The China Geological Survey Series

This Open Access book series systematically presents the outcomes and achievements of regional geological surveys, mineral geological surveys, hydrogeological and other types of geological surveys conducted in various regions of China. The goal of the series is to provide researchers and professional geologists with a substantial knowledge base before they commence investigations in a particular area of China. Accordingly, it includes a wealth of information on maps and cross-sections, past and current models, geophysical investigations, geochemical datasets, economic geology, geotourism (Geoparks), and geo-environmental/ ecological concerns.

Wenchang Li · Guitang Pan ·
Zengqian Hou · Xuanxue Mo ·
Liquan Wang · Xiangfei Zhang

Metallogenic Theory and Exploration Technology of Multi-Arc-Basin-Terrane Collision Orogeny in "Sanjiang" Region, Southwest China

Wenchang Li
Chengdu Center
China Geological Survey
Chengdu, Sichuan, China

Guitang Pan
Chengdu Center
China Geological Survey
Chengdu, Sichuan, China

Zengqian Hou
Institute of Geology
Chinese Academy of Geological Sciences
Beijing, China

Xuanxue Mo
China University of Geosciences
Beijing, China

Liquan Wang
Chengdu Center
China Geological Survey
Chengdu, Sichuan, China

Xiangfei Zhang
Chengdu Center
China Geological Survey
Chengdu, Sichuan, China

ISSN 2662-4923   ISSN 2662-4931   (electronic)
The China Geological Survey Series
ISBN 978-981-99-3654-0   ISBN 978-981-99-3652-6   (eBook)
https://doi.org/10.1007/978-981-99-3652-6

Editorial Board

Introduction

The Tethys Sanjiang (Chi: three rivers, i.e., the Nujiang, Lancangjiang and Jinshajiang) region is one of the most complex orogenic belts in the world. It has not only experienced the Tethyan tectonic evolution, but also effected by the subsequently plate collision and plateau uplifting between India and Eurasia. Owing to the complex geological structures, violent magmatic activities, active ore-bearing fluids, diverse metallogenesis and various deposit types, this region becomes the world-famous nonferrous precious metal metallogenic belt, where multi-episodic tectonic movements make multi-stage structural features widely distributed in every structural belt. Subsequently, the intense compression-shear-nappe of neotectonic movements has formed the current Sanjiang landscape, which is thus called as the natural geological museum. Meanwhile, this region has become the most potential metal enrichment base in China, as a large number of (super-) large deposits have being discovered. Therefore, a series of scientific-technological researches and mineral exploration projects was arranged in this region in the past decades.

The research region of this book covers parts of Tibet, Yunnan, Sichuan Province, which is about 1500 km longitudinally and 420 km latitudinally, with a total area of about 500,000 km^2.

Owing to decades of unremitting efforts and accumulations, both metallogenic theory and exploration practice have been got great breakthrough, and a large number of metal deposits have been discovered and evaluated in the Sanjiang region. The achievements of metallogenic theory innovation and prospecting breakthrough were awarded the First Prize of National Science and Technology Progress Grand of China in 2005.

This book systematically summaries the research and exploration results of many projects, e.g., the "Special prospectiong plan in the Sanjiang region" project jointly undertaken by several provinces of SW China, the "Newly land and resource survey" and "Exploration of nonferrous metal base in the southern Sanjiang region" projects organized by the Ministry of land and resources (China Geological Survey, CGS), several Science and Technology attacking projects granted by the Ministry of Science and Technology, etc. Therefore, this book represents the integration of several generations' hard work in the Sanjiang region. We sincerely appreciate the participating unites, e.g., Chengdu Institute of Geology and Mineral Resources (the predecessor of Chengdu Center, China Geological Survey, CD-CGS), China University of Geosciences, Institutes of Geology and Mineral Resources, Chinese Academy of Geological Sciences (CAGS), Yunnan Geological and mineral exploration and Development Bureau and Yunnan Geological Survey. In summary, this book is an integrating achievement supported by combination of "industry-university-research" units.

1. Tectonic framework of the Sanjiang region

Situated at the oblique joint of the Indian and Eurasian plates, the Sanjiang region has always been a focus of geoscience researches. Many researches have carried out on the geological-tectonic evolutions, metallogenic series and metallogenesis of several ore belts in this region by many geologists worldwide, and they have also made many innovative achievements accordingly. Based on decades of scientific accumulation and continuous dissection, our research team has clarified that the multi-arc-basin-terrane (MABT) structural framework was developed in this region during Paleo-Tethys. In addition, we have carried out systematically researches on the MABT evolution and collisional orogeny of the Cenozoic "Hengduan Mountain" type in this region.

(1) Based on the systematic research on MABT evolution (mainly through dissection of several Ophiolitic mélange belts, island arcs, basins and terranes of different types) in the Sanjiang region, the new MABT structural model was proposed

MABT structure refers to a complex structural system formed by the subduction of Paleo-Tethys Ocean, which is composed of a forward arc and following of a series of island arcs, volcanic arcs, ridges, island chains, seamounts, terranes and the related back-arc oceanic basins, inter-arc-basins and marginal ocean basins. This structural model emphasizes that orogeny is mainly resulted from MABT orogeny process which is triggered by the back-arc-basin reduction (e.g., the oceanic crust subduction or under-thrust, arc-arc collision, arc-terrane collision, block-block collision, etc.). Meanwhile, this model highlights that the Tethyan subduction and closure was caused by a series of multiple ocean subductions and arc-terrane or terrane-continent collisions, rather than a single subduction with a simple structure of trench-arc-basin from ocean to continent. The arc-continent orogeny is mainly caused by the reduction and closure of multiple small-scale oceanic basins, rather than the subduction of the Paleo-Tethys Ocean. Accompanied with MABT evolution, new arcs were emerged and small-scale basins disappeared continually. For example, the crustal reduction in E-W direction was achieved by the reduction of Paleo-Tethys Gantze-Litang, Jinshajiang-Ailaoshan, Lancangjiang and Cangning-Menglian oceans in the center-south Sanjiang region, all of which were closed successively in the Late Triassic, even though the subduction time was slightly different.

(2) Six orogenic belts of different scales and types were identified, and "Hengduan Mountain"-type orogenic model was proposed based on systematic studies on the orogenic process of typical tectonic belts in the Sanjiang region

The Paleo-Tethys Sanjiang, developed under the background of MABT tectonic framework, had undergone subduction orogeny and arc-arc (or arc-continent) collision since the Triassic and the Cenozoic intracontinental strike-slip convergence orogenic process, which formed the composite orogenic belt that contained various orogenic types, namely the "Hengduan Mountain"-type orogenic belt. The "Hengduan Mountain"-type orogenic belt had experienced the Paleo-Tethys MABT evolution and the intracontinental strike-slip—transforming orogeny during Indian-Asian collision process. These geological events were superimposed on the extremely complex six orogenic belts to adjust the transforming tectonic system composed of mainly stress and strain resulted from Indian-Asian collision, which presented as large-scale thrust nappe system, strike-slip fault system and strike-slip-shear system.

2. Study on MABT evolution and metallogenic system of intracontinental tectonic transformation

(1) The MABT metallogenic theory was proposed based on the Paleo-Tethys MABT composition unites evolution and metallogenesis study

Based on the establishment of a new MABT tectonic model, we summarized the metallogenic mechanisms, formation regulations and metallogenic systems accompanying with the MABT evolution as the MABT metallogenic system. This metallogenic system emphasizes the magmatic hydrothermal system, tectonic-hydrothermal system, submarine fluid circulation and other metallogenesis caused by the dynamic processes of continental margin detachment, back-arc subduction, arc-arc and arc-continent collision, post-collisional extension in the specific spatial-temporal structure of the MABT system. Special types of arcs, terranes and small-scale oceanic basins resulted to certain types of metal deposits. For example, Au deposits are distributed in the suture belts of Gantze-Litang and Jinshajing-Ailaoshan, sedimentary exhalative massive sulfide Pb-Zn polymetallic deposits and porphyry Cu (-Mo-Au) deposits were developed in Yidun Island arc, and hydrothermal Pb-Zn-Ag polymetallic deposits were formed in the eastern margin of the Lanping-Pu'er basin, but sedimentary exhalative and volcanic hydrothermal Cu polymetallic deposits developed in the western margin of the Lanping-Pu'er basin. The special evolution background resulted in certain deposit types and controlled the regional distributing regulations.

(2) The "Intracontinental tectonic transformation metallogenic theory" was proposed based on the study of intracontinental tectonic transformation and metallogenesis after oblique amalgamation in the Sanjiang region

Many (super-) large deposits in the Sanjiang region, especially those distributed in the orogenic belts, often have rather young ages, commonly known as "late formation of great instruments". Owing to the studies of intracontinental strike-slip convergence orogenic process, it was recognized that the Cenozoic large-scale metallogenic events in the Sanjiang region developed in the background of intracontinental transformation dynamics, accompanying with the Indian-Asian collision activities resulting in large-scale transformation structure system since 65 Ma. In addition, large-scale strike-slip-shear and thrust nappe systems triggered tectonic transformation, which induced regional extremely magmatic-fluid-metallogenic events. This theoretical system that reveals certain types of deposits resulted from special transforming tectonic system is called as "Intracontinental tectonic transformation metallogenic theory".

3. Research on regional metallogenic regularities and integrated prospecting technologies

(1) Regional metallogenic models were established based on the systematic research on metallogenesis of several ore belts and typical deposits in the Sanjiang region, which will further support prospecting and exploration evaluation in this region.

Three metallogenic giant systems and twelve metallogenic systems were delineated in the Tethyan Sanjiang region. Focusing on the main controlling factors, spatial-temporal structures, deposit types and mineralization associations of each metallogenic system, we defined ten important ore belts. Then based on systematic research on metallogenic background, ore-controlling conditions of each ore belt, it became possible to reveal the main deposit types and metallogenic regularities, dissect the forming conditions of typical deposits, establish regional metallogenic models and point out the prospecting directions, which eventually provide robust support for prospecting and exploration evaluation in this region.

(2) Delineating metallogenic prospecting areas, optimize prospecting target areas and achieving breakthroughs in prospecting practice

Fifty-one metallogenic prospecting areas were delineated, and twenty-one metallogenic target areas were optimized based on multi-information metallogenic prediction. These works provide robust technical support for exploration deployment, and prospecting breakthroughs have been achieved in many prospecting target areas in the Sanjiang region.

(3) Several sets of ore-prospecting integrated techniques were summarized to help achieve breakthroughs in ore-prospecting practice

Based on exploration practice, several sets of effective integrated ore-prospecting techniques were summarized focusing on the special metallogenic backgrounds and deposit types in the Sanjiang region. For example, the integrated technique of "porphyry model + hyperspectral imagery + PIMA + high-precision magnetic survey + IP" was summarized for porphyry Cu deposits, "superimposed metallogenic model + transient electromagnetic method + excitation intensification method" for massive sulfide deposits, "ductile shear zone + geochemical anomalies" for ductile shear zone Au deposits, "structural trap + hydrothermal circulation center + multiple electrical methods" for hydrothermal vein-type Pb-Zn-Ag deposits and "metallogenic system + gravity + magnet + electricity (IP)" for skarn/porphyry concealed deposits. These integrated techniques have played important role in ore-prospecting and exploration evaluation in the Sanjiang region and achieved rapid and efficient ore-prospecting results.

4. Open questions

Though great progress has been achieved, several main questions remain in the Sanjiang region.

(1) What is the main control factor resulting in the ocean-continent, basin-orogeny, crust-mantle transition and the coupling mechanism between MABT evolution and special metallogenesis in the Sanjiang region?

(2) The collisional orogenic effects and composite metallogenic theories should be further studied.

(3) New methods and integrated techniques focusing on ore bodies/deposits location should be summarized in alpine valley areas.

(4) The coupling relationship between Neo-Tethys evolution and metallogenesis should be further studied in the Sanjiang region.

(5) Did the Emei Mountain mantle plume evolution have any effects on metallogenesis in the Sanjiang region?

(6) What effect did the mantle-derived heat flows on metallogenesis in the Sanjiang region?

This book has been revised from the monograph <MABT-Collisional orogenic metallogenic theories and exploration techniques in the Sanjiang region, SW China> published in 2010 by Geological Publishing House. In addition, this book has been introduced many new discoveries, research results and academic literatures. The authors want to express their gratitude to the related units and individuals for their helps.

Contents

Tectonic Framework of Sanjiang Tethyan Metallogenic Domain

Abstract

The Sanjiang Tethyan Metallogenic Domain refers to the Hengduan Mountain Range where the Nujiang River, Lancang River and Jinsha River run side by side, spanning western Yunnan, western Sichuan, eastern Tibet and southern Qinghai, including the eastern Qinghai-Tibet Plateau and the western Yunnan-Guizhou Plateau, with an area of about 50×10^4 km^2. The range belongs to the eastern part of the global Tethys-Himalayan Tectonic Domain structurally, which is in the region where Gondwana and Eurasia collide significantly, and it is also the intersection of Tethyan tectonic domain and Pacific Rim tectonic domain. Influenced by the action of the Indian Ocean Plate, Pacific Plate and Eurasian Plate, it has experienced the complicated process of Tethys formation and evolution, India-Eurasia collision and plateau uplift. Therefore, the Sanjiang geological structure is extremely complex and diverse, with very favorable metallogenic conditions formed. This range is one of the most important metallogenic prospects of nonferrous metals, precious metals and oil and gas in the world and also one of the most critical areas to study the global structure.

1.1 Global Tectonic Background of the Formation of Sanjiang Tethyan Tectonic Domain

1.1.1 Enlightenment from the Formation and Evolution of the Atlantic Ocean, Indian Ocean and Pacific Ocean

It is known that the Atlantic Ocean, Indian Ocean and Pacific Ocean are in different stages of tectonic evolution, with different manifestations and evolutionary characteristics. The Atlantic Ocean is still in the period of spreading, and the passive continental margins on both sides are entering the stage of mature development; the Indian Ocean has begun to subduct unidirectionally in both contraction and spreading, the North Pacific Ocean is in the two-way subduction, and the South Pacific is in the two-way subduction and spreading, but in general has been contracted. Their evolution characteristics are shown as follows.

1.1.1.1 Evolutionary Trend of Closing (Closure), Merging and Transferring of the Ocean Basin

The Tethys tectonics in Mesozoic developed along the boundary between Eurasia and Gondwana is called the Meso-Tethys by Huang and Chen (1987) and the Neo-Tethys by Stocklin (1974) and. Its southern branch—Zagros-Yarlung Zangbo River—has not been completely closed to now, except that the junction zone of India Landmass and Eurasia Landmass was closed at the end of Mesozoic and the beginning of Cenozoic, and there was a subduction zone in the Arabian Sea in the west of India Landmass. A 260 km wide Mokelam subduction accretionary complex system (including fore-arc carbonate rocks, heterogeneous melange and flysch (E_2–N_1)) was formed from southeast Iran to southwest Pakistan, and subduction continues to the present. In the Gulf of Oman, there are still residual oceans between the Arabian Landmass and the Eurasia Landmass (Boulin 1991). Therefore, the Arabian Sea and the Gulf of Oman are both part of Meso-Tethys. Moreover, due to the subduction of the spreading ridge in the Arabian Sea, the ocean basin has stopped spreading, and the oceanic crust of the remaining Arabian Plate has obviously been merged by the Indian Ocean. The subduction of the Indian oceanic crust and the Australian Plate to the Eurasian Plate along the Andaman Islands-Sumatra-Java-Timor Island in the eastern India Landmass is actually a continuation of the subduction of the Meso-Tethys oceanic crust, and there may also be the merger of the Meso-Tethys Ocean to the Indian Ocean (Fig. 1.1). When the Indian Ocean and the Gulf of

W. Li et al., *Metallogenic Theory and Exploration Technology of Multi-Arc-Basin-Terrane Collision Orogeny in "Sanjiang" Region, Southwest China*, The China Geological Survey Series, https://doi.org/10.1007/978-981-99-3652-6_1

Oman are closed, different ocean basins will be closed in the same closed zone. Due to the subduction of the ocean basin, some areas are closed by the oceanic volcanic arc and turned into marginal seas, such as the Philippine Sea. According to Dr. Chen of Taiwan Province, an early intra-oceanic initial island arc was found in Luzon Island. He believes that the South China Sea was originally connected with the Philippine Sea and belonged to the marginal sea of the Pacific Ocean, but then Luzon Island moved northward to separate the South China Sea from the Philippine Sea. Chimei volcanic arc along the coast of Taiwan Province is the northern extension of Luzon Island, and an oblique arc-land collision occurred along the longitudinal valley of Taiwan Province. The Tethys region in Mesozoic, distributed in the area of Lesser Caucasus, is not only a continuation of the Paleo-Tethys, but also merged into the new Tethys Ocean formed in Mesozoic to the west and may be connected with Bangong Lake-Nujiang River to the east. The southern part of the ancient Atlantic Ocean, which was not closed at the end of Early Paleozoic, may have merged into the Paleo-Tethyan Ocean in Late Paleozoic. After the closure of Paleo-Tethyan Ocean, a nearly east–west Hercynian Fold Zone was formed in the southeastern USA, which can be used as an ancient example of ocean basin merging and transformation.

The above statement indicates that the development of an ocean basin has a long and complicated evolution of closing, merging, transforming, connecting the past with the future, which makes the formation and evolution of continental orogenic belts more complicated and changeable, and it is increasingly difficult to reproduce its tectonic evolution history.

1.1.1.2 Destruction of the Oceanic Plate Caused by Three Types of Subductions

The subduction of the Pacific Plate is not the whole subduction, but the subduction of a series of broken small plates, the subduction of the mid-ocean ridge and finally the subduction of residual ocean basins. For example, the eastern margin of the Pacific Ocean has been subducted by a series of small plates such as the Gorda Plate, Cocos Plate and Nazca Plate, as well as by mid-ocean ridges that have subducted under the North American Continent, and the North Pacific Ocean is no longer spreading. The different subduction (or local obduction) speed, angle and direction (vertical and oblique movement) of small plates have caused the different ocean basin closure time on the same tectonic zone or plate junction zone in the orogenic belt and may cause the segmentation, heterochrony and difference of magmatic activities and the characteristics of the same volcanic-magmatic arc zone, for example, the Andes Volcanic Arc. The volcanic-magmatic arc of Jiangda-Weixi-

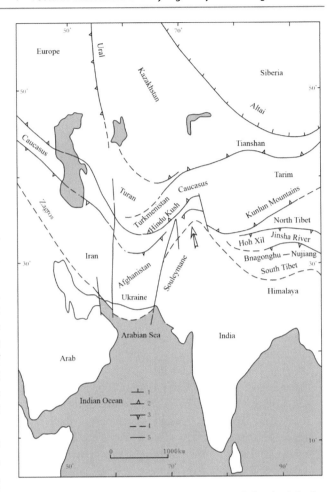

Fig. 1.1 Main suture lines in central Asia and South Asia 1—Caledonian suture zone; 2—Hercynian suture zone; 3—Kimerich suture zone; 4—Alps suture zone; 5—Regional fracture

Lvchun Zone and the volcanic arc of Kaixinling-Nanzuo-South Lancang River Zone in the Sanjiang region of southwest China, which match the Jinsha River Zone and Lancang River Zone, respectively, have the above three characteristics (Liu et al. 1993; Mo et al. 1993). Whether these three characteristics of magmatic activities of the volcanic-magmatic arc can be used in the orogenic belt to invert whether there may be subduction of small plates, or whether it is related to the destruction of multiple back-arc ocean basins, is a topic to be studied in depth with reference to the characteristics of magmatic activities of volcanic-magmatic arc in the eastern and western Pacific Ocean.

1.1.1.3 Coexistence of Subduction, Obduction, Strike-Slip, Accretion and Tectonic Erosion

The oceanic plate is characterized by subduction, obduction, strike-slip, subduction accretion and subduction tectonic erosion. In addition to plate subduction, there are many oblique subductions of micro-landmasses or strike-slip displacement collages of island arc terrane on both sides of the eastern and western Pacific Ocean, such as collage of many island arc terranes in the western USA, and the aforementioned northward slip of Luzon Island. The trench-arc-basin system of the western Pacific Ocean is developed with subduction accretion, and the western North America on the east coast of eastern Pacific Ocean has subduction accretion, such as the Francesco Melange Zone, but no melange zone is found in western South America, which shows the tectonic erosion or cutting of the South American Continent by subduction. Therefore, in an orogenic belt, there is often an intermittent discontinuity in the plate junction zone or ophiolitic melange zone, and the interrupted part is either a tectonic erosion zone or a large strike-slip ductile shear zone.

1.1.1.4 Three Types of Orogenic Belts After Ocean Basin Closure

When a long-developed ocean (such as the Pacific Ocean) finally closes, it may be manifested as arc-arc collision or arc-land collision orogeny caused by subduction of back-arc-basins in the MABT on both sides of the ocean. Professor Xu et al (1994) put forward the back-arc-basin collision orogeny and arc-arc collision orogeny after observing and studying many orogenic belts in the world. But this is only suitable for inter-continental collision orogenic belt after the ocean closure with two-way subduction. For the ocean basin with one-way subduction, the orogeny after closure is the back-arc orogeny on one side of the active margin or the passive margin orogeny, forming a composite orogenic belt with arc-land collision. Many majestic mountains are often formed on the passive continental margin, such as the Himalayas, Longmen Mountain in the west of Sichuan Basin and Gongga Mountain, which is more than 7000 m high. For the inter-continental trough or aulacogen without oceanic crust and subduction, it is purely the intra-continental orogeny by compression, such as Zongwulong Mountain in the northern margin of Qaidam. On the north side of Huaitoutala, the fold basement and unconformity of Early Paleozoic can be found and were merged into the folded mountain system formed from the end of the Late Paleozoic to the beginning of Early Mesozoic together with the sedimentary rocks of the inter-continental deep-sea trough which was further fractured in Late Paleozoic. Therefore, according to the destruction mode of ocean basins or basins, continental orogenic belts can be roughly divided into 3 types: ① Composite orogenic belts formed by two-way subduction and closure of oceanic crust in ocean basins, with arc-arc collision or arc-land collision orogeny on both sides of the active margin and different formation time. Of course, when the back-arc-basin near the continental margin is closed, there will also be arc-land collision orogenic belts, such as the closed orogeny of the Paleo-Asian Ocean in Inner Mongolia and the Bayankala Ocean. ② Composite orogenic belts formed by one-way subduction and closure of oceanic crust in ocean basins, with arc-land collision between active margin and passive margin, such as Himalayas and Gangdise orogenic belt, and the closure of Jinsha River-Ailaoshan and Lancang River. ③ Inter-continental compression orogenic belts formed after the closure of the inter-continental trough or aulacogen, with nonarc-land collision. The back-arc orogeny proposed by Xu et al. (1994) is only one type. As plate subduction is often characterized by oblique subduction and mass strike-slip in the late-collision orogenic stage, the orogenic belt finally formed is mainly a composite orogenic belt caused by subduction, collision and strike-slip, and most of the orogenic belts in Sanjiang region belong to this type.

The complexity of continental orogenic belts depends not only on the complexity of its internal material and structure and the complexity of lithosphere rheological stratification, but also on the complexity of ocean basin evolution. Since the orogeny after the closure of ocean basins is mostly manifested in arc-arc collision orogeny and arc-land collision orogeny, coupled with the subduction of small plates and strike-slip collage of blocks or island arc terranes, the cooling and solidification of the old oceanic crust and the increase in density are mostly abated by subduction and difficult to remain. Thus, in an orogenic belt, it is sometimes difficult to determine the main suture line for the final closure of a long-evolving ocean basin (its closure time may sometimes be earlier than that of some back-arc-basins). It may be a complex tectonic zone composed of not only one line, but two or more subduction accretion zones, such as the Paleo-Asian Ocean with two-way subduction, which has multiple suture zones of different periods with north–south symmetry, and the suture zone for its final closure at the end of Hercynian consists of not one but two zones, interspersed with Xilin Gol, Kiamusze, Mazong Mountain and other blocks of different sizes. The Bayankala Ocean with two-way subduction is finally closed in both north and south, with a "covered" folded thrust orogenic belt formed from the sediments of foreland depression. Several sections may not be closed; thus, there are residual ocean basins with subduction complexes as the base and filled with turbidite.

1.1.1.5 Nonophiolite in Ocean Basins

In the Atlantic Ocean, in addition to ophiolite formed by spreading ridges, in the transition zone between the ocean and land on its edge, the lithosphere can be stretched and thinned, and the underlying pyrolite can be stretched out, or after stretching and thinning, the underlying mantle peridotite can be exposed to the seabed through later thrusting, such as the serpentine mantle lherzolite in the transition zone of oceanic crust in Gahcia Bank section of the western Iberian continental margin (Lemoine and Trümpy 1987;

Whitmarsh and Sawyer 1993), when the Atlantic Ocean is closed, the iherzolite may be easier to preserve in the orogenic belt than ophiolite with spreading ridges. The ultrabasic rocks like ophiolitic melange zone exposed in the south of Daofu in western Sichuan and the Muli-Kangwu area in the south of Ganzi-Litang Zone may belong to this type, because the associated pillow basalt is not oceanic ridge type but is the rift-type alkaline basalt on continental margins. Therefore, when studying orogenic belts, this type of ophiolite should be identified carefully, yet the existence of ocean basin shall not be denied due to the discovery of such nonophiolite.

1.1.1.6 Changes in Oceanic Crusts

Due to the continuous spreading of the mid-ocean ridges and the continuous generation of new oceanic crusts, especially when the oceanic crusts of ocean basins subduct, the old oceanic crust is continuously destructed. In the long-term development and evolution of the ocean basin, the oceanic crust has been changed. Therefore, the ophiolite formation age determined by radiolarians in siliceous rocks associated with oceanic crust basalt may not represent the age of initial ocean basin formation. Nowadays, there are numerous radiolarian siliceous rocks of Paleogene, Neogene and Quaternary in the Pacific Ocean, but it cannot be considered that the Pacific Ocean was formed in Paleogene, Neogene and Quaternary. Moreover, due to the decomposition and tectonic mixing of the ophiolite complex, the radiolarian siliceous rocks found may not be the same as those at the time of the initial ocean basin formation. Therefore, the age of ocean basin formation can be determined more reliably only when it is corroborated by historical data on the formation and evolution of passive continental margins or active margins on both sides.

Some of the evolution features of the three oceans can be used to deepen the understanding of the tectonic evolution of the Sanjiang Tethyan tectonic domain and its global tectonic setting.

1.1.2 Global Tectonic Setting

It can be seen from the geological maps of Europe and Asia that the giant Tethyan tectonic zone, which runs from east to west, is obviously inlaid with strips (orogenic belts) and blocks (landmasses), just like a giant "ductile shear zone" or "tectonic melange zone". This unique tectonic zone, which distributes between the south and north continents, plays an important role in the global tectonic evolution and has always attracted much attention. The Sanjiang orogenic system in southwest China is in the eastern part of the Tethyan tectonic domain—Eastern Tethyan tectonic domain (east of Pamir), and its formation and evolution are closely

related to the formation and evolution of three continental groups, especially Gondwana Continental Group and Pan-Cathaysian Continent Group and their continental margins. The history of global ocean-land evolution since the break-up of the Rodinia super-continent has been characterized by the coexistence of the three major landmass groups, namely Laurasia, Gondwana and Pan-Cathaysian landmasses, and three major oceans, namely Panthalassa Ocean (ancient Atlantic Ocean), Paleo-Asian Ocean and Tethys Ocean (Li et al. 1995; Lu 2004; Pan et al. 1997). The main body of the Eastern Tethyan tectonic domain, including the Sanjiang orogenic system, is in the basic framework —"one ocean and south and north continents"—of Gondwana Continental Group, Pan-Cathaysian Continent Group and the Tethys Ocean between them (Fig. 1.2).

1.1.2.1 Pan-Cathaysian Continent Group

1.1.2.2 Gondwana Continental Group

The Gondwana continent refers to the super-continent composed of several landmasses of East Gondwana (including India, Australia, South Asia, etc.) and West Gondwana (including South America, Africa, etc.) from the end of Neoproterozoic to the beginning of Paleozoic, with the destruction of Mozambique Ocean (Shackleton 1996) and the Pan-Africa orogeny (600–550 Ma) (Kröner et al. 1993; Kriegsman, 1995). The Gondwana continent is also called "the southern continent", and its scope is much larger than that proposed by in *The Face of the Earth*. It includes South America, Africa, Australia, Antarctica, India Peninsula and Arabian Peninsula, as well as Iran, Turkey and Himalayas. In the Gondwana Continent, the Gondwana Rock is developed due to much glacial activities in Carboniferous-Permian, with biological characteristics of cold-water fauna spermatophyte in Carboniferous-Permian, and ferns-Gangamopteris-Glossopter is dominant in Permian and

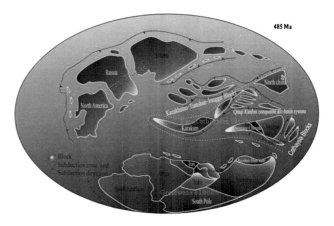

Fig. 1.2 Schematic diagram of global ocean-land pattern of early Paleozoic (Ordovician)

other Gondwana flora. In Late Triassic of Mesozoic, a narrow trench was formed in Madagascar in eastern Africa, and the Neo-Tethys in the northern margin of Gondwana continent spread. From the end of Jurassic to Cretaceous, the India Landmass and the Australian Landmass separated from the Antarctic Landmass, the Indian Ocean began to spread, the South American Landmass separated from the African Landmass, the South Atlantic Ocean began to spread, and Cenozoic gradually migrated to its present location.

1.1.2.3 Tethys Ocean

The Tethys, originally proposed by Suess (1893), refers to the vast ocean in Mesozoic between the ancient land of Angola in the north and Gondwana in the south. The destruction of the ancient ocean and the subsequent uplift formed the magnificent Alps-Himalayas. With the establishment of plate tectonics theory in 1960s, not only the Tethys area is larger than before, but also the formation time of Tethys Ocean dates to Paleozoic. From the time of Tethys evolution, the division of Proto-Tethys, Paleo-Tethys and Neo-Tethys can better reflect the evolution of Tethys spatial–temporal pattern: The Proto-Tethys (Sinian-Silurian) was mainly characterized by the dispersion of Pan-Cathaysian Continent Group and Laurasia Continental Group, the separation of Laurasia Continental Group from Gondwana continent and the spreading of Tethys Ocean (Li et al. 1995; Pan et al. 1997); the combination of "Pan-Cathaysian orogen" at the end of Early Paleozoic formed a unified Pan-Cathaysian continent (Lu et al., 2006) and the basis of the formation of Cathaysia flora (Xie et al., 1994; Pan et al., 1997); the Paleo-Tethys (Devonian-Middle Triassic) was characterized by the convergence of Pan-Cathaysian Continent Group and the Laurasia Continental Group, the connection between Laurasia Continental Group and Gondwana Continental Group, and the shrinking of the Tethys Ocean (Li et al., 1995; Pan et al., 1997), the East Asian continent and its marginal orogenic system were formed by the combination of Indosinian orogen from the end of Late Paleozoic to the Early-Middle Triassic and became an integral part of Pangea super-continent; the Neo-Tethys (Late Triassic-Eocene) was mainly characterized by the break-up of Pangea super-continent and Gondwana continent (Pan et al. 1997); the Tethys Ocean was destructed and transformed into continental lithosphere and entered the period of continental collision and orogeny.

The Sanjiang Tethyan tectonic zone was formed and evolved in global tectonic formation. It is located at the junction of Pan-Cathaysian Continent Group and Gondwana Continental Group and has gone through two times of Pangea break-up and three major development stages—the Proto-Tethys, Paleo-Tethys and Neo-Tethys. Its tectonic evolution is completed by continuous continent break-gathering evolution since Paleozoic, and its dynamic

mechanism may be related to the southward migration of the earth's mass center, the expansion of the southern hemisphere, the destruction of the Pangea and its northward drift. In the mantle convection of the asthenosphere that rotates clockwise from south to north and drifts the continent northward, apart from the vertical mantle convection, there may be horizontal vortices of different sizes, which can only promote the rotation of the landmass, the formation of the Pangea and its immediate destruction (Li et al. 1995; Pan et al. 1997).

1.2 Division of Main Tectonic Units

For the Qinghai-Tibet Plateau and the Sanjiang region, many scholars have divided these areas by tectonic units (Liu et al. 1993; Mo et al. 1993). In view of the improvement of geological survey, the deepening of basic research, the further determination of tectonic environment of some tectonic units, the discovery of some new tectonic units and the demand of metallogenic prediction and resource evaluation, this book is based on the above-mentioned division schemes of many scholars, combined with the latest research and the national division schemes of tectonic units. The main tectonic units in Sanjiang Tethyan Metallogenic Domain and its adjacent areas are divided into 4 first-class tectonic units (namely Yangtze Landmass, Sanjiang MABT, Bangong Lake-Shuanghu-Nujiang River-Changning-Menglian Mage-Suture Zone and Gangdise-Gaoligong Mountain-Tengchong Arc-Basin System) and further divided these units (Fig. 1.3). Also, the reference of "landmass" and "block" is further standardized. A landmass generally refers to a relatively stable area composed of consolidated land crust in the whole geological period, which is generally large in scope, and the paleogeographic features often undergo land-sea changes. A landmass may be an uplift denuded area or a sedimentary basin. A block refers to a small or very small continental crust block, whose geotectonics can be the product of a crack near the edge of a continental plate or an exotic terrane from other tectonic domains.

1.3 Basic Characteristics of Tectonic Units

1.3.1 Yangtze Landmass (I)

The Yangtze Landmass has pre-Sinian System crystalline basement and fold basement and Dahongshan Group, Yanbian Group, Huili Group, Kunyang Group and Ailaoshan Group, etc. in Mesoproterozoic. The basement tectonic layer is formed by arc-basin system development and arc-land collision. The Suxiong "bimodal" volcanic rift is developed in Nanhua. A wide range of carbonate rock plateaus was

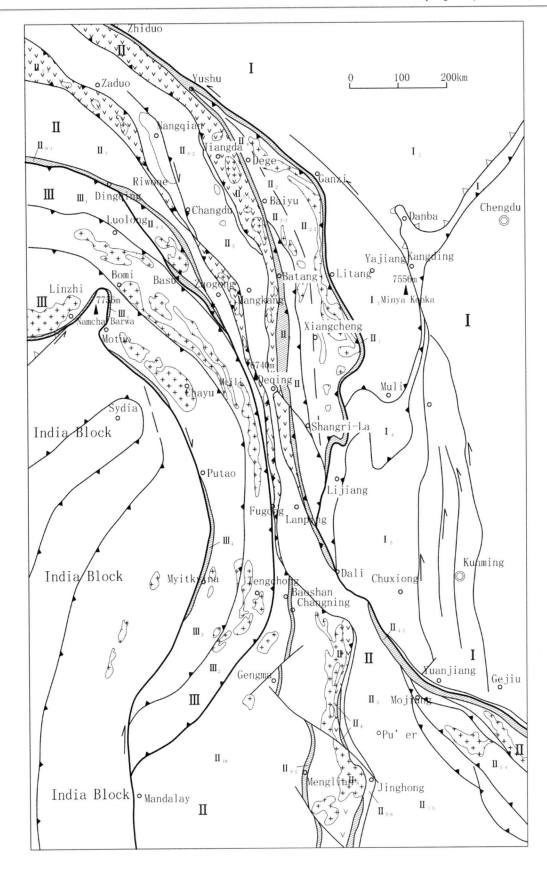

Fig. 1.3 Division of tectonic units in Sanjiang region of Southwest China. I—Yangtze landmass: I_1—Longmen mountain thrust zone, I_2—Bayankala foreland basin, I_3—Yajiang relict basin, I_4—Yanyuan-Lijiang continental margin depression zone, I_5—Chuxiong foreland basin; II—Sanjiang MABT: II_1—Ganzi-Litang junction zone, II_2—Dege-Xiangcheng Island arc (Yidun Island Arc): II_{2-1}—Que'er mountain-Daocheng outer arc zone, II_{2-2}—Jiegu-Yidun back-arc-basin zone, II_3—Zhongza-Shangri-La block, II_4—Jinsha river-Ailaoshan junction zone: II_{4-1}—Jinsha river Ophiolitic Melange zone, II_{4-2}—Ailaoshan Ophiolitic Melange zone, II_5—Qamdo-Pu'er block: II_{5-1}—Jiangda-Jijiading-Weixi continental margin volcanic arc, II_{5-2}—Qamdo-Markam bidirectional back-arc foreland basin, II_{5-3}—Zadoi-Dongda mountain continental margin volcanic arc, II_{5-4}—Mojiang-Lvchun continental margin volcanic arc, II_{5-5}—Lanping-Pu'er bidirectional back-arc foreland basin, II_{5-6}—Yunxian-Jinghong continental margin volcanic arc, II_6—Lancang river junction zone, II_7—Zuogong block, II_8—Lincang magmatic arc, III—Bangong lake-Shuanghu lake-Nujiang river-Changning-Menglian Mage-suture zone: III_1—Bangong lake—Nujiang river junction zone, III_2—Changning-Menglian junction zone, III_3—Jiayu bridge relic arc zone; IV—Gangdise-Gaoligong mountain-tengchong arc-basin system: IV_1—Baoshan block, IV_2—Shading-Luolong fore-arc-basin, IV_3—Bowo-Tengchong magmatic arc, IV_4—Xiachayu magmatic arc, IV_5—Yarlung Zangbo river junction zone

formed in Dengying, and sedimentary covers were formed in the Sinian System-Mesozoic and Cainozoic. Its basement is along Jianchuan-Dali and Ailaoshan fault in Diancang Mountain and Ailaoshan and overthrusts westward on the stratum in Mesozoic in Lanping-Pu'er Depression Zone and Ailaoshan Ophiolitic Zone, forming Diancang Mountain-Ailaoshan Basement Overthrust Zone.

1.3.1.1 Longmen Mountain Thrust Zone (I_1)

Longmen Mountain Thrust Zone is distributed in the northeast and spreads into Shaanxi Province along the northeast. It is cut by northwest-trending Sanhe Fracture to the southwest and enters Jinping Mountain Overthrust Zone in the southwest. This zone is mainly composed of formations in Sinian, Paleozoic, Mesozoic, Paleogene and Neogene, but the overall metamorphic degree of rocks is very low. The thrust nappe in this zone is significant, and thrust pieces of different sizes are developed, with many detached blocks formed. The Jiangyou-Dujiangyan Fracture Zone is the eastern boundary of Longmen Mountain Overthrust Zone. To the northwest, the Longmen Mountain Foreland Overthrust Nappe Zone, the Longmen Mountain Central Fold Nappe Zone and the Longmen Mountain Hinterland Arc Slip-nappe Zone can be further delineated. From northwest to southeast, the three zones show the characteristics of ductile to brittle.

1.3.1.2 Bayankala Foreland Basin (I_2)

The Bayankala Foreland Basin was transformed from the Yangtze passive margin in Paleozoic at the end of Middle Triassic, adjacent to the Sanjiang MABT tectonic zone in the west, and bounded by the Longmen Mountain-Jinping

Mountain Fault in the east and the Yangtze Landmass. The main part of the basin is composed of flysch in the Triassic. In the Paleozoic, this area was a part of the Pan-Yangtze Landmass and from Ordovician to Devonian, with extremely thick clastic rocks and clastic flysch deposits developed. From Early Permian to Late Permian, the Ganzi-Litang Ocean gradually opened, and the carbonate gravity flow accumulation of slope facies was developed, accompanied by extensional basic basalt flow, and the passive continental margin began to form. The passive continental margin continued to develop in the Early Triassic and Middle Triassic, and the extremely thick clastic flysch and turbidite were deposited in the Late Triassic. After being transformed into a foreland basin, the distribution of lithofacies shows that the sediment is gradually deep from east to west.

1.3.1.3 Yajiang Relict Basin (I_3)

Yajiang Relict Basin (I_3) is a secondary basin in the south of Bayankala Basin and is separated by Xianshui River Strike-Slip Fault. Its basic characteristics are similar to those of Bayankala Basin. It entered the residual ocean stage from Late Permian to Middle Triassic and was closed at the end of Triassic. Its main part is the residual basin composed of abyssal sediments, turbidity sediments and neritic flysch sediments of Late Triassic. According to the pillow basalt in Late Permian in Jiulong-Muli area and the oceanic ridge basalt in Luhuo-Daofu area, it is speculated that it is the filling and destruction of the relict basin where the oceanic-continental transitional crust subducted westward. At the end of Late Triassic, with the closure, overall uplift or folding of Ganzi-Litang Ocean, few sedimentary records of Jurassic-Cretaceous can be found. In Paleogene, the right-handed strike-slip pull-apart basin, which only distributed along the narrow strip, accumulated continental molasse deposits. In the intracontinental convergence after the collision (since Jurassic), significant folds and thrusts were formed, and the Longmen Mountain-Jinping Mountain Nappe Zone was formed on the western edge of the Yangtze Block. Metamorphic core complex was formed in the rear edge of the nappe zone. The metamorphism in this area is not significant, mainly low greenschist facies. The magmatic activity is mainly caused by the collisional (Jurassic) terrestrial crust remelting granite except for the basic magmatic activity in the tensional period in Late Permian.

1.3.1.4 Yanyuan-Lijiang Depression Zone (I_4)

This depression zone is located in the southwest margin of Yangtze Landmass. From Sinian to Paleozoic, there are mainly littoral-neritic clastic rocks and carbonate rocks deposited stably. Only in the local depression during Ordovician–Silurian, there are deep-water graptolite shale and siliceous rocks deposited. The basalt in Permian began to erupt at the end of Early Permian. There were still littoral-neritic clastic rocks and carbonate rocks in the Triassic. At the end of Triassic, coal-bearing clastic rocks of marine-land transitional facies are deposited, which shows that the sedimentary is in shallow water, but the depression amplitude is large, and the sedimentary thickness can reach 6000 m. In the Middle Triassic and Late Triassic, in Xiangyun Area on the east edge of the depression zone, due to the fault of Chenghai, a local deep depression was formed, and clastic rocks and turbidites were deposited, accompanied by intermediate-basic pyroclastic rocks. The Jinping gliding nappe sandwiched between Ailaoshan Fault and Adebo Fault shows that the oldest stratum exposed is in Ordovician, and its sedimentary characteristics from Ordovician to Permian are similar to those of Yanyuan-Lijiang Depression Zone, which indicates that both of them are originally connected and belong to the Yangtze Block. The Yangtze Block was torn only because of the left slip along the Ailaoshan Fault, and then (possibly in Triassic) due to the southwest thrust of the Yangtze Block, the Ailaoshan Metamorphic Basement Thrust and exposed, which separated the block from the Yanyuan-Lijiang Depression Zone. The Ailaoshan Group is composed of a sequence of migmatitic gneiss, granulite, amphibolite, schist and marble, and the mylonite is transformed by the former rocks.

1.3.1.5 Chuxiong Foreland Basin (I₅)

Chuxiong Foreland Basin is located in the depression zone of the southwestern margin of Yangtze Block, and it was mainly a fault block depression sedimentary zone before Mesozoic. It was transformed into a foreland basin in the Late Triassic, and after the Jurassic-Cretaceous, especially the collision between India and Eurasia, there was a significant folding deformation. In the Yanyuan-Lijiang Depression Zone and Chuxiong Basin, there were many intermediate-acid rock masses and alkali-rich porphyries invaded in the Himalayan.

The continental overflow basalts and their corresponding intrusive rocks in Permian are mainly developed in the Yangtze Landmass, which are characterized by high TiO_2, Na_2O and K_2O and high enrichment of LREE. They are a part of Emei Mountain Igneous Province and have a genetic relationship with super-mantle plume. According to, the eruption period of Emei Mountain basalt is 262–258 Ma. In addition, in the area from east Erhai Lake to Red River, on both sides of Ailaoshan Fracture Zone, some intraplate deep volcanic rocks and small intrusive masses such as potassium-rich and high-magnesium lamprophyres, alkali-rich porphyry and potassium basalt in Himalayan are distributed intermittently, forming the post-collision alkali-rich porphyry and volcanic zone superimposed on the older tectonic–magmatic zone.

1.3.2 Sanjiang Archipelagic Arc-Basin System (II)

1.3.2.1 Ganzi-Litang Junction Zone (II₁)

The zone runs from western Xiewu Temple in the northwest to Ganzi in the southeast then to the south from Litang and Muli Yazui Ranch to the mouth of the Sanjiang at the junction of Sichuan and Yunnan and then turns to the west and spread to the south along Haba Snow Mountain and the west of Yulong Snow Mountain to Jianchuan, where it meets the Jinsha River Junction Zone extending to the south at the north of Qiaohou. Its northwest end may be connected with the Jinsha River Junction Zone in the west Deng Ke-Yushu. The southern end of the junction zone may be truncated or covered by the westward thrusting of the southwest Yangtze Landmass to the south of Jianchuan. The junction zone is over 500 km long and 5–30 km wide, and it is a tectonic melange zone composed of Late Permian (P₂)-Late Triassic (T₃) oceanic ridge tholeiite, picrite basalt, mafic and ultramafic cumulate, gabbro-diabase wall, serpentinite, radiolarian siliceous rock and flysch. The foreign sedimentary rock blocks are from Ordovician to Triassic, and the matrix is sand-slate and volcanic rocks of Late Permian and Late Triassic. Most ophiolites are decomposed to form ophiolitic melange blocks, but most of the basalts feature pillow structures, and their geochemical characteristics are similar to those of mid-ocean ridge basalts (MORB).

There is a well-preserved ophiolite sequence near Litang, and ophiolite in most places is decomposed. There are amphibole eclogite exposed in the south of Litang and glaucophane schists in Xinlong-Yiji Muli and Sanjiang estuary. Near Manigango in the north and Tuguan Village in the south, pillow basalts, massive basalts and abyssal sedimentary rocks (radiolarian siliceous rocks) are mainly exposed, as well as sporadic gabbro and pyroxenite. Basalt is usually pillow-shaped, characterized by low K_2O (average content of 0.19–0.37%), medium TiO_2 (average content of 1.38–1.63%) and average REE (rare earth elements) distribution pattern, which is similar to mid-ocean ridge basalt (MORB). The K_2O content in basalts in Tuguan Village area in the south is slightly higher, but the rare earth elements (REE) pattern is still average. The basalts in the south member (such as Tuguan Village) were formed in Middle Permian, those in the middle member (Litang) were formed in Early Triassic according to the radiolarian siliceous rock overlaid, and those in the north member (north of Ganzi) were formed in Late Triassic, indicating that the ocean basin opened between Middle Permian and Late Triassic. The ophiolite was localized before the Rhaetian of Late Triassic according to the age of arc volcanic rocks in Changtai-Xiangcheng (T_3^{1-2}) and the age of continental coal measure strata (T_3^3). That is, the Ganzi-Litang Ocean Plate may have subducted since Late Triassic and the ocean basin closed at the end of Late Triassic.

1.3.2.2 Dege-Xiangcheng Island Arc (Yidun Island Arc Zone) (II₂)

This zone is located on the west of Ganzi-Litang Junction Zone, and the main exposed strata are Triassic and a few Paleogene and Neogene. The Middle Triassic and Lower Triassic stratum are clastic rocks mixed with carbonate rocks and siliceous rocks, with a thickness of nearly 5000 m; the lower part of Genlong Formation, Gacun Formation and Miange Formation of Upper Triassic is composed of extremely thick flysch and sand-slate with basic, intermediate-basic and acidic volcanic rocks and carbonate rocks, about 10,000 m thick; from the upper part of Mian Formation to Lamaya Formation, there are neritic clastic rocks and coal-bearing clastic rocks of land-sea transitional facies. To the east of Dingqu Fracture Zone in the south, slump breccia can be seen in the early stage. Most of the breccia are foreign blocks (flysch-like), as well as radiolarian siliceous rocks and back-arc basalts similar to oceanic crust, and ultramafic rocks are exposed. Paleogene and Neogene are molasses deposits in inter-mountain basins.

The tectonic deformation of this zone is significant, and the folding started from the end of Indosinian-Yanshanian, with syn-cleavage folding deformation and left translational ductile shear zone with nearly vertical attitude and the same strike as the tectonic line, and then high-angle positive ductile shear occurred. The Gacun Polymetallic Deposit was controlled by ductile strike-slip and ductile shear and was complicated. Since the Himalayan, there has been left strike-slip intense magmatic activities and well-developed magmatic rocks, which are composed of volcanic formation complex and intrusive rocks in four development stages.

There are crust structures similar to the Yangtze Landmass in the eastern Chas in the south, which are pre-Sinian basement, Sinian and sedimentary cover of later Paleozoic and Triassic, and they are island arc-continental crust basement. The Qias Group in pre-Sinian is composed of metamorphic intermediate-basic volcanic rocks mixed with carbonate rocks and clastic rocks, with a thickness of 2644–2818 m. Guanyinya Formation in Sinian is composed of clastic rocks, and Dengying Formation is composed of carbonate rocks, which are unconformably overlain by the Qias Group and are 300–1200 m thick. The Lower Paleozoic (with the lacuna of Middle and Upper Silurian) is mainly composed of clastic rocks, carbonate rocks mixed with intermediate-basic volcanic rocks and siliceous rocks that evolved from neritic to bathyal deposits and contact Sinian in a parallel unconformity manner. The Upper Paleozoic (with the lacuna of Upper Devonian) series are littoral-neritic clastic rocks and carbonate rocks with basic volcanic rocks and unconformably cover Sinian, Cambrian and Ordovician. The Triassic in Mesozoic is composed of marine clastic rocks, carbonate rocks and volcanic rocks, and the Upper Triassic stratum is composed of marine-continental coal-bearing clastic rocks.

The above stratigraphic characteristics show that Qias Group is equivalent to Hekou Group in Kangdian area of the Yangtze Block, and Sinian is similar to the Yangtze Block, which indicates that they were originally part of the Yangtze Landmass. At the end of Early Permian or from Late Permian to Early Triassic, with the opening of Ganzi-Litang Ocean Basin, this part separated from the Yangtze Landmass and formed the basement of Shaluli-Yidun Island Arc Zone. It also shows that the island arc zone developed on the graben of the continental crust basement of the Yangtze Landmass, and it contains blocks with old basement. This part can also be classified as a secondary tectonic unit-Chas fault uplift.

The Dege-Xiangcheng Magmatic Arc, which is nearly pod-shaped in the north–south direction, is mainly occupied by volcanic-sedimentary rock series in Late Triassic and granite basement of Indosinian-Yanshanian. Roughly bounded by the Yidun-Haizishan line, it can be divided into two arc volcanic-sedimentary basins in the south and north: Baiyu-Changtai Basin and Xiangcheng Basin. Volcanic activities occurred in three periods: the former island arc period, the main arc period and the later arc period. The main arc period can be divided into three stages: early arc-forming stage, mid intra-arc rift stage and late arc-forming stage. Before the formation of the volcanic arc in Early Carnian (T_1), a series of graben and horst is developed in the rock area, resulting in a rift-type alkaline-transition series basalt with high content of TiO_2 or a "bimodal" volcanic assemblage of basalt-rhyolite, which has similar petrogeochemical characteristics to Emei Mountain basalt in Yangtze Plate. The volcanic rocks in the main arc period (cycle in Gacun, T_3^{1-2}–T_3^{2-1}) are dominated by andesite in two arc-forming stages, with a small amount of calc-alkaline basalt and dacite-rhyolite, which have typical characteristics of arc volcanic rocks. Rhyolite-tholeiite "bimodal" volcanic assemblage is developed in the inter-arc rift stage. The development of intra-arc rift and corresponding "bimodal" volcanic rocks is the main feature of Yidun arc, which is different from other magmatic arcs and is also the basic tectonic-volcanic condition for ore-forming and ore-controlling of Gacun massive sulfide polymetallic deposit. This feature is most obvious in the middle and northern segment of magmatic arc. The volcanic rocks (cycle in Miange, T_3^3) in the late arc-forming stage are only developed in the northern part of the arc, which is a "bimodal" assemblage of high-potassium basalt, shoshonite and rhyolite, with rhyolite as the main component, and were not exposed in the southern segment of the arc.

Que'er Mountain-Daocheng Outer Arc Zone (II_{2-1}), mainly composed of adamellite-granodiorite plutonic rock basement, is mainly distributed in the northern member of the magmatic arc. The diagenetic period is in Indosinian (237–195 Ma), but in some huge rock foundations (such as Cuojoma and Dongcuo), there are large xenoliths formed in Hercynian, which are granitoids in the same collision period. Therefore, the granitoids in the outer arc zone are mainly caused by collision, superimposed on the early island magmatic arc zone.

Jiegu-Yidun Back-arc-Basin Zone (II_{2-2}), located in Baiyu Gacun Basin in the north member, is symmetrically distributed with old parts on both sides and new parts in the middle; in the southern Xiangcheng Basin, the zone is mainly distributed in the west of the basin near the Zhongza Block. During the transition from the end of the former island arc period to the early stage of the main arc period, the sub-cyclic volcanic rocks in Chizhong, which are composed of high-MgO pillow tholeiite, massive island arc low-potassium tholeiite and high MgO, high-SiO_2 and extremely low-TiO_2 bonitite andesite, began to form obvious characteristics of arc volcanic rocks, but also retained some characteristics of late rift volcanic rocks. This sequence of volcanic rocks is only exposed in the middle area of Chizhong, Xiangcheng, located on the east side of cyclic volcanic rocks in Genlongya. Changdagou-Pulang Inner Arc Zone is mainly developed with calc-alkaline volcanic assemblage in the main island arc period and the associated intermediate-acid porphyry, which is distributed in Xuejiping-Pulang area of Shangri-La in the southern member of the magmatic arc, with dioritic porphyrite-monzonite porphyry. Its petrochemistry is characterized by low SiO_2, high CaO, high MgO, rich Na and poor K. It is a type I granite, a product of pressure magmatic arc, with porphyry copper deposits and polymetallic mineralization. There is intrusion of intraplate magma after collision.

In addition, there are post-collision granites in Yanshanian-Himalayan in the zone, which are superimposed in the distribution area of early arc magmatic rocks and syn-collision granites.

The Dege-Xiangcheng Island Arc (Yidun Island Arc Zone) has a close temporal and spatial relationship with the Ganzi-Litang Zone. In the northern member of the volcanic arc (Zengke-Changtai), the following spatial configurations can be seen from east to west: the disappeared Ganzi-Litang Ocean (Ganzi-Litang Ridge Volcanic Rock-Ophiolite Zone)—the fore-arc area (arc-ditch hiatus)—is the main arc area of the granite zone and the sedimentary rocks in Late Triassic in Que'er Mountain; the outer arc (east basaltic andesite zone), intra-arc rift ("bimodal" volcanic zone), inner arc (west basaltic andesite zone) and back-arc area (Miange Rhyolite—High-potassium Basalt Zone). The back-arc area is roughly divided from the main arc area by the

Keludong-Dingqu River Fracture, along which there are many serpentinites of tectonic diapiric fold. Volcanic rocks are old in the east and new in the west, and volcanic activity centers migrate from east to west. In the middle-south member of the arc (Xiangcheng Area), the spatial configuration of volcanic rocks is roughly similar to that of the northern member. The difference is that there is no volcanic rock exposed in the late arc, and the back-arc area is occupied by volcanic rocks from the end of the fore-arc period to the beginning of the main arc period. The volcanic activity center tends to move from west to east, which may be due to the different dynamic boundary conditions and subduction mechanisms between the south and north members. It can be seen from the foregoing that the Ganzi-Litang Rock Zone and Yidun Volcanic Arc Zone are also closely connected in time. It can be seen that they are "double zones" organically linked, clearly marking the relative position of the active continental margin of the ancient Ganzi-Litang Ocean and the Zhongza Micro-landmass and pointing out the westward subduction direction of the ancient plate.

To sum up, the Dege-Xiangcheng Island Arc (Yidun Island Arc Zone) was developed since the continental rift grabon-horst system during the short Late Triassic due to the westward subduction of Ganzi-Litang Ocean Plate. Because of its history of alternating tension and compression and the existence of intra-arc rift, it is generally characterized by tension arc in the middle member and compression arc in the south and north members.

1.3.2.3 Zhongza-Shangri-La Block (II_3)

This block is bounded by Jinsha River Junction Zone in the west and Yidun Island Arc Zone in the east and is a long and narrow spindle-shaped fault block. The oldest exposed strata are Shigu Group (south) and Chamashan Group (north) in Sinian, with a thickness of 4800–11,700 m. Shigu Group is a sequence of fine clastic rocks of metamorphic flysch, which is covered by unconformity of Devonian. Chamashan Group is a sequence of metamorphic carbonate rocks and intermediate-basic volcanic rocks.

The exposed strata in the slope zone on the eastern edge of Zhongza-Shangri-La Block are Upper Permian to Triassic, with a total thickness of nearly 10,000 m. The Upper Permian is a sequence of clastic rocks mixed with basic volcanic rocks (mainly alkaline basalt), carbonate rocks, siliceous rocks and siliceous turbidite, and the number of upper basic volcanic rocks increases, with a large number of diabase invading and radiolarian siliceous rocks appearing. The Lower Triassic and Middle Triassic are mainly a sequence of flysch calcareous sand-slates, with multiple beds of muddy limestone and turbidite, and basic volcanic rocks in the lower part. The Upper Triassic is a sequence of flysch, flysch-like sand-slates with carbonate rocks, siliceous rocks and pyroclastic rocks, and slump breccia with more exotic rocks is developed in the middle-south member. It shows that the strata in this block belong to slope-to-uplift or abyssal plain facies from Late Permian to Early Triassic, and the basic volcanic activity is significant. The crust is extensional, which echoes the opening of Ganzi-Litang Ocean. In the Middle Triassic, there was the replacement of debris deposits from slopes to outer continental shelf and the tension and fault depression, as well as the formation of flysch and flysch-like rocks in the Late Triassic, which corresponded to the expansion of Yidun Back-arc-Basin, belonging to the western passive margin zone of the back-arc-basin.

The Lower Paleozoic in Zhongza Plateau is composed of clastic rocks and carbonate rocks mixed with basic and intermediate-acid volcanic rocks, with a thickness of nearly 10,000 m. From Cambrian to Silurian, clastic rocks gradually decreased, carbonate rocks gradually increased, and volcanic rocks changed from lower basic to upper intermediate-acid features. On the east side of Jinsha River Junction Zone, Silurian stratum has deteriorated into schist with significant structural deformation. It shows that the plateau was active and deposited in Early Paleozoic, with significant metamorphism and deformation. In the Late Paleozoic, the plateau was mainly littoral-neritic carbonate rock deposits, with weak metamorphism and deformation, with clastic rocks at the bottom. Devonian and Upper Permian inter-bedded basic volcanic rocks and were about 5000 m thick. Mesozoic is only distributed in the eastern margin of the block in a small amount, mainly clastic rocks and carbonate rocks of Triassic, with intermediate-acid volcanic rocks in the upper part of Triassic.

Shangri-La Plateau was stably developed in Paleozoic, mainly the sedimentary plateau of carbonate rocks. The basic volcanic activities in Devonian and Late Permian may be related to the extensional environment before the opening of Jinsha River Ocean in the west and Ganzi-Litang Ocean in the east, respectively. The strata deformation and metamorphism have been weak since Paleozoic, and the main part is compound anticline in Baiyinchang. The eastern margin of the plateau is thrust on Triassic, and a series of isoclinal folds and thrust faults that reverse westward are developed on the western side of the plateau from Tanglangding, Tuoding to Zhongcun, showing the sector profile of thrust on both sides of the block.

Volcanic activity in this block started from Cambrian. Volcanic rocks in Early Paleozoic are marine basalts with high TiO_2 content and low MgO content, sometimes accompanied by a small amount of dacite-rhyolite, forming a "bimodal" assemblage, formed in clastic rock or carbonate rock formation period and featuring the property of continental intraplate extensional volcanic rocks. Volcanic rocks

in the Late Permian are the most important volcanic rocks in this block, mainly alkalinity continental intraplate extensional basalt with high TiO_2 content, which has many geochemical characteristics similar to Emei Mountain basalt in the Yangtze Landmass, reflecting that the mantle source areas of their magma may be related. According to the characteristics of symbiotic sedimentary facies, volcanic rocks mainly erupt in marginal trough (such as Boge west trough composed of basalt-deep-water carbonate rock-siliceous rock) and neritic plateau or bay (such as Derong-Guxue area, coexisting with neritic carbonate rocks). The above shows that the Zhongza Block was in continental intraplate extension in most time of Paleozoic (especially in the Permian), corresponding to the extension of the southwest Yangtze Plate, and the intraplate extensional alkaline to transitional basalt in the former island arc period of Shaluli-Yidun Arc Magmatic Rock Zone was also produced.

A few square kilometers of calc-alkaline volcanic rocks mainly composed of basaltic andesite and andesite breccia-agglomerate are exposed in the Lana Mountain, Yidun, which is in fault contact with the surrounding rock strata and may be a nappe from the Jiangda-Weixi volcanic arc in the west.

1.3.2.4 Jinsha River-Ailaoshan Junction Zone (II₄)

The northern member of Jinsha River-Ailaoshan Junction Zone may connected with Ganzi-Litang Junction Zone in the west Dengke-Yushu Area, and from then on, it extends to Yanghu and Guozhacuo areas in Northern Tibet to the West Xijir Ulan Lake. Then it goes south through Batang, Benzilan—the west side of Diancang Mountain, turns to Ailaoshan in the southeast longitude and extends out of the border, connecting with the Gemia Zone in Northern Vietnam.

The mafic–ultramafic rocks, carbonate rocks, basic volcanic rocks, slate, siliceous rocks and other melange in the junction zone range from Devonian to Permian, and the matrix is composed of flysch siliceous rock and siliceous rocks of Permian–Triassic. Ultramafic rocks are distributed in groups and strips, and sometimes chromite can be seen, but complete ophiolite profiles are rare. The volcanic rocks in ophiolite are oceanic ridge—quasi-oceanic ridge basalt, indicating the existence of oceanic crust. Ophiolite was formed from Early Carboniferous to Triassic. Significant deformation, fold, overthrust nappe and ductile translational shear are developed. Metamorphism is dominated by greenschist facies, and kyanite and sillimanite, which represent medium-pressure and high-temperature metamorphism, can be found locally. There are some high-pressure and low-temperature blue schists in the Ailaoshan Zone.

(1) Jinsha River Ophiolitic Zone (II$_{4-1}$)

This zone is bounded by the Gaiyu-Zhongza Fault in the east, the narrow area to the east of Yushu Longbao Lake-Aila Mountain-Xiquhe Bridge-Yangla-Ludian Fault in the west, the Zhongza Block in the east, the Jiangda-Weixi Volcanic Arc in the west, Ludian in the south, Ganzi-Litang Zone in the north in Dengke Zone and then the Xijir Ulan Lake-Tongtian River Zone.

This zone is mainly composed of oceanic ridge basalt, quasi-oceanic ridge basalt, serpentinite (the source rock is harzburgite), conglomerate gabbro, diabase wall, radiolarian siliceous rock, etc., which constitutes decomposed ophiolite or ophiolitic melange. The oceanic ridge basalt is only seen near Jiyidu and is the tholeiite with low K_2O content, medium TiO_2 content and flat REE distribution. The value of $w(Mg^{2+})/w(Mg^{2+} + Fe^{2+})$ is low (0.46). It is formed in Early Carboniferous (C_1). Quasi-oceanic ridge basalt is widely exposed in the rock zone, stretching for hundreds of kilometers from Gaiyuxi, Batang, Derong, Guxue, Benzilan, Luosha and Tuoding to the vicinity of Ludian. The geological environment of quasi-oceanic ridge volcanic rocks is similar to that of the oceanic ridge basalt, but it has more characteristics of continental overflow basalt or oceanic island basalt in lithochemistry and geochemistry. Compared with oceanic ridge basalt, its K_2O content is higher (greater than 0.50%), the TiO_2 content is high, the total REE and LREE enrichment are also high, which may come from the enriched mantle source area under the ridge. The quasi-oceanic ridge basalt in this zone is formed in the Early Carboniferous-Early Permian (C_1–P_1), and its lithology is quite stable, indicating that the ocean basin continues to expand at the same speed. The Zhubalong-Gongka Intra-oceanic Arc Volcanic Zone in Permian is located in Zhubalong, Markam, Xiquhe Bridge-Deqin Gongka and Dongzhulin Temple. The volcanic rocks in the main arc period are calc-alkaline andesite, basalt, basaltic andesite, sodium dacite, etc., which are characterized by high Al_2O_3 content and low TiO_2 content. In Gongka and Dongzhulin Temple zones, this zone forms melange with serpentinite, gabbro, diabase walls and radiolarian siliceous rocks. The back-arc volcanic rocks are only exposed between Zhubalong-Xiquhe Bridge and located on the west side of the main arc zone. They are composed of low-TiO_2 tholeiite, dense sill-like diabase wall, radiolarian siliceous rock, flysch, etc. The volcanic rocks in Zhubalong-Xiquhe Bridge were formed in Early Permian according to radiolarian fossils in Xiquhe Bridge Formation, and those in Dongzhulin Temple and Gongka were formed in a period spanning from Sakmarian, Early Permian to Middle Permian.

Given the above facts and the characteristics of Jiangda-Weixi Arc, it can be determined that Jinsha River

Ocean had a prototype of ocean basin in Late Devonian, and it expanded significantly in Early Carboniferous to form an ocean basin. It began to subduct westward at the end of Early Permian and closed and collided from the end of Middle Permian to Early Triassic.

(2) Ophiolitic melange zone of Ailaoshan (II$_{4-2}$)

This zone is bounded by Ailaoshan Fracture in the east and adjacent to Ailaoshan high-grade metamorphic zone in the basement of Yangtze Landmass and bounded by Amo Jiang Fracture in the west and adjacent to Lvchun Volcanic Arc in the eastern margin of Pu'er Micro-landmass, extends to Vietnam in the south and pinches out near Midu in the north, with a length of more than 240 km in China. Most people believe that the Ailaoshan Zone can be connected with the Jinsha River Zone.

The ophiolitic melange zone is composed of basic rock units such as metamorphic peridotite, cumulate, basic lava, radiolarian siliceous rocks, in which sheeted dyke swarms are not developed. Due to their tectonic decomposition, these rock units do not constitute a complete sequence. Metamorphic peridotite is composed of harzburgite representing depleted upper mantle and lherzolite representing primitive upper mantle. Two kinds of primary magma and two sequences of evolved rock series are generated from partial melting of lherzolite in different degrees: one is pyroxene basalt-gabbro-diabase evolved from primary tholeiite magma, and the other is picrite basalt-albitite basalt-basaltic andesite-gabbro diorite evolved from primary picritic basalt magma. That said, there is a close genetic relation among the mantle peridotite, cumulate and basic lava in the ophiolite complex. Most of the basic lava are characterized by oceanic ridge basalt or quasi-oceanic ridge basalt in terms of lithochemistry, trace and geochemistry for rare earth element and petrography, and their composition points are uniformly distributed in oceanic ridge basalt zone or ocean floor basalt zone on various discrimination diagrams of major, trace and rare earth elements. It is judged as the abyssal sediments based on the ecological environment of radiolarian assemblage in siliceous rocks and the characteristics of silicon isotope and rare earth elements in siliceous rocks. This point is further proved by the fact that the fuchsia radiolarian siliceous rocks were found in Lower Carboniferous stratum in Pingzhang, Xinping County. These characteristics indicate that Ailaoshan ophiolite has a property of oceanic crust.

About formation age of ophiolite. Radiolarian siliceous rocks and radiolarian fossils above Bailadu albitite basalt are identified to be formed in the Early Carboniferous. The data obtained by showed that the $^{40}Ar/^{39}Ar$ whole rock age of gabbro in ophiolite is 339 Ma (Early Carboniferous), and the

U–Pb age of single grain zircon from plagioclase granite variant is 256 Ma (Early Permian). The age value obtained from the isotopic dating for gabbro, basalt and siliceous rock in ophiolite made by Yang and Mo (1993), and is between 345 and 320 Ma (Early Carboniferous). Therefore, it can be reasonably considered that the formation age of Ailaoshan ophiolite shall be earlier than the Early Carboniferous. On a regional basis, Yiwanshui Formation in the Upper Triassic stratum unconformably overlaid on ophiolite, and its basal conglomerate contains ophiolite and chromite debris. It is believed that Ailaoshan ophiolite was formed before the sedimentation in the Late Triassic (Yiwanshui Formation). It also matches with the formation time of the arc volcanic zone in Late Permian and the collisional acid volcanic rocks of Lvchun in Late Triassic.

1.3.2.5 Qamdo-Pu'er Block (II₅)

The development of Qamdo Block in the north of Qamdo-Pu'er Block and Pu'er Block in the south of Qamdo-Pu'er Block may be different in Early Paleozoic, but the evolution of these two blocks is basically similar since Late Paleozoic.

Qamdo Block is a double-bed structure of basement and caprock. Gneiss, schists and granulites of Precambrian Ningduo Group and Xiongsong Group are outcropped in Xiariduo, Xiaosumang and Jiangda in the east, and their source rocks are a sequence of clastic rocks and carbonate rocks mixed with basic volcanic rocks. In the Xiaosumang area, Caoqu Group is located above Ningduo Group, which is a sequence of low-grade metamorphic series. The Early Paleozoic stratum is mainly exposed in areas of Gebo, Qingnidong, Haitong of Markam and Duoji of Yanjing in Jiangda. The Gebo Group in the Gebo area may contain Late Precambrian and Early Paleozoic strata, which is a sequence of island-type volcanic-sedimentary rock series; and Middle-Lower Ordovician strata in Qingnidong and Haitong areas are a sequence of continental slope turbidite fan, clastic rocks of the lower continental shelf and upper carbonate rocks. The Lower Devonian stratum of Silurian is mainly found in Duojiban, Yanjing in the south, which is composed of a sequence of clastic rocks and carbonate rocks. It was developed into a stable carbonate plateau in the late period of Middle Ordovician to Devonian without Caledonian folding orogeny, and the Middle Devonian stratum in Qingnidong spread on the folded Ordovician with unconformity contact. Since Devonian, Qamdo Block has entered a stable stage of caprock development. Devonian stratum is composed of continental to neritic clastic rocks and plateau carbonate rocks and is developed into stromatoporoid reef in the south, but is developed into basic and intermediate-acid volcanic rocks in the Middle and Upper Devonian stratum of Jiangda area. Similar to that in Devonian, a sequence of continental clastic rocks and plateau carbonate rocks as well as sponge reefs was developed in Qingnidong area in Carboniferous. The Lower Carboniferous stratum is composed of coal-bearing clastic rock, and the Upper Carboniferous stratum is composed of carbonate rock mixed with volcanic rock on the west side of the Qamdo-Kaixinling area. A sequence of clastic rocks, clastic turbidite and carbonate turbidite belonging to continental shelf to slope fluvial sand bodies and shoreline sand bodies can be found in Deqin area on the east side. In the Permian, island-type volcanic-sedimentary rock series were developed on the east and west sides in Triassic, especially in Late Triassic, the strata on the south landmass correspond to that on north landmass.

The oldest stratum exposed in Pu'er Block is the Silurian stratum, which is found on the west side of Ailaoshan Zone. The Silurian stratum-Devonian stratum are composed of flysch and graptolite shale, Carboniferous stratum in Carboniferous is a sequence of flysch sand-slate, basic volcanic rocks and carbonate rocks, and Permian stratum is an island-type volcanic-sedimentary rock series; and a sequence of flysch sand-slate (turbidite), turbidite limestone, siliceous rocks and slump breccia accumulated due to gravity flow, basic and intermediate-acid volcanic rock is developed in Longdong River area in the west. The stratigraphic age is dominated by the Carboniferous-Permian. Radiolaria in Devonian is found in the Dapingzhang area of Pu'er. This sequence of stratum may contain Devonian stratum. The Devonian stratum is also a sequence of turbidite sand-slate, siliceous rocks, slump breccia and intermediate-acid volcanic rocks in Nanguang Formation of Jinghong in the south. In addition, it may contain Carboniferous stratum-Permian stratum. They may be connected in the north and the south to form deposition in the passive margin zone in the eastern part of the Lancang River in the early stage and then develop into back-arc-basin deposition after in the later stage.

Qamdo-Pu'er Block can be further divided into 6 tertiary tectonic units.

(1) Jiangda-Jijiading-Weixi Continental Margin Volcanic Arc (II₅₋₁)

This arc is located in Jiangda County-Tongpu-Dongdu-Jiaduoling-Deqin-Jijiading-Pantiange. Arc volcanic rock is a marine-continental assemblage of alkaline basalt-andesite-rhyolite-dacite, and the age is from Early Triassic to Carnian in Late Triassic. It consists of three sequences of volcanic rocks, which are partially overlapped in space. The first sequence is collisional rhyolite-dacite with high contents of SiO_2 and Al_2O_3 in Early Triassic (Pubeiqiao Formation and Malasongduo Formation in Jiangda, which are located in Longqiao area of Markam) and Middle Triassic (Walasi Formation and Pantiange formation, which are mainly located in areas of Jiaolongqiao of Markam-Jiading-Pantiange-Shizhongshan), which is exposed in the south of Deqin along with collisional granite intrusion. Volcanic tuffaceous turbidite sand-slate, siliceous rocks and slump breccia, andesite and andesitic volcanic breccia lava can be found in Jiaolongqiao area. The second sequence is an assemblage of post-collisional (lagging) arc volcanic rock-type andesite-dacite porphyry-rhyolitic porphyry in Late Triassic (represented by Jiangda Formation), with marine-continental facies for the two sequences above. Located in Jijiading area in the middle north margin, the third sequence is an assemblage of basalt, basaltic andesite and radiolarian siliceous slate, while it is an assemblage of spilite, quartz keratophyre and radiolarian siliceous rock in Cuiyibi-Jigaiji areas in the south margin, which were formed in the Late Triassic (Cuiyibi Formation) and are characterized by a bathyal-abyssal environment. Volcanic rocks representing the nature of inter-arc rift basin are also formed after the collision. The Carnian stratum (Jiangda Formation) in the Late Triassic is the most developed. The volcanic rocks in Early Triassic were formed after the accumulation of basale molasses in Pushuiqiao Formation in Lower Triassic stratum, while the volcanic rocks of Carnian in Late Triassic were formed after the accumulation of basale molasses in Jiangda Formation in Upper Triassic stratum, so they are typical post-collisional (lagging) arc volcanic rocks. Rift basins such as Shengda-Chesuo Basin, Xuzhong Basin, Luchun-Hongponiuchang Basin and Reshuitang-Cuiyibi are developed in the superimposed back-arc rift basin from north to south; the superimposed back-arc rift valley basin is a "two-peak" assemblage of tholeiite spilite (low content of TiO_2) and rhyolite, with marine facies and pillow structure; moreover, siliceous rock, carbonaceous slate and a small amount of laminal limestone can be found in this basin. Jiangda-Weixi arc are characterized by the segmentation and inhomogeneity along the strike.

The temporal and spatial distribution of volcanic rocks shows that the Jiangda-Weixi arc was formed by the westward subduction of the oceanic plate of Jinsha River. The subduction started in Late Permian, and the collision started in Early Triassic and ended in Late Triassic. After the collision, the magmatic activity with the characteristics of arc magmatic rocks (lagging arc magmatic rocks) occurred.

(2) Qamdo-Markam bidirectional back-arc foreland basin (II_{5-2})

The back-arc foreland basin was formed on a stable block of Late Paleozoic and exposed on Ordovician stratum-Neogene stratum. Only the Lower Paleozoic stratum is exposed in the Lower Ordovician stratum, which is a sequence of flysch sand-slate mixed with carbonate rocks.

Devonian stratum-Permian stratum is a sequence of continental to stable neritic carbonate rocks and clastic rocks mixed with a small amount of volcanic rocks, and the cold and warm water organisms coexisted in Carboniferous-Early Permian, but it is dominated by warm water organisms. The Upper Permian stratum in the Tuoba area is coal-bearing clastic rock, which is rich in the Cathaysia flora. The Lower Triassic stratum is composed of clastic rocks and acid volcanic rocks mixed with carbonate rocks, with a thickness of about 3000 m. The Paleozoic stratum and early, Middle Triassic strata constitute the Qingnidong-Haitong overthrust zone which thrusts westward. The lower part of Upper Triassic stratum is composed of red molasse, the middle part is composed of limestone mixed with clastic rocks, and the upper part is composed of extremely thick coal-bearing clastic rocks, which constitute the main body-back-arc foreland basin with the characteristics of foreland basin deformation. Jurassic stratum-Cretaceous stratum are a sequence of red molasses and copper-bearing prunosus clastic rocks mixed with gypsum-salt, with a thickness of nearly 10,000 m. It is dominated by continental facies but has local marine facies. Distributed in strike-slip pull-apart basins, Paleogene stratum and Neogene stratum are a sequence of red clastic rocks mixed with intermediate-acid volcanic rocks, coal streaks and gypsum-salts. Folds of Mesozoic stratum in Yulong-Markam area are distributed in an echelon on the right side, reflecting the dextral strike-slip orogeny of the faults in Qingnidong-Gongjue area. The Lower Paleozoic stratum is characterized by relatively significant deformation but no significant metamorphism, and it is dominated by greenschist facies but has local amphibolite facies. The deformation of Upper Paleozoic stratum is not significant, and only relatively significant folds are found in the Lingzhihe Bridge area to the east of Haitong on the edge of the basin. Only slight metamorphism is found in the Upper Paleozoic stratum to Lower and Middle Triassic stratum, while no metamorphism is found basically in Upper Triassic stratum and above strata.

Due to the opposite subduction between the oceanic plate of Lancang River and oceanic plate of Jinsha River, some subduction arc volcanic rocks such as an assemblage of andesite and dacite in Late Permian (Jiageding in Markam and Lingzhihe Bridge in Haitong) and andesite-dacite-rhyolite in Middle-Late Permian were formed in the landmass and developed on inward rifted basalt in continental plate of Baoshan Block and Zhongza Block and then lost in the same period; the rifted volcanic rock in Ze'e Formation of Late Triassic and trachyte in Lawula Formation in Neogene occurred in Qamdo Block. The oceanic plate of Jinsha River subducted below the Qamdo Block and remained in the mantle may provide the source conditions for the Yulong porphyry zone and porphyry-type copper deposits in Himalayan.

(3) Zadoi-Dongda Mountain continental margin volcanic arc (II$_{5-3}$)

This arc is distributed in the western margin of Qamdo-Markam Block and controlled by the eastward subduction of the north oceanic plate of Lancang River. It is an assemblage of island-type tholeiite, alkaline basalt, andesite, dacite with prismatic jointing developed, rhyolite and corresponding pyroclastic rocks. The collisional dacitic-rhyolitic volcanic rocks in the Middle Triassic are developed in Zhuka-Yanjing areas along the Lancang River.

(4) Amojiang-Lvchun continental margin volcanic arc (II$_{5-4}$)

This arc is located on the southwest side of the ophiolitic melange zone of Ailaoshan and consists of two sequences of volcanic rocks in Amojiang-Lvchun of Western Yunnan. The first sequence is the Late Permian calc-alkalic basaltic andesite-andesite-dacite (and corresponding pyroclastic rocks) assemblage of transitional facies; the second sequence is the Late Triassic collisional dacite-rhyolite assemblage (Gaoshanzhai Formation) with high contents of SiO_2 and Al_2O_3 and distributed in Lvchun-Yuanyang. The intrusive rock batholith is also distributed in the southern Lvchun area, with the isotopic ages of 230–211 Ma. It mainly consists of type I biotite monzogranite and moyite and also contains a few two-mica granites, all of which are products formed in the same collision period. In particular, a basic volcanic zone with a length of about 30 km is distributed along Bulong-Dalongkai-Wusu-Wannianqing area, which closes to the east side of arc volcanic zone in Late Permian. Its lithology is mainly pillow basalt and contains a few acid volcanic lava. This zone is formed in marine arenopelitic flysch strata in Carboniferous, which is obviously the product of extensional environment, but its exact tectonic properties remain to be investigated.

Collisional granitoids formed in Late Indosinian include Renda, Anmeixi, Jiaren, Baimang Snow-capped Mountain, Ludien, Datuan, Xinanzhai, Bade, etc. from north to south, all of which are distributed in strips along both sides of Jinsha River Junction Zone and belong to normal aluminum-super-saturated series rocks, with an initial value for $^{87}Sr/^{86}Sr$ of 0.7175.

(5) Lanping-Pu'er bidirectional back-arc foreland basin (II$_{5-5}$)

The exposed strata in this basin are Silurian stratum-Neogene stratum. Silurian stratum-Devonian stratum are composed of flysch and graptolite shale, and Carboniferous stratum-Permian stratum are composed of littoral-neritic rocks and carbonate rocks, with coal-bearing clastic rocks in the upper part. The whole Paleozoic stratum is nearly 10,000 m thick. Carboniferous stratum and subsequent strata are similar to strata in Qamdo area in the north, but lack a Lower Triassic stratum. The volcanic rocks and intermediate-acid intrusive rocks in this zone are not developed.

The Mesozoic stratum is characterized by deformation of the foreland basin. Folds and thrusts are developed on the east and west sides of the basin due to the back-arc thrust and the westward thrust nappe of Diancang Mountain and Ailaoshan Zone, but the folds became wide and gentle toward the center of the basin. Nappe tectonic group and some dome structures are developed in the Lanping-Yunlong area. The nappe consists of the overturned succession of strata from Waigucun Formation of Upper Triassic stratum to Jingxing Formation of Lower Cretaceous stratum, indicating that it is formed by deformation and displacement of the basement of overlying strata in Lanping-Jiangcheng depression zone, such as the detachment, folding, thrust nappe and slip detachment. Dome structure may be formed by plastic flow of detachment bed and uptrusion of diapiric folds. The fracture zone at the basement of nappe (or slip nappe) has become an important ore-bearing space for Lanping super-large lead–zinc deposit.

Due to the northward pushing of the Indian Plate and the high compression stress caused by the blocking of Yangtze Landmass during Himalayan, Deqin-Weixi area in the waist is significantly compressed, and Qamdo Block in the north and Pu'er Block in the south are extruded toward both ends, forming a group of conjugate strike-slip fault systems in the depression zone. A right strike-slip pull-apart basin and a left strike-slip pull-apart basin are formed in Gongjue area on the east side and in Nangqian area on the west side of the northern block, respectively. A left strike-slip pull-apart basin (Qiaohou and Weishan area) and a right strike-slip pull-apart basin (such as Lanping-Yunlong Basin, Zhenyuan Basin and Jiangcheng-Mengla Basin) are formed on the east side and on the west side of the southern block, respectively. Most of these strike-slip pull-apart basins are important metallogenic basins.

(6) Yun Country-Jinghong continental margin volcanic arc (II$_{5-6}$)

This arc is mainly distributed on the east side of Lincang granite belt and spread basically along Lancang River Valley, and it is mainly composed of two sequences of volcanic rock series, which are overlapped: ① Permian andesitic-dacitic-rhyolitic volcanic rocks, mixed with Late Carboniferous intermediate-acid volcanic rocks sporadically exposed, are mainly distributed in the southern part of the volcanic arc and the western margin of Pu'er Basin; ② Triassic collisional volcanic rocks, post-collision arc volcanic rocks and post-collision extensional volcanic rocks, including (from old layer to new layer): Middle Triassic (Manghuai Formation) collision dacite-rhyolite assemblage → Late Triassic (Xiaodingxi Formation) post-collision shoshonite-latite assemblage (northern member in Yun County) → assemblage of basalt-andesite-dacite-rhyolite (southern member) → Late Triassic (Manghuihe Formation) post-collision extensional kalisyenite trachybasalt-rhyolite "bimodal" assemblage.

1.3.2.6 Lancang River Junction Zone (II$_6$)

Most scholars believe that the southern member of Lancang River Junction Zone is located on the east side of Lincang magmatic arc, extending southward to Jinghong (border) and connecting with the Nan River zone in Thailand. The north member extends from Caojian to the north, passes through Yingpan of Lanping, the west side of Nanzuo, Meri Snow Mountain and Zhayu of Zuogong and then reaches the Qudeng-Jitang fracture zone, which may be the northwest extension zone of Lancang River Junction Zone. It is connected northwestward with the Ulaan-Uul-Northern Lancang River between Northern Qiangtang and Qamdo Block.

This zone is mainly composed of oceanic ridge basalt, mafic–ultramafic cumulate complex, serpentinite and radiolarian siliceous rock, belonging to ophiolitic melange and representing the remnants of the Lancang River Ocean after its closure.

According to the analysis of the temporal and spatial relationship between this rock zone and the collisional-post-collisional magmatic rock zone (P–T$_3$) of Yun Country-Jinghong arc, Lancang River Ocean was opened in Early Carboniferous, the oceanic crust began to subduct eastward under the Qamdo-Pu'er Block in Early Permian, and the ocean basin was closed and the arc collided with land in the Middle Triassic.

1.3.2.7 Zuogong Block (II$_7$)

This block is located between Lancang River Junction Zone and Nujiang River Junction Zone and is covered by the nappe of Gaoligong Mountain from the south of Chawalong to the east of Bijiang River. There is only one narrow belt in Zuogong-Riwoqê-Yaanduo area in the north of Chawalong.

The lower part of Jitang Group has high-grade metamorphism and local migmatization, with amphibolite facies. The Lower Paleozoic stratum is a sequence of low-grade metamorphic clastic rocks mixed with carbonate rocks and metamorphic volcanic rocks; Devonian stratum and Permian stratum are the passive margin sedimentary zone on the west side of Lancang River Ocean, which is a sequence of neritic-bathyal fine clastic rocks, siliceous rocks mixed with volcanic rocks and carbonate rocks. Coal-bearing clastic rocks are found in Upper Permian stratum, which belong to residual marine sediments. Jiapila Formation in Upper Triassic stratum was formed by molasse accumulation, and Bolila Formation was formed by the marine carbonate rocks and coal-bearing clastic rocks in Adula Formation and Duogaila Formation and unconformably overlaid on underlying strata. The Paleogene stratum and Neogene stratum are composed of continental coal-bearing clastic rocks.

Biluo Snow Mountain-Chongshan Block is distributed in the east of Puladi Fault (ductile strike-slip shear zone in the Nujiang River Junction Zone) and the west of the ductile shear zone of Biluo Snow Mountain. It is exposed in a narrow strip due to intense extrusion and shear deformation. The block is mainly composed of high-grade metamorphic rock series of Chongshan Group Complex in Proterozoic stratum, Mode Formation Complex of Carboniferous stratum, volcanic-sedimentary rocks and acid magma intrusion of Permian stratum.

Chongshan Group Complex can be roughly divided into two sequences of rock assemblages. The first sequence is biotite plagioclase gneiss, Amphibolite granulite and siliceous biotite garnet gneiss with significant migmatization, with the characteristics of parautochthonous anatectic granite intruding migmatized metamorphic supracrustal rocks, in which vein flow folds, rootless intrafolial folds and hook folds are developed, showing the structural feature of deep plastic flow rheology. The other sequence is composed of biotite quartz schist, amphibole plagioclase granulite, marble and hornblende schist, with obvious foliation transposition; S$_2$ foliation is characterized by permeability, which is manifested as bedding shear, development of concealed folds; and the rocks generally show high-grade greenschist facies metamorphism.

Mode Formation Complex is a metamorphic body of Carboniferous relict sediments, and its sedimentary formation is significantly different from that of the Carboniferous stratum in the Nujiang River Zone to the west and is dominated by coarse clastic glutenite mixed with mudstone, siliceous rocks, carbonate rocks and a small amount of volcanic rocks. The lithic quartz graywacke and mudstone form a very frequent sedimentary rhythm, which are a sequence of products of the slope environment at the margin of the block due to rapid accumulation of terrigenous clast.

Jidonglong Formation of Permian stratum is composed of clastic rocks, volcanic rocks and pyroclastic rocks inter-bedded with unstable carbonate rocks. Volcanic rocks are a sequence of basalt-andesite-dacite-rhyolite and tuff assemblages. The environment of sedimentary rocks varies greatly, including abyssal turbidite, adlittoral plateau and slump breccia, which is generally a process of environmental evolution of volcanic arc tectonic facies. ^{40}Ar/^{36}Ar isochron age of granodiorite in Biluo Snow Mountain-Zhazhuqing area in the block is 221.9 Ma (Liu et al. 1999); in addition, the intrusion of granite in Cretaceous was developed.

1.3.2.8 Lincang Magmatic Arc (II₈)

The main body of the magmatic arc basement is the Lancang Group Complex of Neoproterozoic stratum. Most of Damenglong Group Complex of Mesoproterozoic stratum have been covered by intrusive magmatic rocks and distributed sporadically in a form of massive rock. The rock types mainly include biotite plagioclase granulite mixed with biotite plagioclase gneiss. Lancang Group Complex had been reformed by several tectonic thermal events. So, it is generally dominated by the quartz-mica tectonic schist, mixed with shear lenses such as granulite, phyllite, marble and metamorphic basic rocks. The metamorphic strata involved in Suyi blueschist zone adjacent to Changning-Menglian Junction Zone in the west side are mainly Lancang Group Complex in the Neoproterozoic stratum, and its high-pressure metamorphic mineral assemblage is characterized by subduction type kinetic metamorphism, which obviously forms a pair metamorphic zone with the high-temperature metamorphic assemblage on the east side being represented by Lincang granite and andalusite.

The main body of the magmatic arc is Lincang composite granite batholith, with isotopic age of 290–208 Ma. The batholith is exposed for more than 300 km (covering an area of $1 \times 10^4 \text{ km}^2$) and distributed on the east side of Changning-Menglian Junction Zone. It is mainly composed of emplaced tonalite and granodiorite of Permian and is mainly characterized by general gneissic structure and dominated by type "I" granite in its petrological and geochemistry terms. Emplaced granodiorite and monzogranite of Triassic are exposed for large areas and are mainly characterized by complex rock types and dominated by type "S" arc granite in its petrogeochemistry terms.

Most of the magmatic arc caprocks are denuded, and the remaining sedimentary caprock is continental red beds of Huakai Formation of Middle Jurassic stratum. It is worth noting that the thrust nappe of the magmatic arc zone from west to east was restricted by its significant regional inland crust deformation in Paleogene, so the syntectonic monzonitic moyite in Paleogene is developed into vein, lenticular and stock-like emplacement.

1.3.3 Bangong Lake-Shuanghu-Nujiang River-Changning-Menglian Mage-Suture Zone (III)

1.3.3.1 Bangong Lake-Nujiang River Junction Zone (III₁)

The characteristics of the junction zone in the Nujiang River zone south of Chawalong are not obvious in the east member of Bangong Lake-Nujiang River Junction Zone (III₁), and the Gongshan-Gaoligong Mountain Thrust (or Nappe) Zone lies to the west of the Nujiang River Fault. Ophiolitic zone (with abyssal flysch in Late Triassic to Early-Middle Jurassic as matrix) mixed with mafic rocks, ultramafic rocks, limestone, marble, siliceous rocks and abyssal mudstone is found between Dingqing and Chawalong. Relatively complete ophiolite complex is preserved in the Dingqing area, and its ultramafic rocks belong to the magnesian type. Located between Dingqing and Basu, Jiayuqiao Group (III₃) looks like a large tectonic len sandwiched in the junction zone, just like a tectonic terrane.

The ophiolitic zone between Dingqing and Basu is a part of the whole Bangong Lake-Nujiang River Junction Zone, and a relatively complete ophiolite assemblage can be found in Dingqing and its west, in which basalt is of pillow structure and is close to oceanic ridge basalt in composition. Volcanic rocks can be found, and ultramafic rocks (serpentinite) are mainly exposed in the area from the east of Dingqing to Bangda. Ophiolite was formed in the Late Triassic to Early Jurassic. This zone extends southward and then covered by the huge nappe zone of Gaoligong Mountain, so it is formed in the Middle Jurassic stratum and then passes through Chongshan metamorphic deformation zone in the southeast to connect with the Changning-Menglian Junction Zone.

1.3.3.2 Changning-Menglian Junction Zone (III₂)

Changning-Menglian Junction Zone (III₂) starts from Changning and Shuangjiang in the north, passes through Tongchangjie and Laochang to Menglian and extends to Myanmar from the south. Oceanic ridge basalts with N-MORB characteristics were found in Manxin of Menglian and Tongchangjie, and quasi-oceanic ridge basalts were found in Manxin, Yiliu and Tongchangjie, etc. The age of Tongchangjie is Middle Devonian (Zhangqi, isotope age: 385 Ma), while the age of the rest is Early Carboniferous. Associated siliceous rocks are composed of siliceous rocks of pelagic uncompensated basin, in which radiolarias are pelagic abyssal assemblages. This shows that a quite wide Paleo-Tethyan Ocean Basin existed from Carboniferous to Early Permian, with an estimated maximum width of 1367 km. Picrites with similar composition among many beds, lenticular shape and pillow structure are found in oceanic ridge basalts and quasi-oceanic ridge basalts in Manxin and Menglian. They are formed by the condensation of magma with a large number of accumulated olivine crystals ejected from the sea floor due to the rupture of magma chamber under the spreading ridge. In addition, well-developed sheeted dyke swarms were found in areas like Tongchangjie. All these proves the existence of the paleo-oceanographic spreading ridge. Oceanic island basalts in Carboniferous-Permian were exposed in Manxin, Yiliu, Laochang and other areas, located on oceanic ridge and quasi-oceanic ridge basalts in sequence and formed the basalt-limestone assemblage in the ridge together with the limestone strata overlapped on them, with the transitional relationship between them. The discrimination of oceanic island basalt is a difficult problem in the study of rocks and tectonics. It is difficult to distinguish volcanic rocks from enriched basalts formed in other tectonic environments only according to their lithochemistry and geochemistry characteristics, and it is also necessary to combine the analysis data of sedimentary facies. According to the study made by, the limestone in the basalt bed in the areas above does not contain terrigenous clast, but is composed of plateau carbonate rocks far away from the mainland.

In recent years, ophiolite melange and metamorphic rock of high-pressure eclogite-blueschist in Early Paleozoic have been discovered through 1:50,000 regional geological survey and related monographic study. Ophiolitic melange in Early Paleozoic was formed in the Changning-Menglian junction zone in the form of structural "blocks" with various scales and distributed in the areas of Mengyong-Manghong-Nanting River-Ganlongtang-Niujingshan in the nearly north–south direction, mainly composed of serpentinite pyroxenite, serpentinite olivine pyroxenite, metapolycrystal gabbro, metagabbro, metabasalt, plagioclase amphibolite, epidote chlorite actinolite rock, albitite epidote chlorite schist, amygdaloidal basalt, basaltic andesite, siliceous rock and low-grade metamorphic argillaceous siltstone. The U–Pb age of zircon in the cumulate gabbro and gabbro from Nanting River varies from 439.0 to 453.9 Ma, and it can be concluded from geochemistry characteristics of rocks that it was formed in the oceanic ridge (Wang et al. 2013); the U–Pb age of zircon in Niujingshan metagabbro or plagioclase amphibolite (schist) varies from 428.5 to 450.5 Ma, and it can be concluded from geochemistry characteristics of rocks that it was formed in the oceanic ridge (Regional Geological Survey Report on a Scale of 1:50,000 in Shuangjiang County, 2019); the U–Pb age of zircon in the amygdaloidal basaltic andesite in Laonanzhang, Mengyong is 449.3 ± 8.4 Ma, and it is characterized by azores-type oceanic island in geochemistry terms (Sun et al. 2017); the U–Pb age of zircon in metamorphic gabbro of Manxin is 420 Ma (Wang et al. 2018). Newly discovered ophiolite melange in Early Paleozoic and widely distributed ophiolite melange in Late Paleozoic shows the geological history of the continuous evolution of Sanjiang Proto- and Paleo-Tethys; regionally, Sanchahe Formation in Upper Triassic stratum unconformably overlaid on ophiolite melange, which is an important symbol of basin-range transition in Tethyan tectonic zone.

Metamorphic rocks of high-pressure eclogite-blueschis are distributed in the nearly north–south direction for more than 100 km, along which the high-pressure low-temperature blueschist, high-pressure medium-temperature blueschist, high-pressure medium-temperature eclogite, degenerative amphibole eclogite and so on can be found.

Among them, eclogite is exposed from Bingdao, Kongjiao, Genhen River and Bangbing in Mengku Town, Shuangjiang, in the north and extends to Nanpen of Jinghong and Mengsongba in the south through Qianmai of Lancang it is formed in the metamorphic rock series of the "Lancang Group and Damenglong Group" in Precambrian stratum in the form of tectonic len with various scales, and the typical minerals include omphacite, jadeite, lawsonite, coesite, phengite, glaucophane, pyrope and rutile. The restored source rocks are mainly tholeiite with similar geochemistry characteristics as E-MORB, followed by alkaline basalt with similar geochemistry characteristics as OIB (Sun et al. 2017). The U–Pb ages of zircon in eclogite are 801.0 ± 9.8 Ma, 227.0 ± 12 Ma, 447.5 ± 3.6 Ma, 291.7 ± 6.3 Ma, 429 ± 2.4 Ma, 231 ± 2.3 Ma, 254 ± 1.4 Ma and 229.0 ± 1.3 Ma (Regional Geological Survey Report on a Scale of 1: 50,000 in Shuangjiang County, 2019; Sun et al. 2018), and 40Ar/39Ar plateau ages of glaucophane are 409.8 ± 23.6 Ma, 279 ± 1.6 Ma, 215 ± 3.3 Ma and 214 ± 0.9 Ma (Zhai 1990; Zhao et al. 1994), and the evolution history of the subduction and accretion, collision orogeny and decomposition of the Proto- and Paleo-Tethys has been recorded.

1.3.3.3   Yuqiao Residual Arc Zone (III$_3$)

Yuqiao residual arc zone (III$_3$) is dominated by Jiayuqiao Group (Pz$_2$), which constitutes a composite anticline in NW–SE direction. The age of Jiayuqiao Group is still controversial up to now, and then it is determined as a sequence of clastic rocks mixed with carbonate rocks with greenschist facies in Late Paleozoic based on the age of fossil collected and lithologic assemblage characteristics in the regional geological survey on a scale of 1:250,000 (Tibet Institute of Geological Survey, 2007). E'xue Group Complex (C-P) is distributed in the E'xuexiong area, Tongka Township, Basu County in NW–SE direction, which is a sequence of clastic rocks with middle-low greenschist facies mixed with carbonate rocks and basic volcanic rocks. The U–Pb age of SHRIMP zircon in hypoamphibole serpentinite is 267 ± 8 Ma (on a scale of 1:250,000 in Qamdo County, 2007). Zhayu area in Zuogong Country (C–P) is called Rongzhong Group Complex, which is mainly a sequence of clastic rocks with middle and low greenschist facies mixed with assemblage of carbonate rocks, metamorphic basalt, metamorphic rhyolite and siliceous rocks. Faults and ductile shear structures exist between the group complexes and formation complexes above, and significantly deformed muscovite quartz schist, amphibole quartz schist and glaucophane schist sandwiched in greenschist are found in the Bangdashegu Reservoir area; with the tectonic deformation being characterized by large-scale bedding ductile shear zone and bedding concealed fold, it is located in the accretionary complex structure-stratigraphic system.

Most of the residual arc blocks in Mesozoic are uplifted, and littoral-neritic clastic rock and carbonate rock assemblage are only deposited in the marginal zone. The seawater receded in the Late Jurassic and then continental molasse deposition developed. In addition to relatively significant basic volcanic activity in Paleozoic, the intermediate-acid and a small amount of basic volcanic rocks causing magmatic activities of this zone are mainly developed in the lower part of Mesozoic stratum. The intermediate-acid intrusive rocks in Jurassic are boron-rich continental crust remelting granites formed in collide orogenic stage, which may provide a formation environment for arc magmatic rocks related to the southward subduction of Bangong Lake-Nujiang River Ocean.

1.3.4 Gangdise-Gaoligong Mountain-Tengchong Arc-Basin System (IV)

1.3.4.1 Baoshan Block (IV$_1$)

Baoshan Block is located between Nujiang River southern fault and Changning-Menglian Junction Zone, with exposed strata formed in Sinian stratum to Neogene stratum. The Middle and Lower Cambrian stratum in Sinian stratum is composed of flysch sand-slate mixed with volcanic rocks and siliceous rocks, which are characterized by turbidite and relatively active transitional sediment and are gradually developed into sediment of stable block type neritic clastic rocks and carbonate rocks in Late Cambrian to Permian. Coarse clastic rocks and magnesian carbonate rocks are developed in the west of Ordovician stratum to Devonian stratum, and the water body gets gradually deeper while flowing to the east of Devonian stratum, indicating that its western part is close to the provenance and the eastern part is close to the relatively deep basin. Clastic rocks containing glacial boulders and cold-water fauna Eurydesma, etc. as well as eruption of basalt and basaltic andesite were found in the Upper Carboniferous stratum. Carboniferous stratum is lack of Middle stratum, while Permian stratum is lack of Lower stratum. Multi-layered carbonate rocks in Paleozoic are important nonferrous metal ore-hosting formation. Mesozoic stratum unconformably underlaid on the underlying strata of different ages, which is a sequence of clastic rocks mixed with intermediate-basic and intermediate-acid volcanic rocks, with red molasse accumulated at the top. Molasse is mainly distributed on the east and west sides. The Pliocene stratum in Neogene stratum is composed of glutenite and coal-bearing clastic rocks, with limited distribution. Volcanic rocks in this zone mainly include basalt formed in the Late Carboniferous and basalt-rhyolite formed in the Late Triassic. Riebeckite quartz syenite and alkali granite body containing aegirine and riebeckite are exposed

in Muchang Township, Zhenkang Country. The tectonic deformation and metamorphism in Baoshan Block are very weak.

Baoshan Block and Yaando-Gengma passive continental margin in Paleozoic to its east side implicitly reflect the continuous evolution process of Paleozoic Tethys Ocean happened in its east. Volcanic rocks are mainly distributed in Baoshan-Zhenkang area and are mainly composed of continental flood tholeiite in Late Carboniferous. They are characterized by the widespread inclusion of quartz standards mineral molecules, and their geochemistry characteristics are similar to those of basalt in Deccan peninsula, reflecting that their mantle primary development regions have some similarities.

1.3.4.2 Shading-Lhorong Fore-Arc-Basin (IV$_2$)

The depression zone is located in the south of Bangong Lake-Nujiang River Junction Zone and north of Delong-Chongsha Fault, and the exposed strata include Triassic stratum-Neogene stratum. The Triassic stratum is composed of flysch sand-slate (a thickness of nearly 10,000 m) mixed with volcanic rocks and siliceous rocks and the magmatic arc fore-arc accretion sedimentary system of Boshula Ridge in the south of Nujiang River Ocean. For example, areas in the north such as shading area are composed of bathyal-abyssal turbidite and siliceous rocks, while Luolong-Bianba area in the south is composed of littoral-neritic clastic rocks. The lower Jurassic stratum lacks a lower stratum. The lower part of the Middle Jurassic stratum is composed of red, variegated glutenite, sand shale and limestone, partially mixed with basalt; and the upper part is composed of grayish black shale mixed with sandstone and a small amount of limestone, which are formed by fore-arc depression sedimentation. The Cretaceous stratum is composed of coal-bearing clastic rock, which indicates that the depression zone is coming to an end. The Paleogene stratum and Neogene stratum are composed of intermediate-acid and slightly alkaline volcanic rocks and sand shale, which are formed by rift basin sedimentation. Cretaceous stratum, Paleogene stratum and Neogene stratum unconformably contact with underlying strata.

1.3.4.3 Bowo-Tengchong Magmatic Arc (IV$_3$)

This magmatic arc is located in the west of Dingqing-Basu Ophiolite Melange Zone, which is generally considered as the southeast extension of Gangdise-Nyainqentanglha composite island arc zone, bounded by Delong-Chongshao fault in the north and adjacent to Shading-Luolong fore-arc-basin and separated from Xiachayu-Sudian metamorphic magmatic melange zone in the south by Bowo-Xiachayu-Binglang River fault. The exposed strata extend from Precambrian stratum to Neogene stratum. Gaoligong Mountain Group and Guqin Group of Precambrian stratum (mainly composed of Paleozoic stratum) and "Precambrian stratum" in Ridong area are a sequence of schist, gneiss, marble and migmatite, in which many mylonite zones are developed. Its source rock is mainly composed of pelites mixed with carbonate rocks, basic and intermediate-basic volcanic rocks. Lower Ordovician stratum are composed of carbonate rocks, but are lack of Middle and Upper Ordovician stratum and Silurian stratum. Devonian stratum-Permian stratum are a sequence of clastic rocks mixed with carbonate rocks and basic volcanic rocks. The lower part of the Devonian stratum is composed of continental clastic rocks, which overlays on Lower Ordovician stratum in the form of para-unconformity. Gravel-bearing slates with glacial marine sedimentation were developed in Carboniferous stratum-Permian stratum. A sequence of abyssal turbidite and slump breccia was developed in the Lower Carboniferous stratum in the Laigu and Songzong area, showing the geotectonic pattern of alternating uplift with depression. The intermediate-basic-acid volcanic rock assemblage such as metamorphic basalt, andesite, dacite and rhyolite of Laigu Formation in Late Carboniferous and its geochemistry properties show the characteristics of arc volcanic rocks at the convergent edge, which shall be regarded as the information that the southwest side of Tethys Ocean has been transformed into active margin.

Volcanic rocks formed in the Jurassic-Cretaceous are mainly composed of calc-alkalic rock series, such as basalt-andesite-rhyolite and corresponding pyroclastic rock assemblage, which coincide with the subduction of Nujiang River Ocean Plate and the age determined for ophiolite. This zone extends westward and connects with the volcanic arc in the northern part of Gangdise-Nyainqentanglha. The temporal and spatial relationship between this zone and the Dingqing-Basu ophiolite (melange) zone indicates that the Bowo-Tengchong arc may be a continental margin arc formed under Lhasa Block due to the westward subduction of the Nujiang River Ocean Plate. The Tengchong volcanic group is a typical representative of the super-position of post-collisional potassic volcanic rocks in Cainozoic. Paleogene stratum and Neogene stratum were formed by accumulation of fluviatile-lacustrine glutenite, coal-bearing clastic rocks, basic and intermediate volcanic rocks and pyroclastic rocks.

Granites are well developed in the zone, and the northern part of the granite zone is composed of 4 large linear rock zones distributed in parallel in NW direction, with an age limit from Late Triassic, Jurassic, Early Cretaceous to Neogene, mainly between 140 and 40 Ma. The age decreases gradually from northeast to southwest. Type "I + S" granite of Paleogene in Xiachayu near the Yarlung Zangbo River Junction Zone forms a composite batholith, which is quite different from the typical continental margin arc magmatic zone. Among them, type "S" granite is

composed of granite-alkali-feldspar granite, which is the parent rock of Sn and W mineralization in this zone.

1.3.4.4 Xiachayu Magmatic Arc (IV₄)

This magmatic arc zone is located in the southwest of the study area and is mainly a sequence of metamorphic complex and acid intrusive rocks in Himalayan with significant mylonitization and migmatization. The age of this metamorphic complex is unknown. The Rb–Sr age of the migmatitic gneiss in Yingjiang area in the south is between 1102 and 806 Ma. High-grade metamorphic schists and gneiss with unclear contact relationship (which may be Paleozoic stratum or older rock bed) are found below Permian stratum in Chayu area in the north.

Magmatic activity in this zone is very intense, especially the intermediate-acid intrusive rocks. It is generally recognized that Bowo-Tengchong main arc zone turns westward from Bowo and connects with the intermediate-acid rock zone in Gangdise, Tibet. Although both of them were formed from Late Yanshanian to Early Himalayan, the rocks in the two zones are quite different. Gangdise Zone is mainly composed of syntectic type (type I) granites and less of type S granites and is distributed with contemporaneous intermediate-basic volcanic rocks, so it has characteristics of typical continental margin arc magmatic zone. Bowo-Chayu areas are composed of type S, type I and transitional I-S type granites, while this zone in Himalayan is mainly composed of type I granites with primary mantle development characteristics and also includes type S granite formed in collision period. The Tengchong area in the south is mainly composed of type S granite and less of type I granite but is not distributed with contemporaneous volcanic rocks, so it does not have the characteristics of island arc magmatic zone. In addition, some linear batholiths distributed along Gaoligong Mountain have the characteristics of type I-S granite, which may be the products formed in the same collision period.

The tectonic deformation in this zone is relatively significant. The two metamorphic nappes in Gaoligong Mountain and Xiachayu-Sudian area are composed of many small thrust sheets or tectonic slices and are distributed with many mylonitization zones and ductile shear zones. An arc overthrust nappe protruding eastward (Lhari-Chongsha-Gaoligong Mountain-Ruili-Mandalay) forms a left oblique thrust at the north wing of the arc, a right oblique thrust at the south wing, a positive thrust at the top of the arc and a series of extensional (anti-slip) normal faults at its rear margin. A series of basins formed in Cainozoic and parallelly distributed in Tengchong-Yingjiang area forms a tectonic pattern of alternating basin with ridge, and its formation may be related to extensional collapse. Rawu and Laigu areas also have relatively significant fold deformation, but the latter is even more significant than the former, reflecting the difference of the graben-horst zones in tectonic deformation. In the grade of metamorphism, the metamorphism of Preordovician stratum is of high grade, with amphibolite facies generally and significant migmatization; the metamorphism of Ordovician stratum and Upper Paleozoic stratum is of low grade, with low greenschist facies.

Boshula-Gaoligong Mountain nappe was formed in the Jurassic, and Liuku-Mengga foreland depression was formed in the front margin of its south member. During Himalayan, the nappe structure was developed further, the mountain system rose sharply, and extensional collapse occurred in the rear margin under the balance of gravity, which may be related to Cainozoic basin-range structure.

1.3.4.5 Yarlung Zangbo River Junction Zone (IV₅)

Located between Gangdise magmatic arc and the Kangmar-Lhünzê thrust fold zone in the northern Himalayan, the Yarlung Zangbo River Junction Zone is a long and narrow significant tectonic thrust zone. Pelagic-abyssal radiolarian siliceous sediments, mafic volcanic lava, sheeted dyke swarm, cumulate complex and ultramafic rocks, as well as abyssal plains, turbidite fans and a few flysch sedimentary wedges in fore-arc-basins with neritic bed as the main body in Triassic and Cretaceous are developed in this zone. Many ophiolite zones, ophiolitic melange and other plate boundary marker composed of ultrabasic-basic rocks, basic volcanic lava, radiolarian siliceous rocks and flysch are exposed along the Yarlung Zangbo River tectonic zone, representing an important junction zone in Mesozoic Gondwanaland and the traces of the final destruction of the East Tethys Ocean. The junction zone is composed of Yumen ophiolite tectonic thrust sheet in the Triassic, Zedang-Luobusha-Nang County ophiolitic melange tectonic thrust sheet and Langjiexue fore-arc accretionary wedge in the Late Triassic. Yumen ophiolite tectonic thrust sheet in Triassic is the oldest ophiolite in the southern margin of the Yarlung Zangbo River Junction Zone, and it is composed of ultrabasic rocks dominated by lherzolite, basic rocks dominated by altered diabase and basalt and distal turbidite and siliceous rocks. Although the ophiolite in Zedang-Luobusha-Nang County ophiolitic melange tectonic thrust sheet has suffered significant structural damage, the original sequence similar to the typical oceanic crust member of Troodos can be roughly recovered, namely metamorphic peridotite, cumulate complex, gabbro-diabase, pillow basalt, radiolarian siliceous rock and plagioclase granite from bottom to top. Langjiexue fore-arc accretionary wedge in Late Triassic is composed of turbidite fans, abyssal plains and adlittoral basin sediments of Songre Formation in Late Triassic, Jiangxiong Formation in the Norian and Zhangcun Formation in Norian-Rhaetian. From the end of Cretaceous to Oligocene, the ocean basin of Yarlung Zangbo River was closed, Himalayan Landmass collided with Gangdise Landmass in succession, and a large-scale remelting granite intrusion occurred in Gangdise, forming Gangdise syntectonic type S granite zone and related tungsten, tin and pegmatite-type niobium, tantalum and gem minerals.

References

Boulin J (1991) Structures in southwest Asia and evolution of the eastern Tethys. Tectonophysics 196:211–268

Huang JQ, Chen BW (1987) The evolution of the Tethys in China and adjacent regions. Beijing, Geological Publishing House, pp 1–78 (in Chinese with English abstract)

Kriegsman LM (1995) The Pan-African event in East Antarctica: a view from Sri Lanka and the Mozambique Belt. Precambr Res 75:263–277

Kröner A, Zhang GW, Sun Y (1993) Granulites in the Tongbai Area, Qinling Belt, China: geochemistry, petrology, single zircon geochronology, and implications for the tectonic evolution of eastern Asia. Tectonic 12:245–255

Lemoine M, Trümpy R (1987) Pre-oceanic rifting in the Alps. Tectonophysics 133:305–320

Li XZ, Xu XS, Pan GT (1995) Evolution of the Pan-Cathaysian landmass group and Eastern Tethyan tectonic domain. Sediment Geol Tethyan Geol 4:1–13 (in Chinese with English abstract)

Liu DZ, Wang GZ, Li YG, Zhu LD, Tao XF, Xu XH (1999) New progress in the study of isotope geochronology of the northern segment of the Lancang river suture zone. Reg Geol China 18:334–335 (in Chinese with English abstract)

Liu ZQ, Li XZ, Ye QT et al (1993) Dividing of tectono-magmatic belts and distribution of the ore deposits in Sanjiang region. Beijing, Geological Publishing House, pp 1–246 (in Chinese)

Lu SN (2004) Comparison of the Pan-Cathaysian orogeny with the Caledonian and Pan-African orogenies. Geol Bull China 23:952–958 (in Chinese with English abstract)

Lu SN, Yu HF, Li HK, Chen ZH, Wang HC, Zhang CL, Xiang ZQ (2006) Early Paleozoic suture zones and tectonic divisions in the "Central China Orogen" Geol Bull China 25:1368–1380 (in Chinese with English abstract)

Mo XX, Lu FX, Shen SY (1993) Sanjiang Tethys volcanism and related mineralization. Beijing, Geological Publishing House, pp 1–267 (in Chinese with English abstract)

Pan GT, Chen ZL, Li XZ, Yan YJ (1997) Tectonic evolution of the East Tethys geology. Beijing, Geological Publishing House, pp 1–218 (in Chinese with English abstract)

Shackleton RM (1996) The final collision zone between east and west Gondwana: where is it? J Afr Earth Sc 23:271–287

Suess E (1893) Are great ocean depths permanent? Nat Sci 2:180–187

Sun ZB, Zeng WT, Zhou K, Wu JL, Li GJ, Huang L, Zhao JT (2017) Identification of Ordovician oceanic island basalt in the Changning-Menglian suture zone and its tectonic implications: evidence from geochemical and geochronological data. Geol Bull China 36:1760–1771 (in Chinese with English abstract)

Sun ZB, Li J, Zhou K, Zeng WT, Wu JL, Hu SB, Li GJ, Zhao JT (2018) Zircon U-Pb age and geological significance of retro-graded eclogites from Mengku area in western Yunnan province. Geol Bull China 37:2032–2043 (in Chinese with English abstract)

Wang BD, Wang LQ, Pan GT, Yin FG, Wang DB, Tang Y (2013) U-Pb zircon dating of early Paleozoic gabbro from the Nantinghe ophiolite in the Changning-Menglian suture zone and its geological implication. Chin Sci Bull 8:920–930

Wang BD, Wang LQ, Wang DB, Yin FG, He J, Peng ZM, Yan GC (2018) Tectonic evolution of the Changning-Menglian Proto-Paleo Tethys ocean in the Sanjiang area, Southwestern China. Earth Sci 8:2527–2550 (in Chinese with English abstract)

Whitmarsh B, Sawyer D (1993) Upper mantle drilling in the ocean-continent transition west of Iberia. Terra Nova 5:327–331

Xu JH, Cui KR, Shi YS (1994) A new model of tectonic facies and orogeny of arc-back collision. J Nanjing Univ (nat Sci) 3:381–389 (in Chinese with English abstract)

Yang KH, Mo XX (1993) Late Paleozoic rifting-related volcanic rocks and tectonic evolution in southwestern Yunnan. Acta Petrologica Et Mineralogica 12:297–311 (in Chinese with English abstract)

Zhai MG, Cong BL, Qiao GS, Zhang RY (1990) Sm-Nd and Rb-Sr geochronology of metamorphic rocks from SW Yunnan orogenic zones, China. Acta Petrologica Sinica 4:1–11 (in Chinese with English abstract)

Zhao J, Zhong DL, Wang Y (1994) Metamorphism of Lancang metamorphic belt, the western Yunnan and its relation to deformation. Acta Petrologica Sinica 10:27–40 (in Chinese with English abstract)

Basic Characteristics and Evolution of Sanjiang Tethys Archipelagic Arc-Basin System

2

The rise of plate tectonics theory in the 1960s contributed to a revival of the long-neglected idea of continental drift, and revolutionized Earth sciences by shifting people's perception that continents were fixed to a belief that they were in a constant state of motion. The extensive practice of plate tectonics in the world in the past 40 years has proved to be a highly successful theory. It features a simple principle that involves a wide range and conforms to scientific laws, as well as great compatibility, explanatory power and foresight. The charm of plate tectonics: At the level of earth view, the theory was changed to the active theory which saw the significant drift and transformation of the ocean and land on the earth surface from the original fixed theory which believed that the position of the continent was unchangeable; at the level of geo-view, the theory was changed to the holistic, systematic and process geology which is integrated from space to depth and from ocean to continent from the original inclination to the theory of uniform change or sudden change, power curve model, and the theory of vertical movement or horizontal movement and rotary movement; at the level of methodology, both the method of induction focused on observation and facts and the theory of causation focused on deduction and hypothesis should be paid attention to. Most geologists explain geological event groups and their processes in the overall framework of plate tectonic evolution from the continent to the ocean, the method of "observing the present to study the past" to the method of "observing the past to study the present", and the global plate tectonics, completely changing the geoscience concepts. There is no doubt that there are traces of oceanic geological evolution on the continent, that is, the geological records of oceanic lithosphere evolution, development, destruction and transformation into continental orogenic belt in the plate tectonics model can be found and reshaped on the continent. The landing of plate tectonics does not announce the end of this theory, but indicates the beginning of a new journey.

There are obvious differences in lithosphere, thickness, genesis and age between continental zone and oceanic zone. In geoscience research, scholars often question whether it is reasonable to adopt plate tectonics originated from marine geology to know, explain and understand continental geology and dynamic processes. However, the research shows that because the most complete geological records of the earth evolution, the marine sediment records from at least Proterozoic Eon, and the geological event group records of ocean-land interaction and transformation can be found on the continent, the plate tectonics theory has been released successfully, and many marine geologists, together with scientists who stick to continental geology theory, launch the Global Lithosphere Comparison Program and Continental Dynamics Program in all continents and orogenic belts around the world after participating in the geological and geophysical surveys of the submarine sediments in the Mediterranean, Atlantic Ocean, Pacific Ocean and Indian Ocean. The practice of earth science shows that after some limitations have been corrected, the classical plate tectonics theory still has great scientific values and is favored by geosciences. The emerging of plate tectonics initiated the upsurge of research on island arc and back-arc-basin. The study on the island arc and back-arc-basin from the perspective of plate tectonics changed the traditional tectonic pattern of trough-platform structure, and revolutionized the entire earth science system. Therefore, the tectonic evolution of continental margin and the formation and development of continental lithosphere are not only the basic problems of contemporary geological research and the frontier subject of geotectology research, but also the introductory guide for the landing of plate tectonics.

W. Li et al., *Metallogenic Theory and Exploration Technology of Multi-Arc-Basin-Terrane Collision Orogeny in "Sanjiang" Region, Southwest China*, The China Geological Survey Series, https://doi.org/10.1007/978-981-99-3652-6_2

2.1 Basic Characteristics and Main Arguments of Archipelagic Arc-Basin System Tectonics

2.1.1 Proposal of Multi-Arc-Basin-Terrane (MABT) Tectonics

The establishment of sea floor spreading and plate tectonics theory and the landing of plate tectonics ushered in the golden age of Tethys geology, especially the research of Qinghai-Tibet Plateau geology and continental geology. For nearly 30 years, a series of works on the geological evolution of Qinghai-Tibet Plateau have consistently incorporated this research into the general framework of evolution of Tethys Ocean (Fig. 2.1), but there are different opinions on the specific evolution models. To sum up, there are mainly "scissors-stretching", "conveyor belt" and "accordion" models (Huang and Chen 1987). All models are based on the formation of Pangaea and Tethys being a bay in the Pan-thalassa Ocean, the splitting of Gondwana and the accretion of the Asian Continent. Most geologists believe that there are only five suture zones in Qinghai-Tibet Plateau, namely Yarlung Zangbo River, Bangong Lake-Nujiang River, Jinsha River, Kunlun Mountain and Qilian Mountain suture zones, and five accretionary terranes separated from Gondwana. They believe that these suture zones are ocean basins destroyed one after another, and all of them suffered subduction destruction pointing to the continent and dipping northward.

Since the 1980s, when the researchers studied the geological and tectonic evolution of Sanjiang (Jinsha River, Lancang River and Nujiang River) Region in Southwest China, they found that the ocean basins restored from Jinsha River Zone and Ganzi-Litang Ophiolite Melange Zone are only about 1000 km wide, and the adjacent island arcs and land arcs indicate that the ocean crust subducted westward and southward (Li et al. 1991; Mo et al. 1993). The main body of the Bangong Lake-Nujiang River Junction Zone also subducted westward and southward. The study of paleogeography and paleostructure once compared it with the arc-basin system in Southeast Asia, pointing out that the Paleozoic Era-Mesozoic Era Tethys had a paleogeographic pattern of islands and seas. Some scholars put forward the idea that the Sanjiang Paleo-Tethys in Southwest China is a multi-island ocean or a multi-island sea (Liu et al. 1993) discovered that 90% of the orogenic belts were formed by the reduction of back-arc-basins and the arc-land collision through the observation, research and analysis of tectonic evolution of the major orogenic belts in the world. This discovery is undoubtedly of great significance to the study of continental geology and continental orogenic belts. Based on the detailed anatomy of the tectonic facies in the major orogenic belts of Qinghai-Tibet Plateau, the hypothesis of a multi-island tectonic model was formally put forward.

The author has discussed the meaning and evolution of Tethys originally put forward by, the spatio-temporal structure of Tethys, the spatio-temporal framework and evolution relationship of three landmass groups in Laurasia, Gondwana and Pan-China, and studied many ophiolitic melange zones and various types of island arcs and basin systems in East Tethys with Qinghai-Tibet Plateau as the main body. It is found that the early so-called scissors-stretching, conveyor belt and accordion models cannot give a relatively reasonable explanation to the spatial configuration of many ophiolite melange zones and various island arcs and basin systems in Qinghai-Tibet Plateau (1997). The spatial configuration of the arc-basin systems in South Asia and on the west coast of the Pacific Ocean shows that the continental accretion of the southwest Pacific Ocean is completed by the complex of back-arc-basin reduction and island arc orogeny, instead of the continental accretion breaking away from Gondwana and drifting northward. Based on the analysis of more than 20 ophiolite melange zones and related island arcs and basins in the arc-basin system in Southeast Asia and the west coast of the Pacific Ocean and the Qinghai-Tibet Plateau (including the Sanjiang area), the author puts forward an MABT model to explain the formation and evolution of major orogenic belts in Tethys and the Asian Continent (Pan et al. 1997).

2.1.2 Definition of Multi-Arc-Basin-Terrane (MABT) Tectonics

The MABT tectonics at the continental margin (Pan et al. 1997, 2003, 2012): the combination of the frontal arc and a series of island arcs, volcanic arcs, blocks and the corresponding back-arc ocean basins, inter-arc-basins or marginal sea basins formed by the subduction of ocean lithosphere on the ancient continental margin, which as a whole shows a specific composition, structure, function, space distribution and time evolution features in time–space domain between continental lithosphere and ocean lithosphere. The identification and in-depth study of the MABT tectonics at the continental margin can not only dissect the material composition, structure and evolution history of the orogenic belt, but also have important enlightenment for analyzing the formation of the Precambrian continental craton basement. Southeast Asia and the west coast of the Pacific Ocean, which are restricted by the subduction of the Indian Ocean and Pacific Plate, are the most typical areas of Cenozoic MABT tectonics.

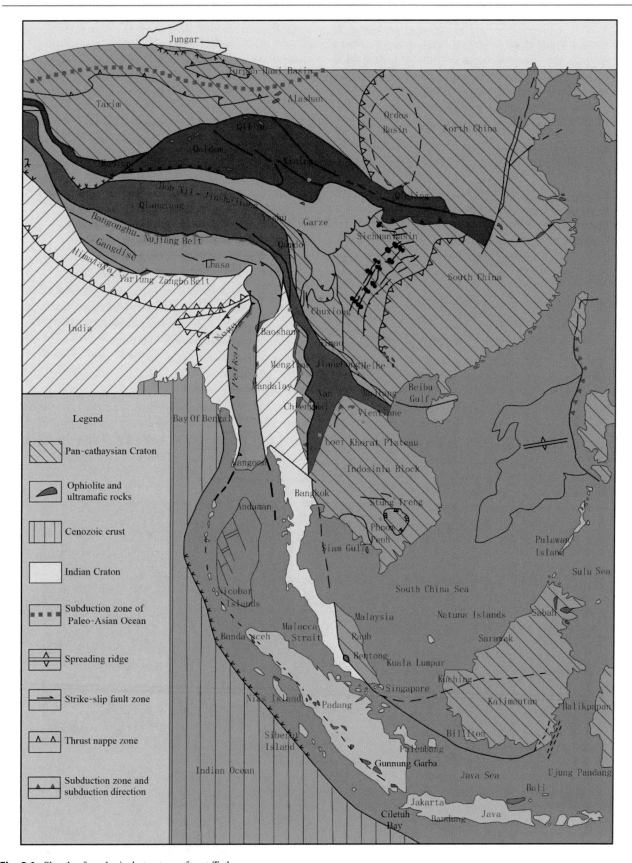

Fig. 2.1 Sketch of geological structure of east Tethys

The formation and evolution of MABT tectonics at the continental margin is an important symbol of the tectonic system transformation from ocean lithosphere to continental lithosphere. The formation of orogenic belt at the continental margin is driven by the reduction of back-arc or inter-arc ocean basin and island arc marginal sea basin, and the continental margin accretion is produced by a series of archipelagic orogeny such as arc-arc collision and arc-land collision. The development of back-arc foreland basin and peripheral foreland basin is an important symbol of basin-range transition. The evolution of the MABT tectonics at the continental margin formed the orogenic system, the destruction of the Tethys Ocean and the splicing of the MABT on both sides of the continental margin into the orogenic system formed mage-suture zones, and finally formed a giant orogenic system (tectonic domain).

2.1.3 Basic Characteristics of Multi-Arc-Basin-Terrane (MABT) Tectonics

According to the analogy analysis of comparative geology, the MABT at the continental margin during ocean-land transition is the continental margin of the Western Pacific Ocean, which concentrates more than 75% of the marginal sea basins in the world, and its formation is closely related to the (super) subduction zone of the Western Pacific Ocean, the most spectacular subduction zone activity in the current global tectonics. Among them, the formation of the MABT tectonics in Southeast Asia is also restricted by the northward subduction of the Indian Ocean Plate, which occupies an important position in global tectonics. It plays a very important role in the comparative study of continental geological history, especially in the formation and evolution of orogenic belts at the continental margin (Pan et al. 1997). Based on the analysis of the material composition, structure, tectonic characteristics and evolution history of the MABT in Southeast Asia, the basic characteristics of the MABT can be summarized as follows.

2.1.3.1 Specific Space–time Structure and Material Composition

The Asian Continental Margin is composed of the Indonesian Island Arc controlled by the one-way subduction of the Indian Ocean and a series of arc-basin systems (Hsü 1994). The Java Trench indicates the subduction of the Indian Ocean Plate, and the Indonesian Island Arc is the frontal arc of the MABT in Southeast Asia. Behind the frontal arc of Indonesia, there are more than ten back-arc-basins, micro-landmasses and island arcs, as well as countless shoals and hundreds of island reefs. Some back-arc-basins have spreading oceanic ridges, oceanic crusts, oceanic islands and seamounts, while others are only marginal seas (Fig. 2.2). The South China Sea Basin is the

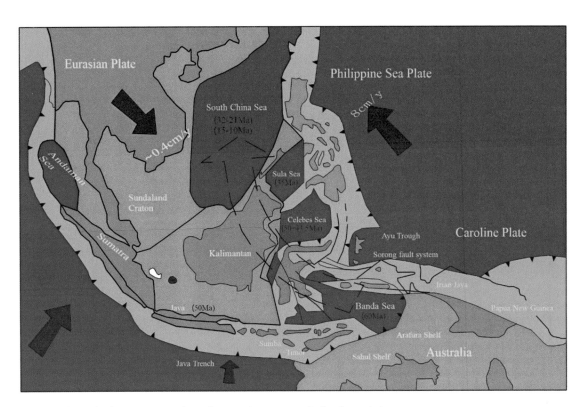

Fig. 2.2 Archipelagic arc-basin system tectonics and space–time structure in Southeast Asia

largest back-arc-basin, with a length of 2600 km from north to south and a width of 1300 km from east to west. Its marginal zone has different tectonics. The basin has a passive margin in the north, sheared margin in the west and an active continental margin in the east. Nansha Islands in the south are located on the residual arc separated from Palawan Trough and are dominated by neritic carbonate deposits. The Izu-Ogasawara-Mariana Intra-oceanic Arc is a frontal arc related to the subduction of Western Pacific Ocean Plate, and there are a series of back-arc-basins, ridges and island arc systems in the west.

Except for Southeast Asia, the Andes Active Margin in Western South America is the continental volcanic-magmatic arc. On the outside of the arc, there is a deep-sea trench, and there is an accretionary wedge on the inner wall of the trench. Some places are fore-arc-basin deposits, which are developed on some landmasses split by extensional structures. When the continental part behind the magmatic arc turns into island land and shallow sea forms islands and back-arc-basins, Andes Margin turns into island arc margin, and the MABT tectonics begins to develop.

2.1.3.2 Three Types of Basement of Island Arcs or Frontal Arcs

In the island arc or frontal arc of the MABT, the basement and composition of different members may be different. Generally speaking, there are at least three different types of basements in the island arc or frontal arc. For example, the frontal arc in Indonesia, the Sumatra Volcanic Arc in the west member, the West Java Volcanic Arc and their back-arc-basins were formed on the continental crust basement; the volcanic arcs of Central Java and East Java in the middle member were formed on the Mesozoic accretionary wedge complex; Sumba Island and Flores Island in the eastern member were formed on the oceanic volcanic arcs and the Neogene back-arc oceanic crust.

2.1.3.3 Three Types of Back-Arc-Basins

A back-arc-basin refers to the marginal sea basin on the continent side of the island arc, which is distributed in many margins of the ocean in the world, and the Western Pacific Ocean Margin is the most typical one. According to the research results of Asia and Southeast Asia, at least rift basin, marginal sea basin and inter-arc-basin can be found according to the development time, tectonic position and evolution characteristics.

(1) Back-arc rift basin: It is behind the volcanic arc, the basement is a stretched and thinned continental crust, and is generally less than 30 km thick, with low negative Bouguer anomaly, mainly neritic or transitional sediment, and "bimodal" volcanic rocks with high heat flow value, such as South Sumatra Basin.

(2) Marginal sea basin: It is located between the active continental margin and the marginal arc, beneath which are the sea floor spreading oceanic crust and marginal oceanic crust. The crust is about 20 km thick, and moderate and deep earthquakes occur, with high heat flow value, such as the South China Sea Basin.

(3) Inter-arc (rift) basin: It is often between the residual arc and the frontal arc, with an oceanic crust about 10 km thick below it, and the upper mantle upwarping is obvious. When the inter-arc-basin expands to a certain extent, it will form an intra-arc rift basin with high positive Bouguer anomaly and extremely high heat flow value, such as the Philippine Sea Basin.

The causes of the back-arc-basins are usually explained by the models of back-arc rifting and sea floor spreading, and the process of back-arc spreading is undoubtedly related to the subduction of the ocean lithosphere.

2.1.3.4 Short Life of Back-Arc-Basins

Compared with the main ocean, the life of the marginal sea basin, back-arc-basin or back-arc ocean basin in the MABT at the continental margin is short, usually only tens of millions of years. The spreading of the back-arc-basin stopped, and then the back-arc-basin was transformed into a residual back-arc-basin. The residual back-arc-basin and back-arc-basin have similar basement, which is usually a new oceanic crust expanded from the sea floor or a thin continental crust. The Caspian Sea and the Black Sea are residual back-arc-basins, while Junggar and Qaidam in China are respectively related to Paleo-Asian Ocean and Paleo-Tethys.

2.1.3.5 Three Different Types of Space–Time Evolution

(1) Back-arc-basins or back-arc ocean basins from the frontal arc inward were formed earlier one by one. For example, the frontal arc of Izu-Ogasawara-Mariana mainly developed Eocene-Miocene volcanic rocks with Pliocene biohermal limestone. Frontal arc inward (from east to west): Mariana Trough is a spreading back-arc-basin. Karig speculated that it was spread in the Pliocene (3 Ma), and its adjacent West Mariana Ridge was formed since 11 Ma; In the west, the Parece Vela Back-arc-Basin was formed by intra-arc spreading, in which there are typical oceanic basalts. Pillow lava, diabase, gabbro and peridotite have been found in Parece Vela Rift, the sea floor spreading occurred in the Oligocene–Miocene, the Kyushu-Palau Ridge in the west is the residual volcanic arc in Oligocene, and the Western Philippines Basin in the west was formed in

the Eocene. It can be predicted that the backward sub-duction of the lithosphere plate in the Western Pacific Ocean will lead to the continuous eastward advance-ment of the arc-basin system in the future, and the split of the arc will separate the new intra-arc rift—back-arc ocean basin by residual arc or new ridge. The ultimate destruction of the Pacific Ocean is not caused by the continental collision but by arc-arc or arc-land collision.

(2) The back-arc-basin with the frontal arc inward was formed earlier one by one. For example, the basin groups after the Indonesia-Timor Frontal Arc (to the north) become new in turn. Banda Sea spread 700 km from east to west, with the estimated sea floor age of about 60 Ma. The northward Sulawesi Basin was formed in the Middle Eocene (50–43.5 Ma). The spreading time of the Sulu Sea floor is from Late Eocene to Oligocene; the first sea floor spreading of the North South China Sea Basin was in Oligocene-Early Miocene (32–21 Ma), and the second one was in Early-Middle Miocene (volcanic rock age 15–10 Ma).

(3) The marginal sea basins parallel to the frontal arc were formed in roughly the same period. For example, Sumatra Basin, Northwest Java Basin, Northeast Java Basin were all formed in 50 Ma, while Japan Sea, East China Sea and South China Sea and other basins were all formed in Oligocene–Miocene.

2.1.3.6 There Are Three Different Types of Collision Orogeny

The present geological and geomorphological features of frontal arcs in Indonesia and its MABT in North Southeast Asia were caused by the interaction of three lithospheric plates: Neogene Philippine Sea Plate (NNW movement, 10 cm/a), Asian Plate (SE movement, about 0.4 cm/a) and Indian Ocean Plate (NNE movement, 7 cm/a). A series of geological event groups (Fig. 2.3) occurred during the for-mation of MABT in Southeast Asia, such as subduction, accretion, volcanic arc formation, back-arc thrust, strike-slip fracture, back-arc spreading, separation and formation of micro-landmasses, arc-arc collision, arc—micro-landmass collision, arc-land-collision, and the formation of foreland fold thrust zone, obduction and uplift of mountain chain zones. The prominent manifestation is the orogeny restricted by three different dynamic mechanisms:

(1) Island arc orogeny restricted by oceanic lithosphere subduction: different island arc orogeny styles may be shown due to different subduction convergence direc-tions. For example, Sumatra (volcanic arc) orogeny restricted by the oblique subduction of Indian Ocean Plate and Java (volcanic arc) orogeny restricted by the forward subduction of the Indian Ocean Plate. The subduction caused accretion complexes and fore-arc-basins. The volcanic arc is located on the pre-Paleogene continental crust basement, with the Barisan Right-lateral Strike-slip Fracture Zone devel-oped, and the Tertiary back-arc-basin developed at the north side of the volcanic arc.

(2) Collision orogeny restricted by the subsidence of the back-arc ocean basin: different orogenic styles caused by arc-arc collision or arc-land collision. Maluku arc-arc collision orogeny is the only example of arc-arc collision orogeny in the global tectonics at present. The Maluku Back-arc Ocean Basin is subducted to the west by the Sangihe Volcanic Arc and to the east by the Halmahera Arc. Their fore-arc zones collided, the subduction complex formed between the two arcs with opposite polarities, and Talaud Ridge was formed. Their fore-arc zones thrust backward on the attached arcs. The Sulawesi arc-land collision orogeny was caused by the collision between two micro-landmasses (Buton-Tukangbesi and Banggai-Sula) separated from the Australian Continental Plate, and the eastern part of the Sulawesi Volcanic-Magmatic Arc. The collision resulted in ophiolite thrusting upward on the micro-landmasses, and the metamorphic zone in Middle Sulawesi thrusting westward on the volcanic-magmatic arc, resulting in the formation of foreland thrust fold zone in Tertiary.

(3) Collision orogeny restricted by continental craton sub-duction: for example, Banda land-arc collision orogeny is caused by the subduction of the passive margin of Northern Australia under Banda Volcanic Arc and its fore-deep complex in Timor Trough. In the southern deformation zone of Timor Island, Australia, the Permian-Pliocene strata were folded and imbricated, and the accretionary complex and ophiolite in Timor Trough thrust upward on the foreland. The fore-arc zone of Timor arc is 100 km wide. It spread westward to the east of Sumba Island, with a 400 km wide fore-arc zone. The collision just started recently. It spread eastward in the northeast of Timor Island, with a 40 km wide fore-arc zone, and then spread northward. The volcanic arc and the nearly disappeared fore-arc zone overthrust northward on the submarine oceanic crust of Banda Sea, and Banda Sea Floor subducted southward into the Australian Continental Margin.

The Melanesia land-arc collision orogeny in New Guinea resulted from the subduction of the Australian Continent under the Tertiary Volcanic Arc and the collision with the oceanic island arc which is considered as New Guinea, and

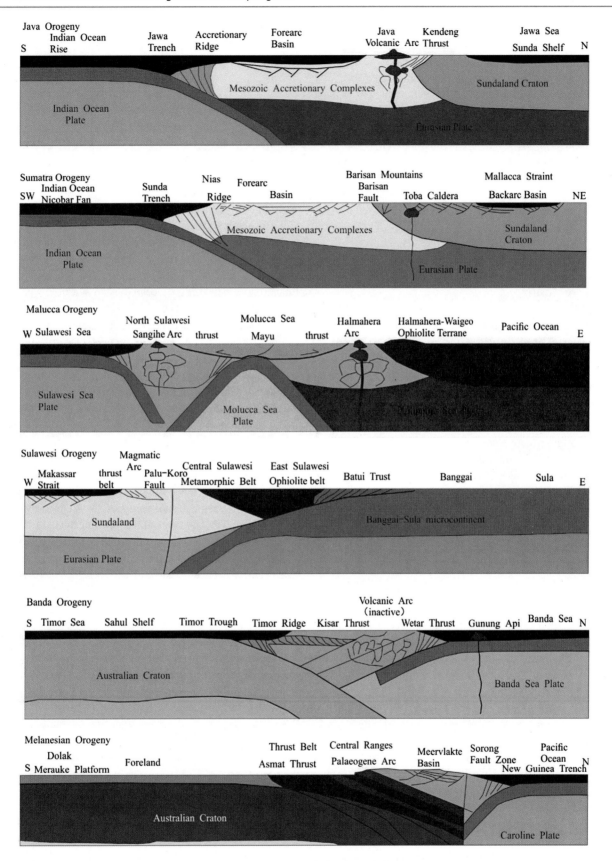

Fig. 2.3 Orogeny during the formation of archipelagic arc-basin system in Southeast Asia

the ophiolite melange zone was obducted and uplifted into the central mountain zone. A foreland fold thrust zone was formed in the northern margin of Australia. The volcanic island arc thrust northward on the Caroline Crust. The transverse strike-slip shear zone in Northern New Guinea was caused by the westward movement of the Caroline Plate at a rate of 12.5 cm/a relative to the Australian Plate.

The study of the MABT tectonics and their evolution characteristics in Southeast Asia is of great enlightenment for researchers to study the continental geology, especially the formation and evolution of Qinghai-Tibet Plateau with complex orogenic belts (especially Sanjiang Orogenic Belt).

2.1.4 Main Arguments of Multi-Arc-Basin-Terrane (MABT) Tectonics

2.1.4.1 Spatial–temporal Pattern of Coexistence of Three Landmasses and Three Oceans

In the history of global ocean-land evolution, there have been more than three spatial–temporal patterns of ocean-land distribution where landmasses and oceans coexist at least since late Neoproterozoic. Since late Precambrian, the world can be divided into Laurasia Landmass Group, Gondwana Landmass Group and Pan-Huaxia Landmass Group (North China, Yangtze, South China, Indo-China landmasses and other landmasses) (see Fig. 2.2), which have their own systems and unique geological evolution. At least in the Paleozoic, there were Paleo-Atlantic Ocean, Paleo-Asian Ocean (including the Ural Ocean) and Tethys Ocean all over the world. Since Mesozoic, there has been the present spatial and temporal distribution pattern of oceans and continents.

2.1.4.2 Long Period of Ocean Evolution

The Atlantic Ocean, Indian Ocean and Pacific Ocean are in different tectonic evolution stages of the ocean lithosphere. Currently, the Pacific Ocean has a history of spreading at least 200 Ma. Since the mid-ocean ridge has dived under the North American Continent and transformed into the San Andres strike-slip fracture, there is no sea floor spreading in the North Pacific Ocean. In the Western Pacific Ocean, it shows the evolution of the MABT in East Asia and Southeast Asia. According to GPS measurement, the distance between Shanghai and San Francisco is shortened by 5 cm every year, thus the Pacific Ocean will be destroyed after 400 Ma. In Tethys tectonic domain, the spreading basins restored by many ophiolite zones have only been "opened and closed" for several tens to millions of years. The period of tens to millions of years is at most the lifetime of marginal sea basins or back-arc-basins. Therefore, there is a long evolution process for an ocean from its occurrence,

spreading to its destruction. For example, the Tethys Ocean has a lifetime of at least 600 Ma from its occurrence, development to its destruction (from Late Precambrian to Eocene).

2.1.4.3 Similar Original Scale of the Tethys Ocean and the Present Pacific Ocean

When the Tethys was evolved out of the "Pangaea and Panthalassa Ocean", the prototype of the Tethys Orogenic Belt, including the Qinghai-Tibet Plateau and its adjacent areas, was restored to a paleo-ocean, instead of a vast Pangaea Bay or shallow Tethys Sea, which led to the quantitative problem of the width of this primitive ocean. If we consider the formation and evolution of the global ocean lithosphere, the Paleo-Tethys Ocean is a link of the global ocean-land transition and evolution historical chain in Phanerozoic. The Tethys Ocean existed at a time when there was no Pacific Ocean. The opening, formation and evolution of the Pacific Ocean is exactly the shrinking and destruction of the Tethys Ocean, i.e., the Tethys Ocean, which was destroyed and closed at the end of Mesozoic, was originally as large as the Pacific Ocean today. The research results of Gondwana Continental Group show that the India Landmass was a part of Gondwana Continent in the early Mesozoic, and it began to split and advance northward from the end of Jurassic and the beginning of Cretaceous. Currently, the distance between the India Landmass and the Antarctic Pole is more than 12,000 km. The width of the Tethys Ocean disappeared along the north–south direction, not less than 12,000 km. The great shortening of the Tethys Himalayas and Qinghai-Tibet Plateau along the north–south direction generally only reflects the relationship between Laurasia and Gondwana Continental Group, while Pan-Cathaysian Continent Group wedged westward between the two continents to form a series of mountain chains (Hengduan Mountain and Helan Mountain, etc.) spreading in south-north direction, which also showed great compression in east–west direction. The spreading of the Pacific Ocean is the sum of the shortening range of the Pan-Huaxia Continent and the shortening range of the nearly north–south orogenic belts to realize the balanced compensation of the horizontal movement of the lithosphere; otherwise, the formation of the East Tethys MABT orogenic belts cannot be explained.

2.1.4.4 Continental Margin Volcanic-Magmatic Arc and Multi-Arc-Basin-Terrane (MABT) Formed by the Two-Way Subduction of Ocean Lithosphere

The two-way subduction of the lithosphere of the Pacific Ocean formed the Cordillera MABT and collision orogenic belt on its east side in Late Mesozoic, and the Cenozoic Andes continental margin arc. The evolution of the

Paleo-Asian Ocean formed the MABT of Kazakhstan-Altai-Greater Khingan and a series of arc-arc, arc-land collision orogenic belts from the southwest margin to the southeast margin of Siberian Traps, while the continental margin volcanic-magmatic arc was the main structure in Tianshan and North China Margin (including the early back-arc-basin subduction complex zone). This asymmetric evolution of the oceanic lithosphere also shows the same characteristics in the evolution of Tethys Oceanic lithosphere. The asymmetric destruction of the North Pacific Ocean is characterized by the destruction of the old oceanic crust, the emergence of the new oceanic floor, and then the spreading of the mid-ocean ridge. When the continent crosses and buries the spreading ridge, the oceanic crust will destruct in reverse order, that is, it will destruct from the new oceanic crust to the old oceanic crust. This feature is very important for understanding the occurrence of ophiolite in the same ancient suture zone in the continental orogenic belt at different times.

2.1.4.5 Existence of Ocean Basins Indicated by the Remnant Arc

The existence of remnant arc indicates the existence of ocean basin, and the MABT is the symbol of the evolution of ocean lithosphere from occurrence and development to shrinkage and reduction. Many ophiolite melange zones composed of ophiolite and subduction complex have been found in Qinghai-Tibet Plateau. Most of the "trinity" ophiolites are "small ocean basin", back-arc-basin and island arc marginal sea type. On the north side of Tethys Ocean, there are the Early Paleozoic Qinling-Qilian-Kunlun MABT and the Late Paleozoic-Triassic Qiangtang-Sanjiang MABT, as well as the Gangdise MABT on the south side of Mesozoic Tethys Ocean. The existence of MABT indicates the existence, reduction and transformation of ocean lithosphere. The lithosphere of Tethys Ocean experienced a long-term continuous and complex evolution from occurrence, development to shrinkage and destruction at least from Paleozoic to Mesozoic. The Paleo-Tethys was the inheritance and development of the Proto-Tethys, and the Mesozoic Tethys did not reopen after the destruction of the Paleo-Tethys Ocean. Some Tethys Oceanic crusts could be merged by the subsequent Indian Ocean. Bangong Lake-Shuanghu Lake-Nujiang River-Changning-Menglian Mage-suture Zone should be essentially a relic of the destruction of Tethys Ocean.

2.1.4.6 Action Mode of the Transformation of Oceanic-Continental Lithosphere Tectonic Systems

The transformation of oceanic-continental lithosphere tectonic systems occurred through the island arc orogeny of oceanic lithosphere subduction, back-arc-basin reduction, arc-arc collision, and arc-land collision. According to the geological characteristics of Qinghai-Tibet Plateau, a series of marginal basins, island arc-basin systems and sediments and other rock assemblages on the western margin of the Pan-Cathaysian Continent Group in the reduction of Tethys Ocean have been continuously involved in orogeny since Paleozoic, and finally become a part of the Pan-Huaxia Continent. The lithosphere of the Tethys Ocean has been formed as an island arc orogenic belt from northeast to southwest, and the shrinkage of the back-arc-basin of the ancient marginal arc was not caused by "soft collision" or "collision without orogeny" during the horizontal tectonic movement of the lithosphere transformed into the continent of Qinghai-Tibet Plateau, but was caused by the island arc orogeny of arc-arc collision and arc-land collision. The transformation of back-arc foreland basin and continental margin basin into peripheral foreland basin is the geological record and important symbol of basin-mountain transformation. The mechanism of the ocean floor renewal, shrinkage and destruction is unknown yet, but it is obvious that it is important to enlighten how to identify and understand the formation and evolution of paleo-oceans in continental geological research after the transformation of oceanic lithosphere into continental lithosphere through island arc orogeny.

2.1.4.7 Qinghai-Tibet Plateau Crust Mainly Composed of Remnant Arc Orogenic Belts

The Qinghai-Tibet Plateau is composed of crustal materials formed and modified by complex and diverse geological processes, with the tectonic characteristics of a mosaic of strips and blocks, which is now manifested as a series of island arc orogenic belts of different scales, types and periods formed at different times, and generally manifested a complex tectonic domain composed of remnant arcs of different sizes and their back-arc, inter-arc or small ocean basin remnants and collision zones. The research shows that the tectonic framework of Qinghai-Tibet Plateau is bounded by the ophiolitic melange zone in the southern margin of Kunlun and the Bangong Lake-Nujiang River Ophiolitic Melange Zone, which can be divided into three tectonic regions: the Early Paleozoic Qinling-Qilian-Kunlun MABT in the northern plateau, the Late Paleozoic-Triassic Qiangtang-Sanjiang MABT in the central and eastern plateau, and the Mesozoic Himalayan-Gangdise MABT in the southern plateau. It is the formation and evolution of the three MABTs that restrict the Cenozoic tectonic deformation history and stress state of Qinghai-Tibet Plateau and its adjacent areas, and control the uplift of Qinghai-Tibet Plateau. The composition of the MABT is the essential reason for the uplift of Qinghai-Tibet Plateau.

2.1.4.8 Three Important Tectonic Processes of Qinghai-Tibet Plateau

The main composition of Qinghai-Tibet Plateau is not the split terrane of Gondwana Continent, but the MABT formed by the two-way subduction of Tethys Ocean in different periods of Phanerozoic, which was combined and converged by multi-stage collision orogeny of multiple island arcs. The original geological bodies of Qinghai-Tibet Plateau were constantly changed and transformed, and new ones were constantly formed, which are manifested in three important tectonic processes: First, the shrinking and destruction of the lithosphere of Tethys Ocean and the arc-arc or arc-land collision orogeny; second, the plateau is growing with the basin-mountain transformation and the continuous formation of orogenic belts; third, the substantial shortening and thickening of the earth crust with intracontinental convergence and plateau uplift. Due to the frequent transformation and evolution of the paleostructure and paleogeographic environment of Qinghai-Tibet Plateau, the multiple transformation and repositioning of material and energy flows are conducive to the formation of various mineral resources.

2.2 Space–Time Structure of Sanjiang Tethys Multi-Arc-Basin-Terrane (MABT)

2.2.1 Space–time Structure and Evolution of Yidun Arc-Basin System

The Yidun Arc-Basin System is mainly composed of Yidun Island Arc Zone on the west side of Ganzi-Litang Ophiolitic Melange Zone. As the product of the westward subduction of Ganzi-Litang Oceanic Crust, Yidun Island Arc Zone is mainly formed on a long-term spread and thinned continental crust. Volcanic-magmatic arc chains are distributed in parallel on the west side of Ganzi-Litang Junction Zone, running through the north and south, and stretching intermittently for 500 km. Yidun Island Arc Zone is not only a typical volcanic-magmatic island arc zone, but also an important polymetallic metallogenic zone of tungsten, tin, molybdenum, copper, gold, lead, zinc and silver.

2.2.1.1 Space–time Structure of Yidun Arc-Basin System

The Yidun Island Arc Zone (Fig. 2.4), which lies between Zhongza-Shangri-La Block in the west and Ganzi-Litang Junction Zone in the east, started in Late Carnian of Late Triassic, and was initially formed on a partially spread oceanic crust or transitional crust. Its main structure was built on the thinned continental crust with the characteristics of graben-horst system, and it experienced a complicated development history of compression-spreading alternation.

Formation of Ganzi-Litang Ocean Basin and Development of Yidun Island Arc

The limestone in Ganzi-Litang Melange Zone contains Late Permian fossils. Radiolarians from Early Triassic to early Late Triassic have been found in siliceous rocks associated with oceanic ridge basalt. Early Triassic radiolarians (*Yanagia Chinensis* Feng, *Paurinella Sinensis* Feng); Middle Triassic-early Late Triassic radiolarians (*Triassocampe cf. nova* Yao, *Pseudostylo-sphaera nazarovi* Kozur, *Squinabolella*? sp., *Hinedorclls* sp., *Muelleritortis cochleata tumidospina* Kozur, *Pseudostylosphaera nazarovi* (Kozur et Mostler), *Triassocampe coronata* Bragin, *Astrocentrus Pulcher* Kozur et Mostler et al. (1:200,000 Gongling Map Sheet, 1984; 1:200,000 Litang Map Sheet, 1984). The $^{40}Ar/^{39}Ar$ plateau age of Panyong ocean ridge pillow basalt in Ganzi-Litang Melange Zone is (231.3 ± 6.7) Ma, which is equivalent to the transition period from Middle Triassic to Late Triassic. It shows that Ganzi-Litang Ocean Basin was formed in Permian-early Late Triassic (the peak of the spreading of Ganzi-Litang Ocean Basin), with a width of about 480 km. Ganzi-Litang Ocean Basin was further spread on the basis of the deep-water rift basin deposited in Carboniferous.

Ganzi-Litang Oceanic Crust began to subduct westward in the middle Late Triassic under Zhongza-Shangri-La Block, forming a typical supporting pattern of Yidun Island Arc-back-arc-basin system on the west side, and its closure and filling destructed at the end of Late Triassic (Liu et al. 1993; Mo et al. 1993). Yidun Island Arc is mainly occupied by volcanic-sedimentary rock series formed in Late Triassic and granite basement formed in Late Indo-China—Yanshanian and is characterized by volcanic-intrusive complex combination developed. The subduction arc-forming occurred in 238–210 Ma, which is equivalent to the formation time of arc volcanic rocks in the Late Triassic. The collision arc-forming (collision orogeny) occurred in 208–138 Ma, which is closely related or distributed with volcanic arc granite in Late Triassic. The back-collision spreading occurred in 138–75 Ma, which is equivalent to the formation time of type A granite in Late Yanshan.

Spatial Distribution of Yidun Arc-Basin System

Yidun Arc-Basin System is developed with Ganzi-Litang Ophiolite Melange Zone, Que'er Mountain-Daocheng Outer Volcanic-Magmatic Arc Zone, Changtai-Xiangcheng Intra-arc Rift Basin Volcanic Zone, Dege-Changdagou Inner Volcanic-Magmatic Arc Zone, Baiyu Kongma Temple-Mianlong Back-arc Spreading Basin Volcanic Zone and Gaogong-Cuomolong Back-arc Inter-plate Volcanic-Magmatic Zone from east to west (Fig. 2.4).

Fig. 2.4 Tectonic framework and deposit distribution of Yidun Island arc zone

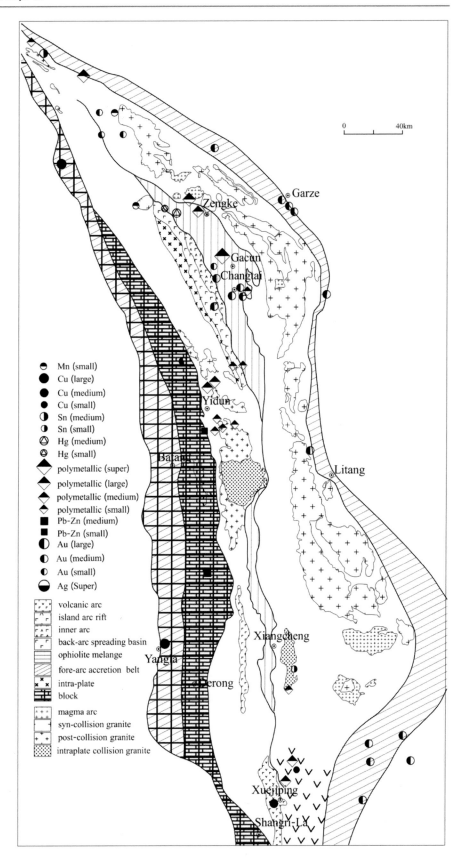

(1) Ganzi-Litang Ophiolite Melange Zone

It is an indisputable fact that the westward subduction of Ganzi-Litang Oceanic Crust is the fundamental factor leading to the formation and development of Yidun Island Arc (Liu et al. 1983; Luo et al. 1998; Xu et al. 1992). Ganzi-Litang ophiolite intermittently appears along Ganzi-Litang Junction Zone in the areas of Yushu Zhishimen, Yulong, Ganzi, Litang and West Muli. Most of them have been dismembered into ophiolite melange due to multiple structural deformations. In the range of 5–20 km in width, mafic rocks, ultramafic rocks, basic basaltic volcanic rocks, gabbro-diabase sheeted dike swarms, abyssal radiolarian siliceous rocks and turbidites are mixed with tectonic slices and melange formed by tectonism and slump. The mafic–ultramafic rocks in the zone are mostly mixed in Triassic basalt series, which are not too much, generally 100–500 m long and 10–500 m wide. The gabbro-diabase body is small in quantity but large in area, and its bedding intrusion is located in the basalt series of Quga Temple Formation. The basic basalt series is the most widely distributed volcanic rock in the Late Triassic, and pillow basalt is developed. In the area north of Ganzi, pillow basalt is dominated by Late Triassic Qugasi Formation, which is widely distributed along the fracture zone. In the Litang Area, the basic volcanic rocks were mainly formed in the Middle and Early Triassic, and in Tuguan Village in the southern member, the basic volcanic rocks were formed in the Late Permian (Mo et al. 1993). The thickness of the whole basalt system along the fracture zone is relatively stable, generally 600–1500 m. The sequence of ophiolite complex in Litang, Muli and other places is clear and complete.

Ganzi-Litang Ocean Basin was formed in Late Permian-early Late Triassic, and subducted westward in the middle Late Triassic under Zhongza-Shangri-La Block, forming a typical supporting pattern of Yidun Island Arc-back-arc-basin system on the west side, and its closure and filling destroyed at the end of Late Triassic. Ganzi-Litang Ophiolite Melange Zone is an important gold metallogenic zone, for example, Cuo'a Gold Deposit, Gala Gold Deposit, Niduocun Gold Deposit, Madake Gold Deposit, Xionglongxi Gold Deposit and other gold deposits.

(2) Que'er Mountain-Daochengwai Volcanic-Magmatic Arc Zone

The outer-arc volcanic-intrusive complex zone is located on the west side of Ganzi-Litang Ophiolitic Melange Zone, which is composed of the first sub-cycle neutral volcanic rocks and its adjacent intermediate-acid plutonic rocks-subvolcanic rocks. The outer-arc volcanic rocks, mainly composed of andesite and andesitic pyroclastic rocks, are caused by the first arcing activity in Carnian of Late Triassic, with an average Rb–Sr isotope age of 220 Ma.

Intermediate-acid plutonic rocks and subvolcanic rocks appear in pairs with andesite line, and they are distributed in dependence, forming the Niyajiangcuo-Daocheng Granite Zone. This zone starts from Niyajiangcuo in the north, passes through Cuojiaoma, Yongjie, Changtaishan to Gongbala and Jiangcuo in Daocheng, and stretches for hundreds of kilometers. There are dozens of rock masses, which are distributed in a strip in SN direction. These rock masses are complex massif formed in multiple stages and caused for many reasons, and the rocks are composed of diorite, quartz diorite, granodiorite, biotite granite and biotite monzogranite. The volcanic-magmatic arc was mainly formed in the early and late Indo-China, with an average age of 226 Ma and 204 Ma, respectively. The representative rock masses in early Indo-China include Niyajiangcuo rock mass (K–Ar age of 221.1 Ma) (the regional survey report of Sichuan Geology and Mineral Bureau Regional Geological Survey Team), Jiaduocuo rock mass (K–Ar age of 224–225 Ma) (the regional survey report of Sichuan Geology and Mineral Bureau Regional Geological Survey Team), part of Cuojiaoma rock mass (K–Ar age of 227 Ma) and part of lithofacies of Dongcuo rock mass. The representative rock masses in late Indo-China include part of Dongcuo rock mass (K–Ar age of 206 Ma; Rb–Sr age of 208 Ma; U–Pb age of 200 Ma), Ajisenduo rock mass (K–Ar age of 201 Ma) (the regional survey report of Sichuan Geology and Mineral Bureau Regional Geological Survey Team) and Ranxigong rock mass (K–Ar age of 207 Ma) (the regional survey report of Sichuan Geology and Mineral Bureau Regional Geological Survey Team). The above data show that the diorites in early Indo-China, which are absolutely dominant in the granite zone, are located in Middle Carnian of Late Triassic, and their genetic type is mainly type I. The location period is basically the same as the formation time of the volcanic rocks on the west side, which indicates that subduction and arcing volcanism-magmatism started in the early and Middle Carnian of Late Triassic. The granitoids in late Indo-China are mainly light-colored granitoids, and the petrogenetic type is type S, which indicates that the arc-land collision and arcing volcanism-magmatism occurred in Rhaetian of Late Triassic (Fig. 2.5). Minerals in the outer-arc volcanic-intrusive complex zone are concentrated in Shangri-La Arc Area at the southern end of the island arc, mainly porphyry and skarn copper, molybdenum and gold polymetallic deposits related to intermediate-acid subvolcanic rocks, for example, porphyry copper, molybdenum and gold deposits in Pulang, Chundu, Xuejiping and Lannitang, and skarn copper polymetallic deposits in Hongshan and Langdu.

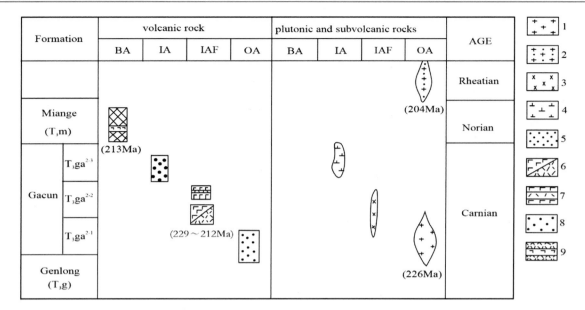

Fig. 2.5 Volcanic-intrusive complexes in different secondary tectonic units of Yidun Island Arc. BA—back-arc area; IA—inner arc; IAF—island arc rift; OA—outer arc. 1—Arc granite; 2—Collision granite; 3—Diabase sheeted dike swarm; 4—Diorite; 5—Arc andesite; 6— Basalt-rhyolite bimodal assemblage (coexisting); 7—Basalt-rhyolite bimodal assemblage (interbedded); 8—Arc andesite; 9—Shoshonite-rhyolite bimodal assemblage

(3) Volcanic zone of Changtai-Xiangcheng Intra-arc Rift Basin

The volcanic zone of the island rift basin is located on the west side of the outer volcanic-magmatic arc zone, and the volcanic-intrusive complex is composed of "bimodal" volcanic rocks and diabase sheeted dike swarms associated with them, which are intermittently distributed in the axial zone of the island rift. The bimodal rock assemblage was caused by the second sub-cycle volcanic activity in Gacun, which occurred in the middle part of Gacun Formation. The Rb–Sr isochron age of basalt is 217 Ma, and that of ore-bearing rhyolite series in Gacun Deposit is 232–203 Ma. The strongly altered ore-bearing dacite volcanic rocks in Gayiqiong Deposit have a K–Ar age of 221–210 Ma, and the ore lead isotope model age of the rock ore is 229–211.8 Ma. Therefore, it can be inferred that the age of acid volcanic rocks in bimodal rock assemblage is 229–212 Ma, which indicates that the "bimodal" volcanic activity occurred in the middle and late Carnian of Late Triassic, which followed the outer arc volcano-intrusive complex. The diabase sheeted dike swarms mainly developed in the northern member of island arc rift zone, that is, between Changtai and Zengke Volcanic Rock Development Area, and penetrated into volcanic-sedimentary rock series of Genlong Formation and the lower member of Gacun Formation in a bedded manner. The intra-arc rift basin has been an important ore-bearing basin for Sedex (black ore) silver-lead–zinc polymetallic deposits, including Gacun Silver Polymetallic Deposit, Gayiqiong Silver Polymetallic Deposit, Shengmolong Sliver Polymetallic Deposit, Quekailongba Silver Polymetallic Deposit, Dongshanji Silver Polymetallic Deposit, etc.

(4) Dege-Chandagou Inner Volcanic-Magmatic Arc Zone

The inner volcanic-magmatic arc zone is located on the west side of the volcanic zone in the intra-arc rift basin. It is composed of third sub-cycle intermediate-acid volcanic rocks (Norian in Late Triassic), diorite-diorite porphyry and biotite granite-granite porphyry associated, and is the main mass of the inner arc, with dacite-andesite dacite and pyroclastic rocks as the main mass.

The intermediate-acid volcanic rocks are mainly composed of dacite-andesite dacite and pyroclastic rocks. Compared with the outer arc "andesite line", the dacite line is much smaller in scale, and its acidity is obviously higher, with a large number of lava domes and subvolcanic rocks accompanying it. Diorite-diorite porphyrite is distributed in groups and zones, concentrated in Zengke Area, mainly in small rock strains or lumps and mainly distributed on the west side of the volcanic rock series, and closely coexisted with them. Some of them intruded into the volcanic rock series, and some of them intruded into and penetrated the gabbro-diabase in the island arc rift. Most of the rock masses are covered by the strata of Miange Formation, which revealed that its formation period was the same as that of the third sub-cycle volcanic activity of Gacun Formation, but it might be a little later, indicating the island arc environment of original arc-land collision.

(5) Volcanic zone of Baiyu Zhama Temple-Mianlonggou Back-arc Spreading Basin

The volcanic zone of the back-arc spreading basin is generally located on the west side of the inner volcanic-magmatic arc zone, with local overlap. The back-arc volcanic rocks in Yidun Island Arc belong to Miange Formation in the Upper Triassic, and the volcanic-stratigraphic horizon is equivalent to that of Lana Mountain Formation (T$_3$) previously divided. The volcanic rock series ranges from Denglong Town, Baiyu County in the north to Dulonggou in the south, with a width of 5–10 km, spreading for about 120 km along the strike in a strip manner in NNW direction, parallel to Changtai Volcanic Arc. The back-arc volcanic rock series is composed of upper and lower members. The lower member is composed of basaltic lava and basaltic tuff, which are small in volume and not spread far away, and are concentrated in Miange, Labagou and other places. The upper member is composed of an acidic volcanic rock series, which consists of dacite lava and dacite clastic rocks in the lower part and rhyolitic welded tuff in the upper part, mainly rhyolite rock series. The scale of acid volcanic rock series is much larger than that of basic rock series, with a thickness of several hundred meters, and it is covered by the overlying black slate series. Generally speaking, the back-arc area is characterized by "bimodal" volcanic activities.

The bimodal volcanic rock series in the back-arc-basin of Yidun Island Arc developed in a significant extensional tectonic environment on the continental crust basement, which has the tectonic conditions for the formation of volcanic rock hygroscopic low-temperature gold and silver polymetallic deposits. Along the volcanic zone (Miange Formation), besides the middle-sized gold-silver polymetallic deposit in Nongduke and the large mercury deposit in Kongma Temple, the regional geochemical exploration and microwave remote sensing data also show a number of comprehensive mineralization anomalies in Tage, Darike and Dulonggou, which shows a good metallogenic prospect in this volcanic zone.

(6) Gaogong-Cuomolong Back-Arc Inter-Plate Volcanic-Magmatic Zone

The back-arc inter-plate volcanic-magmatic zone is located on the west side (continental side) of the volcanic zone in the back-arc spreading basin, which is distributed in a strip shape and roughly parallel to the volcanic rocks in the back-arc-basin, and is separated by the strata of Lana

Mountain, which is dominated by psammite formed later. The inter-plate volcanic strata belong to the Tumugou Formation (T$_3$t) formed in the Late Triassic, which is similar in horizon to the Miange Formation in the back-arc-basin. The volcanic rocks are rhyolite. The inter-plate magmatic rocks developed in the narrow area between the Keludong-Xiangcheng Fracture Zone and the Aila-Riyu Fracture Zone, starting from Gaogong, Shiqu in the north and reaching Batuoren, Batang in the south, and constitute the second large granite zone, namely Gaogong-Cuomolong Granite Zone. The rock mass intrudes in Tumugou Formation of Upper Triassic, showing obvious intrusive contact; most of the rock masses are complex rock masses, with porphyritic moyite in early stage and porphyritic biotite monzogranite and moyite in late stage. The genetic type of rock is type A, which reflects the stress system of continental crust uplift and spreading after arc-land collision, and indicates the formation environment after collision orogeny; the duration of magma varies from 8 to 29 Ma. For example, the early diagenesis of Gaogong Rock Mass is 115.8 Ma, and the late diagenesis is 86.9 Ma; the early diagenesis of the Cuomolong Rock Mass is 84.8 Ma, and the late diagenesis is 76.8 Ma. The peak of the magmatic event is around 87 Ma.

The rock mass in the back-arc inter-plate volcanic-magmatic zone is often accompanied by tin and silver polymetallic mineralization, and many tin and silver polymetallic deposits have been found, such as medium-sized Lianlong (Xizhigou) Tin, Silver and Bismuth Polymetallic Deposit and large-sized Xiasai Silver Polymetallic Deposit, forming a large Tin and Silver Polymetallic Mineralized Granite Zone. In addition, the large pyrite-type silver, gold, copper and lead polymetallic geophysical and geochemical anomalies related to inter-plate volcanic rocks show an extremely broad mineral exploration prospect.

2.2.1.2 Formation and Evolution of Yidun Arc-Basin System

Characteristics and Properties of Yidun Island Arc "Basement"

The ancient metamorphic basement in Yidun is composed of pre-Sinian Qias Group, exposed in "Qias Fault Uplift", which is equivalent to the Hekou Group on Yangtze Platform. The overlying strata are deposited by platform-type Sinian Guanyinya Formation and Dengying Formation, which indicates that Qias Fault Uplift was once a part of the western margin of Yangtze Platform. The sedimentary facies characteristics and biological features of Ordovician-Lower Permian stratigraphic units are similar to those of Yangtze Platform, indicating that the Zhongza Block and Yidun Area are the components of the western margin of Yangtze Platform. Carbonate and clastic sediments of Upper Permian-Middle Triassic reveal that they were formed in the continental margin basin or slope environment. The Qugasi Formation formed in the Upper Triassic changed from conglomerate and coarse sandstone to sand shale from bottom to top, reflecting the evolution of its formation environment from shoreland to continental shelf, and its basic volcanic rocks also showed the affinity of inter-plate basalt. Therefore, Yidun was once on the western margin of Yangtze Platform. Estimated that the thickness of the basement continental crust of Yidun Island Arc was 20–25 km according to the restriction of magma density on magma eruption and intrusion, which indicated that Yidun Island Arc had the characteristics of thin continental crust basement.

The island arc basement is a continental crust basement that has been stretched for a long time. It is further stretched and thinned in Xiangcheng Area, and a new oceanic crust is produced by local stretching and cracking. The spreading center is located in the Panyong-Baisong in the western part of Xiangcheng, and the age of the oceanic crust is about 231 Ma. Ocean crust fragments marked by mafic–ultramafic rock masses, gabbro-diabase sheeted dike swarms, pillow-massive basalt, abyssal radiolarian siliceous rocks, etc., are widely found in Chizhong-Muyu in the east and Fanyong-Baisong in the west. With the opening of Jinsha River and the westward subduction of the oceanic plate, the passive continental margin of Yangtze River in Yidun developed a graben-horst system under the tension, which controlled the stratigraphic distribution and sedimentary facies characteristics. The graben of the graben-horst system in Yidun Area is located on the east side of Zhongza Landmass and Ganzi-Litang Zone, respectively, and the horst is located on the west side of Ganzi-Litang Zone and Zhongza Block, respectively; the horst has carbonate platform deposits, while the interior of the graben is composed of fine terrigenous clastic deposits and a small amount of siliceous rocks. In Early Carnian of Late Triassic, the graben-horst tectonic pattern in Yidun remained basically unchanged, but a secondary graben-horst system developed in the graben basin (Fig. 2.6).

Formation and Evolution of Yidun Arc-Basin System

On the basis of the ocean basin formed by the spreading of Ganzi-Litang Zone in Late Permian-early Late Triassic, the strata subducted westward under Zhongza-Shangri-La Micro-landmass in the middle Late Triassic. With the westward subduction of Ganzi-Litang Oceanic Plate, Yidun Area entered a new development period on the basis of the graben-horst tectonic pattern in the fore-island arc period, and started the generation, development and evolution of Yidun Arc-Basin System. Generally, it experienced the following processes (Fig. 2.7).

Period of Subduction Orogeny (238–210 Ma)

The products of subduction orogeny include Ganzi-Litang ophiolite melange zone, volcanic arc granite zone, main volcanic arc andesite zone, volcanic zone of intra-arc rift basin, back-arc bimodal rock assemblage zone and back-arc intraplate volcanic-magmatic rock zone, etc. They are distributed from east to west across Yidun Island Arc Zone, and can be divided into the following stages:

(1) Formation period of outer volcanic-magmatic arc (early and middle Carnian in Late Triassic). With the westward subduction of Ganzi-Litang Oceanic Crust, the metasomatism of dehydrated fluid from subduction plate to mantle source induced mantle rock melting, forming a low-density and low-viscosity incipient melting zone below the frontal volcanic zone, and the diapiric fold rising and magma segregation formed calc-alkalic magma, followed by crystallization differentiation to form calc-alkalic volcanic rock series, which constituted the so-called east "andesite line".

In Xiangcheng in the southern member of the area, the crust was relatively thin and magma differentiation was poor, forming a volcanic arc dominated by andesite-basaltic andesite; in Changtai in the northern member of the area, the crust was relatively thick and magma differentiation is sufficient, forming the volcanic arc dominated by andesite-dacite. On both sides of the volcanic arc running through the whole area, sedimentary rock series characterized by turbidite fans were developed. On the outer side (the

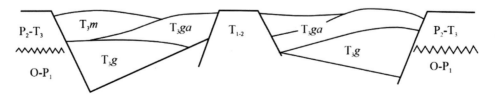

Fig. 2.6 Graben-horst tectonic pattern and sediment distribution in fore-island arc period in Changtai area

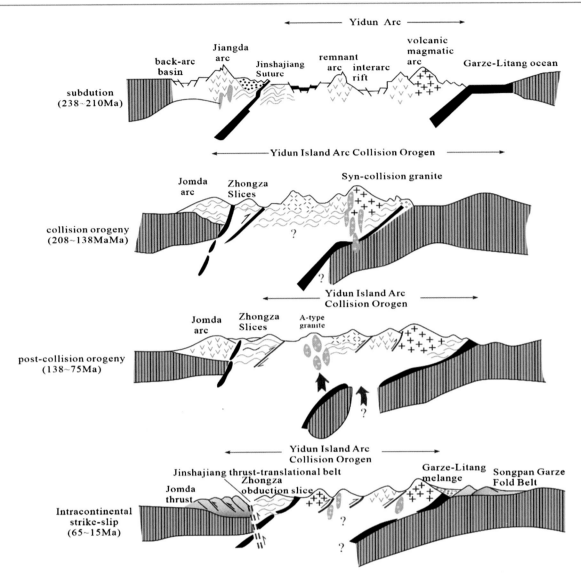

Fig. 2.7 Tectonic evolution of Yidun arc-basin system

east side) of the frontal volcanic arc, the mantle rocks and even the rocks in the crust-mantle transition zone, which are metasomatism by dehydrated fluid and mixed with SZC, partially melt, forming the volcanic arc intermediate-acid magma, which was located on the outer side of the volcanic arc, forming Niyajiangcuo-Daocheng Intermediate-acid Intrusive Rock Zone, and forming the magmatic arc coexisting with the volcanic arc. The age of magmatic granite is 237–208 Ma, and volcanic arc andesite occurred in Tumugou Formation in Upper Triassic, which have similar REE distribution patterns and geochemical characteristics of trace elements. The latter was mainly caused by the partial melting of mantle wedge contaminated by SZC metasomatism, while the former was caused by large quantity of crust-derived materials. The development of granite-andesite volcanic-magmatic arc marks the formation of Yidun

Island Arc Orogenic Belt, suggesting that Ganzi-Litang Ocean Basin was reduced and closed in early Late Triassic.

(2) Development stage of island arc (inter-arc) rift basin (middle and late Carnian of Late Triassic) Basic characteristics of island rift: ① Typical "bimodal" volcanic activities occurred. The "bimodal" volcanic activity occurred after the early arc volcanic activity and was strictly limited to the narrow zone sandwiched by inner and outer volcanic-magmatic arcs. ② With "bimodal" volcanic activities, the significant emplacement of basic magma from shallow to ultrashallow occurred, forming the diabase sheeted dike swarms closely associated with bimodal rock assemblage. ③ There are two units of bimodal rock assemblages, the basic rock of the lower unit is composed of tholeiite with high Fe content and

low Mg content, and that of the upper unit is composed of tholeiite with high Mg content and low Ti content. ④ In the island arc rift, the faulted basins are mostly graben or semi-graben, and the basin margin fractures are mostly extensional fractures in the fore-island arc period, spreading in NNW direction. Currently, Zengke Basin, Gacun Basin and Changtai Basin have been identified, which constitute a fracture subsidence zone with different fracture subsidence time, scale and depth. ⑤ Deposition in the island arc rift zone shows clastic rock facies on the slope margin of volcanos, sand shale facies in rift basin and mudstone facies in silled basin. ⑥ The silled basin or depression in the island arc rift zone is an important ore-bearing place for the exhalative-sedimentary volcanic-associated massive sulfide deposit. Formation mechanism of island arc rift: The development of rift depends on the thermal structure of crust and mantle below it. In the continental rift, the temperature near Moho is about 500–700 °C; in the island arc zone, the crust and mantle are in a high thermal state, and the ground temperature at Moho is as high as 900 °C. According to the paleotemperature of its adjacent area—West Panzhihua, the paleotemperature of Yidun Island Arc Zone is close to that of the paleo-ocean. The power source of the arc rift may be related to the subduction angle of the plate and the mode of action between the upper and lower plates. The oceanic plate subducted steeply at a large angle, and decoupling occurred between the obducted plate and the subducted plate. The tensile stress field appeared in the back-arc zone. The steep and deep plate subduction often naturally leads to subsidence, breaking and seaward movement of subducted plates. The oceanic trench may cause the island arc plate to be in a tensile state, forming an arc rift or even a rift basin, and accepting the mudstone facies of silled basin or sand shale facies of trough basin.

(3) Formation of inner volcanic-magmatic arc (Late Carnian-Early Norian in Late Triassic). Due to the regional compression, Ganzi-Litang Oceanic Crust Plate, which once stopped subducting or moving seaward, subducted westward again at the end of Carnian in Late Triassic. As a result, the magma source rock of the island arc rift stopped melting due to the pressure increase, and the island arc rift was destroyed, while its dehydrated fluid and SZC were metasomatism and mixed mantle rocks again. Due to the intervention of water fluid, mantle rocks melted, forming calc-alkalic magma, which was separated and crystallized, possibly mixed with magma or mixed with the crust, forming an inner arc volcanic rock series. Perhaps due to the thickening of the crust, some calc-alkalic magma was partially emplaced from shallow to ultrashallow,

forming a large number of subvolcanic rocks and diorite-diorite porphyrite, which constituted the main mass of volcanic-magmatic arc. Although the assemblage of inner arc and outer arc volcanic rocks formed by the early and late arcing volcanism is similar, their geochemical and geodynamic backgrounds are not completely the same. The early outer arc volcanic-magmatic rocks are subducted, and the late inner arc volcanic-magmatic rocks may have been formed in a short period after Ganzi-Litang Ocean closed and started small-scale arcing (Yidun Island Arc-land (Yangtze Plate) collision), which represents the island arc environment during the initial arc-land collision.

(4) Development stage of back-arc spreading basin (middle and late Norian of Late Triassic) Geophysical research shows that after the initial collision between the continent and the island arc, the oceanic plate will undergo subsequent subduction. In Yidun Island Arc, after the closure of Ganzi-Litang Ocean and even the initial arc-land collision, the subduction of the subducted oceanic plate may also take place. On the one hand, the SZC component mixed with mantle rock and the water fluid from the subduction zone metasomatism mantle rock, on the other hand, the tensile stress field appeared in the back-arc area, and the back-arc spreading basin was formed, which was mainly manifested by the development of the shoshonite-rhyolite bimodal rock assemblage and the sedimentation of the black sand-slate series in the faulted basin. On the west side of Changtai Island Arc in the north member, Miange Back-arc Spreading Basin was developed, the spreading center was located on the line of Kongma Temple-Miange-Nongduke, and developed "bimodal" volcanic activities, forming a high-potassium rhyolite-shoshonite assemblage, which formed the same Rb–Sr isochron, with an age of 213.7 Ma, indicating that Miange Back-arc-Basin was formed in the middle and late Norian of Late Triassic. Wanrong-Qingdarou Back-arc Spreading Basin is developed on the west side of Xiangcheng Island Arc. The basin was mainly filled with clastic rock series of Tumugou (Gacun) Formation and Lamaya Formation, with little or no volcanic activities. According to the stratigraphic correlation of biological fossils and strata, the basin filling sequence was formed in Norian in Late Triassic. On the west side (continental side) of the back-arc spreading basin, the back-arc inner-plate volcanic zone is developed, which is mainly a sequence of acidic volcanic rock series, which is distributed in a strip shape and roughly parallel to the volcanic rocks in the back-arc spreading basin. The volcanic rock stratum belongs to the Tumugou Formation (T_3t) formed in the Late Triassic, and its horizon is equivalent to that of Miange Formation in the

back-arc spreading basin. The volcanic rocks have typical characteristics of inter-plate rift magmatism. The back-arc spreading basin is another important ore-bearing place of volcanic shallow-low temperature gold-silver polymetallic deposits. Along the volcanic zone (Miange Formation), besides the middle-sized gold-silver polymetallic deposit in Nongduke and the large mercury deposit in Kongma Temple, the regional geochemical exploration and microwave remote sensing data also show a number of comprehensive mineralization anomalies in Tage, Darike and Dulonggou, which shows a good metallogenic prospect in this volcanic zone.

Collision Orogeny Period (208–138 Ma)

The collision orogeny period is marked by the successive development of collisional granite, orogenic uplift and late granite, accompanied by complex deformation tectonics formed by island arc and crust compression and contraction and shear strain. The magmatic activity during collision orogeny period is interdependent or closely associated with the granite zone of 237–208 Ma. The rock masses of interdependent distribution are distributed at the north and south ends of the Cuojiaoma-Daocheng Granite Zone, and the closely associated rock masses are small in scale and often intrude into the interior and edge of the granite base of 237–208 Ma.

At the end of Late Triassic, Ganzi-Litang Ocean Basin contracted into a residual basin along the subduction zone, and the main mass of Yidun Island Arc rose, partially accepting delta distributary channel facies and coastal swamp facies deposits, forming the molasse-miscellaneous debris tectonics. The high-density oceanic crust dragged the low-density continental crust to subduct along the subduction zone, resulting in the shortening of the continental crust and the arc-continent convergence collision. With the downward subduction and the change of P–t conditions, significant dehydration occurred, which on the one hand resulted in the double thickening of the collision zone crust, and on the other hand formed the collision granite with crust source, which penetrates into or was located in the volcanic arc granite zone and its vicinity. With significant shortening of the continental crust and the double thickening of the crust, the collision zone rose greatly, and the island arc magma in the crust, which may have been "frozen" in the island arc orogenic belt, was "activated" again, resulting in hypabyssal rock—ultrahypabyssal rock emplacement.

The syn-collision granites were relatively large in quantity, interdependent or closely associated with the 237–208 Ma granite zone, and distributed at north and south ends of Cuojiaoma-Daocheng Granite Zone. The main rocks were two-mica granite and moyite, followed by monzonitic granite and monzodiorite, which were the continental crust remelting granite. Late orogenic granites mainly occurred in the western granite zone of Shangri-La, and the dating data of granites vary greatly, ranging from 138 to 200 Ma, indicating that the granites were formed and located after the syn-collision granites. The magma evolution sequence is quartz monzonite diorite porphyrite → quartz dioritic porphyrite → quartz monzonitic porphyry → monzonitic granite porphyry, which belongs to crust-mantle granite. Minerals in collision orogenic period are concentrated in concentrated in Shangri-La Arc Area at the southern end of the island arc, mainly porphyry and skarn copper, molybdenum and gold polymetallic deposits related to intermediate-acid subvolcanic rocks, for example, porphyry copper, molybdenum and gold deposits in Chundu, Xuejiping and Lannitang, and skarn copper polymetallic deposits in Hongshan and Langdu.

With significant shortening of the continental crust, a series of axial vertical syn-cleavage folds in NWW-SN to south-north direction were formed, and the shear strain from west to east occurred, forming an arc-shaped ductile shear zone and phyllonite zone with gentle dip to the west in nearly SN direction, accompanied by the stretching lineation of recumbent folds and vertical shear zones.

Post-Orogenic Extension Period (138–75 Ma)

The typical magmatic product of post-orogenic extension is type A granite in Late Yanshanian, which developed in the narrow area between Keludong-Xiangcheng Fault Zone and Aila-Riyu Fault Zone, and was spatially distributed in the back-arc zone, forming another important granite zone in this area, namely Gaogong-Cuomolong Granite Zone. Type A granites formed in Late Yanshanian are often accompanied by tin and silver polymetallic mineralization, and many tin and silver polymetallic deposits have been found, such as medium-sized Lianlong Tin, Silver and Bismuth Polymetallic Deposit and large-sized Xiasai Silver Polymetallic Deposit, forming a large Tin and Silver Polymetallic Mineralized Granite Zone. A significant tin-bearing granite zone has been formed in the back-arc area of Yidun Island Arc, showing a broad mineral exploration prospect.

Since the concept of type A granite was put forward, its connotation and extension have changed greatly, and its genetic types and tectonic environments are also varied. Eby divided type A granites into two types, namely type A1 and type A2, based on the geochemical characteristics and tectonic environment of type A granites all over the world. The former type is geochemically similar to the magmatic hearth mantle of ocean island basalts (OIB), and formed in the continental rift and mantle plume (hot point). The latter type is geochemically similar to continental and island arc basalts, and the magma is derived from the crust or island arc and formed in the extensional tectonic environment after collision or orogeny. Gaogong-Cuomolong Granite Zone reflects that it is equivalent to inter-plate extensional granite, that is, type A granite defined by Pearce generally shows the geochemical characteristics of type A2 granite.

Due to double thickening of the crust, rock metamorphism in the lower crust led to high-density mineral assemblages (eclogite facies), and shortening of the continental crust led to thickening of the crust and lithosphere, resulting in the temperature of the mountain root being relatively lower than that of the asthenosphere. The hot material structure produced potential gravity instability, which led to rooting or lower crust subsidence. As a result, the hot asthenosphere at the lower part upwelled greatly, replacing the cold lithosphere, and led to partial melting of the crust, resulting in significant magmatism, that is, the significant development of type A granite. The main mass of post-orogenic spreading in this area occurred in 138–75 Ma, and the peak of spreading may be 80 Ma.

Himalayan Intracontinental Orogeny (65–15 Ma)

The main Himalayan intracontinental orogeny occurred in Qinghai-Tibet Plateau, manifested as the collision uplift of Tibetan Plateau. Its long-range effect in Yidun Island Arc Collision Orogenic Belt showed thrust nappe structure and significant strike-slip translation, as well as the formation of pull-apart basin and Himalayan granite emplacement.

Yidun Island Arc Collision Orogenic Belt has basically been shaped after the subduction orogeny in Late Triassic, collision orogeny in Jurassic-Cretaceous and subsequent post-orogenic spreading. In Cenozoic, due to the northward wedging collision of the Indian Plate and its intracontinental convergence orogeny, the main mass of Qinghai-Tibet Plateau uplifted and was orogenic, while the post-orogenic spreading of Yidun Island Arc Collision Orogenic Belt stopped and then vertically uplifted rapidly, and significant thrust nappe occurred. The significant translational strike-slip accompanied formed a series of beaded strike-slip pull-apart basins, accepting the Paleogene and Neogene red bed deposits. Himalayan granitic magmatic activity occurred along the nearly SN direction translational fracture, forming a nearly north–south granite belt. Although this sequence of granite is also type A granite, which shows the characteristics of intraplate extensional granite, the geochemical difference between this sequence of granite and type A granite during post-orogenic spreading indicates that they may have different magma sources and dynamic backgrounds. Himalayan granitic magma shows more characteristics of crust-mantle mixed source, which is mainly controlled by Dege-Xiangcheng Deep Fault (which changed to strike-slip

translation fracture in Himalayan); the granite magma in Late Yanshanian mainly originated from continental crust remelting and was restricted by the spreading of regional orogenic belts.

Himalayan granite emplacement was the latest magmatic event of the Yidun Island Arc Collision Orogenic Zone. The exposed rock masses in this area include Batang Genie rock mass, Xiangcheng Cilincuo-Riyong rock mass, Changtong rock mass, Shangri-La Chuliang rock mass, etc. Among them, Genie rock mass is distributed in Gaogong-Cuomolong Granite Zone, and other rock masses constitute the main mass of Cilincuo-Chuliang Granite Zone. Genie rock mass intruded in Triassic sand-slate series and Hagela rock mass (39–81 Ma), and its main mass is porphyritic medium-fine moyite, with a $^{40}Ar/^{39}Ar$ plateau age of 15 Ma. Cilincuo-Riyong rock mass intruded in the sand-slate system of Lamaya Formation in Upper Triassic, and is the compound batholith. The main rocks are porphyritic fine-grained monzonitic granite and hornblende biotite monzogranite. The age of Cilincuo rock mass is 65 Ma, while that of Riyong sheeted dike is 60 Ma. Generally speaking, the Himalayan magmatism lasted for a long time (50 Ma), but the period of magmatism in each rock mass was short.

All ore deposits formed in Yidun Island Arc Zone in different ages were shaped in the process of intracontinental orogeny, and were superimposed and reformed by tectonism and magmatic activity to varying degrees, showing the characteristics of multiple minerals and complex and diversified deposit types in an ore deposit.

2.2.2 Temporal and Spatial Structure and Its Evolution of Jinsha River Arc-Basin System

On a regional basis, the Jinsha River tectonic zone and Ailaoshan tectonic zone belong to the ocean basin subduction and arc-land collision junction zone. Both of them are comparable in terms of the formation, subduction and destruction of ocean basin and the evolution history of the continental margins on both sides. Thus, they belong to the same arc-land collision junction zone and are also an important polymetallic metallogenic zone of copper, gold, lead and zinc.

2.2.2.1 Temporal and Spatial Structure of Jinsha River Arc-Basin System

Jinsha River Tectonic Zone (Fig. 2.8) between Qamdo-Lanping Landmass in the west and Zhongza-Shangri-La Landmass in the east experienced the splitting of Devonian - Carboniferous strata, spreading of Late Carboniferous-Early Permian Ocean Basin, subduction and destruction of

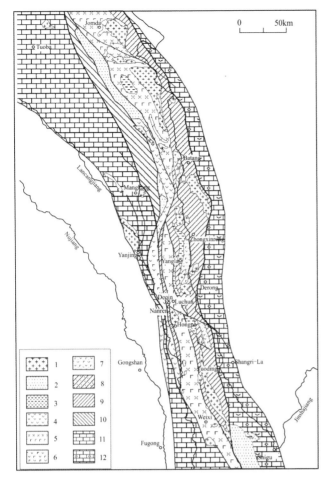

Fig. 2.8 Distribution of tectonic–magmatic rocks of Jinsha river zone. 1—Granite (γ4-5); 2—red sandstone distribution area (J-K); 3—clastic rock sedimentation area (T3); 4—intermediate-acid volcanic zone in continental margin arc (T1-2); 5—volcanic zone of Jiangda-Deqin-Weixi superimposed rift basin ($T2^2$–$T3^1$); 6—intermediate-basic volcanic zone of Ardenge-Nanzuo continental margin arc ($P1^2$–P2); 7—intermediate basic volcanic zone of Zhubalong-Benzilan intra-oceanic arc ($P1^2$–P2); 8—Deqin-Shimianchang tectonic melange zone; 9—Jinsha river tectonic melange zone; 10—Qingnidong-Haitong overthrust zone; 11—Qamdo-Lanping block; 12—Zhongza-Shangri-La block

Permian stratum, and collisional orogeny of Triassic stratum. The strata in the zone suffered from significant tectonic deformation and metamorphism, which destroyed the original sequence and relationship of strata. Except for the volcanic arc zone and stable landmass on both sides that are relatively complete, the extrusion fracture, schistosity and mylonitization were extremely developed in Jinsha River Zone. Tectonic melanges such as mafic–ultramafic rocks, carbonate rocks, siliceous rocks and basic volcanic rocks were found everywhere in the period from Devonian to Permian; the matrix was flysch sand-slate and intermediate-basic pyroclastic rock in Permian stratum-Triassic stratum.

Formation Age of Jinsha River Back-Arc Ocean Basin

In recent years, radiolarians formed in Late Devonian-Early Permian have been found in siliceous rocks associated with oceanic ridge basalt in Jinsha River Melange Zone. Radiolarians in Early Carboniferous, such as *Albaillella paradoxa defladree*, *Astroentactinia multispinisa* Won; radiolarians in Early Permian, such as *Albaillella* sp., *Pseudoalbailla* sp., etc.; radiolarians in Late Devonian-Early Carboniferous, such as *Entactinia* sp., *Entactinosphera* sp., *Entactinia parva* Won, *E. tortispina* Ormiston et Lane, *Entactinosphera foremanae* Ormiston et Lane, *En. cometes* Foreman, *En.deqinensis* Feng, *Belowea varibilis* (Ormiston et Lane), *Astroentactinia multispiosa* (Won) etc. The U–Pb age of zircon in oceanic ridge-quasi-oceanic ridge basalt in Jinsha River melange zone is (361.6 ± 8.5) Ma indicating that zircon was nearly formed in Early Carboniferous. The Rb–Sr isochron age of cumulate in Jiyidu is (264 ± 18) Ma (Mo et al. 1993), indicating that cumulate was nearly formed in Early Maokou of Early Permian. It shows Jinsha River back-arc ocean basin was formed in Early Carboniferous, and Jinsha River back-arc rift basin in Late Devonian was developed into the late development stage with the embryonic form of back-arc ocean basin. Jinsha River back-arc ocean basin was developed into the peak stage in the early stage of Early Permian, with a width of about 1800 km (Mo et al. 1993).

Jinsha River Ocean Basin began to subduct westward in the late stage of Early Permian and then destruct below Qamdo-Lanping Landmass, forming an intra-oceanic arc in Zhubalong-Yangla-Dongzhulin and a back-arc-basin of Xiquhe-Xueyayangkou-Jiyidu-Gongnong in the west of the intra-oceanic arc (basement of oceanic crust, $P1^2$–P_2) from east to west (basement of oceanic crust, $P1^2$-P_2); Jiangda-Deqin-Weixi continental margin volcanic arc and Qamdo-Lanping back-arc-basin (base bed of oceanic crust, $P1^2$–P_2) in the west of the continental margin arc. The Early and Middle Triassic formed Jiangda-Deqin-Weixi volcanic-magmatic arcs and the Qamdo-Lanping arc-back foreland basin (T_{1-2}) on the west side of the volcanic arc through oblique subduction collision; the superimposed rift basin of Chesuo Township-Xu Zhong-Luchun-Hongpo-Cuiyi was formed in the late stage of Middle Triassic to the early stage of Triassic (T_2^2-T_3^1).

The so-called angular unconformity between Upper and Lower Permian stratum in Gerongna, Batang was once considered as the product of oceanic crust destruction and ocean basin close to Jinsha River Paleo-Tethyan Ocean in Early Permian and "Late Variscan Orogeny". After investigation and tracing for three lines, it was determined as a slip nappe composed of sandshale-limestone formed in Late Permian mixed with volcanic rocks from east to west, which overlaid on the melange with garnet mica quartz schist as the matrix.

Spatial Distribution of Jinsha River Arc-Basin System

Jinsha River Arc-Basin System can be divided into the following zones in space from east to west: passive margin fold-thrust zone, ophiolite tectonic melange zone, volcanic zone of superimposed rift basin, Shimianchang Tectonic Melange Zone and continental margin arc volcanic zone (Fig. 2.8).

Passive Margin Fold-Thrust Zone

The passive margin fold-thrust zone is located on the western margin of Zhongza-Shangri-La Block, with developed fault tectonics, significant thrust nappe-extensional detachment tectonics, and multi-stage tectonic transformation and transformation, which is characterized by tectonic slices of thrust in different stratigraphic ages overlaying westward in an imbricate way, or are limited to the tight folds in shear fracture zones. With its unique limit Smith stratigraphic configuration, this zone overthrusts westward on the non-Smith stratigraphic system of Jinsha River Tectonic Melange Zone, while it is dominated by the nappe-detachment fault in the east, which is different from the Smith stratigraphic sequence in Zhongza-Shangri-La Block. Silurian stratum-Devonian stratum are composed of carbonate rocks mixed with clastic rocks and are developed into stromatoporoid reef in Middle and Late Devonian; Carboniferous stratum-Permian stratum are a sequence of collapse accumulation of carbonate rocks, turbidite mixed with basalt, basaltic andesite, pyroclastic rocks and carbonate rock, which are characterized by stratigraphic sequence and rock assemblage in the passive margin rift basin in the western part of Zhongzan-Shangri-La Block.

The western passive margin fold-thrust zone in Zhongza-Shangri-La Block starts from Dongpu in Dege Country in the north, and extends to Nixi, Tuoding and Shigu in Shangri-La in the south through Batang, Zhongza and Derong. The main body of this zone extends from north to south along the east bank of Jinsha River, and is characterized by developed imbricate thrust nappe, extensional detachment fault and detached block tectonics, accompanied by significant tectonic mylonitization, rheid fold, flow cleavage and dynamic metamorphism. It is not only a huge nappe-detachment tectonic zone with a length of 600 km, but also a metallogenic zone of copper, lead, zinc and other nonferrous metals. Tuoding Copper Ore Deposit (medium-sized carbonate rock in Devonian stratum), Gelan Copper Ore Deposit in Shangjiang (Pt_2 small-sized low-grade metamorphic rock), Sanjia Village Lead–Zinc Ore Deposit (medium-sized carbonate rock in Cambrian stratum) and Najiao Lead–Zinc Ore Deposit (medium-sized carbonate rock in Cambrian stratum).

Ophiolite Tectonic Melange Zone

The ophiolite melange zone is on the west side of the passive margin fold-thrust zone, and is the main body of Jinsha River Junction Zone. The reason why the ophiolite melange zone is described here is that in addition to ocean basin sediments, many sedimentary blocks in the passive continental margin in the west of Zhongza-Shangri-La Block are also mixed in the ophiolite melange zone. For example, the study on Yangla Copper Ore Deposit showed that a sequence of continental slope clastic rock formed in Devonian stratum, mixed with carbonate rock sediments was developed between oceanic ridge basalt in Late Carboniferous and intermediate-basic volcanic rocks in Permian, which was an overturned fold and was in contact with surrounding rocks in a way of fault. The similar condition is common on both sides of Jinsha River Valley, of which the distribution is consistent with that of the ophiolite melange zone, namely, a north–south distribution. The ophiolitic melange zone can be roughly divided into two subzones as seen from the surface. The eastern subzone extends from Bengzha to Derong in Batang, generally including E'aqin Group, some of Zhongxinrong Group and Gajinxueshan Group divided by the regional geological mapping with an original scale of 1:200,000. From the perspective of lithologic assemblage, it is a sequence of metamorphic basic volcanic rocks, ultrabasic rocks and oceanic ridge basalt; the main body is a sequence of mafic and ultramafic rock assemblage of the lower oceanic crust. The western subzone extends through Zhubalong-Yangla-Gongka and is composed of the upper oceanic crust substances with the characteristics of intra-oceanic arc, such as basaltic andesite, amphibolite basaltic andesite, siliceous slate, fine clastic rock, siliceous rock, etc. Taking Xiquhe Bridge and Yangla Copper Deposit as an example, it is composed of Gajinxueshan Group and the members a + b of the mining area, which are mainly composed of sedimentary rocks, mafic rocks and andesite, and its volcanic rocks have the characteristics of island arc volcanic rocks. Generally speaking, the oceanic crust is mainly composed of the lower oceanic crust with a large thickness, while the members of the upper oceanic crust are composed of sedimentary rocks, oceanic island volcanic rocks and island arc volcanic rocks. At the rapidly spreading oceanic ridge, the thickness of the lower oceanic crust is relatively large and has certain sequences; at the slowly spreading oceanic ridge, the thickness of the lower oceanic crust is very small and has no sequence. Compared with the distribution characteristics of the ophiolite melange zone in Jinsha River, it can be seen that Jinsha River Ocean Basin is a slow-spreading tecnotics. Besides the later tectonic reworking, the three groups mentioned above shall be regarded the same in composition, and the difference is that the eastern zone is dominated by lower oceanic crust substances while the island arc (intra-oceanic arc) volcanic rocks are accumulated in the western zone.

According to the study of the surrounding rock of Yangla Copper Deposit, the age of oceanic tholeiite is 361.6 Ma (zircon U–Pb method), indicating that Jinsha River Ocean Basin has spread into ocean in Early Carboniferous, and began to subduct in the late stage of Early Permian, so intra-oceanic arc volcanic rocks (257 Ma, amphibole K–Ar method) were found in the west side of the ocean basin; and radiolarians in Maokou were formed in the associated siliceous rock in Zhubalong intra-oceanic arc volcanic rocks (Peng Xingjie); continental margin arc volcanic rocks in the late stage of Early Permian were formed in Deqin-Weixi area on the western margin of ocean basin, rocks in Maokou were formed in limestone associated with arc volcanic rocks, and 220 Ma (biotite K–Ar method, Yunnan Bureau of Geology and Mineral Resources) to 280 Ma (zircon U–Pb method) island-type granodiorites were formed in Baimangxueshan area, all of which are the evidence that the Jinsha River Ocean Basin began to subduct and destruct westward. It can be seen that Jinsha River Ocean Basin was mainly spread in the early stage of Carboniferous-Early Permian, and subducted and destructed from the late stage of Early Permian to Early and Middle Triassic. Jinsha River-Ailaoshan Ophiolitic Melange Zone is an important metallogenic zone of precious and nonferrous metals such as gold, chromium and copper, including Ailaoshan Super-large Gold Ore Deposit (ophiolitic melange), Yangla Large Copper Ore Deposit (intra-oceanic arc volcanic-sedimentary rocks) and a sequence of copper and gold ore occurrences.

Volcanic Zone of the Superimposed Rift Basin

The volcanic zone of the superimposed rift basin is on the west side of the ophiolite melange zone, distributed in Xuzhong of Yanjing County, Luchun-Jijiading area of Deqin County, Reshuitang-Sizhuangzi Bridge area, Pantiange-Qiaohou area of Weixi, and developed in the continental margin volcanic arc of Jiangda-Weixi and its marginal zones. It is a superimposed rift basin formed by extension in the post-collisional island arc orogenic belt, with the formation age from the late stage of Middle Triassic to the early stage of Late Triassic. There are volcanic-sedimentary rocks with well-developed geological sections in Luchun-Jijiading area of Deqin County, Reshuitang-Sizhuangzi Bridge area of Deqin County and Pantiange-Cuiyibi area of Weixi County. In the basin, a sequence of tholeiite and alkaline rhyolite assemblage with the characteristics of "bimodal" volcanic rocks and a large number of gabbro-diabase dyke groups under extensional background are developed, accompanied by bathyal

siliceous rocks, radiolarian siliceous rocks, tuffaceous turbidite, tuffaceous-siliceous turbidite, arenopelitic flysch, fine clastic rocks and limestone lens. The initial value for 87Sr/86Sr of basalt in Luchun Deposit varies from 0.7065 to 0.7194, and the Rb–Sr isochron age is 236 Ma. The initial value for $^{87}Sr/^{86}Sr$ of rhyolite varies from 0.7099 to 0.7213, and the Rb–Sr isochron age varies from 224 to 238 Ma (on a scale of 1: 200,000 in Deqin country, 1985); the initial value for $^{87}Sr/^{86}Sr$ of rhyolite in Pantiange area of Weixi is 0.7074, and the Rb–Sr isochron age is 235 Ma. It can be seen that the "bimodal" volcanic rocks (Cuiyibi Formation) in the Pantiange area of Weixi and the "bimodal" volcanic rocks (Renzhixueshan Formation) in Luchun Deposit in Deqin have the same geotectonic background and volcanic rock development sequence. Basalt and rhyolite are from the same magma source, and the volcanic rocks were formed in the early stage of Late Triassic.

The superimposed rift basin has become an important ore-bearing basin for sedimentary exhalative massive sulfide copper–gold-silver-lead–zinc polymetallic ore deposits, such as Zuna Lead–Zinc (Silver) Ore Deposit in Shengda-Chesuo Basin, Zhaokalong Large Siderite Type Silver-Rich Polymetallic Ore Deposit and Dingqinnong Copper–Gold (Lead–Zinc-Silver) Ore Deposit; Luchun Zinc-Copper-Lead (Silver) Polymetallic Ore Deposit in Luchun-Hongpo Basin, Hongponiuchang Copper–Gold (Lead–Zinc) Polymetallic Ore Deposit and large gypsum ore deposit in Lirenka-Bamei area; Laojunshan Medium-sized Lead–Zinc (Silver) Ore Deposit and Chugezha Large Siderite Type Ore Deposit in Reshuitang-Cuiyibi Basin.

Shimianchang Tectonic Melange Zone

Deqin-Shimianchang Tectonic Melange Zone is on the west side of the volcanic zone in the superimposed rift basin, and distributed on the west side of Xuzhong and passes through Deqin County, Shimianchang-Yanmen area. It is the product of the destruction of the rift valley basin on the margin of Qamdo Landmass and its post-arc-land collision in Carboniferous-the early stage of Early Permian. This zone in Shimianchang area is a melange mainly composed of serpentinite ultrabasic rocks, schistose basalt, gabbro-diabase, siliceous rocks and limestone as blocks and mica schist, mica quartz schist and chlorite quartz schist as matrix. Slumping blocks composed of continental shelf carbonate rock, clastic rock and slope carbonate rock, as well as basinal turbidite, contourite, radiolarian siliceous rock (P₁), ultrabasic rock and gabbro-diabase were developed in Early Permian from west to east on the geological section of Yonghongrongqiu, Deqin, showing the sequence characteristics of continental margin rift (valley) basin. Shimianchang asbestos ore deposit related to serpentinite ultrabasic rocks has been found in Shimianchang Tectonic Melange Zone.

Continental Margin Arc Volcanic Zone

As an important part of Jinsha River Arc-Basin System, the continental margin arc volcanic zone is located on the west side of Shimianchang Tectonic Melange Zone and on the east margin of Qamdo-Lanping Landmass. In spatial terms, its main body is distributed on the east side of Zongla Mountain Pass in Markam County, Qamdo and area between Nanren-Bucun-Nanzuo in Deqin County and Badi-Yezhi area in Weixi County, which is called Jiangda-Deqin-Weixi Continental Margin Arc in Permian, including the upper part of Jidonglong Formation and Shamu Formation overlaid on it. The earliest arc volcanic activity was found in Nanren-Feilaisi area of Deqin County in the late stage of Early Permian, and *Neomisellina aff.douvillei* (Gubber), *N. aff.sichuanesis* Yang, *Kahlerina* sp., *Reichelina* sp. were found in the limestone containing bioclast in the upper thin bed of Jidonglong Formation, which was an arc volcanic activity happened in the late stage of Maokou of Early Permian and lasted until Late Permian. Tholeiite → calc-alkaline → potassium basalt rocks were developed in the volcanic rocks from morning to night. The rock types include quartz tholeiite, intermediate potassium andesite, dacite, rhyolite and its pyroclastic rocks. The properties of volcanic rock mark the complete process of generation, development and maturity of island arc (Mo et al. 1993).

Columnar joints of basaltic andesite formed in Permian were developed on the east side of Zongla Mountain Pass in Markam County, Qamdo, which belongs to continental eruption. The volcanic rocks in Adengge, Deqin are mixed with siltstone and carbonate rocks, forming a semideep-neritic environment. Very developed columnar joints of basaltic andesite were found on the west side of Feilaisi, Deqin, which belongs to continental eruption. The volcanic rock-carbonate rock assemblage with pillow structure developed in the volcanic rock in the Nanzuo-Bucun area belongs to a semideep-neritic environment. Volcanic rocks in the Shamu area are associated with sandshale containing plant fossils and brachiopod fragments, reflecting the marine-continental environment. Volcanic turbidite and volcanic source turbidite of submarine fan facies are developed in Yanmen Township. In Badi-Yezhi area, Weixi Country, a sequence of flysch sand-slate, metamorphic volcanic rocks and conglomerate, intermediate-acid volcanic breccia, slump breccia, marlstone and sedimentary sand-slate with incomplete Bouma sequence were found in Kangpu, Weixi and Xidagou, Jicha, indicating that arc volcanic rocks have entered the marginal slope-basinal deep-water environment. It shows that the distribution environment of arc volcanic activities is quite different in space. Arc volcanic rocks have varied lithofacies and diverse sedimentary types in space, and the topography of island arc fluctuates greatly.

There is a land with terrestrial plants and columnar joints outcropped from the water surface, and there are also carbonate rock plateau and abyssal valleys grown underwater. There are various sedimentary facies and types of sediments, varying from continental facies-marine-continental transitional facies-neritic facies-plateau slope facies-basinal deep-water facies, so as to form a tectonic paleogeographic pattern in an island-chained distribution way.

Jiaduoling-Dongka Medium-sized Iron-Copper Ore Deposit, Renda Copper Ore Deposit, Azhong Gold Ore Deposit, Nanzuo Medium-sized Lead–Zinc (Copper, Silver) Ore Deposit, Lirenka Large-scale Lead–Zinc (Copper, Silver) Ore Deposit, and a sequence of ore occurrences and gold anomaly areas have been found in Jiangda-Deqin-Weixi Continental Margin Arc Volcanic Zone in Permian, with various ore deposit types.

2.2.2.2 Formation and Evolution of Jinsha River Arc-Basin System

Geological and Paleogeographic Features of "Metamorphic Soft Basement" in Pre-Devonian

Regionally, Yangpo Formation of the former Shigu Group in Shigu area of Zhongza-Shangri-La Block on the east side of Jinsha River Zone is a sequence of metamorphic rocks of high-grade greenschist facies-amphibolite facies, among which the Nd model age of plagioclase amphibolite is between 1343.9 and 1369.9 Ma, and its metamorphic Rb–Sr isochron age is (996.1 ± 33.7) Ma which is quite consistent with the curing time limit (800–900 Ma, Jingningian) of the metamorphic basement in Yangtze Landmass. Therefore, it is the basement of Yangtze Landmass, and the quite stable plateau-type sediments formed in Paleozoic remain on it. In Qamdo-Lanping Block on the west side of Jinsha River, the age of single mineral zircon in the gneiss of Ningduo Group in Changqingke area, Yushu, Qinghai Province is 1870 Ma, which can be compared with that of Qias Group in Western Sichuan and Ailaoshan Group in Western Yunnan. The age values of the two types of granites intruded are 1780 Ma and 1680 Ma, respectively. There are tillites in the upper part of the Caoqu Group and basalts in its lower part, both of which are two important characteristics of plateau-type sediments in Sinian and Yangtze Landmass in the south extending into the Sanjiang. The main metamorphic age of Ningduo Group and Caoqu Group is 640 Ma, which is equivalent to the time limit of Chengjiang Orogeny during the final consolidation of the basement of Yangtze Landmass. Therefore, Qamdo-Lanping Block shall also be the basement of Yangtze Plateau, that is, Qamdo-Lanping Block and Zhongza-Shangri-La Block on both sides of Jinsha River were a unified block during Sinian- "Pan-Yangtze Landmass".

In Early Paleozoic, the areas of Western Sichuan, Western Yunnan and Eastern Tibet on the unified landmass had the "sedimentary cover formed in Sinian and later age" (Liu et al. 1993). Aulacogen was formed along Jinsha River and its two sides in Caledonian of Early Paleozoic, and its representative sediments are mica phyllite, mica-quartz phyllite mixed with marble and amphibolite (schist) in Longba Formation of original Shigu Group, and the representative source rocks are composed of semi-pelagic siliceous and arenopelitic flysch and semi-pelagic flysch clastic rocks in Qingnidong Group of Middle and Lower Ordovician stratum. Qamdo-Lanping Landmass and Zhongza-Shangri-La Block on both sides of aulacogen of Jinsha River maintain a relatively stable plateau-type sedimentary environment; in Early and Middle Ordovician, neritic carbonate rocks and clastic rocks were deposited in Yanjing and Markam areas of Eastern Tibet, while the Upper Ordovician stratum is lacking, and the Silurian stratum is composed of neritic sandy arenopelitic sediments. Apart from clastic rock and carbonate rock sediments of trilobites in Cambrian, there are also continuous biocarbonate rock sediments of Ordovician stratum-Silurian stratum in Zhongza-Shangri-La area. The fossils are characterized by neritic coral and reef-building coral of bryozoa and stromatopora; there are also other abundant benthos. The neritic plateau type sediments of Lower Paleozoic stratum "can be compared with that of Yangtze Plateau in both biological and sedimentary characteristics" (on a scale of 1: 200,000 in Shangri-La 1985).

The aulacogen of Jinsha River was closed and rose to form land along with the stable landmass on both sides at the end of Caledonian, which is characterized by Devonian stratum being unconformably overlaid on or being disconformably overlaid on underlying strata (Middle Ordovician stratum and Upper Silurian stratum) regionally. At the same time, Caledonian orogeny caused extensive and intense deformation and metamorphism of Pre-Devonian stratum. In Early Devonian, it began to enter the formation, development and evolution stage of Paleo-Tethys of Jinsha River on "metamorphic soft basement" in Pre-Devonian.

Formation and Evolution of Jinsha River Arc-Basin System

By taking Qiangtang-Jitang-Chongshan-Lancang Residual Arc on the west side of Qamdo-Lanping Block as the frontal arc, Jinsha River Arc-Basin System entered a new development period at the beginning of Devonian on the basis of the metamorphic "soft basement" of Early Paleozoic, namely the generation, development and evolution of Paleo-Tethyan Jinsha River Arc-Basin System (Fig. 2.9). It generally includes several stages as below.

Rift (Valley) Basin Stage (D)

In Early Devonian, Western Sichuan, Western Yunnan and Eastern Tibet had been connected with each other and formed the plateau type sediments on the "Pan-Yangtze Landmass", and seawater flows into the Sanjiang area from the south and north of Jinsha River Zone; except for the Jitang-Chongshan-Lancang Remnant Arc Zone, where Lincang archicontinent and Riwoqê archicontinent (island arc mountain system) were formed at the north and south ends respectively, the areas on both sides of Jinsha River Zone generally descend to be sedimented, forming terrigenous alluvial, alluvial-proluvial and diluvial clastic sediments first and then developing into littoral-neritic clastic rock and carbonate rock sediments. In Jinsha River Zone, Early Devonian stratum may be continuously deposited on Silurian underlying strata, which was formed by neritic continental-shelf carbonate-clastic rocks.

In Middle Devonian, the transgressive area was spread, and local extension and rift occurred on the background of the extension of neritic continental shelf in Early Devonian. In the Yangla-Benzhilan-Xiaruo-Tacheng area, the neritic-semi-pelagic carbonate rocks and siliceous-arenopelitic flysch formation were formed and intermediate-basic volcanic rocks erupted in the rift basin. In Late Devonian, the basin was extended and rifted significantly. In the rift basin with Yangla-Dongzhulin-Shigu area as the central axis, the bathyal sediments represented by radiolarian siliceous rocks-thick-heavy limestone-sandstone-mudstone assemblage are developed, accompanied by the eruption of extensional continental tholeiite and intermediate-basic volcanic rocks, showing that the extension and rifting have deepened the seawater and developed volcanic activities in the rift basin, namely rift basin stage formed by the thinning and rift of the continental crust. The stable landmasses on both sides of Jinsha River (Qamdo Landmass on the west and Zhongza-Shangri-La Landmass on the east) are composed of plateau carbonate rock-clastic rock formations in epicontinental neritic basins.

Ocean Basin Formation Stage (C_1–P_1)[1]

The period from Carboniferous to the early stage of Early Permian is an important period for the spreading of Jinsha River Back-arc Ocean Basin. On the basis of the rift basin in Late Devonian, it further spread to form an ocean basin in Carboniferous-early stage of Early Permian, that is, the evolution and development of Jinsha River Ocean Basin began. With the formation of Jinsha River Ocean Basin from Carboniferous to the early stage of Early Permian, the Qamdo Landmass split off from the "Pan-Yangtze Landmass" to form an independent micro-block. In Early-Middle Carboniferous and Late Devonian, rift basins continued to

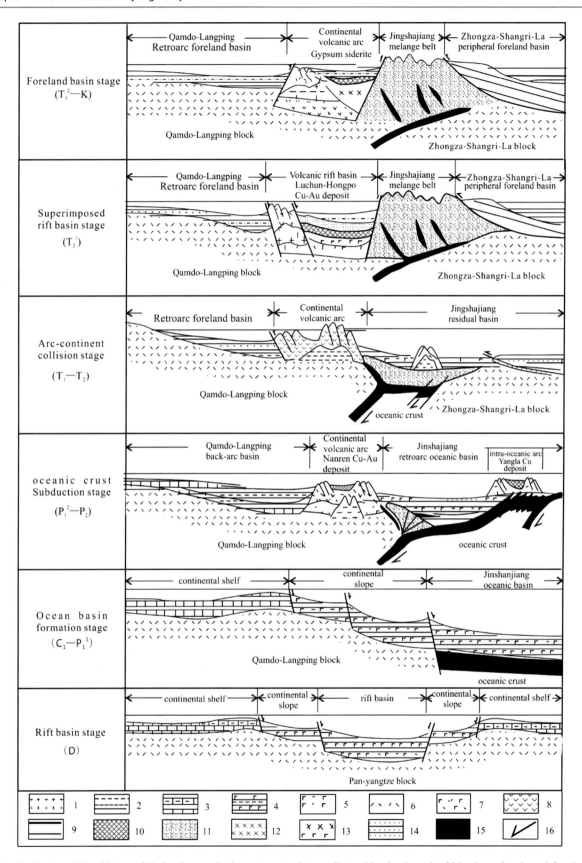

Fig. 2.9 Geologic evolution history of Jinsha river arc-basin system. 1—acid intrusive rock; 2—marginal clastic rock; 3—plateau carbonate rock; 4—basinal volcanic rock; 5—basalt; 6—continental crust basement; 7—intermediate-basic volcanic rock of island arc; 8—intermediate-acid volcanic rock of island arc; 9—abyssal flysch; 10—ore body; 11—tectonic melange; 12—rhyolite; 13—"bimodal" volcanic rock; 14—plateau clastic rock; 15—oceanic crust; 16—subduction direction

spread and the continental crust split off, forming Jinsha River Initial Ocean Basin in Early-Middle Carboniferous, in which volcanic, endogenous and terrigenous low-density turbidites represented by radiolarian siliceous rock-thick bedded limestone-black mudstone-tuff assemblage were deposited and oceanic ridge-oceanic island basalt erupted, belonging to semi-pelagic-abyssal carbonate rocks, volcanic rocks and arenopelitic-siliceous flysch formation.

In Late Carboniferous-early stage of Early Permian, with the increase of crustal splitting intensity, Jinsha River Ocean Basin spread rapidly in an asymmetric tectonic pattern, forming a mature ocean basin. The lithofacies in the ocean basin are composed of mid-ocean ridge mafic–ultramafic rocks, oceanic ridge-oceanic island tholeiite assemblages, and form the ocean basin ophiolite complex in Gajinxueshan, Gongka, Jiyidu and other places together with radiolarian siliceous rocks, along with which the abyssal sediments represented by the arenaceous-siliceous flysch formation composed of radiolarian siliceous rocks, siltstone, silty mudstone, black arbonaceous shale assemblages (rich in iron and manganese) in Gajinxueshan, Yangla, Gongka, Jiyidu-Xiaruo and Xinzhu area in Late Carboniferous-Early Permian were deposited in basins, so it is a sequence of volcanic, endogenous and terrigenous low-density turbidite.

The main body of Qamdo Block on the west side of Jinsha River Ocean Basin is a metastable plateau type sediment formed by carbonate rocks and clastic rocks mixed with intermediate-basic volcanic rocks in epicontinental neritic basins. Closing to the ocean basin, the eastern margin of Qamdo Block formed marginal rift (valley) basin in the Jiangda-Deqin-Shimianchang area with the significant spreading of Jinsha River Ocean Basin from Middle and Late Carboniferous to the early stage of Early Permian, in which slope carbonate rock slump accumulation at the margin of the continental shelf, submarine turbidite fan sedimentation, semi-pelagic basinal turbidite, radiolarian siliceous rock, contourite, as well as extensional basic and intermediate-basic volcanic rocks were developed. By the early stage of Early Permian, volcanic rocks had the characteristics of "bimodal" assemblage (Mo et al. 1993), showing the sequence characteristics of the rift (valley) basin at the margin of the continental shelf. The main body of Zhongza-Shangri-La Landmass on the east side of Jinsha River Ocean Basin is stable plateau type sediment formed by carbonate rocks-clastic rocks in epicontinental neritic basins, and the continental shelf marginal facies near the basin is composed of metastable carbonate rock-clastic rock mixed with intermediate-basic volcanic rocks.

Oceanic Crust Subduction and Destruction Stage (P_1^2–P_2)

The tectonic geology background of Jinsha River Zone changed greatly from the late stage of Early Permian to Late Permian. On the basis of the ocean basin formed by

spreading in Late Paleozoic, Early Carboniferous-the early stage of Early Permian, Jinsha River Basin subducted and destructed westward significantly in the late stage of Early Permian, marking the development of volcanic rocks in the intra-oceanic arc and the formation of back-arc-basin.

Due to the subduction and destruction between oceanic crusts (its formation mechanism is related to fracture and detachment and subduction inside the oceanic crust plate), Xiquhe Bridge-Xueyayangkou-Dongzhulin-Jiyidu-Gongnong Back-arc-Basin (oceanic crust basement) on the west side of the volcanic arc and its intra-oceanic arc in Zhubalong-Yangla-Dongzhulin area in the late stage of Early Permian-Late Permian were formed respectively in the central axis area of Jinsha River Ocean Basin. Neritic carbonate rocks and clastic rocks and semi-pelagic arenaceous-siliceous flysch formation were deposited in Zhubalong-Yangla-Dongzhulin intra-oceanic arc environment. Island-type volcanic rocks composed of quartz tholeiite-basaltic andesite-andesite-dacite (a small amount) were developed in the volcanic arc from morning to night. The middle part (oceanic crust basement) of Xiquhe-Xueyayangkou-Dongzhulin-Jiyidu-Gongnong Back-arc-Basin is on the west side of the intra-oceanic volcanic arc. The semi-pelagic-abyssal siliceous-arenopelitic flysch formation is formed in the back-arc spreading basin, accompanied by the diabase-sheeted dyke swarm in the back-arc spreading environment and the assemblage of tholeiite and basaltic tuff on it.

The oceanic crust subducted and destructed westward under Qamdo-Lanping Block on the west side of Jinsha River Ocean Basin, and then Jiangda-Deqin-Weixi continental margin volcanic arc of Permian and Qamdo Back-arc-Basin (continental crust basement) on the west side of the continental margin arc were formed respectively in the late stage of Early Permian to Late Permian. Volcanic-sedimentary rocks in the continental margin volcanic arc have varied lithofacies and diverse sedimentary types in space, and the topography of island arc fluctuates greatly. There is a land with terrestrial plants and columnar joints outcropped from the water surface, and there are also carbonate rock plateau and abyssal valleys grown underwater. There are various sedimentary facies and types of sediments, varying from continental facies-land-sea transitional facies-neritic facies-plateau slope facies-abyssal basin, so as to form a tectonic paleogeographic pattern in an island-chained distribution way. Tholeiite → calc-alkaline → potassium basalt volcanic rocks were developed in the arc volcanic rocks from morning to night. The properties of volcanic rock marks the complete process of generation, development and maturity of island arc (Mo et al. 1993). Qamdo Back-arc-Basin (continental crust basement) is on the west side of Jiangda-Deqin-Weixi continental margin volcanic arc, in which the metastable marine-continental biterrigenous coal-bearing clastic rock and volcanic rock formation, littoral biterrigenous clastic rock and volcanic rock formation, and neritic carbonate rock, clastic rock and volcanic rock formation are deposited.

The main body of Zhongza-Shangri-La Block on the east side of Jinsha River Ocean Basin maintains neritic carbonate rock sediments of plateau facies, the western margin of Zhongza-Shangri-La Block corresponds to the island arc-basin system on the eastern active margin of Qamdo Landmass; Permian is the evolution and development period of passive continental margin rift basin; the slope carbonate rock slump accumulation at the margin of the continental shelf, submarine turbidite fan sedimentation, semi-pelagic arenopelitic-siliceous shale flysch, and extensional basic and intermediate-basic volcanic rocks were developed in the passive continental margin rift basin in Zhiyong-Fulong Bridge-Nixi-Tuoding, and the lithogeochemistry characteristics of volcanic rocks and sandstones show that they are passive continental margin environment (Mo et al. 1993). The formation of the rift basin on the western margin of Zhongza-Shangri-La Block corresponds to the development of the island arc-back-arc-basin system on the active margin of Jiangda-Deqin-Weixi Block in Permian on the eastern margin of Qamdo Landmass.

Due to the westward subduction and destruction of the oceanic crust of Jinsha River, the spatial configuration structure of Jinsha River Arc-Basin System was formed during the period from the late stage of Early Permian to Late Permian, namely the intra-oceanic volcanic arc-back-arc-basin and the continental margin volcanic arc-back-arc-basin. This process is not only the conversion of the oceanic-continental lithosphere, but also the process of significant adjustment, recombination and conversion of material constituents.

Arc-Land Collision Stage (T_1–$T_2{}^1$)

In the Early and Middle Triassic, the tectonic and sedimentary environment of Jinsha River Arc-Basin System, Qamdo Block and Zhongza-Shangri-La Block on the east and west sides changed dramatically. Jinsha River Ocean Basin was destroyed and closed, and then the oceanic crust disappeared at the end of Late Permian. Jinsha River Zone turned into an arc-land collision development stage in Early and Middle Triassic, which is marked by the development of Jiangda-Deqin collisional continental margin volcanic arc and the formation of Qamdo Back-arc Foreland Basin and the residual marine basin (marginal sea) of the Jinsha River.

Jinsha River Ocean Basin was subducted and destroyed in Permian and then was closed at the end of Late Permian, with the arc-land collision and land-land docking. On the basis of the original Jinsha River Ocean Basin of Permian, Jinsha River Ocean Basin entered into the development stage of the residual marine basin (marginal sea) in the Early and Middle Triassic, in which the semi-pelagic fine clastic turbidite composed of endogenous and volcanic sources and mixed with spillite-keratophyre, radiolarian siliceous rock and marlstone assemblages was formed, belonging to carbonate rock, siliceous-arenopelitic flysch and volcanic rock formation. Collisional island arc intermediate-acid volcanic rocks, subvolcanic rocks and intrusion appeared in Shusong-Tongyou, which resulted from the subsequent development of the intra-oceanic arc from Permian to Middle Triassic. The main body of Zhongza-Shangri-La Landmass on the east side of Jinsha River Residual Marine Basin is dominated by the uplift and denudation area, but is lack of Middle Triassic stratum and its subsequent strata. Clastic rock sediments were formed in the epicontinental neritic basin in the Early Triassic, the basal fluvial conglomerate disconformably overlaid on or unconformably overlaid on Permian stratum, and the middle and upper part is composed of littoral-neritic carbonate rock and clastic rock formation.

The Jiangda-Gebo-Xuzhong area on the eastern margin of Qamdo Landmass is on the west side of Jinsha River Residual Marine Basin. Due to the arc-land collision and land-land docking, collisional continental margin volcanic arc was formed in Early-Middle Triassic and superimposed on the subduction type continental margin volcanic arc in Permian, and the volcanic rock assemblage of basaltic andesite-andesite-dacite-rhyolite series with island arc nature was developed. The arc volcanic rock was formed following the piedmont purplish red conglomerate in the Early Triassic, and consists of facies assemblage of alluvial-proluvial facies $\rightarrow$ littoral-neritic facies $\rightarrow$ marginal slope facies $\rightarrow$ basin facies from bottom to top and from west to east. It was turned into the spatial pattern of the fore-arc, inter-arc and back-arc-basins in the Middle Triassic, and bathyal volcanic rocks, terrigenous and volcanic turbidite were developed in the basins.

Qamdo Landmass on the west side of Jiangda-Gebo-Xuzhong collisional continental margin volcanic arc in Early and Middle Triassic changed from the back-arc-basin in Permian into the back-arc foreland basin. Most areas were uplifted due to the arc-land collision, and lack of Early and Middle Triassic strata. In the early stage of Early Triassic, fluvial and littoral clastic rock and intermediate-acid volcanic rock formations were formed in the marginal zone of the back-arc foreland basin only in Markam area near the island arc in the east of the block, and disconformably overlaid on underlying strata.

Superimposed Rift Basin Stage ($T_2{}^2$–$T3{}^1$)

Jinsha River Residual Marine Basin (marginal sea) disappeared during the period from the late stage of Middle Triassic to the early stage of Late Triassic, and the geological pattern of Jiangda-Deqin-Weixi Island Arc Orogenic Belt changed from extrusion into extension, and the possible conversion mechanism of its mechanical properties - lithosphere delamination, which caused continental crust to be stretched due to the extensional collapse caused by thinning; the superimposed rift basin split mainly formed in the early stage of Late Triassic in the original volcanic arc and its marginal zone due to extension and splitting; it was formed after the subduction and destruction, arc-land collision and land-land docking and collision of Jinsha River Ocean Basin and before the large-scale and large-area accumulation of molasse formation in Jinsha River in time; the upper main body is superimposed on Jiangda-Deqin-Weixi Island Arc Orogenic Belt in space, belonging to the post-collision extensional tectonic background.

From the late stage of Middle Triassic to the early stage of Late Triassic, the superimposed rift basin was characterized by the development of semi-pelagic volcanic turbidite, tuffaceous turbidite, tuffaceous-siliceous turbidite and argillaceous flysch, as well as "bimodal" volcanic rocks and gabbro-diabase dykes and dyke swarm composed of basalt and rhyolite assemblage. The geochemistry characteristics of volcanic rocks show that they are rift basin environments under extensional background. In the early development stage, the rift basin was composed of the assemblage of neritic to semi-pelagic basalt, basaltic tuff, sandstone, sandy mudstone, tuffaceous siliceous rock and marlstone, in which a large number of gabbro-diabase dykes and dyke swarm were developed. In the middle development stage, the rift basin was composed of the assemblage of semi-pelagic basalt, basaltic tuff, rhyolite, rhyolite tuff, rhyolite volcanic breccia, sandy mudstone, tuffaceous siliceous rock and marlstone, in which a large number of gabbro-diabase dykes and dyke swarm were developed. In the late development stage, the rift basin was composed of the assemblage of semi-pelagic and neritic (continental rock had appeared in some areas) rhyolite, rhyolite tuff, rhyolite volcanic breccia, sandstone, sandy mudstone and marlstone, in which gabbro-diabase dykes and dyke swarm were distributed. At the end of its development, the geological pattern of the rift basin changed from extension and rift into extrusion, the basin gradually shrank and destructed, forming the littoral-neritic clastic rocks (with the nature of molasse) mixed with intermediate-intermediate-acid volcanic rocks and pyroclastic rocks, in which a large number of gypsum-salt sediments composed of gypsum, barite and siderite deposits were developed.

From the late stage of Middle Triassic to the early stage of Late Triassic, the tectonic paleogeographic environment of the superimposed rift basin had changed greatly both in time and space. In terms of time, in the early and middle development stage, the rift basin had significant extension and rift, volcanic activity erupted in the deep water, and the gabbro-diabase dykes and dyke swarm in the extensional

tectonic background as well as subvolcanic gabbro porphyrite were developed. In the late development stage, the rift basin had slight extension and rift, and was closed with acidic volcanic activity, volcanic rocks were formed in the shallow water, and even continental eruption and columnar joints occurred. At the end of its development, it turned into an extrusion environment, and ended by the appearance of intermediate-intermediate-acid volcanic rocks. Moreover, a large number of gypsum-salt sediments were developed. The rift basin is composed of continental-adlittoral volcanic islands and faulted extensional basins spatially, with abyssal sediments developed, forming a tectonic paleogeographic pattern dominated by "graben and horst". Shengda-Chesuo Township-Xialaxiu volcanic-sedimentary basin, Xuzhong-Luchun-Hongpo volcanic-sedimentary basin and Reshuitang-Cuiyibi-Shanglan volcanic-sedimentary basin can be roughly found from north to south.

By taking geochronologic scale as the age standard and based on the correlation diagram (Fig. 2.10) drawn depending on data about K_2O and rifting and extension speed in basalt of Ethiopia provided by Mohr and Berberi et al. and data about K_2O and rifting and extension speed in basalt of Kenya-Tanzania provided by Mohr et al. and Villims, the extension and rifting distance of three volcanic-sedimentary basins were measured. As for Shengda-Chesuo Township-Xialaxiu volcanic-sedimentary basin in the north member, the average value (1.43%) of w (K_2O) from six basalts is adopted, which is set at point A in the figure, with the rifting velocity (V_P) of 0.27 cm/a and the rifting distance (D) of 63 km; as for Xuzhong-Luchun-Hongpo Volcanic-Sedimentary Basin in the middle member, the average value (0.48%) of w (K_2O) from 10 basalts is adopted, which is set

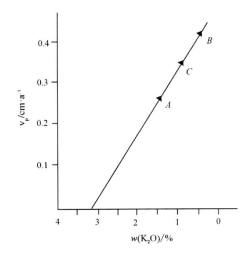

Fig. 2.10 Relationship between Basalt w (K_2O) and rifting speed in rift basin. A—rifting speed of Shengda-Chesuo township-Xialaxiu volcanic-sedimentary basin; B—rifting speed of Xuzhong-cLuchun-Hongpo volcanic-sedimentary basin; C—rifting speed of Reshuitang-Cuiyibi-Shanglan volcanic-sedimentary basin

at point B in the figure, with the rifting speed (Vp) of 0.43 cm/a and the rifting distance (d) of 140 km. As for Reshuitang-Cuiyibi-Shanglan volcanic-sedimentary basin in the south member, the average value (0.81%) of w (K_2O) from four basalts is adopted, which is set at point C in the figure, with the rifting speed (Vp) of 0.36 cm/a and the rifting distance (d) of 116 km. Mo et al. (1993) estimated that the spreading width of three volcanic-sedimentary basins is 49.5 km in Chesuo Basin, 113 km in Jiaojiading Basin and 81 km in Cuiyibi Basin, with similar results.

According to the critical spreading rate (0.5–0.9 cm/a) proposed by Sleep and Kuzmir, at which the magma under the ridge is formed, the rifting speed of Chesuo Basin in the north member (0.27 cm/a) is lower than that of Jijiading Basin in the middle member and that (0.43 cm/a and 0.36 cm/a) of Cuiyibi Basin in the south member. Therefore, the rifting intensity of Jijiading Basin in the middle member and Cuiyibi Basin in the south member is relatively high, with "bimodal" volcanic rock assemblage appearing in the basin, while the rifting intensity of Chesuo Basin in the north member is relatively low, with only tholeiite appearing in the basin but no "bimodal" volcanic rock. In addition, the rifting speed (0.27 cm/a, 0.43 cm/a, 0.36 cm/a) of the three basins in the north, middle and south members are all lower than the critical spreading rate (0.5–0.9 cm/a) of the magma under the ridge, so the spreading ridge ophiolite assemblage was not formed.

Foreland Basin Stage ($T_3{}^2$–K)

In the middle and late stage of Late Triassic, Jinsha River Zone entered the comprehensive intracontinental collision orogeny stage. Clastic rocks, molasses and coal-bearing formations were formed in marginal foreland basin in Jinsha River Orogenic Belt and its rear margin due to accumulation, and unconformably superimposed on Jinsha River Tectonic Melange. In Qamdo Landmass on the west side of Jinsha River Junction Zone, the fluvial-littoral clastic molasse formation was formed in the back-arc foreland basin in the early stage of Late Triassic, and unconformably covered and underlaid on the strata of different ages; its back-arc foreland basin continued to develop and evolve to form the formation varying neritic carbonate rocks from marine-continental coal-bearing clastic rocks in the middle and late stage of Late Triassic. By the Late Cretaceous, the foreland basin gradually shrank and destructed.

Intracontinental Convergence Stage (E–Q)

Cainozoic is the formation and uplift period of the Sanjiang area and even Qinghai-Tibet Plateau. The last orogeny formed a large-scale overthrust nappe, a large-scale strike-slip and the layered detachment of the surface and lithosphere formed by overthrust nappe and stretching. On the one hand, some fracture, depression, strike-slip,

stretching and pull-apart basins were formed; on the other hand, the mountains formed in the early stage were superimposed and reworked, and the crust was thickened significantly, accompanied by significant magmatism, metamorphism, tectonism and mineralization of nonferrous metals and precious metals. All ore deposits formed in Jinsha River Zone in different ages were shaped in the process of intracontinental orogeny, and were superimposed and reformed by tectonism and magmatic activity to varying degrees, showing the characteristics of multiple minerals and complex and diversified deposit types in an ore deposit.

2.2.3 Temporal and Spatial Structure and Its Evolution of Zhongza-Shangri-La Block

Zhongza-Shangri-La Block lies between Jinsha River Junction Zone on the west side and Yidun-Xiaqiaotou Back-arc-Basin Orogenic Belt on the east side. In the Paleozoic, it was a part of the western passive margin of Yangtze Continent and a horst on the west side of Muli-Haidong Graben. In the middle and late stage of Late Paleozoic, it split from Yangtze Landmass with movement of Dege-Shangri-La Micro-landmass due to the opening of Ganzi-Litang Ocean, forming a stable block in the micro-landmass. Before splitting, this block moved up and down with the movement of Yangtze Landmass, so that its movement is related to but has difference with the sedimentary characteristics of Yangtze Landmass. As a stable block, Zhongza-Shangri-La Block was formed in the Middle and Late Cambrian. This block experienced three major development stages. That is, basement formation stage, stable block formation stage and reverse polarity orogenic stage of block fold and uplift (Li et al. 2002).

2.2.3.1 Basement Formation Stage

This block has a typical dual structure of basement and caprock. The basement is a metamorphic crystalline basement, with the first rock formation of the primary Shigu Group in the south member and Chamashan Group in the north member. The caprock is the Paleozoic stratum-Triassic stratum.

The basement (Shigu Group) in the south member of the block is mainly distributed in Shigu, Lijiang-Tacheng outside Weixi in the west of Yinchanggou Anticline in Shangri-La Block and in the east of Ludien Granite Zone in Jinsha River Tectonic Zone. It is divided into three groups from bottom to top: Yangpo Group, Longba Group and Tacheng Group. The study on Shigu Group is unknown and its stratigraphic division and formation age are still controversial. By our observation, Shigu Group can be roughly divided into three rock formations according to the degree and characteristics of deformation. The first rock formation is Shigu Group outcropped in Liming Township, Lijiang, which is the lower member or lower part of Yangpo Formation. With high-grade metamorphism, it belongs to high-grade greenschist-low amphibolite facies, and granite veins formed by partial melting are seen locally. The deformation level is also relatively deep, S_0 has been completely replaced by S_1 and many folds are seen locally. The second rock formation is Shigu Group outcropped in the east of Taiping Bridge, Judian-Lamaluo, Shigu and in the west of Baita Village, Judian-Wuhou Village-Hetaoping, Shigu; with low-grade metamorphism and belonging to the middle-grade greenschist facies, so it is equivalent to the upper part of Yangpo Formation and Longba Formation. The deformation level is relatively low, S_0 has not been completely replaced by S_1 and it is characterized by flysch sedimentation. A series of sharp edges or congruous folds that reverse eastward is formed. It is separated from the first rock formation by the mylonite zone with ductile shear. The third rock formation is distributed in the east of Baita Village-Hetaoping, which is equivalent to marginal Tacheng Formation. It is mainly composed of mica schist and mica quartz schist, so its metamorphism is lower than that of the second rock formation and the layers are still visible, forming a series of folds that reverse westward and contacting with the second rock formation by fault, which is just opposite to the second rock formation. In addition, a large sequence of metamorphic basic volcanic rocks is found in the rock formation, and siliceous rocks are found locally. This is very similar to the Carboniferous stratum and Permian stratum in the passive margin zone of Jinsha River in Tacheng-Tuoding in the north. Therefore, we consider that the first rock formation belongs to the real metamorphic crystalline basement of Shangri-La Block; the second rock formation belongs to the miogeosynclinal sedimentation at the passive margin, may also belong to the contemporaneous heterotopical sedimentation on the block, which is formed by the stable block-type shallow water in Cambrian-Silurian changing into the abyssal environment on the western margin, and may also contain some Precambrian strata. The third rock formation is regionally connected with Late Paleozoic stratum in Tacheng-Tuoding in the north and distributed in zones from north to south, forming the passive margin sedimentary zone in the eastern part of Paleo-Tethyan Ocean in Jinsha River. The lithogeochemistry characteristics of its basic volcanic rocks show that it is continental margin rift type (Li et al. 1999a; b), so the third rock formation may be Late Paleozoic stratum.

The basement of the northern member of the block is Chamashan Group. It is a sequence of metamorphic carbonate rocks and intermediate-basic volcanic rocks, located under the Cambrian stratum and in contact with Cambrian stratum by faults. It should be pointed out that some suspect that the Chamashan Group was formed in Late Paleozoic due to upward thrust of Cambrian stratum and distributed on the passive margin of Jinsha River. As the strata on the Zhongza-Shangri-La Block overthrust westward on Jinsha River Junction Zone, the thrust superposition relationship mentioned above is reasonable. In the case of the above statement, Chamashan Group is not the real basement of the block.

To sum up, up to now, the direct contact relationship between the caprock and the basement of the block has not been found. However, in terms of metamorphism and deformation, as there was a tectonic change between them, it is reasonable to take the first rock formation-Shigu Group as its crystalline basement. It shows that the block is kept stable after a tectonic event.

2.2.3.2 Stable Block Formation Stage

The stable block was formed in Middle and Late Cambrian and ended at the end of Permian. The whole Paleozoic stratum is an assemblage of clastic rocks and carbonate rocks, but the Lower Paleozoic stratum is composed of basic and intermediate-acid volcanic rock, which is relatively active, while the Upper Paleozoic stratum is relatively stable due to no volcanic rocks mixed. Therefore, its evolution can be divided into two stages, namely Early Paleozoic and Late Paleozoic.

Evolution Stage in Early Paleozoic

From Middle and Late Cambrian to Silurian, the block is mainly composed of the sedimentary sequence of carbonate rocks-clastic rocks-carbonatite, which generally shows the stable coastal-continental shelf plateau-type sedimentary environment.

The lower bed of Middle Cambrian stratum is composed of dark-gray thin laminated dolomitic silty slate, containing trilobite locally, which shows an offshore environment; the middle bed is composed of feldspathic quartz sandstone mixed with dolomitic debris, sandstone mixed with dolomitic debris, and feldspathic quartz sandstone, in which swash bedding can be found, so it shows a coastal environment; the upper bed is composed of dolomite mixed with sandstone, which shows a littoral-neritic environment. The Upper Cambrian stratum is composed of secondary dolomite and contains biodetritus, belonging to a neritic carbonate rock gentle slope environment. The analysis on sedimentary facies shows that Middle and Late Cambrian strata in Zhongza-Shangri-La Block were located in the shore and offshore tectonics, with shallow water but high energy. Paleocurrent data (wave ridge strike: 160°, 150°) indicates that the coastline at that time generally extended from north to south. Considering that the west of this area is Proto-Tethyan Jinsha River, it is speculated that the west at that time should be a front slope zone of plateau and continental slope zone inclined to the west, while this area was kept in the marginal shoal environment of plateau. The main lithology of Ordovician stratum is quartz sandstone or feldspathic quartz sandstone mixed with sandy slate, with dolomitic limestone lens in the lower bed. Abraded wave ripple, cross bedding and parallel bedding were developed, showing a littoral-neritic environment. Early sediments of Early Silurian are mainly composed of gray, medium-layer medium-grained feldspathic quartz sandstone, showing a foreshore environment. In the middle and late stages, the stratum is composed of siltstone and slate. It is an inshore-offshore environment. The Middle and Late Silurian stratum is composed of crystal powder dolomite, showing a tidal flat environment. In a word, the Early Paleozoic stratum has high energy and shallow sea water and is generally located in a stable environment on the plateau margin.

Late Paleozoic Evolution Stage

The Early Paleozoic stratum in this area is a part of the western margin of Yangtze Landmass, and Caledonian orogeny still has an influence on it. It is mainly characterized by the conformity of Devonian stratum and underlying stratum. Lugu Lake-Erhai Lake on the east side is a depression zone with continuous sediments and deepening water body, and the block still has the tectonic pattern of early horst-shaped uplift.

In the early stage of Early Devonian, the stratum is composed of feldspathic quartz sandstone sediments, with plate-like cross bedding and lenticular basal conglomerate, belonging to a fluvial environment. In the middle and late stage, the stratum is a thin layer of sediments composed of quartz sandstone and silty shale, with well-developed horizontal bedding, belonging to a littoral-neritic environment. The Middle and Late Devonian stratum is composed of limestone and bioclastic reef rock with fore-reef collapse. It is a bordered carbonate rock plateau environment. The sea level dropped at the end of Late Silurian, causing the extensive exposure of the western land margin of Yangtze Landmass and the uplift of carbonate rocks on the plateau margin. Aulacogen was developed in Lugu Lake-Erhai Lake on its west side, in which the carbonate turbidity current and abyssal radiolarian siliceous rocks can be found. Therefore, the paleogeographic pattern of Zhongza-Shangri-La Block at that time was dominated by a high-energy zone on the margin of a continental shelf, with the fore-reef and back-reef colluvium developed on the west side. There is a continental shelf lake in the east of the high-energy zone;

with synsedimentary fracture, the sea water is extremely deep and changes suddenly. The Lower Carboniferous stratum is composed of light gray and dark gray, thin-medium-thick micritic limestone and silty limestone, some of which have become crystalline limestone, belonging to an open plateau environment. The Middle and Upper Carboniferous stratum is composed of light gray bioclastic limestone and oolitic limestone. In the northern Zhongza area, the Lower Carboniferous stratum is also composed of bioclastic limestone and oolitic limestone. Therefore, it shows that Zhongza-Shangri-La Block in the whole Carboniferous is a marginal shoal environment of carbonate plateau, and the local area is an open plateau environment. The Carboniferous stratum on the east side of the block is dominated by limestone, but also contains argillaceous limestone, flint limestone and siliceous rocks; the sea water is relatively deep, especially in Lugu Lake-Erhai Lake area, where an aulacogen was formed and manganese ore and radiolarian rock can also be found. Like the Carboniferous stratum, the Permian stratum is also a neritic carbonate rock environment. There is also an aulacogen in the east, in which siliceous rocks and corrosive fluid sediments can be found. In Late Permian, basalt erupted on the east side of the block, which marked the complete splitting of the block from the Yangtze Landmass and formed an independent block.

It can be seen from Paleozoic sedimentary evolution that Zhongza-Shangri-La Block is a part of the western margin of Yangtze Landmass, which constitutes a complete plateau and passive continental margin. This passive continental margin is not a simple epicontinental sea inclined to the ocean, but always maintains a graben-horst pattern of two highs and one low, two shallows and one deep, namely marginal reef (shoal) of continental shelf, marginal moraine (shoal) of land and marginal slope of plateau. In several areas with sea level drop, shoal areas are nearly exposed and subject to erosion and even incised valley sedimentation. This pattern lasted until the late stage of Paleozoic, and ended upon the formation of Ganzi-Litang Back-arc Ocean Basin.

2.2.3.3 Reverse Polarity Orogenic Stage

The uplift of Zhongza-Shangri-La Block in Late Permian (local uplift to form land) shows that Jinsha River Oceanic Crust had been destroyed, marking that it has entered the land-arc and land-land collision stage and the foreland uplift has appeared. Coarse clastic molasses are accumulated in some members of the Late Triassic block, such as the northern Lalashan and the southern Tuoding area, which are characterized by sedimentation of the post-orogenic foreland basin. However, due to the tectonic inversion of Jinsha River Zone, that is, the back-thrust spreading of the junction zone and the passive margin zone (see Sect. 2 below), the strata on Zhongza-Shangri-La Block thrust westward on the passive margin zone and even the junction zone of Jinsha River Tectonic Zone, resulting in the tectonic inversion of the foreland basin (Liu et al. 1993), forming a back-thrust spreading foreland basin of Late Triassic in Qamdo-Pu'er Block, so the development of foreland basin in Zhongza-Shangri-La Block was quickly ended and only a trace of the initial embryonic form of foreland basin was left.

The collision between Zhongza-Shangri-La Block and Qamdo Landmass also led to the fold deformation of Paleozoic stratum on the block, forming a series of nearly north–south short-axis geanticlinal tectonics, and making Late Triassic stratum unconformably cover on it. Its tectonic deformation pattern is from the center to the west of the block and from the wide and gentle isopachous fold without cleavage to the overturned fold with the close cleavage and same inclination, showing a change from weak to strong and a kind of reversed orogeny. This reverse polarity orogeny overthrusts Zhongza-Shangri-La Block westward, forming an important regional thrust sheet on the east side of Sanjiang area. Due to the large-scale westward thrust nappe of the block, a thrust fault also appeared at its rear margin, which causes the strata on the block (such as Devonian stratum) to thrust eastward on Mesozoic stratum on the east margin of the block, showing a horst on a cross section.

Reverse polarity orogeny of Late Triassic in Zhongza-Shangri-La Block may be further intensified during Yanshanian-Himalayan. This is because the overthrust nappe on the east side of the whole Qamdo-Pu'er Basin reached a peak in Yanshanian-Himalayan, which ended the development of the back-thrust spreading foreland basin but started the orogeny. Moreover, this may also be proved by the fact that Zhongza-Shangri-La Block had not sedimented since the Late Triassic.

We believe that Zhongza-Shangri-La Landmass was always in a horst state in the graben-horst tectonic pattern on the western margin of Yangtze River from Ordovician–Silurian to Early Carboniferous and was separated from Yangtze Landmass with the opening of Ganzi-Litang Ocean until Carboniferous to Permian. The Late Permian uplift is related to the closure of Jinsha River Ocean, which is characterized by foreland uplift. It was a sequence of plateau-type adlittoral sediments from Middle and Late Cambrian to Permian, showing the inherent characteristics of stable blocks, while its tectonic deformation was relatively significant and the reverse polarity orogeny occurred, which was different from the normal blocks. This seems to be the common characteristic of different blocks in the Sanjiang area in terms of deformation characteristics, which may be related to the significant extrusion applied on the Sanjiang" area.

2.2.4 Temporal and Spatial Structure and Evolution of Qamdo Basin

Located between Jinsha River Junction Zone and the Northern Lancang River Fracture Zone, Qamdo Basin is a composite back-arc foreland basin system developed and formed on Proterozoic-Lower Paleozoic basement in Late Paleozoic and Mesozoic as the same name, and is Paleo-Tethyan Basin in the northern member of Sanjiang area in southwest China.

2.2.4.1 Composition and Characteristics of the Upper Crust Tectonic Bed

The upper crust of Qamdo Block has the tectonic characteristics of "double basements" and "three caprocks", the double basements refer to Proterozoic crystalline basement and Lower Paleozoic fold basement and the three caprocks are composed of Upper Paleozoic, Mesozoic and Cainozoic caprocks. According to the tectonic-rock assemblage and tectonic orogeny cycle, Qamdo Block can be divided into five tectonic beds from bottom to top (Table 2.1).

Basement Tectonic Bed

It is composed of Proterozoic crystalline basement (I) and Lower Paleozoic fold basement (II). The Proterozoic crystalline basement is outcropped in Xiaosumang and Xiariduo in the northeast of Qamdo Basin, which is composed of the Paleo-Mesoproterozoic Ningduo Group and Neoproterozoic Caoqu Group and may also include a part of metamorphic rocks in the lower bed of Xiongsong Group. The rock assemblage in Ningduo Group is composed of (garnet) biotite plagioclase gneiss, plagioclase amphibolite, biotite/binary quartz schist, granulite, etc. The metamorphic grade is of low amphibolite facies and the rock has significant multi-stage tectonic deformation. Caoqu Group is a sequence of metamorphic rocks of high-grade greenschist facies, with the lithology of schist mixed with metamorphic volcanic rocks; and its lower bed contains metamorphic conglomerate. There is no direct contact relationship between Caoqu Group and Ningduo Group, but they are obviously different in lithologic assemblage and metamorphic grade, so they may be unconformable.

Proterozoic Xiongsong Group in Boluo-Xiongsong on the west side of Jinsha River is composed of gneiss and schist members in the lower bed and marble members in the upper bed. The gneiss in the lower bed has the same lithology as that of Ningduo Group. Abundant colonial coral, bivalves, lamellibranch and crinoidal biological fossils are found in the marble in the upper bed of continuous Xiongsong Group in Xiugeshan on the west side of Xiongsong. Among them, the coral is identified as Lithostrotion, which was formed in Early Carboniferous.

Table 2.1 Tectonic bed characteristics of Qamdo block

Tectonic bed	Tectonic period	Age	Sedimentary formation and magmatic activity	Deformation and metamorphism	Tectonic environment	Tectonic stage
V	Himalaya Layaqi	N	Red/variegated gypsum-salt-bearing molasses sediments. Mantle source alkaline magma intrusion and inland alkaline-calc-alkaline volcanic eruption	Strata tilting, simple fold, brittle fracture, and no metamorphism	Intracontinental convergence and extrusion: thrust, nappe, strike-slip, stretching or tension	Intracontinental convergence and extrusion deformation stage
		E				
IV	Yanshanian	K	Marine-continental and continental red (locally variegated) clastic rocks	Open-isoclinic fold, simple fold, brittle fracture, no metamorphism	Foreland-intracontinental depression basin	Foreland-intracontinental depression basin stage
		J	Sedimentation and intermediate-acid I–S type magma intrusion developed in orogenic belts on both sides of the block			Tectonic collapse and extension (east) stage of collisional orogenic belt
	Indosinian	T_3	Molasse clastic rock, carbonate rock, coal-bearing terrigenous clast Calcium-alkali volcanic rocks and intermediate-acid I–S type magma intrusion developed in the east	Wide and gentle fold, brittle fracture, no metamorphism	Post-collision stretching: the eastern part is composed of a superimposed volcanic-sedimentary basin and a rift basin; the western part is composed of a foreland basin	Collisional orogenic belt
		T_1–T_2	Molasse clastic rock, carbonate rock, clastic turbidite. Calcium-alkali volcanic rocks and intermediate-acid magma intrusion developed in the east		Collision and extrusion: crustal uplift, ancient land in the west; collisional orogeny in the east, small tectonic basin	
III	Hercynian	P	Neritic carbonate rock, transitional coal-bearing terrigenous clast; flysch and abyssal turbidite (on both sides of the block).	Wide and gentle fold, brittle fracture, no metamorphism basically, local greenschist facies	Active terrestrial source (east), back-arc-basin (middle) Epicontinental Ocean Basin and continental margin rift basin (on both sides)	Arc-basin system stage Block splitting stage
		C				
		D	Basic/intermediate-acid volcanic rocks and a small amount of I–S type magma intrusion			
II	Caledonian	C–S (?)	Flyschoid clastic rocks and carbonate rocks Type I plagioclase granite invasion	Complex folds such as tight-inclined folds, syn-cleavage fold, general low-grade metamorphism	Orogenic belt (?)	Fold and basement formation stage
I	Jingningian-Pre-Jingningian	Z	Lack of geological records	Significant ductile deformation, plastic flowing deformation, mylonitization, foliation (local)	Ancient orogenic belt	Crystalline basement formation stage
		Pt_2–Pt_3	Chlorite-sericite schist, dolomite quartz schist (Pt_3), schist and gneiss mixed with marble (Pt_{1-2}), the source rock is flysch clastic rock mixed with carbonate rock, and also contains basic volcanic rocks			

The source rock age of Ningduo Group is 1680–2200 Ma (zircon U–Pb method, based on data of a scale of 1: 200,000 in Dengke and Laduo respectively), and the age of plagioclase amphibolite (source rock is volcanic rock) in the lower bed of Xiongsong Group is 1594 Ma (SM–Nd whole rock isochron age, based on data of a scale of 1: 200,000 in Wanxiong), which was basically formed in (Paleo) Mesoproterozoic. The volcanic rocks in Caoqu Group are earlier than 876–999 Ma (rock U–Pb method, based on data of a scale of 1: 200,000 in Dengke), which was basically formed in Mesoproterozoic and Neoproterozoic. The gneiss in the lower bed of Proterozoic Xiongsong Group obtained two metamorphic ages of 857.4 Ma $\pm$ 143 Ma (Rb–Sr method) and 611–669 Ma (Rb–Sr method, on a scale of 1: 200,000 in Xiongsong), respectively, which were equivalent to the period from Jingningian to Chengjiang, indicating that Jingningian orogeny and Chengjiang orogeny occurred during the formation period of the crystalline basement of Qamdo Block. The fold basement on the crystalline basement, namely the second tectonic bed (II), is composed of Middle-Lower Ordovician stratum of Lower Paleozoic and a small amount of Silurian stratum. Middle and Lower Ordovician strata are mainly outcropped in the Qingnidong-Haitong-Duojiban, and mainly composed of low-grade metamorphic sand-slate mixed with carbonate rocks, with significant deformation and development trendline as the same as that of tight-inclined folds. The Lower Devonian stratum obviously unconformably covers it in Jiangda and Jueyong. The fold basement was formed in Caledonian, which not only caused folds in Lower Paleozoic stratum, but also caused the plagioclase granite intrusion with the age of 462 Ma in Jiefang Township, Jiangda.

Upper Paleozoic Tectonic Bed

The Upper Paleozoic tectonic bed (III) (including Devonian stratum, Carboniferous stratum and Permian stratum) is well developed and is the first stable caprock sediment in Qamdo Block. In the interior of the block, it is a sequence of stable neritic plateau-transitional carbonate rock and clastic rock sediments, with the thickness between 3000 and 5000 m, and the east and west margins of the block are dominated by metastable-active sediments with large sediment thickness. There is no regional tectonic unconformity interface in the Upper Paleozoic tectonic bed except the local parallel unconformity interface between Devonian and Carboniferous strata and Upper and Lower Permian strata, which represents a complete regional transgressive–regressive sedimentary cycle.

Hercynian orogeny ended the development of this tectonic bed. Wide and gentle folds and brittle deformation are mainly developed in the Upper Paleozoic stratum, and the metamorphic grade is generally low. However, there is significant deformation and metamorphism locally in the orogenic belts on the east and west sides of the block.

Mesozoic Tectonic Bed

Mesozoic tectonic bed (IV) is the second caprock developed on Qamdo Block, which was formed in the reversed foreland-intracontinental basin after Hercynian-Early and Middle Indosinian orogeny. Except for Lower and Middle Triassic strata, the main body of Mesozoic stratum was composed of continuous Upper Triassic stratum to Cretaceous stratum, which constitute the sedimentary sequence of foreland basin, and are fully overlaid on Proterozoic, Paleozoic and Lower and Middle Triassic strata, with almost more than 80% of Qamdo Block. The relationship between Lower Triassic stratum and Upper Paleozoic stratum, and between Upper Triassic stratum and Lower (Middle) Triassic stratum in this tectonic bed is tectonic unconformity.

Mesozoic tectonic bed was formed on the erosion surface after Hercynian-Early and Indosinian orogenies, and was composed of transgressive–regressive cycles developed in the eastern part of Early and Middle Triassic strata and the regional transgressive–regressive cycles in Late Triassic-Cretaceous strata, which were the evolution history of Qamdo Basin developed into foreland basin. Influenced by the tectonic activities of orogenic belts on both sides of the block, the eastern part of the Mesozoic stratum is dominated by the Triassic stratum and is characterized by the development of a large number of volcanic rocks. On the basis of the Upper Paleozoic stratum, the central and western parts are foreland basin filling sequences starting from Late Triassic molasse formation, which are dominated by stable sedimentary assemblages. After the end of the late stage of Late Triassic, the foreland basin moved westward, and Jurassic-Cretaceous strata were extremely thick molasse sediments accumulated in a significant intracontinental depression.

Rocks of the Mesozoic stratum were not metamorphic and dominated by simple, wide and gentle folds. Significant superficial brittle deformation occurred during Himalayan intracontinental extrusion.

Cainozoic Tectonic Bed

Cainozoic tectonic bed (V) is a strike-slip pull-apart basin sediment formed in Cainozoic intracontinental convergence stage and mainly distributed in Paleogene/Neogene basins such as Gongjue, Nangqian, Jiqu and Lawu. Paleogene stratum in Gongjue and Nangqian basins is a sequence of red molasses with a thickness of 4000 m, and also contains gypsum-salt, accompanied by the intrusion of intermediate-acid volcanic rocks and alkaline porphyry/vein rocks. The Paleogene/Neogene stratum in Jiqu Basin is mainly a sequence of fluviatile-lacustrine red clastic rocks; the lower bed of Neogene stratum in Lawu Basin is a sequence of latitic-trachytic volcanic rocks, and the upper bed is a sequence of fluviatile-swampy coal-bearing clastic rock sediments. Cainozoic tectonic beds are limited in distribution, formed before the large-scale uplift of the plateau, and mainly composed of Paleogene stratum. There is an unconformity relationship between the Paleogene stratum and the underlying strata of different ages as well as between the Paleogene stratum and the Neogene stratum. Influenced by the intracontinental convergence, the Paleogene/Neogene stratum is mainly characterized by tectonic deformation such as stratigraphic tilting, local folds and strike-slip faults.

2.2.4.2 Temporal and Spatial Structure of Qamdo Basin System

The Qamdo Composite Basin System is represented by the above two tectonic beds of the Upper Paleozoic and Mesozoic, both of which belong to different types of basins controlled by different tectonic stages and mechanisms, and have experienced different development and evolution processes, respectively, with different temporal and spatial structures.

Late Paleozoic Lithofacies Paleogeography and Basin Structure

Late Paleozoic basin was formed on the erosion surface of fold basement formed by Caledonian orogeny in Qamdo Block, and Lower Devonian stratum unconformably overlaid the underlying Lower Ordovician stratum, on which the stable Devonian, Carboniferous and Permian sedimentary assemblages were continuously developed, but volcanic active formations representing rift basins were developed in different degrees on the east and west sides.

Mesozoic Lithofacies Paleogeography and (Back-Arc) Foreland Basin Structure

The Mesozoic Qamdo Basin was formed on the erosion surface of the above-mentioned Late Paleozoic basin after the Hercynian-Middle and Early Indosinian orogeny. Mesozoic basin was located on the subduction plate on the west side of Jinsha River Subduction Orogenic Belt, in which the foreland basin filling sequence starting from Piedmont molasse formation was developed. Therefore, the basin has the nature of a back-arc foreland basin. The Mesozoic basin was formed in the Early and Middle Triassic, reached the peak of development in the Late Triassic, and then shrank in the Jurassic, and finally closed in the Cretaceous. It is characterized by long evolution time, obvious dynamic change, diverse basin types and complicated temporal and spatial structure.

Early and Middle Triassic Lithofacies Paleogeography and Basin Properties

During the Early and Middle Triassic, Qamdo Block was generally in an uplifting state, and only a few small depressed basins existed in the eastern orogenic belt of the block. Jiangda-Walasi area is the only basin with continuous Early and Middle Triassic sediments. Pushuiqiao Formation in the lower bed of Lower Triassic stratum unconformably overlaid on Hercynian granite body by eluvial purplish red granitic basal conglomerate, in which a sequence of continental, littoral and continental shelf coarse sandstone containing gravels, sandstone mixed with oolitic limestone, andesitic volcanic rocks and dacite volcanic rocks was deposited, and then was developed into a sequence of shoal-tidal flat assemblage (Quxianong Formation) with the upper transition, which is composed of oolitic limestone, sand or calcarenite and micrite and mixed with few volcanic rocks; in addition, littoral-neritic → slope brecciated limestone, volcanic turbidite, oolitic limestone, calcarenite, siliceous rocks mixed with intermediate-acid volcanic rocks were developed in the upper bed (Serongsi Formation). In the Middle Triassic Walasi Formation, a sequence of volcanic, endogenous and terrigenous turbidites with a thickness of 2000–2500 m containing intermediate-acid volcanic rock interbeds developed, which vertically constituted repeated regressive and progressive submarine fan superposition. It shows that the basin had sustained subsidence from the Early Triassic to the Middle Triassic. There are a series of small basins of Early Triassic distributed in the northwest narrow strip along Xiariduo-Malasongduo-Jiaolongqiao at the western rear edge of the orogenic belt. The lower member of Lower Triassic (Malasongduo Formation) is littoral-neritic grayish black, variegated sandstone, siltstone mixed with marl and limestone, which is unconformity with Carboniferous-Permian, and the upper member is a sequence transitional-continental of grayish green and locally purplish red acidic dacite-rhyolitic volcanic rock mixed with quartz sandstone and black shale erupted. The basins may have been squeezed by the eastern orogenic belt and soon destroyed at the end of Early Triassic.

The above basins were formed on the erosion surface of Hercynian movement, all of which were isolated small basins, which were not connected with each, and had different filling sequences and evolution histories. They belong to tectonic basins trapped in orogenic belts and thrust depressions on the western edge. Judging from the difference in the time of basin destruction and the general existence of collision-type intermediate-acid volcanic rocks, the collision orogeny in the eastern part of Qamdo Block in the Early and Middle Triassic was significant, and the difference between the uplift and depression of the crust was quite obvious. The small basins and intermediate-acid volcanic activities in this period were just the response of collision orogeny.

Late Triassic Lithofacies Paleogeography and Basin Properties
After the general uplift in Early and Middle Triassic, the Qamdo Block began to sink integrally in Late Triassic, resulting in the complete overlap of molasse formation in the lower part of the Upper Triassic, and the sedimentary sequence of littoral-neritic clastic rocks, shelf carbonate rocks and land-sea transitional clastic rocks was successively deposited upward, forming a complete transgression-regression cycle. In Carnian-Norian of Late Triassic, except for two NW-NNW parallel ancient lands in Boluo-Xiongsong on the west of Jinsha River and Xiariduo-Qingnidong-Haitong on the west, the whole Qamdo Block is a wide sea basin. After the Norian, the eastern margin of the block rose, and Boluo-Xiongsong Ancient Land spread to the north and south and to the west. At the end of the Late Triassic, the eastern part of the whole basin rose further from east to west, and the sedimentary area of the Late Triassic became land.

The Late Triassic Carnian was bounded by Xiariduo-Qingnidong-Haitong Ancient Land in the northwest direction. The east and west sides were affected by the different tectonic activities, so the paleogeographic environment, basin types and sedimentary assemblages are quite different. Post-collision tectonic spread occurred in the eastern orogenic belt, forming a superimposed volcanic-sedimentary basin, and on its west side, a rift basin was formed by stretching along the back-margin fracture of the main arc of the early continental margin. The central and western part of the block belongs to an intracontinental neritic basin.

In Late Triassic Carnian, Jiangda-Gebo superimposed volcanic-sedimentary basins were distributed on the early continental margin arc, that is, Gelashan-Jiangda-Gebo, and spread southward to Xuzhong-Deqin, Yunnan. A sequence of active volcanic-sedimentary assemblages was developed in Carnian, which consisted of Dongdu Formation, Gongyenong Formation and Dongka Formation from bottom to top, which was equivalent to Jiapila Formation in the same period in the central and western parts of Qamdo Block, so it was also called Large Jiapila Formation. The Dongdu Formation in the lower part was a sequence of fluvial and littoral amaranth-variegated clastic rock molasse, with a small amount of intermediate-acid volcanic rocks locally, which was pseudo-integrated with the Middle Triassic, and featured angular unconformity with Proterozoic and Paleozoic. It contained multiple layers of conglomerate mainly composed of quartzose mylonite and locally composed of metamorphic rocks, obviously its provenance was mainly from the orogenic belt at that time. The Gongyenong Formation in the middle part was composed of neritic-shelf medium-thin to thick limestone; the Dongka Formation in the upper part was mainly composed of neritic-transitional sandstone and siltstone and locally composed of volcanic turbidite and intermediate-acid volcanic rocks.

The research on the fine structure of the basin in the early Late Triassic shows that the basin was formed on the continental margin arc with island chain structure in the early stage, with an undulating basement and complicated paleogeographic environment. In the northwest, the basin was generally a shallow sea basin except the Xiaosumang-Caoqugulong zone composed of Proterozoic stratum, and in the southeast, it was Kagong-Songxi-Gebo bay basin connected with the shallow sea in the northwest between Boluo-Xiongsong Ancient Land and Xiariduo-Qingnidong-Haitong Ancient Land in the west. Massive volcanic activities in the basin occurred in Carnian, and the magmatic activities quickly ended at the end of Carnian, and then the volcanic-sedimentary basin was transformed into a stable neritic basin. In Norian-Rhaetian (Bolila-Adula-Duogaila) in Late Triassic, stable neritic carbonate rocks to transitional clastic rocks developed in this zone, and the sedimentary facies of the whole block tended to be consistent during this period.

Shengda-Chesuo Rift Basin was located in the west of the volcanic-sedimentary basin mentioned above. The Shengda-Chesuo-Songxi zone to the east of Xiariduo-Qingnidong-Haitong Uplift spread in NW direction for about 300 km, with a width of only 5–10 km. It is a series of limited rift basins in the Late Triassic, stretching and intermittently developing along Chesuo-Deqin Fracture on the west side of Jiangda-Mangling continental margin arc orogenic belt in Late Paleozoic, and is similar in nature to Deqinluchun Rift Basin in the south. As the Upper Triassic in the southwest of the basin overlaps the Devonian and Carboniferous-Permian, the basement of the basin may be mainly the Upper Paleozoic. In the northern part of the basin, along Xialaxiu-Moditan-Shengda, the Upper Triassic is called Shengda Group (T_{3Sd}), and the lower part is clastic rocks with a small amount of fluvial-lacustrine to neritic basic basalt (equivalent to Quezhika Formation and Chayekou Formation), and neritic limestone/marl (equivalent to Niangken Formation, Nilenong Formation and Mianda Formation-Luose Formation). The middle and upper parts are a sequence of grey-dark grey low-grade metamorphic sand-slate, sandwiched with multiple layers of alkaline olivine basalt (equivalent to Caijunka Formation, Bama Formation, Zhaga Formation and Zaileda Formation). In Ridanguo, Zaileda and other places, alkaline olivine basalt can be found in multiple interlayers, with a thickness of about 1500 m, accompanied by the intrusion of basic and ultrabasic rocks such as olivine diabase. The Bouma series is widely developed in clastic rocks, belonging to flysch turbidite formation dominated by land debris, with a total thickness of 6000 m (top and bottom not seen). In the slope zone on the east side of the basin, according to the observation of Shengda and other places, a sequence of transitional deposits has been developed, which is composed of angle gravelly slump limestone, channel deposits and cal-careous turbidite fans of slope-marginal basin facies, and there are load casts in the sand-slate.

The southward rift basins are small. In Chesuo and its northwest area, the Upper Triassic in the same period is a sequence of mottled/dark grey clastic rocks from shallow water to semi-deep water, with a sedimentary thickness of nearly 4000 m. A large number of pillow-shaped breccias, conglomerates and lava mainly composed of (intermediate) basic volcanic lava are developed in the upper part, with a volcanic rock thickness of nearly 2300 m, accompanied by diabase vein intrusion. The thickness of volcanic cycle from bottom to top is increasing, which indicates that the exten-sion of the basin is increasing. In Songxi-Chetai area in the northeast of Gongjue County, a sequence of volcanic tur-bidite is developed in the lower part of the Upper Triassic (Jingu Formation-Waqu Formation), and a large number of grayish green-grayish black block-bedded, pillow-shaped, almond-shaped intermediate-basic volcanic lava, agglomer-ate and breccia are developed in the upper part, with andesite at the top, and the volcanic rock thickness is up to 2000 m. A large number of diabase and gabbro intrusion occurred at the same period.

In the west of Xiariduo-Qingnidong-Haitong ancient land, the Carnian of Late Triassic started to sink and trans-gress as a whole from the ancient land to a vast shallow sea basin, and a sequence of sub-stable molasses to stable littoral-neritic carbonate rocks and clastic rocks was devel-oped, which is called Jiapila Formation. At the bottom of Jiapila Formation, there are gray-purple composite con-glomerates and gravelly sandstones deposited by piedmont debris flow and fluvial alluvial fan, and they transit upward to littoral-neritic medium coarse to coarse-grained sand-stones, which are unconformity with all old strata. The middle and upper parts are composed of fuchsia to gray-purple, grayish-yellow to gray-green debris of quartz sandstone, siltstone and mudstone of different thickness, interbedded with limestone or lens, and belong to littoral-neritic sedimentation.

The Upper Triassic Jiapila Formation, also known as Small Jiapila Formation, developed in the area west of Xiariduo-Qingnidong-Haitong Ancient Land, is at the same horizon as the so-called "Large Jiapila Formation" in the eastern part of the block, that is, Dongdu Formation, Gon-gyenong Formation and Dongka Formation. They are iso-chronous and heterogeneous. Compared with the active "Large Jiapila Formation" in the eastern area, the sedimen-tary assemblage is very different, and the Small Jiapila Formation in the western area is the sub-stable to stable sedimentation, with simple lithology and lithofacies and almost no volcanic rocks. The sedimentary characteristics of Jiapila Formation show that the Jiapila period is the primary stage of transgression, with abundant material sources,

extremely fast deposition and poor sorting. The eastern part of the basin is dominated by clastic sediments, and the grain size of sediments becomes smaller from east to west and from bottom to top. The structure is dominated by conti-nental facies to littoral-neritic facies, which indicates that the paleogeographic environment experienced changes from Piedmont river, alluvial plain to clastic shelf from east to west and from bottom to top. However, the grain size of sediments in the western part of the basin is relatively fine, with the littoral-neritic gray layer accounting for a significant proportion, and there are more carbonate rock deposits in the middle and upper parts, which may have experienced the evolution from alluvial fan and alluvial plain to clastic rock, carbonate rock shelf and plateau. The provenance filling from the eastern orogenic belt may mainly control the sed-imentary basin. The analysis of the huge difference in the sedimentary thickness of Jiapila Formation reveals that the basin was formed on the undulating basement in this period, with Xiariduo-Qingnidong-Haitong Uplift in the east, Dangbala-Machala Uplift in the west and Tuoba Uplift in the middle. The uplift-depression pattern of the basin basement may have been strengthened and enlarged by tectonic activity in the Jiapila period, and the deposition thickness of Jiapila Formation in the depression (such as Dangga) between Tuoba Uplift and Machala Uplift has doubled, which proves this aspect. The sporadic volcanic rocks in the middle and lower horizons of Jiapi Formation coincide with the volcanic rocks developed in the contemporaneous active volcanic-sedimentary basin and rift basin in the east, which reflected the contemporaneous tectonic–magmatic activity in the western basin. From the volcanic rock assemblage, the eastern part of the basin is dominated by (alkaline) basalt, while the western part is dominated by intermediate-acid volcanic rocks. Further studies regarding whether magmatic activity is related to extensional tectonic mechanism are required.

In Norian (Bolila) of Late Triassic, the spread of the eastern orogenic belt of Qamdo Block stopped and the dif-ference of EW tectonic activity disappeared, which was also the period of the largest transgression range and the most stable sediment in Qamdo Basin. On the vast clastic shelf formed by the accumulation and filling of terrigenous clastic materials in the Jiapila Period, a wide sea carbonate shelf plateau with gentle inclination from east to west was formed, and a sequence of stable neritic-platform carbonate deposits (called Bolila Formation) was developed, integrated on Dongka Formation (east) or Jiapila Formation (west), cov-ering almost the whole Qamdo Block. The Bolila Formation has a single lithology, consisting of gray-dark gray lime-stone, arenaceous limestone mixed with micrite-silty lime-stone, nodular limestone, arenaceous limestone, bioclastic limestone and dolomite, mixed with a small amount of argillaceous siltstoue. The upper and lower parts are mainly thin and medium-thick layers, and the middle part is mainly thick blocks. The thickness of the western part is generally greater than that of the eastern part, and the thickness of the eastern edge drops sharply to only tens of meters, which is directly unconformity with the underlying strata. It shows that the sea water overlaps from west to east.

During Late Norian-Rhaetian of Late Triassic (Adula-Duogaila), the orogenic belts on the east and west sides of Qamdo Block rapidly uplifted, the erosion rate was accelerated, the material supply was sufficient, and a large amount of terrigenous clastic materials were injected into the basin at an accelerated rate and accumulated on the early carbonate shelf, resulting in the rapid filling and silting of the basin and the relative lowering of sea level. It mainly formed a regressive sedimentary sequence, which was composed of Adula Formation (lower part) and Duogaila Formation (upper part), also called Bagong Formation. Adula Formation was in conformity contact with Bolila Formation in the lower part and Duogai Formation in the upper part. The sedimentary assemblages of Adula-Duogaila Formation are similar, mainly belonging to the neritic-shelf to transitional terrigenous clastic deposits, with shed coal (layers) locally, which are widely distributed with little change, and the total thickness is generally 500–1000 m. According to the stratum distribution of Adula-Duogai Formation, the deposits at that time mainly occurred in the west of Xiariduo-Qingnidong Uplift, and there were only some connected to semi-connected local depressions or finger-shaped bays in the east. Except that the local depressions were deep and the deposition thickness reached 3000 m, the deposition in most other areas was not thick. During the same period, the difference between the internal basins was obvious, with significant local persistent depression and large accumulation. For example, the thickness of Adula-Duogai Formation in Niaonong Area between Tuoba Uplift and Qingnidong Uplift surged to over 3000 m.

Since Early Jurassic, the tectonic pattern of Qamdo Basin has changed greatly. On the one hand, the Late Triassic sedimentary area in the eastern part of the basin has been uplifted into land, and the parallel orogenic belt in the eastern part of the basin has retreated westward and moved to the west of Maozhuang-Tuoba-Juelong; on the other hand, the sea water withdrew gradually and entered the development stage of intracontinental depression basin. Because there was continuous sedimentary transition between Jurassic and Upper Triassic, and no new tectonic sequence has been found, the transition occurred by gradual "peaceful evolution". During Early Jurassic-Middle Jurassic, the basin was still in a relatively open environment. With the change of sea level, seawater flooded in from time to time, and continental and transitional deposits alternately appeared. In Late Jurassic, all seawater finally withdrew and

completely evolved into a continental basin, with the development of fluvial-lacustrine deposits representing the basin destruction.

Jurassic-Cretaceous stratum is a series of transitional to continental clastic rocks deposited continuously on Upper Triassic. The deposition thickness is generally between 1500 and 3000 m, the thickness in the east is small and gradually increases to the west. Jurassic stratum is composed of lower Chalangga Formation, middle Tutuo Formation, Dongdaqiao Formation and upper Xiaosuoka Formation.

E'aipu Fracture in the east of the basin (the western boundary fracture of Tuoba Thrust) played an important role in controlling the whole Jurassic sedimentation. The northeast side of the fracture is basically the tidal flat, mainly thinner red clastic rock deposits, while the southwest side of the fracture is mainly silty, muddy and marlaceous lacustrine deposits with a dark color. A delta is formed in Xiangdui Area in southwest of the southeast member of the fracture. The Jurassic deposits are over 4000 m thick, and the Cretaceous deposits integrated on it are over 5000 m thick. It shows that the northeast member of the fracture is in an active state of relatively rising and the southwest member is relatively falling, and the foreland thrust activity on the east side of the basin started from the end of Late Triassic to the west and lasted until Jurassic. In addition, in Randui-Angchong to the east of Lancang River on the western edge of the basin, a large number of composite conglomerates and sandy conglomerates deposited in the alluvial foothill debris flow-river appeared in the Middle Jurassic to Upper Jurassic. It shows that the eastward thrust uplift of the orogenic belt in the western margin of the basin was also very significant.

Cretaceous stratum is mainly distributed in Cuowa-East Markam in the southeast of the basin, which indicates that the basin has shrunk from northwest to southeast. The Cretaceous stratum is in conformity with the underlying Jurassic stratum, mainly developing a sequence of red continental molasse deposits upward coarsening, which generally shows the characteristics of coarse grain size of sediments on both sides of the basin and relatively small grain size of sediments in the center. It shows that at that time, with the accelerated speed of mountains on both sides of the basin, a large number of bioerosion was formed and transported to the depressed basin, resulting in the final siltation and closure of the basin. The Cretaceous stratum is well exposed in Markam area, and the lower series (namely Cuowa Formation, Laoran Formation or Xiangdui Formation) is mainly composed of purple sandstone, mixed with light gray, grayish yellow-grayish green calcareous sandstone containing argionite, calcareous sandstone, siltstone and fine conglomerate. From bottom to top, the reverse cycles of siltstone $\rightarrow$ sandstone $\rightarrow$ conglomerate form an upward coarsening sedimentary sequence. The upper series (namely Markam Formation and Hutoushan Formation) is composed of fuchsia fine and medium grained quartz sandstone, feldspar quartz sandstone, argillaceous siltstone, mudstone mixed with conglomerate and glutenite, and oblique bedding and cross-bedding are well developed. The above Cretaceous strata are continental basin fluvial-lacustrine deposits.

2.2.4.3 Evolution and Orogeny of Qamdo Basin

Based on the above analysis of the space–time structure and sedimentation of Qamdo Basin, the tectonic evolution and orogeny process of the reshaped basin are as follows.

Evolution of Epicontinental Sea Basin During the Cleavage of Paleo-Tethyan Ocean in Devonian-Early (Middle) Carboniferous

From Early Devonian to Early Carboniferous, a geodynamic mechanism of extension caused by mantle uplift opened the prelude to the tectonic evolution of the Paleo-Tethys. The development of Jinsha River Rift split Qamdo Block from Yangtze Plate, and Qamdo Block was brought into the Paleo-Tethyan ocean-land system. In the same period, there are indications that the west side of the block is roughly along Tienailie-Denglongnong and Leiwuqi Gangzi-Kagong areas in Nangqian County, Qinghai Province, and the North Lancang River rift or ocean basin also developed.

Compared with the active continental margin slope-deep water basin deposits developed in the rift basins on the east and west margins, the Qamdo Landmass located between the two oceans is a shallow sea basin, which is mainly composed of stable neritic-shelf deposits.

Evolution of Arc-Basin System Under the Subduction of Permian Paleo-Tethyan Ocean

According to the typical calc-alkaline series arc volcanic rocks in Permian, Jinsha River Ocean may have started to subduct under Qamdo Block in the late period of Early Permian. The eastern part of Qamdo Block forms an active continental margin arc-basin, and the western part is a contemporaneous back-arc-basin, which generally constitutes a complete arc-basin system. At that time, the subduction trench in the northern member may be very close to the continental margin and subducted directly under Qamdo Block. The Permian oceanic crust subduction in the south of Gobo occurred in the sea east of the continental margin, belonging to the intra-oceanic subduction. Therefore, the continental margin island arc in the eastern part of the block developed on the basis of different crust, and there are obvious differences in the properties and crustal structure between the southern and northern arc members.

Collision Orogeny in Early and Middle Triassic

With the westward subduction of Jinsha River Oceanic Basin closed at the end of Late Permian, the evolution of Late Paleozoic basin ended. Collision orogeny in Early and Middle Triassic started. The important basis is as follows: First, Qamdo Block generally uplifted at the end of Late Permian, which is reflected in the upward coarsening of Permian stratigraphic sequence as a regression sequence; second, in the eastern part of Qamdo Block, there are sporadic Lower (Middle) Triassic unconformity overlying the Upper Paleozoic or Hercynian intrusive rocks, followed by the overall overlapping of Late Triassic deposits, and the lower part of Lower Triassic and Upper Triassic are molasse deposits; third, in Jiangda in the eastern area and Xiariduo, Malasongduo, Zhiba, Reyong and Xuzhong in the western area, high-silica, high-potassium dacite-rhyolite and homologous intrusive S-type granites of Early (Middle) Triassic are developed.

As the east and west sides of Qamdo Block in Early Triassic and Middle Triassic are dominated by collision orogeny, Qamdo Block is uplifting as a whole under the tectonic compression.

Development and Evolution of Mesozoic Composite Foreland Basin in Qamdo Block

Extensional Tectonics After Collision in the Early Period of Late Triassic

After the collision orogeny and uplift in Early Triassic and Middle Triassic, Qamdo Block was once again subjected to transgression and sedimentation in the early period of Late Triassic. In the eastern part of the block, with the detachment of mountain roots and upwelling of mantle, the orogenic belt was stretched and collapsed. In Carnian of Late Triassic, a superimposed volcanic-sedimentary basin was formed in the orogenic belt, and a limited rift basin was developed in Shengda-Chesuo-Songxi on the west side of the basin.

Evolution of the Composite (Back-Arc) Foreland Basin in Qamdo Block in Late Triassic-Cretaceous

In the early period of Late Triassic, Qamdo Block was transformed into a composite (back-arc) foreland basin located between the Jinsha River Collision Orogenic Belt in the east and Leiwuqi-Dongda Mountain Orogenic Belt in the west on the basis of back-arc-basin formed in early stage, except for the active volcanic-sedimentary basin formed by spreading in the east. During the whole Late Triassic, the basin was mainly descended and uplifted, and from east to west, it experienced a tectonic cycle from slow descending to slow uplift. At the end of Late Triassic, the dynamic mechanism of basin evolution began to change, and the eastern margin arc orogenic belt of Qamdo Block was further squeezed and uplifted, and pushed westward, so that the Upper Triassic stratum began to be drawn in the orogenic belt. During the Jurassic-Cretaceous, the foreland basin also moved westward parallel to the orogenic belt, and the sedimentation center was roughly located on Markam, Qamdo. In the same period, westward intracontinental subduction occurred along the North Lancang River Fracture, and a significant relative depression appeared in the northeast side. The general tectonic framework of Jurassic-Cretaceous is that the mountain chains on both sides of the basin are constantly uplifting and pushing backward into the basin, while the central part of the basin is continuously depressed, with continuous thick deposits from Upper Triassic to Cretaceous. During the Cretaceous, the basin gradually retreated and destructed from northwest to southeast. Under the continuous compression and collision, the orogenic belts on both sides of the basin developed significant intrusion of S-type intermediate-acid magma, forming Dongda Mountain and Dizhong deep-seated granite bases, respectively.

Intracontinental Convergence and Its Tectonic Response in Himalayan

Since Late Yanshanian-Himalayan, with the subduction and closure of Neo-Tethyan Ocean and the continuous significant jacking of Indian Plate into Eurasia, the whole Qinghai-Tibet Plateau started the comprehensive intracontinental convergence and compression and significant uplift. Qamdo Basin is located in the eastern part of Qinghai-Tibet Plateau. During Himalayan, it was generally in a tectonic system dominated by compression. With the readjustment of massive crustal materials, the tectonic deformation of the block occurred from the early transition from strike-slip to extension to the late massive horizontal shortening and the sharp uplift and thickening of the crust.

Massive Right Strike-Slip Structure

During Late Yanshanian-Early Himalayan, given the N-NNE orogeny of Indian Plate and the relative S-SW movement of Yangtze Plate, Sanjiang composite orogenic belt between the two plates became a giant strike-slip shear system. Qamdo Block is located on the southwest side of the strike-slip system, with Jinsha River Fracture and deep NNW fractures such as Chesuo-Deqin Fracture as boundaries. It is mainly controlled by the near NS stress field from Indian Plate, which is mainly manifested in the right-handed orogeny between the main body of the block and its eastern orogenic belt along the above fractures, resulting in a series of NW-NNW folds and alternate compressional and torsional fracture structures in the middle-east of Qamdo-Markam Composite Basin in Upper Triassic. Its strike intersects with the NNW main strike-slip fracture in an acute angle in the

shape of Chinese character "人" (Fig. 2.11), among which Xiariduo-Yulong-Haitong Oblique Fold Tectonic Zone developed on the west side of Chesuo fracture is the most representative, and it plays an important role in controlling the distribution of porphyry in Himalayan.

Formation of Strike-Slip Basin and Deep-Seated Magmatism in Paleogene

During the Paleogene Eocene, strike-slip extensional tectonic activities took place in Eastern Tibet, and a series of strike-slip extensional basins were formed on the surface, such as Gongjue, Nangqian, Jiqu and Lawu basins. Paleogene crustal extension was coupled with mantle uplift, and crustal decompression led to the melting of lower crust or upper mantle. Regional extensional fractures promoted magma to rapidly rise into the crust, and at the same time as the formation of strike-slip extensional basins on the surface, deep volcanic-magmatic activities, which were mainly alkaline, developed. Volcanic activities mainly occurred in basins with significant strike-slip tension. In the Gongjue, Nangqian and Lawu basins, there were Paleogene-Neogene volcanic rocks with rough-coarse texture in varying degrees. In the same period, the relative uplift of the surface, such as Xiariduo-Haitong Belt and Gaoji-Tuoba Belt, has become a favorable environment for the emplacement of intrusive rocks due to its tectonic dilatancy, forming small porphyry bodies distributed in groups and strips, constituting important tectonic intrusive rock belts, which were closely related to the formation of porphyry copper deposits.

Hedging System with the Giant Thrust Belt of North Lancang River as the Main Structure

The giant thrust belt of North Lancang River was mainly formed in Late Yanshan-Himalayan, which had an important influence on the tectonic deformation in Qamdo Block. During Himalayan, Leiwuqi-Dongda Mountain composite island arc orogenic belt on the west side took on the huge stress from Indian Plate in N-NE direction, which not only caused its own significant tectonic deformation, forming a series of imbricated sheets, but also took the North Lancang River Junction Zone as the main thrust belt, resulting in an overall thrust nappe to northeast (Fig. 2.12). Due to the significant thrust nappe, a part of the west side of Qamdo Basin has been swallowed up by overthrust fracture. Due to the transmission and application of tectonic stress to NE-NNE Qamdo-Markam Basin, Qamdo Basin has been greatly shortened and narrowed in NE-NNE direction, and the significant tectonic deformation zone has been developed at the front edge of the main thrust fracture, with a series of axial planes inclined to the W-SW and closed deflection-overturned fold and a series of secondary thrust fractures of Upper Triassic and Jurassic-Cretaceous. A large

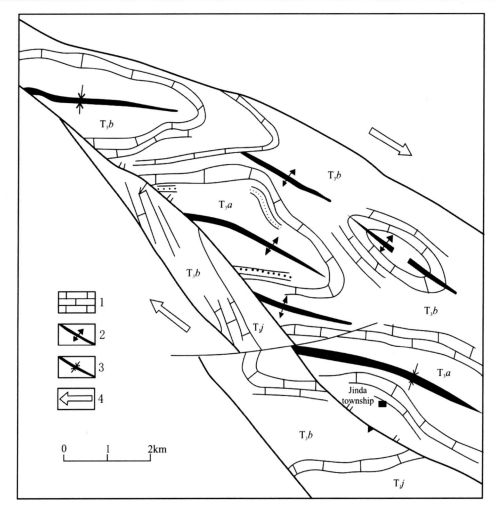

Fig. 2.11 Secondary fold reflecting dextrorotation in upper triassic on the east side of Tuoba fracture zone. 1—Limestone; 2—Anticline; 3—Syncline; 4—Twist direction

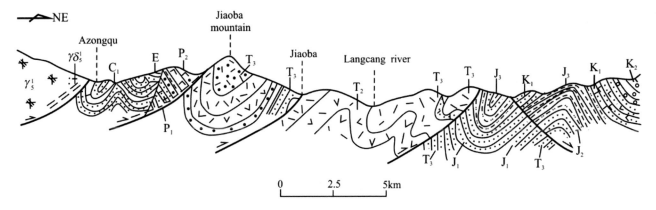

Fig. 2.12 Jiaoba Mountain-Zhuka lancang river Nappe zone overthrust eastward over Jurassic-Cretaceous stratum

number of detached blocks pushed from the west were developed. The horizontal compression range of the basin increased from northwest to southeast, and tended to tip out near the Sanjiang waist in the south of Yanjing. It shows that the amplitude and intensity of thrust nappe compression in Qamdo-Markam Basin in Leiwuqi-Dongda Mountain Island Arc Zone increased from northwest to southeast, and the nappe direction also changed from NE-NEE direction to nearly EW direction, mainly nearly EW compression in Yanjing. It is precisely because the island arc zone completely squeezed the Qamdo-Markam Basin that the strike-slip extensional basin formed in Paleogene quickly destroyed, and some of them were further transformed into depressed basins. Corresponding to Lancang River on the west side, the eastern orogenic system has further thrust westward on the basis of the foreland thrust zone (Fig. 2.13). Its frontal zone was Tuoba-E'aipu Fracture Zone in the north and Laoran Yanjing Fracture in the south. The two-sided thrust system with the basin as the central axis has been formed (Fig. 2.14), which has become the most spectacular tectonic deformation in Himalayan intracontinental convergence of Qamdo Basin.

Nearly EW Tectonic

The EW tectonic in Qamdo Block is hidden but exists widely. It is composed of a series of small NWW to nearly EW left-lateral shear fracture bundles, which cut through all strata including Paleogene and Neogene, mainly belonging to brittle surface and formed in Himalayan. There are two important zones on the surface: First, Tongtian River located on the northern edge of Qamdo Block and its south area. The Tongtian River Fracture in the region is connected with Xianshui River Fracture in Western Sichuan, and it strikes in the NWW direction. Since Himalayan, there has been a

significant left-hand strike-slip, which has significantly transformed Yushu-Luoxu Member of Jinsha River Junction Zone. It is also clearly reflected in Qamdo Block on the south side, and a series of secondary EW fractures have been developed. Second, the structure is located in Xiaya-Chaya-Latuo, spreading in the NWW direction with different characteristics in the east and west members. The west member shows NWW left-hand orogeny in Xiaya and its north and south sides, which results in the dislocation of Luwuqi-Dongda Mountain Island Arc Zone and the east or NEE arc protrusion of the main thrust fracture along Wokadi-Zhuka-Yanjing.

In Chaya-Latuo, the basement block is dislocated, and the traction tectonics formed on the surface layer makes the NW-NNW fracture and stratigraphic strike line change to NWW-SEE direction obviously, and then merges with NW-NNW tectonics in the south of Latuo, and the connected Gongjue-Mangcuo Paleogene Gongjue Basin may be disconnected due to dislocation compression. According to the observation of the Paleogene-Neogene EW left strike-slip fractures between Gongjue County and Latuo, the horizontal dislocation distance of a single fracture reached tens to nearly hundreds of meters, and the continuous dislocation displacement caused by group faults resulted in a large total displacement in the whole zone. Most of the EW left strike-slip fractures took the regional NW fracture

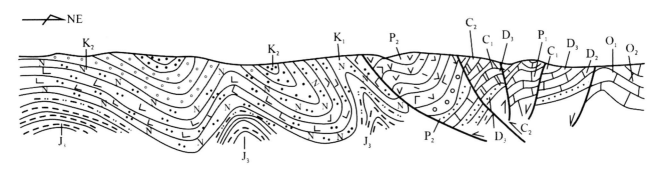

Fig. 2.13 Haitong-Markam Nappe thrust westward over Jurassic-Cretaceous stratum

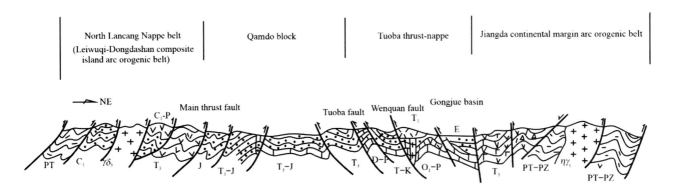

Fig. 2.14 Profile of hedging system on two sides of Qamdo basin

tectonic as the boundary line, and occurred in the divided blocks and strips, and a few of them cut the NW fractures. The understanding of the nearly EW tectonic in Qamdo Block is consistent with the remote sensing of Sanjiang Region recently made by the Institute of Mineral Deposits of Chinese Academy of Geological Sciences (according to the "Ninth Five Year Plan", Sanjiang and "Research on Yidun Island Arc Tectonic Evolution and Metallogenic Regularity"), and the two large EW linear tectonic zones interpreted for Eastern Tibet coincide with the above generally: First, it extends from Shangka in the west to the east through Latuo, Ziga and Ailashan passes in Western Sichuan. It has obvious aeromagnetic anomalies (treated by ΔT polarization) and gravity anomalies (treated by residual gravity anomalies), showing frequent Himalayan activities. Second, it is located in Chayu-Bitu-Daocheng in the south. In addition, a series of small and medium EW linear tectonics have been found in Eastern Tibet. The EW left strike-slip fracture is the result of the interaction of the inhomogeneity of materials in the block and the asymmetric compression stress in EW. It is the result of adjustment and transformation between crustal materials and stress field after the NS compression was blocked in Himalayan, and it has become an important manifestation of Himalayan tectonic deformation in Qamdo Block, which is also in line with the characteristics of tectonic stress field in this area.

2.2.5 Space–Time Structure and Evolution of Ailaoshan Arc-Basin System

For many years, it has been considered that Ailaoshan Arc-Basin System is roughly in the same back-arc ocean basin subduction—arc-land collision orogenic belt as Jinsha River Arc-Basin System, which is mainly based on the fact that both systems are located on the east side of Qamdo-Lanping Block. However, the east side of Ailaoshan is directly adjacent to the southwest continental margin of Yangtze Continent, and the east side of Jinsha River is adjacent to Zhongza-Shangri-La Block which is divided from Yangtze Continent.

2.2.5.1 Space–Time Structure of Ailaoshan Arc-Basin System

Ailaoshan Orogenic Belt is mainly composed of (from northeast to southwest) Red River Fracture, Ailaoshan Fracture, Tengtiao River Fracture, Jiujia-Anding Fracture and Amo Jiang-Lixian River Fracture, and their basement high grade metamorphic zone, arc-land collision ophiolite melange zone (commonly called low-grade metamorphic zone), Mojiang-Lvchun Arc Volcanic Zone and foreland thrust belt. Generally speaking, the belt converges to the northwest and spreads to the southeast (Fig. 2.15).

2.2.5.2 Formation Period of Ailaoshan Back-Arc Ocean Basin and Lvchun Continental Margin Arc

The main Ailaoshan low-grade metamorphic zone is the ophiolite melange zone. Early Carboniferous radiolarian *Albaillella paradoxa* (?), *Deffaadree*, *Astroentactinia multispinisa* Won. can be found in siliceous rocks. The silicon isotope (δ^{30}Si) content of radiolarian siliceous rocks in Laowangzhai is 0.2‰, which is close to the 0.6 ‰ to 0.18 ‰ (average 0.16‰) of that in abyssal environment (δ^{30}Si) classified by Ding. The ΣREE of siliceous rare earth elements in Devonian and Carboniferous is (47.56–137.93) $\times$ 10^{-6}, the fractionation of light and heavy rare earth elements is obvious, the LREE is 7.07–9.46, and the (La/Yb)$_N$ is 0.61–10.63. The rare earth partition is light rare earth enrichment right, and the negative δEu is abnormal. δEu ranges 0.58–0.79, indicating the abyssal to abyssal environment, which can represent the characteristics of abyssal siliceous rocks in the arc ocean basin. Fuchsia siliceous rocks (C_1) can be found in Pingzhang, Xinping County. It is generally believed that fuchsia radiolarian siliceous rocks represent the vast ocean deposit. In this zone, ocean ridge volcanic rocks can be also identified, and gabbro and plagioclase granite are formed in Shuanggou Ophiolite Formation. The U–Pb age of this zone is 328–362 Ma, which is basically consistent with the U–Pb age of Shusong plagioclase granite, Jinsha River Zone, which is 340–294 Ma. It reveals the initial formation period of ocean floor expansion at the turn of Devonian/Carboniferous, which represents the development period of Ailaoshan Back-arc Ocean Basin from Early Carboniferous to Early Permian.

The volcanic rocks of Panjiazhai in Late Permian, which occurred in the flysch sedimentary strata composed of metamorphic sandstone, phyllite and quartz schist on the northeast side of Ailaoshan low-grade metamorphic zone, have been metamorphosed into green schist and "blue schist". Its lithochemistry and geochemistry characteristics indicate that it is composed of continental rift rough basalt and basalt. It is characterized by low Si content and high Ti and alkali content. The total amount of rare earths is high, and the fractionation of light and heavy rare earths is obvious; The REE distribution is of right enrichment, similar to Emei Mountain basalt. Therefore, Panjiazhai metamorphic volcanic rocks belong to continental rift volcanic rocks. However, they have been involved in the block of melange, and the glaucophane in the blue schist is arfvedsonite and blue tremolite. In addition, according to the Middle Triassic Ganbatang flysch sedimentation with metamorphic deformation in the ophiolite melange zone, it shows that the back-arc ocean basin finally destructed and the melange was formed in Middle Triassic, and it was covered by Late Triassic molasse tectonic unconformity.

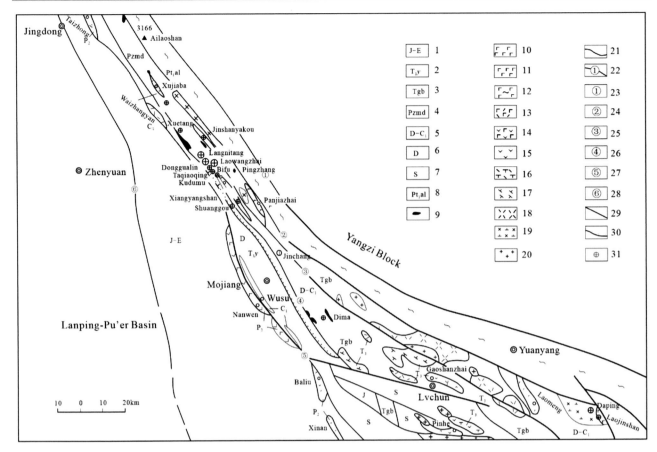

Fig. 2.15 Sketch of geological tectonics of Ailaoshan. 1—Jurassic-Paleogene continental sedimentation; 2—Yiwanshui formation in upper Triassic; 3—Triassic Ganbatang group; 4—Paleozoic Madeng group complex; 5—Devonian-lower carboniferous (low-grade metamorphism); 6—Devonian; 7—Silurian (not metamorphic); 8—Proterozoic Ailaoshan group complex; 9—Ultrabasic rock; 10—Albitite basalt; 11—Pyroxene basalt; 12—Metamorphic basalt; 13—"Bimodal" volcanic rocks; 14—Basalt-andesite; 15—Andesite; 16—Latite; 17—Rhyolite porphyry; 18—Rhyolite; 19—Gabbro-diorite; 20—Granite; 21—Geological boundary; 22—Main fracture and its number; 23—Red river fracture; 24—Ailaoshan fracture; 25—Tengtiao river fracture; 26—Jiujia-Anding fracture; 27—Amo Jiang-Lixian river fracture; 28—Babian river fracture; 29—Other fractures; 30—Unconformity boundary; 31—Gold deposit (point)

The Ailaoshan Back-arc Ocean Basin began to subduct southwest in the early period of Late Permian, and the eastern edge of Lanping-Pu'er Block in the southwest formed the Mojiang-Lvchun Continental Margin Arc. Mojiang-Lvchun Continental Margin Arc is mainly characterized by volcanic-sedimentary rocks formed in Late Permian-Late Triassic and granite intrusion formed in the late period of Late Triassic. The subduction arc was formed before 260–250 Ma, which is equivalent to the time of arc volcanic rocks formed in Late Permian; the collision arc was formed before 240–210 Ma, which is consistent with the development age of arc volcanic rocks formed in the Late Triassic.

2.2.5.3 Spatial Distribution of Ailaoshan Arc-Basin System

The spatial structure and tectonic pattern of Ailaoshan Arc-Basin System have been significantly reformed by the significant collision between India Plate and Eurasia since the end of Cretaceous and Paleogene. The rotation and slip of the landmass and the overlapping displacement of different tectonic stratigraphic units lead to the blurring of boundaries of some tectonic stratigraphic units, but the spatial configuration of volcanic arc and ophiolite melange zone can basically be identified.

Active Basement Thrust Zone in the Southwest Margin of Yangtze Continent

Since the Red River Fracture and Ailaoshan Strike-slip-Thrust Ductile Shear Zone are mainly composed of the high grade metamorphic zone of Ailaoshan and Jinping overlying fracture block units, according to the research of Wang during the "Eighth Five-Year Plan", Ailaoshan Group Complex is divided into Xiaoyangjie Rock Formation, Qingshui River Rock Formation and Along Rock Formation. Xiaoyangjie Rock Formation is an aluminum-rich rock series assemblage, it is mainly composed of biotite plagioclase gneiss, garnet fibrolite biotite plagioclase gneiss, disthene biotite plagioclase gneiss, etc., sandwiched with granulite, and locally sandwiched with olivine dolomite marble lens and boudin. Amphibolites are rare. Folding layers, bedding recumbent folds, synclinal inverted fold and plunging fold are well developed. Flow fold and rootless fold formed by plastic rheology are more common, and they are mostly in the form of boudin migmatite, shadow migmatite and homogeneous migmatite. It shows that the rock has experienced local melting under deep conditions, and even formed in-situ/quasi-in-situ low-intrusion remelting granite rock mass. In Yuanyang and Langdi, the intrusion of magnesian ultramafic rock mass is also found in the rock formations, which indicates that there was the intrusion of anatectic magma in the upper mantle. It shows that this tectonic-rock stratum was once a part of the deep crust or at least the middle crust.

Qingshui River Rock Formation and Xiaoyangjie Rock Formation are separated by a regional tectonic detachment surface, mainly composed of plagioclase amphibolite gneiss, amphibolite plagioclase gneiss and granulite, with biotite plagioclase gneiss, granite gneiss and heronite, and amphibolite and marble in the lower part. The strata are characterized by the widespread development of folding layers, bedding recumbent folds, and bedding ductile shear zones. The plastic rheological characteristics of deep local melting in this tectonic-rock stratum are rare. Generally, the grade of metamorphism can reach high greenschist facies, with amphibolite facies locally, and its source rock is volcanic-sedimentary rock assemblage, which shows obvious difference from Xiaoyangjie Rock Formation. Along Rock Formation is a calcium-rich rock assemblage, mainly composed of marble, with a small amount of amphibolite, amphibolite plagioclase gneiss and calcium silicate, etc. The intrusion of mafic dikes and granite veins has resulted in metamorphism and deformation. Marble is characterized by folding layers and bedding shear zones, with obvious solid plastic rheology, which is actually recrystallized carbonate mylonite. The metamorphism and deformation of this rock formation is similar to that of Qingshui River Rock Formation.

Up to now, no valuable isotopic age data has been obtained in Ailaoshan Group Complex. Only the Institute of Geology and Geophysics, Chinese Academy of Sciences has measured the whole-rock apparent isochron age of Rb–Sr (839 ± 0.739) Ma in this group complex, and the K–Ar age of gneiss and granulite in Along Rock Formation includes three ages in Late Paleozoic, Mesozoic (171–85 Ma) and Cenozoic (47.7–11.5 Ma). It may reflect the superposition of tectonic thermal events such as the formation of Ailaoshan Arc-Basin System, arc-land collision and post-orogenic spreading and exposure.

The Jinping Unit above Ailaoshan Metamorphic Group Complex is characterized by the development of Paleozoic-Triassic sedimentation, and it should be an integral part of the overlying strata at the western margin of Yangtze Plate. The oldest exposed stratum is the Ordovician stratum, which is mainly sand-mud formation with flysch-like characteristics, with slight metamorphism. It is covered by parallel unconformity of Middle Silurian, and the Middle and Upper Silurian stratum is basically a sequence of magnesium carbonate rocks. The Devonian stratum lacks the lower series, and the middle and upper series are carbonate rocks mixed with dolomite, argillaceous siliceous rocks and shale. The Carboniferous stratum is composed of limestone and biological limestone. The Lower Permian stratum is composed of biological limestone and cryptocrystalline limestone. The lower part of the Upper Permian is Emei Mountain basalt and pyroclastic rock equivalent to the western margin of Yangtze Plate, which is eruption unconformity with the lower series, with a thickness of 4536 m; the upper part is the transitional coal-bearing stratum. The Lower Triassic is in lacuna, and only the Middle and Upper Triassic strata are exposed. The Middle Triassic is dominated by limestone and dolomite, with tuffaceous sandstone, shale or basalt occasionally, directly overlying the Upper Permian basalt; the Upper Triassic is dominated by sand shale, locally mixed with limestone and dolomite, containing coal and rich biological fossils. The strata exposure, contact relationship and lithologic and lithofacies characteristics of Jinping Overlaying Unit are similar to those of Haidong, Dali. It seems that it was originally a part of the same tectonic unit and was dismembered into two parts by the left-handed strike-slip tectonic of Ailaoshan-Red River Fracture in the later period, with a strike-slip distance of about 350 km. Therefore, together with Ailaoshan Group, it formed the early passive margin subduction unit of orogenic belt and has been transformed into an overlapping block unit since the end of Late Triassic.

Ailaoshan Ophiolite Melange Zone

Ailaoshan Ophiolite Melange Zone is located in the low-grade metamorphic zone between Jiujia-Anding Fracture Zone and Ailaoshan Fracture (Fig. 2.15). There are many studies on this zone. In recent years, more oceanic ridge basalt, pyroxene-diorite cumulate, lherzolite and radiolarian siliceous rock (C_1) have been found in studies. Based on previous studies, a relatively complete ophiolite sequence can basically be established, with metamorphic peridotite (including lherzolite and harzburgite), cumulate complex (including pyroxenite, gabbro, gabbro diorite and plagioclase granite), diabase, basic lava (including albitite basalt and pyroxene basalt) and radiolarian siliceous rock from bottom to top. Compared with the typical ophiolite in the world, the layered gabbro and diabase sheeted dike swarms in the cumulate are undeveloped.

The mafic–ultramafic rocks are distributed in groups and strips in a melange zone with a length of 400 km and a width of several kilometers along the NW strike (Fig. 2.15). Basic volcanic rocks, mafic–ultramafic rocks, and radiolarian siliceous rocks are mixed with each other in the muddy flysch matrix in various sizes. There is obvious tectonic contact between rocks with different genesis and composition and their matrix, showing the appearance of ophiolite melange. Generally, the matrix is mainly sandy shale with turbidite sedimentary, but the significantly deformed limestone blocks can be seen directly in the sandy shale layer, and their attitudes are very discordant, the rocks formed in different environments are mixed, and some limestone blocks may be slumping rocks on the slope edge. There are biological fossils in limestone blocks, including conodonts of Early Carboniferous and Ordovician–Silurian. In Pingzhang, Kudumu and other places, Carboniferous-Permian biogenic limestone of shallow plateau origin with block length greater than 500 m can be seen in lower Carboniferous argillaceous limestone, fuchsia radiolarian siliceous rocks and black sandy argillaceous rocks mixed with siliceous rocks deposited in deep water, showing obvious mixing tectonics. The exposed strata along this zone, which are relatively complete and have the basis of biological fossils, such as the lower Carboniferous black thin-bedded argillaceous limestone or nodular limestone, are exposed in sporadic fragments, and the rocks are significantly squeezed and deformed, forming complex folds. However, the source rock features are well preserved, and there are abundant conodont fossils. In the aleuropelitic turbidity sediment along the east side of Jiujia-Anding Fracture Zone, a sequence of conglomerates with the same composition as the sediments are intermittently exposed in the range of about 50 km from Yakou Street, Heping to Xiangyang Mountain, Shuanggou, with poor sorting and grinding, mixed size, different gravel size, different shapes and single composition. Such sediment is the gravity sediment of submarine fan channel in slopes.

According to Wang Yizhao's observation, the macroscopic tectonic deformation of this zone is mainly characterized by lateral tectonic displacement, and the formation of tectonic community related to bedding shear, which is characterized by bedding recumbent folds, bedding shear zones and folding layers, and by the extensive development of S1 lateral displacement of S0 in the region. Generally, the replacement is thorough, forming a regional permeable foliation composed of S1. Due to the significant compression during the orogenic period, S1 and S0 are widely developed in the area, forming a regional tectonic style composed of synclinal inverted folds and imbricate thrust fractures associated with thrust fractures. The axial plane of the inverted fold generally falls to the SW and tends to the NE, reflecting

that the thrust nappe orogeny is in NE-SW direction. The tectonic characteristics of metamorphic rock formations and mineral paragenesis assemblage show that the two foliation replacements are basically in the tectonic deformation environment of low greenschist facies. Near Ailaoshan Fracture Zone in the northern member, the low-grade metamorphic rock belt formed phyllonite and tectonic schist due to significant tectonic deformation, and the significant strain zone developed. The low-grade metamorphic rocks generally reach the low greenschist facies, and it shows the single-phase dynamic metamorphism with no progressive metamorphic zones, except that it can reach the high greenschist facies locally. It is worth noting that the source rock of the albite chlorite schist containing glaucophane, which is exposed from the tectonic block at 244 km between Mosha and Malutang, is almond-shaped basalt, which was formed by high-pressure and low-temperature metamorphism. The green schist obviously belongs to mylonite formed by ductile shear deformation of basic volcanic rocks. The rock has obvious S-C fabric. The ductile shear zone represented by this green schist has been severely damaged and reformed by the later fractures, which means that it developed into an important part of the thrust tectonic melange zone during the arc-land collision on the basis of the oceanic subduction.

Mojiang-Lvchun Continental Margin Arc Zone

This zone is located in Mojiang-Lvchun, which is in the west of Jiujia-Anding Fracture Zone on the southwest boundary of Ailaoshan Ophiolite Melange Zone, and in the east of Amo Jiang-Lixian River Oblique Thrust Fracture Zone. In the orogenic stage of the arc-land collision, this zone was basically involved in the overlapping landmass unit of the orogenic belt. It consists of Paleozoic and Mesozoic strata and Late Triassic granite. Silurian-Lower Carboniferous strata developed graptolite shale, clastic turbidite, radiolarian siliceous rocks, thin argillaceous limestone and reticulate limestone. In Early Carboniferous, a sequence of "bimodal" volcanic rocks, sandstone, siliceous shale and siliceous rocks, which are composed of basic volcanic rocks and acidic volcanic rocks, developed in Bulong-Wusu of Mojiang, and were characterized by the passive margin rift on the east side of Pu'er Block. The Middle and Upper Carboniferous to Lower Permian strata are a sequence of unstable plateau carbonate rocks and coal-bearing clastic rocks, showing the uplift of fracture blocks. In the Late Permian, a sequence of arc volcanic rocks developed along the Taizhong-Lixian River. In the Late Triassic, a sequence of collision acidic volcanic rocks and lag arc volcanic rocks developed in Gaoshanzhai and Keping of Lvchun, respectively.

The "bimodal" volcanic rock assemblage in Mojiang-Wusu shows two eruptive cycles (Fig. 2.16). Both

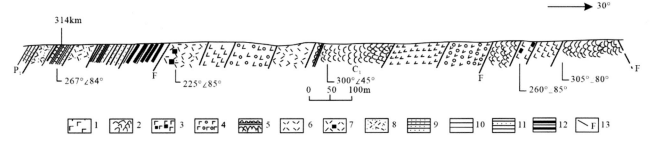

Fig. 2.16 Section of Mojiang-Wusu at 314 km. 1—Massive basalt; 2—Pillow basalt; 3—Pyritized basalt; 4—Almond-shaped basalt; 5—Jasper filled pillow basalt; 6—Rhyolite; 7—Pyritized rhyolite; 8— Rhyolitic tuff; 9—Sandstone; 10—Feldspar coarse sandstone; 11—Silty siliceous rocks; 12—Siliceous shale; 13—Fracture

eruptive rhythms of the lower cycle are all composed of lava, with basalt first, rhyolite later, and basalt pillow tectonic developed; both eruptive rhythms of the upper cycle are rhyolitic tuff, indicating the action of volcanic eruption. There is a sequence of rhythmic sediments of siliceous shale-sandstone-siliceous rock between two volcanic activities. The top and bottom of the member are deep-water siliceous rocks and flysch-like sand shale, and the petrochemical and geochemical characteristics of the volcanic rocks are double rift type "bimodal" volcanic rocks

(Fig. 2.17). The basalt has high TiO_2 content, which mainly falls in the intraplate basalt area in ATK diagram, indicating that the "bimodal" volcanic rocks were formed in the continental margin on the east side of Pu'er Block.

The spatial overlap of Late Permian arc volcanic zone with Mojiang-Wusu "bimodal" rift zone can be divided into three types of volcanic rocks: main arc volcanic rocks, collision volcanic rocks and lagged arc volcanic rocks.

Volcanic rocks are widely exposed in the main arc period, represented by Taizhong volcanic rocks, Nanwenqiao

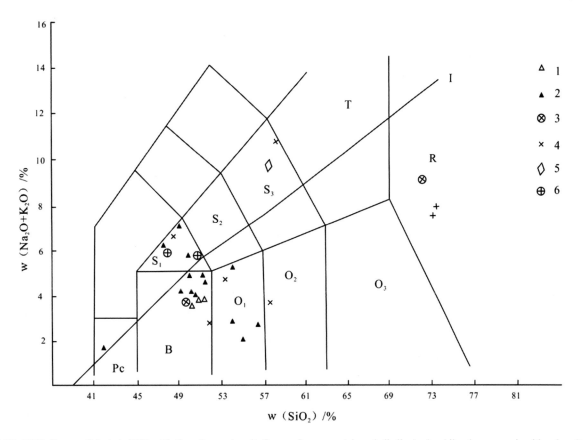

Fig. 2.17 TAS diagram (Li et al. 1999a, b). S_1—Coarse basalt; S_2—Basaltic andesite; S_3—Coarse andesite; T—Trachyte; Pc—Picrite basalt; B—Basalt; O_1—Basaltic andesite; O_2—Andesite; O_3—Dacite; R—Rhyolite; I—Irvine boundary (the upper part is alkaline and the lower part is subalkaline); 1—Ailaoshan oceanic ridge basalt; 2—Ailaoshan quasi-oceanic ridge basalt; 3—Wusu-Bulong volcanic rocks; 4—Main arc volcanic rocks; 5—Collision volcanic rocks; 6—Lagged arc volcanic rocks

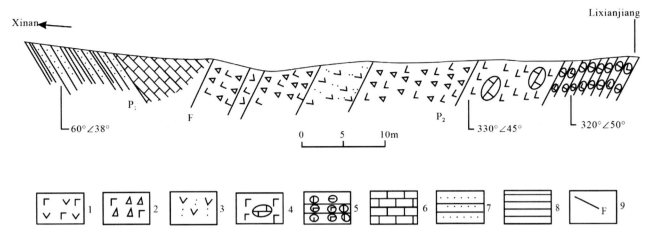

Fig. 2.18 Profile of Lixian River-Xin'an Volcanic Arc. 1—Basaltic andesite; 2—Basaltic volcanic breccia; 3—Andesite tuff; 4—Limestone-bearing breccia basalt; 5—Slump sedimentary rock; 6—Limestone; 7—Sandstone; 8—Shale; 9—Fracture

volcanic rocks and Lixian River volcanic rocks (Fig. 2.18) formed in the Late Permian. There are mainly almond-shaped rough basalt, plagioclase basalt, basaltic andesite, biotite andesite and pyroclastic rocks. Basalt interbedded with andesite. The tholeiite series coexisted with the calcium alkali series. The content of Ti of clinopyroxene in basalt is obviously lower than that of pyroxene in continental tholeiite. The chemical composition and petrochemical composition are shown in the ATK diagram (Fig. 2.19) and $lg\tau$-$lg\sigma$ (Fig. 2.20), almost all of which fall in the island

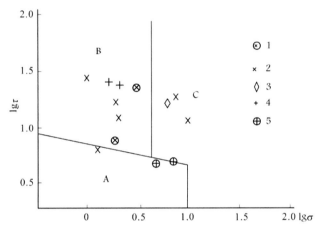

Fig. 2.20 $lg\tau$-$lg\sigma$ Illustration. A—Volcanic rocks in intraplate stable area; B—Volcanic rocks in subduction zone; C—Alkaline volcanic rocks evolved in areas A and B 1—Bulong-Wusu volcanic rocks; 2—Taizhong-Lixian River main arc volcanic rocks; 3—Lagged volcanic rocks; 4—Collision volcanic rocks; 5—Panjiazhai volcanic rocks

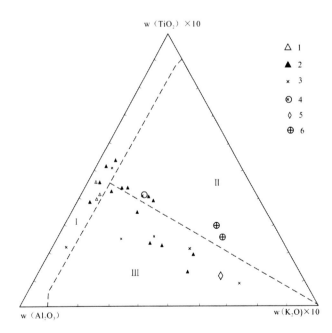

Fig. 2.19 ATK Diagram. I—Oceanic basalt area; II—Continental basalt and andesite area; III—Island arc, basalt and andesite area in orogenic belt. 1—Ailaoshan oceanic ridge basalt; 2—Ailaoshan quasi-oceanic ridge basalt; 3—Main arc volcanic rocks in Taizhong-Lixian River; 4—Wusu volcanic rocks; 5—Pinghe latite; 6—Panjiazhai metamorphic basalt

arc orogenic belt, and the rare earth pattern and trace element pattern are similar to those of island arc volcanic rocks (Figs. 2.21 and 2.22). According to the spatial distribution, from east to west, the general trend is from tholeiite series to calc-alkali series, and the content of Al_2O_3 and (K_2O + Na_2O) gradually increased, reflecting the westward subduction of the ocean crust.

The collision volcanic rocks, represented by Lvchun Gaoshanzhai volcanic rocks, are a sequence of acidic (rhyolite porphyry) assemblages, dating from the Late Triassic. Their petrochemistry is characterized by high SiO_2 (73.99%) content and K_2O (5.20%) content (Fig. 2.17), which is completely the same as that of the collision volcanic rocks in Sanjiang Region (Mo et al. 1993), and also similar to the collision rhyolite in the eastern zone of western United States.

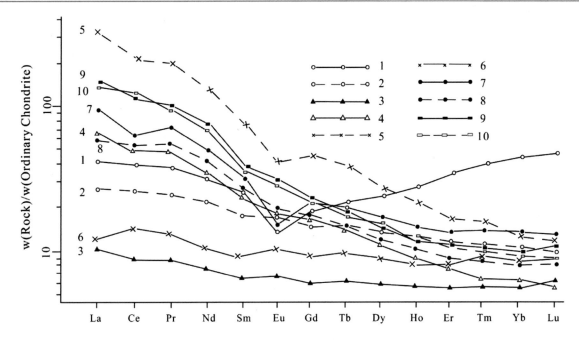

Fig. 2.21 REE distribution pattern in volcanic rocks in Taizhong-Lixian River. 1—Rhyolite (Wusu); 2—Tholeiite (Wusu); 3 —Dolerite (Nanwen Bridge); 4—Basaltic andesite (Lixian River); 5— Biotite andesite (Taizhong); 6—Plagioclase basalt (Baliu); 7—Rhyolite porphyry (Gaoshanzhai); 8—Latite (Pinghe); 9 and 10—Metamorphic basalt (Panjiazhai)

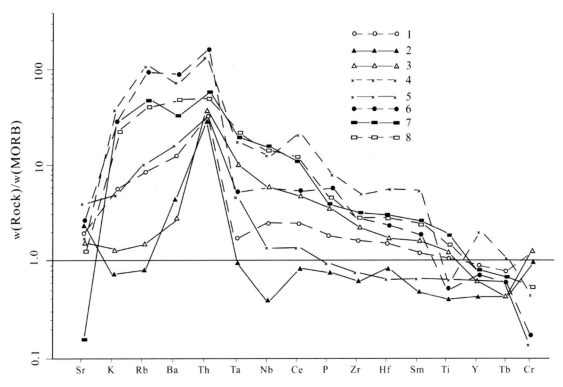

Fig. 2.22 Geochemical distribution pattern of trace elements in volcanic rocks in Taizhong-Lixian River. 1—Tholeiite (Wusu); 2— Dolerite (Nanwen Bridge); 3—Basaltic andesite (Lixian River); 4— Biotite andesite (Taizhong); 5—Plagioclase basalt (Baliu); 6—Latite (Pinghe); 7 and 8—Metamorphic basalt (Panjiazhai)

Post-collision extensional volcanic rocks, represented by Late Triassic volcanic rocks in Lvchun-Pinghe, are mainly a sequence of neutral to intermediate-acid pyroclastic rocks mixed with some lava (alkaline andesite, Fig. 2.17), with a volcanic explosivity index Ep of 0.7. They are a series of high-potassium volcanic rocks with petrochemical characteristics, which can be compared with T32 latite in the South Lancang River. The REE distribution pattern (Fig. 2.21) is similar to that of continental volcanic rocks, and the trace element distribution pattern is similar to that of island arc volcanic rocks (Fig. 2.22), which indicates that it was far away from plate boundaries and close to continental volcanic island arcs.

2.2.5.4 Formation and Evolution of Ailaoshan Arc-Basin System and Arc-Land Collision Orogenic Belt

The Ailaoshan zone experienced the transformation of different tectonic systems and tectonic deformation of different tectonic levels from Paleozoic to Mesozoic and Cenozoic. According to the space–time structure, rock composition and deformation characteristics of the above-mentioned tectonic units, they generally experienced the following stages.

Ocean-Land Transition Stage

The expansion and contraction of ocean lithosphere is marked by the appearance of arc-basin system. Ailaoshan Arc-Basin System and Jinsha River Arc-Basin System developed in the same way, with Jitang-Chongshan-Lancang residual arc on the west side of Qamdo-Lanping-Pu'er Block as the front of the northeastern Tethyan Ocean, and entered a new development period at the beginning of Devonian on the basis of Early Paleozoic subduction complex or metamorphic "soft basement" deposited on the western margin of Yangtze Continent. In Middle-Late Devonian neritic shelf carbonate plateau background, local tension occurred. On the west side of Jiujia-Anding Fracture Zone, the Middle-Upper Devonian semi-pelagic mud-sand mixed with siliceous sediment were deposited in the rift basin. Ocean crust represented by oceanic ridge basalt appeared in Shuanggou-Pingzhang-Laowangzhai in the Early Carboniferous. On the west side of the back-arc ocean basin, Bulong, Mojiang-Wusu experienced a "double-peak" volcanic eruption in the rift zone that split off the edge of the landmass. Carboniferous-Early Permian is the period of further expansion and finalization of Ailaoshan back-arc ocean basin. Carbonate rocks in Carboniferous-Early Permian plateau near the land margin were distributed in isolated island chains, with siliceous flysch mixed with basalt. Lanping-Pu'er Block split from the Yangtze Landmass to form an independent micro-landmass.

In Late Permian, the back-arc ocean basin stopped spreading and turned into significant subduction and reduction to the west Lanping-Pu'er Block, forming the Mojiang-Lvchun continental margin volcanic arc.

Left Strike-Slip Arc-Land Collision Orogenic Stage

At the end of Late Paleozoic-Late Triassic, Yangtze Landmass wedged westward, and Ailaoshan Back-arc Ocean Basin was the earliest orogenic belt formed by the arc-land collision after the subduction and destruction of Hoh Xil-Jinsha River-Ailaoshan Zone. As a result of the arc-land collision, Taizhong-Lixian River collision continental margin volcanic arc was superimposed on or on the east side of Late Permian subduction continental margin volcanic arc, and a sequence of basaltic andesite-dacite-rhyolite and other calc-alkaline volcanic rock assemblages with island arc properties developed.

During the Middle-Late Triassic, Lanping-Pu'er Basin began to deposit on the side of Jinsha River-Ailaoshan, receiving the material source supplied by the low-grade metamorphic belt. At the bottom of Yiwanshui Formation in the Upper Triassic, a large amount of gravel from the low-grade metamorphic zone can be found, including basic-ultrabasic rock gravel. Yiwanshui Formation also contained red sand and mudstone gravel deposited in its early stage. As mentioned above, Yangtze Plate subducted westward, and the foreland basins should have developed on the edge of Yangtze Block, but they all developed on the eastern edge of Lanping-Pu'er Block, and the tectonic position of the foreland basin was reversed. In the low-grade metamorphic zone and the whole ophiolite melange zone, the westward obduction nappe collision orogeny can be found anywhere. At this time, Ailaoshan Thrust Nappe Fracture has been formed, and a group of imbricate thrust nappe tectonics developed in the Ailaoshan Group, then the Paleozoic strata in Jinping in the lower plate were significantly reversed westward. In Fenggang, Yuanyang, Ailaoshan Group foliation, pegmatite and arizonite developed along the previous foliation also formed westward inverted folds, and the fragments in mylonite belt also indicated the upward thrust of the upper plate. The foreland folds and Permian heavy folds in Permian–Triassic strata in Changdong, Zhendong Village, Zhenyuan may have been formed in this period. The thrust stacked mountains moved into the foreland basin, which also resulted in the denudation and redeposition of the lower or early sediments of Yiwanshui Formation. However, the Ailaoshan Group of high grade metamorphic zone was not denuded out of the surface at that time, because no gravel from the high grade metamorphic zone was found in Triassic-Jurassic strata.

Stage of Lithosphere Extension and Basement Stripping

At this stage, the mountain may rise sharply due to the significant orogeny of the overthrust in the previous stage, and the reverse slip or spreading may be formed at the rear edge of the overthrust zone due to the imbalance of gravity, so that the metamorphic basement of Ailaoshan Group and Cangshan Group was gradually exposed by the tectonics, resulting in a series of tectonic deformations found in the metamorphic core complex. In Xinjian-Jiayin in the southern member of Ailaoshan, along the contact interface between Xiaoyangjie Rock Formation and Qingshui River Rock Formation, as well as within Qingshuihe Rock Formation and Along Rock Formation, regional or secondary detachment fault zones occurred, resulting in bedding (foliation) slip, forming pudding/brecciation zones, bedding ductile shear mylonite zones, folds, sheath folds and domino tectonics, all these reflect the spreading and stripping in northeast direction. Its spreading may occur at the same time as that of Cangshan, that is, after Late Triassic and Jurassic. Through extensional tectonic stripping and erosion of overlying rocks, Cangshan Group and Ailaoshan Group are exposed on the surface and subjected to erosion, resulting in gravel in high grade metamorphic zones in subsequent strata. The exposure mechanism of Ailaoshan Group and Cangshan Group is similar to that of the overthrust of Kangma metamorphic core complex and Ximeng metamorphic core complex at the rear edge of Himalayas and Dashan, Gengma.

Overthrust and Left Strike-Slip Stage

At the end of Late Cretaceous to Paleogene, due to the closure of Yarlung Zangbo River Ocean Basin and the northward compression of Indian Plate, the whole Qinghai-Tibet Plateau and even the Sanjiang Region entered a stage of comprehensive intracontinental convergence, and the northeast horn of Indian Landmass and Yangtze Landmass collided with each other in Sanjiang Region, forming a transverse wasp waist-shaped tectonic knot. As the northeast horn of India Landmass was slightly offset to the north relative to the southwest corner of Yangtze Landmass, both twisted during the collision, which made the overall strike of the tectonic belt in Sanjiang Region reverse in an "S" shape. What's more, the above-mentioned Xilaping-Diancang Mountain Overthrust of Yangtze Landmass appeared in Sanjiang wasp waist zone. There is a significant left strike-slip in Ailaoshan Zone in the south wing of the arc overthrust zone, and vertical foliation zone, nearly horizontal tensile lineation and inclined vertical folds were developed in the high and low-grade metamorphic zones. However, in Diancang Mountain, which is located at the top of the nappe arc, there is no sign of left strike-slip, but mylonite foliation and westward inverted folds formed in Cangshan marble.

The Triassic nappe of Lanping Lead–Zinc Deposit is overlaid on the Paleocene Yunlong Formation and covered by its sediments, while the Cangshan nappe overthrust on the Paleogene and Neogene strata of Lanping-Pu'er Basin, indicating that its overthrust nappe occurred in the Paleogene and later.

The whole-rock K–Ar age of felsic mylonite from the northeast of Jinshan Bealock in the northern member of Ailaoshan is 18.7 ± 1.9 Ma. Schearer et al. measured the U–Pb isotopic age of yttrium ore and monazite with two samples of light-colored granite in Ailaoshan Metamorphic Zone and found that the result was the same (23.0 Ma $\pm$ 0.2 Ma). These two ages can represent the upper limit time of the end of left strike-slip ductile shear. This left strike-slip process transformed the early tectonics of Ailaoshan Zone more thoroughly in the northwest member than in the southeast member. Some members in the south of Yuanyang in southeast member still have traces of early thrust and extension, and it is difficult to find early deformation traces in Jinshan Bealock in middle member. The mylonite foliation is steep or inclined to the northeast, and the tensile lineation is nearly horizontal, with a direction angle of 327°, which is consistent with the direction of the regional tectonic line. The interbreeding of striped mylonite and large and small eyeball mylonite reflects the interbreeding of relatively strong and weak strain zones. The porphyroclast pressure shadow shows the left strike-slip characteristics of the northeast plate slipping to the northwest, and then to northwest, according to the fact that the ophiolite melange zone and low-grade metamorphic zone gradually destructed and the high-grade metamorphic zone narrowed, it seems that the closer it is to the nappe front, namely Diancang Mountain, the more obvious overthrust than strike-slip, which was probably a westward oblique thrust.

Right Strike-Slip Fracture Stage of Red River

The overthrust tectonics made Diancang Mountain and Ailaoshan rise to the present height of more than 3000 m. After Miocene, due to gravity, anti-slip normal fractures were formed along Red River Fracture, which was accompanied by right strike-slip orogeny. The formation of the right strike-slip fracture may be caused by the India Landmass pushing into the Qinghai-Tibet Plateau, resulting in the uplift of the plateau, the eastward movement of the materials in the plateau, forcing the Yangtze Landmass to rotate counterclockwise and sliding eastward along Red River Fracture. In Middle Pliocene sediments developed in Red River Valley, there are a large number of gravels of high-grade metamorphic rock series in Ailaoshan Group, which indicates that it is in the stage of intense rising and denudation, which is similar to the time of rapid cooling of

20–19 Ma, and it also indicates that it is closely related to normal faulting along Red River Fracture.

2.2.6 Temporal and Spatial Structure and Evolution of Lanping Basin

The Lanping Basin is located in the south of Chaya-Jiangcheng Depression Zone in Mesozoic, sandwiched between Jiangda-Weixi Lvchun volcanic arc zone (east) and Yunxian-Jinghong volcanic arc zone (west). The basin has experienced two important evolution stages since its formation in Mesozoic: the back-arc foreland basin stage (Mesozoic) and the strike-slip pull-apart basin stage (Cenozoic).

Before the back-arc foreland basin was formed, the Qamdo-Pu'er Block was Jinsha River-Ailaoshan Ocean in the east and Lancang River Ocean in the west. From Late Carboniferous to Permian, the two ocean crusts began to subduct under the landmass between them, forming volcanic arcs on both sides of the landmass and a common back-arc-basin in the middle. During Triassic collision orogeny, the tectonic reversal occurred, which caused the rocks of Dongda Mountain-Lincang Landmass and Yangtze Landmass on the other side of the above two ocean basins to thrust on Qamdo-Pu'er Block, and started the foreland basin stage.

In Cenozoic, as the India Plate continued to push northward and northeasterly after the Indus River-Yarlung Zangbo River Ocean destructed, and the Yangtze Landmass pushed westward possibly due to the subduction of the Pacific Ocean, the two facies collided and squeezed, and the wasp waist zone contracted. All Mesozoic basins in Sanjiang Region contracted in a beaded shape, and the blocks crawled northward and southward or were squeezed, forming a massive strike-slip, which made the basin enter an evolution stage dominated by the development of strike-slip pull-apart basins.

2.2.6.1 Back-Arc Foreland Basin Stage (Mesozoic)

Lanping Basin lacks the Lower Triassic stratum, the Middle Triassic stratum is mainly developed in the Xiaomojiang Area between Jinggu and Pu'er in the basin except in the volcanic arc zones on both sides, and in Xiapotou Village of Yunxian in Pu'er and Laogongzhai of Batang Village in Zhendong Town, respectively, which is angular unconformity with C_3 and P_2, reflecting that the basement of the basin experienced significant tectonic deformation and denudation after Permian. Xiapotou Formation (T_2x) and Dashuijing Mountain Formation (T_2d) are a sequence of littoral-neritic clastic rocks and carbonate rocks, with composite conglomerate at the bottom. In Xiapotou Cundi conglomerate, the gravel mainly comes from the siliceous rock, sandstone,

mudstone and volcanic rock of the underlying Longdong River Formation (C_3l), which is supported by matrix, without mud and paleosoil at the bottom. It is a kind of transgression overlying sediment that has been cleaned by water. The Choushui Formation (T_2c) is composed of mudstone and carbonate rocks, and some carbonate rocks may be large sliding blocks with turbidite limestone and sand-slate on them, reflecting the deepening of the water body in the sedimentary basin, the appearance of a slope zone, and the overlapping to the north and east. The basin uplifted and suffered erosion after this period. During the Late Triassic, a sequence of clastic rocks (feldspar lithic sandstone, siltstone and mudstone, calcareous mudstone) and marl were deposited on the parallel unconformity, with basal conglomerate, fine conglomerate containing volcanic debris and boulder conglomerate at the bottom. In the area of Yiwanshui on the eastern edge of the basin, there was a huge thick molasse deposit at the bottom of Yiwanshui Formation in the Upper Triassic, and fluvial sediment appeared on the composite molasse deposit. On the unconformity surface of the other section, a sequence of thick conglomerate accumulated by fuchsia sand shale gravel can be seen, which indicates that the Mesozoic red beds developed before the deposition of Yiwanshui Formation. This fully shows the significant orogeny of Ailaoshan Zone and the movement of orogenic belt to basin. After that, the basin sediment changed from continental facies to marine facies, and a sequence of neritic clastic rocks and carbonate rocks were deposited. The Upper Triassic stratum is obviously divided into three parts, the lower part is continental-marine coarse clastic-fine clastic rock, the middle part is neritic limestone, the upper part is delta front sand shale and delta wedge sandstone, and finally coal-bearing clastic rocks. The Jurassic stratum is still a sequence of sand shale mixed with limestone in neritic tidal flat environment, occasionally deposited in continental sedimentation, and the sea area in Middle Jurassic once expanded, with transgressive overlapping to the east and west sides; The Cretaceous stratum is composed of a sequence of typical fluvial-lacustrine sand shale and conglomerate, and the basin obviously shrinks from the east and west sides to the center. Besides the east–west change, the sedimentary thickness also changes from the north to the south, reflecting that the sedimentary basin shrinks into a beaded shape. Due to the continuous migration of orogeny on both sides to the center of the basin, strike-slip pull-apart basins were superimposed in the center and edge of the basin in Paleogene and Neogene, and Paleogene and Neogene red clastic rocks, gypsum-salt and coal measures were deposited.

The unconformity between Upper Eocene Baoxiangsi Formation (E_2b) and Lower-Middle Eocene Guolang Formation occurred in the basin. This is the first regional unconformity in the Mesozoic basin since the Late Triassic, which is considered as the beginning of the first act of the

Himalayan Orogeny. However, except for the sea area on both sides of the basin expanded in the Middle Jurassic and the sea water invaded the orogenic belts on both sides, the mountain zones on both sides were kept in an orogeny status and moved toward the basin, which mainly occurred in the Late Triassic and Cretaceous. Late Triassic is characterized by the early basin sediments being involved in orogenic belts and then being uplifted, denuded and re-deposited; Cretaceous is mainly characterized by basin contraction. Fluvial sandstone and coarse conglomerate can be found near the orogenic belt (Ailaoshan Zone). Due to denudation caused by mountain uplift and the overthrust toward the basin, no unconformity between Cretaceous internal strata and the underlying strata was found. But the unconformity between different horizons of Paleogene and Neogene strata and Cretaceous and Jurassic strata can be found. Pliocene Sanying Formation (N_2s) and Cretaceous Jingxing Formation (K_1j) (Liantie Township, Eryuan Country), Pliocene (N_2) strata unconformably underlaid on Middle Jurassic Huakaizuo Formation, and the unconformity between Paleocene Yunlong Formation (E_1y) and Cretaceous Hutousi Formation (K_2h) can be even seen in Longtang, Lanping Country. However, Yunlong Formation conformably underlaid on Cretaceous strata, and it is difficult to divide them completely in Yunlong and other basins. This reflects the gradual progradation of orogeny from the margin to the center of the basin, but does not indicate that no Yanshanian orogeny occurred in the basin. Regionally, Yanshanian magmatic activity indicates that Yanshanian orogeny occurred, while the sedimentary records show that there is a gradual progradation process between Yanshanian and Himalayan orogeny.

Generally speaking, the evolution process of the foreland basin includes the early abyssal-bathyal flysch sedimentation stage, the middle marine molasse sedimentation stage and the late continental molasse sedimentation stage.

According to the study on foreland basins in the Sanjiang area of Southwest China, foreland basins can be divided into two types, namely marginal foreland basins and back-arc composite foreland basins. The former can be divided into early marginal foreland basin (marginal marine basin after the ocean basin disappears) and late marginal foreland basin (orogenic foreland basin) depending on its two significant development stages. The early marginal foreland basin/residual marine basin were developed between the passive continental margin of the western part of Yangtze Landmass and Changtai-Xiangcheng Volcanic Arc Zone, that is, the marginal marine basin after Ganzi-Litang Ocean disappeared in the late stage of Late Triassic. The marginal sea changed from the early deep-water flysch to the late shallow-water sedimentation, and was finally silted up and folded for orogeny, so that the post-orogenic Sichuan foreland basin developed on the eastern part of Yangtze Block

(Liu et al. 1993). It should be noted that the formation and evolution of Sichuan foreland basin, especially in its early stage, are mainly controlled by the northern and northwestern Qinling-Qilian-Kunlun Orogenic Belts, and may also be influenced and controlled by the passive continental marginal orogenic belt in the western margin of Yangtze Landmass after the closure of Ganzi-Litang Ocean at the end of Late Triassic and in Jurassic. Due to lack of sediments formed in the Jurassic and later ages in the passive continental marginal zone of the western margin of Yangtze Landmass, a large number of continental molasses sediments had developed in Longmen Mountain Foreland Basin since Jurassic. Marginal foreland basins with the same evolution model can also be found in Baoshan Block in the west of Changning-Menglian Junction Zone in Yunnan in the Sanjiang area. After the Changning-Menglian Ocean disappeared in Late Permian, the residual sea was formed on the passive continental margin in the east of Baoshan Block, which was located in the foreland uplift part and was lacking sediments formed in Late Permian. Its sedimentary evolution process generally develops from the early deep-water flysch, carbonate rock and radiolarian siliceous rock (T_1) to shallow-water sediments, and post-orogenic Shuizhai-Muchang Mesozoic foreland basin/depression zone (1999a, b; Liu et al. 1993) were developed in the east of Baoshan Block since Late Triassic.

However, Lanping Basin is a special back-arc composite foreland basin, which does not completely conform to the evolution model of this normal foreland basin. Lanping Basin was developed on the obduction plate with opposite subduction between ocean basins and oceanic crusts on both sides, and its main body is located on the back-arc-basin zone, but it is not completely similar to Dickson back-arc foreland basin (1974). Especially in the early stage of Middle Triassic, the sedimentary basins were mainly in the forearc zones on both sides, which are consistent with the early marginal foreland basins in terms of tectonics, that is, there should be an early abyssal-bathyal flysch sedimentation stage. No typical deep-water flysch sedimentation was found in Shanglan Formation or Manghuai Formation in the west of the basin, or Xiapotou Formation and Dashuijingshan Formation in the central Pu'er, that is, there was no foredeep development, and the carbonate rocks deposited by turbidity current were found only in Choushui Formation in the upper part of Middle Triassic strata in Yunxian, Pu'er-Zhendong, showing the change of water body from shallow to deep, rather than the overall change from deep to shallow (the sedimentary changes from shallow to deep also occur in the area where the depression center moves to the foreland). This may be the reversed orogenic polarity tectonics in the collision orogenic belt due to subduction on both sides, that is, the orogenic action of the passive margin mountain zone of the subduction plates on both sides has no progradation to the foreland in both east and west sides, but has progradation to the obduction plate-Lanping-Pu'er Block located between them in the reversed polarity, forming the orogenic pattern of the opposite overthrust nappe, resulting in Late Paleozoic volcanic arc zone and back-arc-basin zone on both sides in the early stage being developed into Mesozoic foreland basin as the basement for the development of foreland basin. Moreover, the central axis of the basin is characterized by a foreland uplift, with foreland depression and foreland thrust zone on both sides, forming a typical foreland basin tectonic pattern. Due to reversed orogenic polarity, the reverse orogenic overthrust nappe covered the early abyssal-bathyal flysch sediments at the front margin, and caused the sediments to be overlaid on the volcanic arc zone in the opposite direction, resulting in unconformity with the volcanic arc zone. The provenance is mainly from the volcanic arc zone, for example, the gravel of the basal conglomerate of Longdi Village, Eryuan is mainly supplied by the volcanic arc zone. At the same time, the depression center of the basin moves to the middle of the basin over time and allodapic limestone of Choushui Formation is deposited in Yunxian, Pu'er-Zhendong. The appearance of allodapic limestone also indicates that there is a marginal carbonate rock plateau environment on the margin of the foreland zone.

Although the evolution of Lanping Foreland Basin has certain particularity, its evolution history can still be discussed from the perspective of sedimentology.

Abyssal-Bathyal Flysch Sedimentation Stage

In this stage, when the thrust wedge emerged from the water surface or had not reached its maximum height, it lagged behind sedimentation due to sedimentation; the sedimentation rate of the foreland basin is higher than the accumulation rate of sediments, so the basin is in a starved state, mainly forming a deep-water sedimentation type and also forming a gradually shallowing sedimentary system toward the foreland uplift at the same time. In the early development stage of foreland basin, the thrust wedge was thrust on the margin of Craton, where the crust thickness decreased rapidly and the stiffness decreased significantly. Under the action of gravity and horizontal extrusion, the crust flexed downward greatly. Therefore, the formed foredeep depression is deep and narrow and is often covered by the late overthrust nappe in the evolution process.

The sediments that occurred in this stage shall belong to Shanglan Formation and Manghuai Formation of the Middle Triassic in terms of the spatial development tectonics of the basin and are generally located between the volcanic arc zone and the passive margin zone, namely the tectonic position of the residual sea. However, just as mentioned above, the sedimentary characteristics show that the sedimentary characteristics of this stage can be found only in Shanglan Formation in the northeast of the basin, and the early abyssal flysch sediments in the lower part of the rest of the area may be covered (may be covered by (T_1) due to the reversed orogenic polarity. No sedimentary evolution sequence from deep to shallow is found, while the evolution from shallow to deep appears, which may just reflect the conversion period of reversal in orogenic polarity, so that Lanping Basin was depressed on both sides and had a shared foreland uplift in the middle during the Middle Triassic, bringing the evolution of foreland basin into the second evolution stage with the reversal in orogenic polarity.

Marine Molasse Sedimentation Stage

In this stage, when the thrust orogenic belt crossed the hinge line of the continental slope to the thick rigid Cratonic crust, the flexural settlement formed by the tectonics gradually decreased compared with the previous stage, so a shallow and wide foredeep depression was formed and the thrust orogenic belt was finally developed to its maximum height and remained in a relatively stable status for a long term. At this time, a retrograde orogeny occurred in the provenance area of the foreland basin, developing from the sediments with high maturity mainly from Craton in the early stage into the sediments with low maturity mainly from the thrust zone in this stage. When the flexural settlement is fixed or gradually reduced, and the basin gradually rises to the sea level and keeps in the stable state for a long term, stable environmental sediments with high maturity will appear. The development period of Lanping-Pu'er Composite Foreland Basin in this stage varies roughly from the sedimentation period of Choushui Formation/Pantiange Formation to the sedimentation period of Huakaizuo Formation. Before the thrust orogenic belt is formed, this stage is mainly continental alluvial sedimentation, such as alluvial fan and fan delta sedimentation of Waigu Village Formation. Littoral-neritic coarse clastic rock sediments, such as littoral-neritic quartz sand or delta sediments of Maichuqing Formation and Huakaizuo Formation; carbonate rock and evaporite sediments in the euxinic environment, such as limestone of Sanhedong Formation.

In the development process of the foreland basin, there may be three ways for the thrust of the fold orogenic belt to Cratonic Block: ① slow climbing; ② fast parallel overthrust; ③ skipping overthrust. The thrust block of Lanping-Pu'er Composite Foreland Basin may belong to skipping overthrust, resulting in several cycles from sea to land: Shanglan Formation (T_2s)—Waigu Village Formation (T_3w), Sanhedong Formation (T_3s)—Yangjiang Formation (J_1y); Huakaizuo Formation (J_2h)—Bazhulu Formation (J_3b), Jingxing Formation (K_1j)—Nanxin Formation (K_1n) and above. Due to the skipping and suddenness of thrust and the lag of sedimentary facies behind settlement, many fast deepening starved sedimentary events have been caused, such as turbidite of Choushui Formation, black shale of Waigu Village Formation, turbidite and carbonaceous shale of Walu Formation, and even the dark gray bed at the top of Huakaizuo Formation. This phenomenon also reflects the result of the continuous skipping migration of the fore-deep into Craton, and many unconformities have been formed at the margin of the basin and many "piggyback" basin covers have also been formed due to this thrust form.

Continental Molasse Stage
With the further migration of the thrust orogenic belt toward Craton, most areas of the basin of Early Cretaceous (Nanxin Formation K_1n) were occupied by an alluvial environment. Rivers from the thrust orogenic belt flowed into the catchment area of the foreland basin, forming a lake environment at the catchment center of the basin and a delta environment at the margin of the basin. It should be noted that the foreland depressions on both sides of the basin are not completely evolved synchronously. However, due to the accuracy of stratigraphic correlation and the incompleteness of stratigraphic preservation, the corresponding relationship between the two sides in terms of evolution shall be further studied.

2.2.6.2 Strike-Slip Pull-Apart Basin Development Stage (Cenozoic)

Since Paleogene, Lanping Basin had entered the strike-slip pull-apart basin development stage. This type of basin is generally an axial basin distributed along a large strike-slip fracture zone, which is mostly rhombic or rectangular, and is a deep basin. Therefore, the sedimentary rock series is extremely thick, and the basin has multiple filled facies and wide varieties, such as clastic rock piles, landslides, alluvial fans, braided rivers, meandering rivers, fan deltas, seashores, shallow lakes and deep lakes, turbidity currents, chemical sediments, and algal limestones. Because of the continuous lateral movement of faults at the margin of the basin, the provenance also changes over time, so the filled facies are complex and diverse. At the same time, synsedimentary deformation and unconformity are also very common. The basin is distributed along the margin of synsedimentary translational motion in a significant asymmetric way, with alluvial fans dominated by small clastic flows, which contain coarse breccia and conglomerate sediments. While large-scale alluvial and fluvial sediments are developed along the inactive or slightly active margins, including fine conglomerate, but few breccia.

Yunlong Formation (E_1y) is the earliest sedimentary stratum in the strike-slip pull-apart basin and forms a sedimentary cycle from coarse to fine with Guolang Formation (E_2g^{1-2}). Due to the reworking caused by the late strike-slip fault, the original appearance of the whole basin cannot be restored, and the alluvial fan facies, fluvial facies, shallow lake facies and semi-deep lake facies can be found in terms of the sedimentary facies. Based on a scale of 1: 200,000 in Lanping, Yongping and Weixi or other regional survey data, the thickness of Yunlong Formation varies greatly, reaching 2025 m in Shijing, Lanping, but decreasing to 266 m in Laomujing. The sedimentary basement is also different. It contacts the underlying Hutousi Formation (K_1h) in a parallel unconformity or unconformity manner in Yunlong, Baofeng and Longtang Village, Lanping. It overlies on Nanxin Formation (K_1n) and Shanglan Formation (T_2s) respectively in Lanping and Qiaoshi, and even overlies on Cangshan Group in the east. At that time, the basin was actually composed of a sequence of deep aulacogens from north to south. For example, in Shunchuanjing-Yanqu Village-Mishajing-Qiaohoujing, Jianchuan County, a large number of broken rocks (mainly limestone) on both sides of the fault fell down, forming angular gravels of different sizes. The basement of this formation is mainly composed of conglomerate, of which the composition is mostly mudstone, siltstone and shale, and few marlstone. Conglomerate is round or sub-angular, and its particle size changes greatly. For example, the particle diameter on Laomujing section in Lanping County is 0.5–1 cm, is 3–5 cm in Lajing, Lanping County, and the largest one can reach tens of centimeters, all of which are supported by matrix. It is worth noting that even mudstone and siltstone with poor abrasion resistance and conglomerate are angular, indicating that conglomerate is not formed through long distance motion but may be deposited by collapse or alluvial fan. The basal conglomerate of Yunlong Formation in Longtang Village, Lanping County is quite special, and the conglomerate is well ground and oval in shape, with different sizes (varying from several centimeters to tens of centimeters in diameter) and it is mainly supported by particles. The interstitial materials are coarse sand with thin mudstone sandwiched in the middle, and the conglomerate is composed of sandstone, limestone, flint, etc., and also contains glutenite, sandstone and mudstone (extremely thin) in the upward beds, forming a sequence of upward-thinning cycles, namely, braided fluvial sediments. There is lenticular feldspathic quartz sandstone on it, in which the cross bedding can be found and the sedimentary characteristics of braided river can also be found.

The middle part of Yunlong Formation is mainly composed of purplish-red fine sandstone and siltstone mixed with mudstone. In the purplish-red sandstone near Guquan Bridge in Lajing, Lanping County, abraded wave ripple and fluvial unidirectional current ripples and launder can be found, and even planar cross bedding composed of fine conglomerate and coarse sandstone can be found, which should be fluvial-deltaic or littoral-neritic sediments. The upper bed of Yunlong Formation is mainly composed of purplish-red siltstone, mudstone and marlstone, mixed with a small amount of medium and thin layers of fine sandstone. Yangcen in Jianchuan County is more representative. The observation reveals that this sequence of stratum is the product vertically aggraded in the center of the basin. Some intervals are composed of purplish-red silty mudstone mixed with yellowish-green shale, occasionally mixed with 1–2 cm thick fine sandstone. Some horizons contain many layers of 10–20 cm thick graded bedding varying from argillaceous fine sandstone and argillaceous siltstone to mudstone, belonging to low density turbidite sediments in the lake basin. Some yellowish-green shales are sandwiched with 1–2 cm thick flat marlstone lens or thin marlstone. A 0.2 mm thick horizontally laminated bed can be found on the limestone bed section, and the horizontally laminated bed can also be found in shale. The lower part of the laminated bed is rich in debris (silt-sized) and the upper part is rich in mud (clay or marl), which is varves sediments. As the yellowish-green shale was formed in the period with low debris injection, which is conducive to calcium precipitation, it coexists with marlstone; however, purplish-red siltstone mixed with fine sandstone was formed in the period with high debris injection, which is not conducive to calcium precipitation, so it does not coexist with marlstone. It can be seen that the yellowish-green shale and purplish-red siltstone appears alternately on the section. Each cycle appears as purplish-red siltstone and ends with yellowish-green shale and off-white marlstone. To sum up, it should be sediments of deep lake-semi-deep lake facies.

The overlying Guolang Formation is in conformable contact with Yunlong Formation, the lower bed is interbedded with sand and mud, and the upper bed is mainly composed of sandstone, with cross bedding and few gravels, belonging to the product generated due to lake water gradual shallowing. The top stratum of this formation is incomplete and covered by the overlying Baoxiangsi Formation in an unconformity manner.

New definition of Baoxiangsi Formation: ① similar to Lijiang Formation; ② the upper bed is the original Baoxiangsi Formation and the lower bed is the original Meile Formation (Yunnan Bureau of Geology and Mineral Resources, 1996). Therefore, Baoxiangsi Formation actually includes two cycles from coarse to fine. Most of the lower strata of Baoxiangsi Formation are in unconformable contact with Shigu Group. The bottom bed is a thick layer of grayish-purple and purplish-red massive gravel bed. The gravel is mainly composed of limestone and quartz schist, followed by vein quartz, purplish-red siltstone and basic rock, with extremely poor separation and moderate rounding. The lower bed is breccia, which gets smaller from

bottom to top, the middle-upper bed is brick-red and off-white sandstone containing feldspar, with well-developed large plate-like cross bedding, and a thickness of strata series of 1–5 m and a cross bedding dip angle of up to 30°, and no fossils were found (Regional Geological Survey Report on a scale of 1: 200,000 in Weixi). In Liming Township, Lijiang County, the bottom bed is composed of pebbly conglomerate, containing a lot of quartzite gravel, with well rounding but poor separation. In Labazhi Copper Ore Deposit, the basal conglomerate is composed of a thick layer of massive conglomerate bed, varying from 1–2 cm to 10–20 cm in diameter. With complex composition, it is mainly composed of limestone and few quartzite and schist, etc., mixed with coarse sandstone lenses or wedges. The former is a fluvial sediment on the inactive or relatively inactive margin in the strike-slip pull-apart basin, while the latter is an alluvial fan sediment on the active margin in the strike-slip pull-apart basin. The middle and upper strata are generally considered as large fluvial cross bedding. It is found through observation of the section of Liming Township that the sandstone is characterized by good degree of separation and extremely high roundness and is free of mud and mica sheets. In addition, the thickness of cross bedding is large, and the dip angle of foreset bed is steep (30°) and the direction of dip is stable, all of which are the typical characteristic of eolian dune. Therefore, it may be the desert sediments. It may be aeolian dunes on the marginal banks of rivers and lakes, as seen on both sides of Yarlung Zangbo River Valley in Tibet at present.

In whole Neogene, the fault activity had been greatly reduced, the scale and number of strike-slip pull-apart basins had also been greatly reduced, and the sedimentation range had been reduced, while small basins dominated by rivers and lakes had appeared and swamps had developed in large numbers, forming important coal-bearing strata. The Miocene stratum is the Shuanghe Formation, Pliocene stratum is Jianchuan Formation and Sanying Formation, all of which are fluvial and lacustrine sediments. For example, Jinding Township, Lanping is a small north–south nearly rhombic basin, in which Sanyingxian fluviatile-lacustrine sediments were deposited and it unconformably overlaid on the underlying Paleogene Yunlong Formation and its previous strata, with the top being outcropped or covered by Quaternary strata.

2.2.7 Temporal and Spatial Structure and Its Evolution of Tenasserim Arc-Basin System

Bangong Lake-Shuanghu-Nujiang River-Changning-Menglian Mage-suture Zone is a relic of the destruction of Tethys Ocean. The eastern part of this zone roughly includes Jitang Residual Arc, Northern Lancang River Junction Zone,

Riwoqê-Dongda Mountain Magmatic Arc, Lincang Magmatic Arc, Southern Lancang River Junction Zone, Yunxian County-Jinghong Volcanic Arc, etc. all of which constitute Tenasserim MABT, and Qamdo-Lanping-Pu'er Block is located in its east side.

2.2.7.1 Temporal and Spatial Structure of Tenasserim Arc-Basin System

The understanding of the temporal and spatial structure and formation process of Bangong Lake-Shuanghu-Nujiang River-Changning-Menglian mage-suture zone plays a very important role in the study of the formation and evolution of geological structures in Eastern Tethys. Due to differences in perspectives, collected data and understanding, the nature and tectonic attributes of ocean basin restored in this zone have long been a subject of debate in the study on Sanjiang Orogenic Belt.

Ocean Basin Formation Age and Tenasserim Islands Development Age

Tectonic stratigraphic units of Devonian to Middle Triassic radiolarian siliceous rocks have been found in Menglian, Changning, including radiolarian rock represented by the earliest *Monograptus uniformis zone* of Devonian, *Archocyrtium menglianesis* Wu. Ar. delicatum et al. (C_1). There is a sequence of siliceous rocks deposited by deep-water flysch on the western slope of Meri Snow Mountain, such as Early Carboniferous *Palaeory phosty.lus uar spina*. Late Carboniferous-Permian *Albaillea* SP., *Pseudea I bailla* sp. were found in the siliceous rocks in Zhayu-Bitu. Interdisciplinary comprehensive research on sedimentary geochemistry of siliceous rocks, REE, stable isotopes and radiolarian paleoecology (Liu et al. 1993) showed the sedimentary environment of the abyssal ocean basin. Radiolarian siliceous rocks of Devonian, Carboniferous and Early Permian in Manxin, Menglian, Gengmanongba and other places have three characteristics of associated rock assemblage, namely flysch sandwiched with siliceous rock, limestone and mudstone, sandwich bed in ocean floor basalt or exposed caprock above pillow basalt, and sandwich bed in basalt above oceanic island volcanic rocks, which form siliceous rock-basalt-limestone assemblage with limestone without terrigenous clast on oceanic island. The amphibole K–Ar isochron age of gabbro in the Tongchangjie ophiolite is 385 Ma (Cong et al. 1993), which represents the geology of Middle Devonian. Therefore, most scholars believe that the Paleo-Tethyan Ocean in Changning-Menglian was formed since the spreading in Devonian stratum. However, the following important geological records allow us to reconsider the age of Tethys ancient ocean basin represented by Dingqing-Bitu-Changning-Menglian zone (Pan et al. 1997, 2004).

(1) The landmasses on both sides of Changning-Menglian have different basements. The lowest outcropped stratum in Baoshan Block is the Cambrian Gongyanghe Group, with fluvial submarine fan-slope environment in its lower bed and basinal siltstone, shale sandwiched with siliceous rock in its upper bed. Significant tectonic activities at the end of Cambrian and significant intermediate-acid magma intrusion occurred in this area, which represented by Pinghe granite body, with isotope age between 495 and 525 Ma and uneven metamorphism and deformation. Baoshan Block is actually the northern extension member of the Shan Landmass. Mogok gneiss, Shan Landmass and Tengchong Landmass of Gaoligong Mountain Group have similar basement to India Landmass, which can be regarded as the accretion part of the northern margin of Gondwana affected by Pan-African Events. Lincang Landmass, Jitang Block and Qamdo-Lanping Block in the east of Changlian-Menglian Zone all have the same basement characteristics with that of island arc accretion in Neoproterozoic Rodinia supercontinent convergence stage, and have the characteristics of volcanic magma accretion in Early Paleozoic as they are all split fragments in the western margin of Yangtze Continent. However, granite intrusion of 750–900 Ma, metamorphic event of 800–900 Ma, sedimentation of Yangtze plateau in New Nanhua-Sinian, glacial sedimentation, etc. on Yangtze Continent unconformably contact with the Mesoproterozoic underlying tectonic strata, as well as the phosphorus-bearing sediments of Early Cambrian are obviously different from Shan Landmass which is close to Gondwana. It shows that Shan (including Baoshan) Landmass has been separated from Yangtze Continent by ocean at least in Neoproterozoic and later. The analysis of paleomagnetic data by Zhuang reveals that the latitude difference between Baoshan Block and Yangtze Landmass of Cambrian-Ordovician was great in Early Paleozoic. The former was around 18° south latitude and the latter was near the equator, which was consistent with the environment reflected by geological records. Shan-Baoshan Block is not the spreading block of Changning-Menglian Ocean Basin due to splitting from Lincang or Qamdo-Pu'er Block on the west side of Yangtze Landmass in Devonian, but the tectonic relationship formed by shrinkage, final subduction and destruction of Tethys due to northward drift of Shan-Baoshan Block.

(2) Subduction event of Tenasserim Frontal Arc in Early Paleozoic. Yong research on Tenasserim Mountain on the west side of the North Lancang River showed that a sequence of metamorphic rock with greenschist facies in Youxi Group of Lower Paleozoic stratum was

decomposed from Jitang Group of Precambrian, with the source rock of basalt and dacite mixed with are-nopelitic debris, among which dacite Rb–Sr age was (371 ± 50) Ma, being the evidence of metamorphic age and active continental margin island arc environment obtained after geochemistry research on rocks. $^{40}Ar/^{39}Ar$ plateau age of granodiorite mineral of Biluo Snow Mountain is between 418 and 424 Ma, and $^{40}Ar/^{39}Ar$ plateau age of the glaucophane of the meta-morphic basic volcanic rocks in Lancang Glaucophane High-pressure Metamorphic Zone is 410 Ma (Zhang et al. 1990). 433 Ma coarse-grained biotite granite and 422 Ma fine-grained biotite granite are found in Lin-cang Granite Basement. Most of the geological records of these tectonic thermal events in Jitang Group Com-plex, Chongshan Group and Lancang Group in Pre-cambrian show the remnants of accretionary wedge under the island arc, such as volcanic-magmatic arc formed by the subduction of Tethys Ocean in Early Paleozoic toward the northeast (present position). This continental crust zone with soft basement may have been originally a series of mountains along the south-west margin of Yangtze Landmass of Pan-Huaxia Continental Group. At the beginning of Devonian, this continental margin arc chain entered the formation and evolution stage of the MABT of Qiangtang-Sanjiang in the form of Japan-Ryukyu Islands splitting, and Jitang Residual Arc-Chongshan Residual Arc-Lancang Residual Arc are referred to as Tenasserim Frontal Arc, of which the southwest side is the residual Proto-Tethys.

(3) Sedimentary records of Baoshan Block. Cambrian to Permian strata in Baoshan Block had kept the passive margin strata of continental shelf-plateau, and the strata formed in each era are mostly in conformable contact with each other, except for any hiatus between Lower Carboniferous stratum and Devonian stratum. It lacks the Middle Carboniferous stratum. The upper Car-boniferous stratum is composed of glacial debris flow at the continental margin, siltstone and shale in the lower bed, and intraplate rift basalt mixed with limestone lens in the upper bed (see the section on Structure and Evolution of Baoshan Block for details). These sedi-mentary records also reflect the continuity of Paleo-Tethyan Ocean from one aspect, which restricts the overall continuity of Paleozoic passive margin strata in the Shan-Baoshan Block in the southwest of the ocean.

Therefore, Shan-Baoshan Block may have become an independent block in Paleozoic (when Rodinia supercontinent decomposed) and was separated from Yangtze Landmass by the Tethyan Ocean. The Paleo-Tethyan Ocean in Changning-Menglian was not reopened after the Proto-Tethyan Ocean was destroyed, but developed and evolved based on the Proto-Tethyan Ocean.

Spatial Distribution of Tenasserim Arc-Basin System
According to the present tectonic pattern of Hengduan Mountain, it is divided into two members (namely, southern member and northern member), which are briefly described from west to east in turn.

Dingqing-Zhayu-Bitu Junction Zone
This zone is located in the curved turning part (extending from NWW-SEE to NS) of the eastern part of Bangong Lake-Nujiang River Junction Zone. According to the avail-able data, Carboniferous-Permian ophiolite, Triassic ophio-lite, Early Jurassic ophiolite and radiolarian siliceous rock have been found in Suruka, Dingqing Country, Nuxilaka, Luolong Country and Zhayu, Zuogong Country, which are mixed with various greenschists or flysch sand-slate of Carboniferous-Permian and Late Triassic-Early Jurassic, respectively. The main remnants of Carboniferous and Per-mian ophiolite are distributed in Nuxilaka on the east side of Jiayu Bridge Residual Arc and Suruka-Tongka in the middle zone. At present, only metamorphic peridotite (serpentinite), gabbro-cumulate, metamorphic lava (with pillow rock locally) and overlying abyssal turbidite mixed with recrys-tallized metamorphic siliceous slate are found. Ophiolite has been significantly decomposed, with lenses of different sizes and in contact with the matrix in form of faults. The ophi-olitic melange zone in Suruka is of low greenschist facies, with metamorphism; jadeite + quartz mineral assemblage are found in quartzite, and Permian spore fossil are found in fine-grained limestone in the tectonic stratum. Late Triassic-Early Jurassic ophiolite in Dingqing area has been widely studied. Although the trinity ophiolite, such as metamorphic peridotite, cumulate, mafic complex, dike swarm and pillow lava, has been descomposed, its remnant is complete. In addition, many radiolarian siliceous rocks in Late Triassic-Early Jurassic are mixed in the sedimentary deformation bed of Quehala Turbidite Fan representing the outer arc ridge slope sediment. A considerable part of Quehala Assemblage should be classified as the fore-arc accretionary wedge in Jiayu Bridge Residual Arc. The so-called Meng'axiong Carbonate Rock Plateau is only a limited fore-arc plateau between the inner side of accre-tionary wedge and Jiayu Bridge Island Arc. Restoration of Tethyan Ocean Basin by siliceous rocks, Carboniferous and Permian ophiolite and Late Triassic-Early Jurassic ophiolite in accretionary wedge is a continuous evolution process. Dingqing Ophiolite Melange Zone bends toward the south-east arc and then overlaps with Suruka-Nuxilaka-Tongka

Ophiolite Melange Zone in Xialinka through Jizhong and Bangdaxi, and finally connects with Zhayu-Bitu Ophiolite Melange Zone. The cumulate gabbro and diabase in Late Triassic-Early Jurassic ophiolite belong to glassy andesite, and were formed in the intra-oceanic arc environment and oceanic intraplate environment of the ocean ridge. The existence of the intra-oceanic arc also indicates that the ocean represented by Dingqing ophiolite had existed much earlier than the Triassic. Middle Jurassic molasse unconformably overlies on the whole ophiolitic melange unit, which indicates that Dingqing-Nujiang River Zone started subduction and destruction of oceanic crust in Late Triassic and then collided and closed at the end of Early Jurassic.

Dingqing ophiolite melange zone bends toward the southeast arc and then overlaps with Suruka-Nuxilaka-Tongka ophiolite melange zone in Xialinka through Jizhong and Bangdaxi, and finally connects with Zhayu-Bitu Ophiolite Melange Zone.

Located between Chayu Landmass and Qamdo Landmass, Bitu Junction Zone is a tectonic melange zone composed of some tectonic blocks, with a width of about tens of kilometers. According to its tectonic characteristics, it can be roughly divided into three parts. Roughly distributed along both banks of Nujiang River, the western part is dominated by a thrust zone composed of Late Devonian-Early Carboniferous sand-slate, marble and siliceous rocks, in which the rock metamorphism generally reach low greenschist facies, and flysch rhythms, flute cast and abyssal cellular arenicolite are developed, showing the characteristics of turbidite sedimentary fans on the continental slope and continental basement, which may be the accretionary wedge of flysch on the eastern margin of Chayu Landmass. The middle part, such as Lawu Village, Zhayu Town, is dominated by a large rock mass composed of neritic continental shelf carbonate rocks of Middle and Upper Devonian and Carboniferous, on which Carboniferous limestone is mixed with a small amount of basalt and limestone is rich in corals, brachiopods and fusulinid. The east and south parts are dominated by a sequence of melange composed of Carboniferous and Permian flysch sand-slate, phyllite, siliceous rock, tholeiite and ultramafic rock. Early Carboniferous *Palaeoryphosty.*, *Lus uar spina* are found in a sequence of siliceous rocks deposited by abyssal flysch on the west slope of Meri Snow Mountain; Late Carboniferous-Permian *Albaillella* sp, *Psiudeal bailla* sp. and Carboniferous conodonts are found in the siliceous rocks in Zhayu-Bitu; basalt is a sequence of sub-alkaline tholeiite, with w (SiO_2) between 47.57 and 50.73%, w (TiO_2) between 1.18 and 2.96%, w (K_2O) between 0.36 and 0.98%, $w(P_2O_5)$ between 0.14 and 0.36%, w (Rb) between 9 and 27×10^{-6}, w (Sr) between 150 and 226×10^{-6}, w(Ba) between 75 and 229×10^{-6}, w (Th) between 1.85 and 9.15×10^{-6}; REE is distributed flatly with weak enrichment in ΣLREE, with ΣREE of $(55.32-79.85) \times 10^{-6}$, w (Sm)/w (Nd) between 0.23 and 0.36, w (Ce)/w(Yb) between N1.01 and 2.05, Eu between 1.04 and 1.35; it is included in ocean ridge and ocean island basalts on the diagram of $TiO_2–P_2O_5$ and $Zr-TiO_2$.

The above basalts is generally of low-grade metamorphism and the metamorphism in the south of Bitu reaches the grade of greenschist facies. Pumpellyite is found in altered basalts and metamorphic basalts in Zhayu and Bitu, accompanied by ultramafic rocks formed by tectonic intrusion, so it may be a transitional type of medium–high pressure facies series. The research made by Zhang on the phengite in Jiayu Bridge Metamorphic Zone shows that $bo = 0.9032$ nm (based on 20 chlorite samples) and RM = 0.09, both of which are all projected in the high-pressure zone II in MgO-RM diagram. In addition, pumpellyite-actinolite assemblage is also found in Rutog-Gêrzê Metamorphic Zone in the west of Jiayu Bridge, forming the medium–high pressure tectonic metamorphic zone in Bangong Lake-Dingqing-Bitu accompanied by ophiolite melange zone.

The eastern part of Bitu Junction Zone is dominated by a large-scale ductile shear zone and a large nappe thrust eastward. Among them, Bitu and Wapu-Chawalong Melange Nappe is the largest in scale.

Jitang Residual Arc and Caprock

As the important part of Tenasserim Frontal arc, Jitang Residual Arc is composed of Jitang Group, Chongshan Group and Early Paleozoic Youxi Group with high greenschist facies-amphibolite facies. The arc is bounded by Northern Lancang River Fault in the east and Bangong Lake-Nujiang River Junction Zone in the west (east member), respectively, and is adjacent to Southern Qiangtang Basin in the north being divided by Sanduo-Duocai NNW Fault, in which there are few stratigraphic units and metamorphic rocks of Tenasserim Mountain and northern Lancang River are outcropped in its east, while a large area of Upper Triassic stratum and few Jurassic stratum are outcropped in its west.

Metamorphic rocks in North Lancang River Zone can be divided into Precambrian Jitang Group Complex and Lower Paleozoic Youxi Group. The Jitang Group Complex is composed of biotite plagioclase gneiss of amphibolite facies, biotite granulite, amphibolite plagioclase gneiss, banded (striated) migmatite and local mixed tonalite mixed with quartz schist, sillimanite garnet plagioclase schist, biotite feldspar schist, plagioclase amphibolite and marble, with an apparent thickness of more than 4000 m. Being reworked by multiple metamorphism and deformation, some sections are characterized by significant plastic flow deformation. The age of gneissic granitic intrusive rocks is 338 Ma, and the latest migmatization age is 212.5 Ma (K–Ar method).

The source rock is composed of a sequence of arenopelitic rocks mixed with island-type intermediate-acid and basic volcanic rocks. Youxi Group is composed of various schists and multi-layer composite conglomerates. The source rock of the schist is arenopelitic rock and intermediate-acid arc volcanic rock mixed with basic volcanic rock formed by dynamothermal metamorphism of greenschist facies at the active continental margin. However, it is of low-grade metamorphism in the west of Dongda Mountain and mainly composed of thin interlayer of quartz sandstone, slate, schist and marble, with an apparent thickness of more than 3000 m, a whole rock Rb–Sr metamorphic age of 371.1 Ma and angular unconformity with Jitang Group, so it may be an accretionary wedge formed by Paleo-Tethyan subduction and destruction.

A number of Permian granitoids have been found in places such as Lajiang, Riwoqê County, which are composed of "tonalite, granodiorite-monzogranite", and are intruded into Jitang Group. The U–Pb age of monzogranite is (269 ± 18) Ma, which may be related to the early island arc magma series restricted by the northeastward subduction of the Tethyan Oceanic Crust of North Lancang River and the intrusion of alkali granite in the later stage (113–90 Ma).

A sequence of volcanic-sedimentary metamorphic complex is outcropped in strips along Biluo Snow Mountain-Chongshan, and is located between Lancang River Fault and Biluo Snow Mountain Fault. Among them, the metamorphic rocks are composed of schist, granulite, gneiss mixed with marble and plagioclase amphibolite (schist), which is called Chongshan Group Complex (Pt2). Its age-dating by Sm–Nd and K–Ar varies from 1100 to 1000 Ma, 956 Ma and 738–72 Ma, and multiple sets of data. Tectonic strata show products of Rudinia supercontinent convergence stage and decomposition stage, and also show collisional monzogranite intrusion in Late Triassic.

Only the upper stratum is found in the Triassic stratum of Zuogong Basin in the western part of the residual arc, and it can be divided into Dongdacun Formation, Jiapila Formation, Bolila Formation and Bagong Formation from bottom to top. Dongcun Formation is distributed eastward along Zuogong County and ended by Wuqida in the north, with the middle and lower beds of 120–150 m thick composing of sandshale and limestone in three rhythms; that is, the upper bed is composed of sandshale mixed with a thin layer of limestone, and the bottom bed is composed of 0.6–3 m thick composite conglomerate and is rich in Carnian corals and bivalve fossils, with a thickness of 750–1000 m and overlaying on the metamorphic rocks of Youxi Group in a way of angular unconformity. The Jiapila Formation is composed of a littoral-neritic purplish-red sandshale mixed with few gray-green rocks, with two sequences of thick limestone in the middle bed and multi-layer 1000–1400 m thick conglomerate assemblages in the middle and lower beds, which

is in a conformable and transitional relationship with the underlying Dongdacun Formation. Bolila Formation on it is a thick layer of limestone mixed with dolomitic limestone and silty mudstone, partly mixed with andesite and rich in Carnian-Norian fossils, with a thickness of 200–500 m. The top bed (e.g., Bagong Formation) is dominated by rhythm interbeddings with greyish-black and black sandshale and grey and green fine sandstone and siltstone; the semideep sedimentary tectonics can be found in the lower bed; sandstone and lithic sandstone are relatively common and phytoclasts are found in the upper bed, with a thickness of more than 3000 m. It is considered as the foreland depression product formed by the westward oblique subduction of destructed residual arc landmass of Dingqing-Bitu Ocean Basin.

The Jurassic stratum is only distributed on a small scale in the south and north of Xiaya and the east of Tiantuo in Zuogong County. Jurassic strata in Xiaya-Jingtang-Beishan belong to the middle strata, which is composed of littoral purplish-red sandstone, mudstone mixed with variegated silty mudstone and unstable limestone and unconformably overlays on Triassic strata, and the thickness of about 1000 m. The Jurassic strata in southeast Xiaya are all purplish-red sandy mudstone sandwiched in the fracture zone, which should be the south extension member of the Middle Jurassic stratum in Beishan, Guojing. Preserved on the fault block between faults, Jurassic strata in the northeast of Tiantuo is also purplish-red clastic rock without top and bottom beds.

Late Triassic-Middle Jurassic sedimentary strata in Zuogong may be the foreland basin product formed by destruction of Dingqing-Bitu Ocean Basin and the southwestward oblique subduction of destructed residual arc landmass, and the fold deformation is characterized by the tectonic pattern of foreland fold-thrust zone.

Northern Lancang River Junction Zone

This zone runs from the northwest of Ulaan-Uul Lake in the northwest end to Zhayu-Bitu Junction Zone in Zhayu, Zuogong County in the south through the east side of Jitang Residual Arc in the southeast, and the Southern Lancang River Zone runs along the east side of Meri Snow Mountain-Biluo Snow Mountain Residual Arc in Deqin County, Baijixun, Weixi County and Yingpan, Lanping County, and runs along Lancang River Fault in its south and then extends out of the frontier southward through Banpo, Jinggu Country, Yakou, Lancang County and Jinghong Country, with the whole length of about 1400 km. Lancang River Zone bends westward in the north member and roughly extends to the southwest of Rola Kangri, which may be connected to Jinsha River Junction Zone. According to the outcropping and spatial distribution characteristics of basic and ultrabasic rocks and tectonic strata as well as the assemblage sequence of geological bodies, this zone can be divided into two members, namely Northern Lancang River Zone and Southern Lancang River Zone.

The Northern Lancang River Zone is distributed from NNW to NW. Jadeite and glaucophane are found in Shitou Mountain and Heixiong Mountain, etc. in the northwest of Ulaan-Uul Lake in the northern part of this zone, and ultramafic rocks (on a scale of 1: 200,000 in Riwoqê 1992) and mid-ocean ridge basalt invaded in Carboniferous stratum are found in Riwoqê. Most members are covered by Permian-Middle Triassic volcanic arc in Zadoi-Riwoqê-Dongda Mountain on the east side due to their westward thrust, and no ophiolitic melange/melange has been out-cropped yet.

In the middle-south member, such as Riwoqê and Jitang, turbidite of abyssal sedimentary basins is found in the upper bed of Riazenong Formation and Majunong Formation in Carboniferous Kagong Group defined by predecessors, which is a sequence of siliceous-calcareous-argillaceous sediments and is coexisted with tholeiite-rhyolite assemblage, with different tectonics from stable coal-bearing formations that have no volcanic rocks of original Kagong Group in Qamdo Stratigraphic Zone on the northeastern side or east side. Riazenong Formation is composed of a "bimodal" volcanic rock series, which is the product of the rift valley (back-arc) where the crust is stretched and thinned. The incompatible elements of basalt conform to the distribution pattern curve of single uplift, and the rare earth elements are of nearly flat distribution pattern and slightly enriched in light rare earths. Vi/V value (between 20 and 40) and the results of multiple element discrimination show that it has the characteristics of E-MORB. Therefore, it should be the product of oceanic crust formed by back-arc spreading. The lithochemistry data of basalt in Junmanong Formation indicate that it belongs to an oceanic island with an intra-oceanic hot spot environment.

The so-called Kagong Group in Northern Lancang River Junction Zone is significantly deformed and is a stratified disordered group complex, in which S0 has been replaced by tectonic foliation and axial flow cleavage has developed. Therefore, it is the product of subduction, collision and extrusion mechanism. This deformation is closely related to the arc-arc collision between Jitang Residual Arc and Dongda Mountain-Zhuka Continental Margin Arc in the western margin of Qamdo Landmass. Late Triassic arc volcanic rocks unconformably cover the stratum in Northern Lancang River Junction Zone (such as Junda-Jueyong). The east side of the junction zone is composed of clastic rock-dacite-rhyolite arc volcanic rock assemblage from Jiaoba Mountain to Zhukabing Station and Late Triassic Type-S granite, and similar arc volcanic rock assemblage is still found in Riwoqê and Jitang on the west side of the junction zone, indicating that subduction and destruction of

the oceanic crust of Northern Lancang River is characterized by two-way subduction. Similar to the two-way subduction of oceanic crust that occurred in Maluku Sea in Southeast Asia today, this two-way subduction is the only example of arc-arc collision occurring in the global tectonics.

A significant superposition of left-lateral ductile shear occurred in the Northern Lancang River Zone in Early Cenozoic.

Volcanic-Magmatic Arc in Riwoqê-Dongda Mountain

Controlled by two-way subduction of Northern Lancang River Ocean Basin in the east and west directions, volcanic-magmatic arc in Riwoqê-Dongda Mountain was developed on the western margin of Qamdo Landmass, and the molasses in the lower bed of Middle Triassic stratum unconformably overlaid on the island arc volcanic rocks dominated by Upper Permian basalt and andesite. And then, a sequence of Middle Triassic rhyolite, dacite and few pyroclastic rocks was developed. With w (SiO_2) of 67.26–76.91%, w (TiO_2) of 0.13–0.6%, w (Al_2O_3) of 11.96–14.52%, w (CaO) of 0.61–1.83%, w (K_2O) of 3.19–6.17%, w (K_2O) > w (Na_2O), it is highly rich in potassium, and rocks is of aluminum supersaturated type and calc alkali series. With W (Rb) of (143–190) × 10^{-6}, w(Sr) of (35–295) × 10^{-6}, w(Ba) of (227–1039) × 10^{-6}, w(Th) of (15–37) × 10^{-6}, $\sum$REE of (154.81–680.46) × 10^{-6}, it is rich in LREE; with w(Sm)/w(Nd) of 0.15–0.22, it is similar to that of Type-S granites. Watering Formation in Upper Triassic stratum is composed of a sequence of andesite, alkaline basalt, potassium basalt and few trachyte and single-spotted chalcocite. With w (SiO_2) of 48.36–72.07%, w (K_2O) + w (Na_2O) of 6.81–12.1%, w (K_2O) > w (Na_2O), w(Al_2O_3) of 11.00–16.10%, w(TiO_2) 0.55–0.90%, it is enriched in Rb, Sr, Ba, with $\sum$ REE of 439.7 × 10^{-6}, w (La)/w (Yb) of 21.33, w (Sm)/w (Nd) of 0.16, belonging to LREE-significantly enriched type and the product of the late continental margin collision orogeny.

Dongda Mountain granite batholith is characterized by type-I subducted granite, and quartz monzogranite, quartz diorite and tonalite and other small rock masses in Late Permian to Early-Middle Triassic stratum are distributed in the main granite batholith of Late Triassic and its margins in scattered shape. Late Triassic granite batholith is composed of 3 units, namely plagioclase granite, granodiorite and monzogranite. The Rb–Sr age of granodiorite is between 215.5 and 219.6 Ma and the K–Ar age of monzogranite is 194 Ma. With $w(K_2O)/w(Na_2O)$ > 1.5, A/CNK of 1.1–1.2 and standard mineral C of 0.7–4.97%, the rock is of the aluminum supersaturated series. They all are included in the Type-S granite zone in the A-C-F diagram. w (Sm)/w (Nd) is between 0.14 and 0.17 and the initial value of $^{87}Sr/^{86}Sr$ is between 0.7125 and 0.7233. All characteristics above indicate that the main granite of Dongda Mountain is Type-S granite after syn-collision. A series of tectonic patterns inclined westward and thrust eastward, such as tectonic foliation, pinacoidal surface of overturned fold with same inclination, and ductile shear zone, can be found in Dongda Mountain-Zhuka.

Changning-Menglian Junction Zone

It can be divided into three different tectonic-lithofacies zones, different melange degrees and different deformed and metamorphic zones from west to east (Fig. 2.23).

(1) **Gengma-Cangyuan Tectonic Deformation Zone**.

This zone is mainly composed of Devonian-Triassic strata. The Devonian Wenquan Rock Formation is mainly composed of a sequence of turbidite sediments of continental uplift slope submarine fan, such as graptolite shale and arenopelitic-siliceous rocks. The lower bed of Early Carboniferous Pingzhang Formation is dominated by andesite basalt, and the upper bed is composed of three sedimentary environments: turbidite channel sediments, carbonate rock plateau sediments without terrigenous clast on ocean islands, and carbonate rock platform of landward seamount. The Late Carboniferous-Early Permian Yutangzhai Formation is dominated by oceanic island carbonate rocks regionally, but locally overlaps on Pingzhang Formation or Early Devonian Wenquan Formation. The Upper Permian Nanpihe Formation is composed of a sequence of arenopelitic clastic rocks, siliceous rocks, argillaceous-siliceous rocks, carbonaceous shale, thin limestone and breccia limestone. However, as Permian radiolarians are developed in siliceous rocks, the age of Nanpihe Formation is still controversial. Radiolarians from Late Devonian to Early Permian are found in clastic siliceous rocks in Kenong in the southeast of Mengsheng Town. Sporopollen from Late Devonian to Early Carboniferous is found through analysis of the prophyte fossils. Recently, we have found Early Carboniferous radiolarian Albaillella sp., Entactinosphaera foremanae and Schrfenbergia turgida in the siliceous rocks in the eastern member of the section. The area around Sipaishan, Gengma is complex in tectonics and developed in shear thrust tectonics, so it is not a continuous section. Nanpihe Formation may originally include part of Late Devonian and Early Carboniferous strata. Papai Formation containing Late Permian and Early Triassic radiolarians, Early Triassic bivalves and ammonites is similar to Nanpihe Formation in lithologic assemblage and sedimentary characteristics. Devonian, Carboniferous and Early Permian radiolarians are also included in the siliceous debris of clastic siliceous rocks. Its horizon may be partly equivalent to that of Nanpihe Formation. The clastic flysch, limestone and thin argillaceous limestone and siliceous rocks outcropped in Ximeng, Cangyuan, Gengmadashan-

Mengtong are a sequence of passive continental margin sediments from the end of Precambrian to Early Paleozoic, and no unconformity relationship with the overlying Late Paleozoic strata has been found so far. It seems that the east side of Baoshan Block has a long history of passive margin development from Early Paleozoic to the end of Late Paleozoic. Parallel unconformity of Upper Permian and the underlying strata and the inclusion of D-P_1 radiolarian siliceous debris with different ages in the clastic siliceous rocks from Late Permian to Early Triassic indicate that Late Permian Changning-Menglian Ocean Basin has been subducted and destructed into the development stage of the foredeep depression, and the eastern basinal zone has been partially uplifted and denuded. Nanpihe Formation and Papai Formation may be formed in this foreland depression zone, because the only source of siliceous rock debris containing D-P_1 radiolarian can be the eastern basin facies zone, and no D-P_1 radiolarian siliceous rock sediments are found in Baoshan Block on the west side.

(2) **Ophiolite Melange Zone in Niujingshan, Shuangjiang-Manxin, Menglian**.

The zone is narrow in the north but wide in the south, and has been outcropped in Changning-Menglian, and then extended out of the frontier southward. Most strata are in a disorderly and mixed state. In Manxin, Menglian, the stratigraphic sequence in some areas is relatively well preserved. It was basically a sequence of arenopelitic-siliceous rock formation from Devonian to Middle Triassic, and radiolarian fossils D₁-P₁, *Archocyrtium menglianesis* Wu, *Ar. delicatum* Cheng (C_1) were found in a sequence of not-too-thick strata, which indicates that most of the basins were in a starved state or some of them were faulted. At the same time, a sequence of ocean ridge/quasi-ocean ridge basalt (C_1), ocean island basalt and plateau-type carbonate rocks (C_1-P_1) formed on oceanic islands are outcropped in the zone. They can form the basin lithofacies zone together with the above arenopelitic-siliceous rocks. Changning-Menglian Ophiolite Zone is mainly distributed in Tongchangjie-Manxin and Yingpan-Yutangzhai-Gengma on its west side. It is further confirmed through relatively systematic research that there are not only ocean ridge/quasi-ocean ridge volcanic rocks, but also oceanic island type and continental margin rift-type basic volcanic rocks in this zone (Fig. 2.24), and the temporal and spatial distribution relationship between them is preliminarily clarified. Ocean ridge volcanic rocks are mainly found in Tongchangjie and Manxin, continental margin rift basalt is mainly found in Yingpan, Yutangzhai and Gengma, and ocean island basalt is found in Tongchangjie and Yiliu. Ocean ridge volcanic rocks are dominated by tholeiite series,

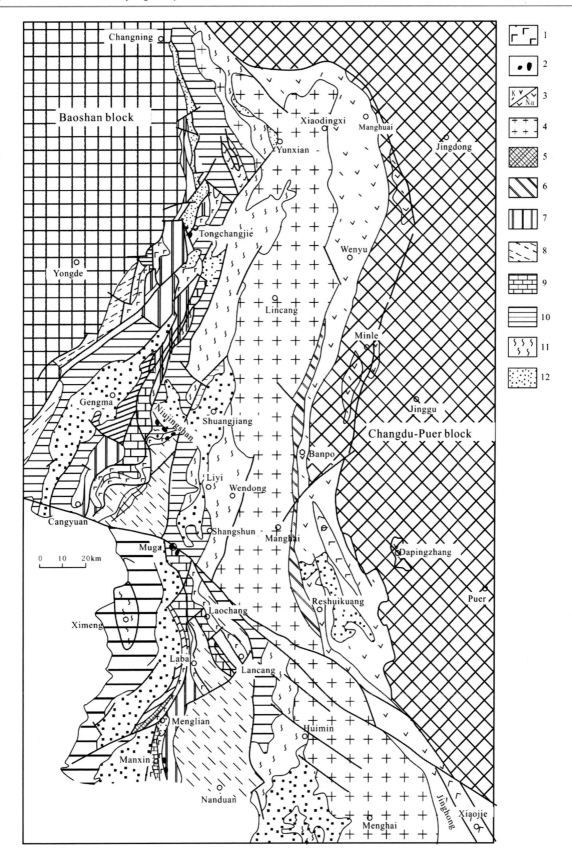

Fig. 2.23 Geological sketch of changning-Menglian-Southern Lancang River zone. 1—Ocean ridge and ocean island basalt; 2—Ultramafic rock; 3—Potassium/sodium arc volcanic rocks; 4—Magmatic arc granite; 5—Bathyal-abyssal facies of landmass and passive margin; 6—Back-arc-basin facies (turbidite); 7—Deep-water facies of ocean basin; 8—Bathyal-abyssal facies; 9—Ocean Island and carbonate rock plateau; 10—Accretionary wedge; 11—Significant metamorphism and deformation zone; 12—T₃-Q stratum

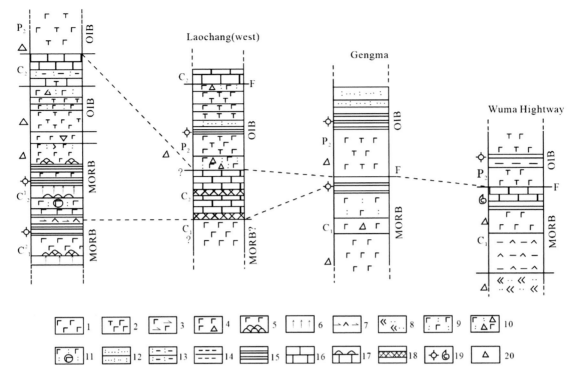

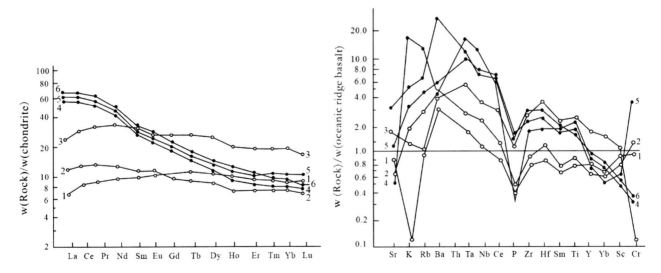

Fig. 2.24 Sequence comparison of late Paleozoic volcanic rocks in changning-Menglian zone (Shen 2002). 1—Basalt; 2—Trachytebasalt; 3—Olivine basalt; 4—Amygdaloidal basalt; 5—Pillow basalt; 6—Picrite; 7—Cumulate; 8—Metamorphic peridotite; 9—Basaltic tuff; 10—Basaltic tuffaceous breccia; 11—Basaltic agglomerate; 12—Tuffaceous siltstone; 13—Tuffaceous mudstone; 14—Mudstone and shale; 15—Siliceous rock; 16—Limestone; 17—Paleo-karst surface; 18—Mine; 19—Radiolarians and other fossils; 20—Sampling position. OIB—ocean island basalt; MORB—mid-ocean ridge basalt; F—fault

including quartz tholeiite, olivine tholeiite and alkaline basalt containing hypersthene, which are associated with metamorphic peridotite (serpentinite), metamorphic cumulate (metamorphic pyroxenite, pyroxenite), metamorphic diabase and radiolarian siliceous rock. Lithochemistry and geochemistry characteristics indicate that these basalts belong to ocean ridge basalts (Figs. 2.25 and 2.26). REE distribution patterns of olivine tholeiite and quartz tholeiite in Manxin, metamorphic olivine tholeiite and quartz tholeiite in Tongchangjie are nearly flat, with $w(La)/w(Yb)_N$ of 0.84–

Fig. 2.25 Distribution pattern of rare earth elements in volcanic rocks and spider diagram of trace element in changning-Menglian. 1–2 refer to ocean ridge basalts; 3–6 refer to ocean island basalt

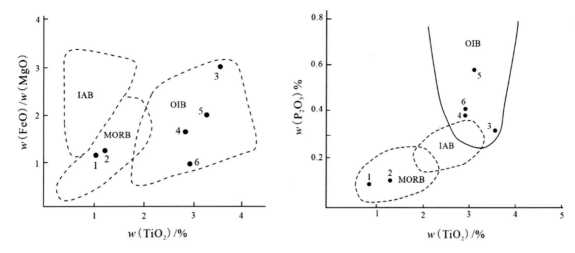

Fig. 2.26 Identification map of volcanic rock tectonic environment in changning-Menglian (Shen 2002). IAB- island arc basalt; MORB—mid-ocean ridge basalt; OIB—ocean island basalt 1–2 refer to ocean ridge basalt; 3–6 refer to ocean island basalt

1.87 and no Eu anomaly, so it basically belongs to T-MORB type; sub-alkaline picrite basalt, olivine tholeiite (the horizon may be slightly higher than before) and alkaline basalt containing Hy in Manxin are of moderately-enriched LREE type, with $w(La)/w(Yb)_N$ of 5.18–9.32 and no Eu anomaly, so it basically belongs to E-MORB type. Oceanic island volcanic rock is composed of alkaline basalt series, including alkaline olivine basalt, picrite basalt, basanite, potassic trachybasalt and sodium trachyte basalt, with lithochemistry characteristics of ocean island basalt (Figs. 2.25 and 2.26). Since the lithochemistry and geochemistry characteristics of ocean island basalt are similar to those of continental overflow basalt, only oceanic island carbonate rocks (no terrigenous clast) developed on the basalt in Yiliu and the basalt associated with ocean ridge/quasi-ocean ridge basalt in Manxin are identified as ocean island basalt. Although alkaline basalts distributed in Yingpan, Yutangzhai, Mengsheng-Xiaoheijiang River area in the west have similar lithochemistry and geochemistry characteristics to ocean island basalts and some of them are covered by carbonate rocks, it is also necessary to make a further study to confirm whether they are ocean island basalts. They are mainly distributed in the passive margin zone of the eastern margin of Baoshan Block, associated with clastic flysch, siliceous rocks and carbonate rocks in the passive margin zone, so they are temporarily classified as continental margin rift type. As for the alkaline basalts on Wuma Highway in Laochang and Tongchangjie, which have been determined by radiolarians to be of Late Permian, from the geological development history of the whole area, the tectonic background of their formation is different from that of oceanic island and passive continental margin rift, which may be formed in the period of subduction and forthcoming destruction of oceanic crust and in the period of foreland

fault block uplift on the original passive margin. The formation age of volcanic rocks in this zone was controversial before. It can be confirmed by observing the section and studying the biological fossils that ocean ridge/quasi-ocean ridge basalt was formed in Early Carboniferous. Early Carboniferous radiolarian fossils are found in the associated siliceous rocks in Manxin, and basalt is locally found under Middle and Late Carboniferous strata. The metamorphic volcanic rocks in Tongchangjie area were originally identified to be formed in Early Carboniferous, and the K–Ar age of amphibole in gabbro is 383 Ma (Zhang et al. 1990), which indicates that it existed until Late Devonian. Metamorphic basalt in the borehole in Laochang area may be similar to that in Tongchangjie area, and Middle Carboniferous conodont fossils *Niognathodus symmetricus* (Lane), *Hindeodilla* sp. are found in the limestone on it. It can also be determined that the ocean island basalt is covered with carbonate rocks of Middle Carboniferous in Yiliu, so it was undoubtedly formed in Early-Middle Carboniferous. The volcanic rocks in the rift valley of the continent on the west side were mainly formed in Early Carboniferous, with basalt being underlaid the limestone in Lower-Middle Carboniferous strata at the place 14–15 km away from Cangyuan (Mengsheng) Highway. Alkaline basalts in Laochang and Wuma Highway were formed in Late Permian just as mentioned above. In addition, the gravity flow slump breccia limestone is agglutinated or wrapped by alkaline basalts locally at the end of Early Permian in Manxin area. This basalt may also be formed in the Late Permian. All types of basic rocks, serpentinite, metamorphic cumulate, peridotite, basic lava, pillow basalt, amphibolite schist, glaucophane, two-mica quartz schist, phyllite, various limestones, siliceous rocks of different ages and other rock assemblages with different sizes, different genetic types and different metamorphic deformation characteristics are in contact with each other by taking tectonic shear as the interface, and disorderly and fragmented melange can be found in the field. It is particularly worth pointing out that after the predecessors discovered the high-pressure blueschist in the late 1980s, Yunnan Institute of Geological Survey carried out a survey on a scale of 1:250,000 in Lincang County and Gunlong in recent years, and found and determined three high-pressure glaucophane schist zones in Changning-Menglian Subduction Complex Zone for the first time, and confirmed the existence of Ophiolite in Niujingshan.

(3) **Fore-Arc Accretionary Wedge Thrust Zone**.

This zone is mainly composed of proximal turbidite in the southern formation of Carboniferous and fore-arc slope turbidites of Permian Laba Formation. The turbidite and siliceous rock sedimentary bed of Laba Formation generally contains intermediate-basic and intermediate-acid pyroclastic rocks and tuff interbedded with basaltic andesite cuttings; moreover, calcirudyte formed by gravity flow is found in many places. The limestone breccia contains C_2-P_1 fossils, and the siliceous rocks contain Late Permian radiolarian *Follicucullus* assemblage and *Neoalballella optima* assemblage. There are also Late Permian ammonites and bivalves buried in situ, and Middle Triassic radiolarians are found in the upper siliceous rocks. In the southern formation, it unconformably overlaps on Lancang Group. The data above show that Lincang-Jinghong Remnant Arc Zone in the east was in a continental uplift state in Carboniferous, with C_2-P_1 slope formed, slope turbidite developed and carbonate gravity flow accumulated in the west margin of the island arc, and intermediate-acid volcano erupted in Permian. This tectonic lithofacies zone may have extended to Shuangjiang area to the north, and then outcropped in Xiaoheijiang River Bridge as a sequence of sandstone, slate, and carbonate rock and breccia field stone (no fossils are found) accumulated by gravity flow in gully distributed from north to south; the sedimentary characteristics are similar to those of the southern member and Laba Group, indicating that the eastern member of Changning-Menglian area was adjacent to the margin of continental margin arc in Carboniferous, and reworked into fore-arc-basin in Permian, and then reworked into fore-arc accretionary wedge in the Early Middle Triassic.

Lincang Magmatic Arc

The Stratum between the Lancang River Tectonic Zone and Changning-Menglian Tectonic Zone is characterized by wide outcropping of intermediate-acid intrusive rocks, forming Lincang Composite Batholith Zone with the largest scale in Yunnan. The batholith zone generally extends from north to south, ranging from 10 to 48 km from east to west, with an average width of 22.5 km. The continuous outcropping length from north to south in Yunnan Province is 350 km, forming a very striking tectonic magma zone. Lincang granite is a wedge inclined to the west and slightly concave in the middle. Generally, the extension depth is relatively shallow in the east, about 0.5–2 km; while the extension depth is relatively deep in the west, about 8–12 km. There is a huge ductile shear zone with thrust nappe between the east side of the batholith and Upper Paleozoic and Triassic strata. The west side of the batholith is mostly in intrusive contact with the Mesoproterozoic Lancang Group. In addition, xenoliths containing metamorphic rocks of Lancang Group in granite can be found in many places on Douge-Mengku Highway. The batholith is often unconformably covered by Huakaizuo Formation of Middle Jurassic stratum. For many years, a lot of isotopic dating data have been collected: The Rb–Sr isochron age of the whole rock is between 236 and 348 Ma (Yunnan Bureau of Geology and Mineral Resources 1990), mainly varying from 279 to 297 Ma (Early Permian), and the U–Pb age of zircon is between 212 and 254 Ma (Early Middle Triassic), and the K–Ar age of single minerals is between 223 and 288. The batholith is mainly composed of medium-grained-coarse-grained biotite monzogranite, followed by biotite granodiorite. The chemical composition of rocks is relatively stable, and most of them are of the aluminum supersaturated series, namely faintly acid calc-alkaline rocks (some of them are of the normal series). Their acidity and alkalinity are lower than the average value of similar rocks, while the contents of compositions magnesium, iron and calcium are higher than the average value, indicating that the rocks are intermediate. Lincang composite granite batholith is dominated by Type-S granite, and also mixed with Type-I granite (Regional Geology of Yunnan 1990), and the middle member (Lincang member) is characterized by the coexistence of Type-S granite and Type-I granite. The characteristic values of lithochemistry vary widely, indicating that the provenance of batholith is characterized by polyphyly and heterogeneity. In addition to substances in the upper and lower crust, there may also contain substances from the upper mantle, reflecting at least that the depth of the provenance area where the partially remelted granite were formed during the long-term diagenesis is different. Intrusions formed by two mechanisms: one is the anatectic highly-intrusive rock formed under the control of dike spreading, that is, hypomagma rises through the fracture to the upper magma reservoir and then cools down and crystallizes (there may be "thermal ballooning" magma intrusions locally); the other is that with the repeated opening activities of faults, magma migrates from the deep crust to the shallow crust for many times, finally forming a whole rock masses with the rock mass extending along a certain direction, namely a large pluton with irregular plane and uniform internal tectonics.

The early stage of batholith formation was the subduction arc-building period (P_1), during which the lower crust was partially melted to form intermediate granitic magma possibly mixed with substances from the upper mantle, and then a rock mass dominated by Type-I granite (this rock mass contains two-pyroxene granulite, hypersthene granulite or granulite enclaves) was formed with spreading and rising to the middle and upper crust of the fault. With subsequent intensification of the arc-land collision and extrusion, melting occurred along different horizons of the crust (mainly in the middle and upper bed of the crust) and along the ductile shear zone, forming a large-scale arc-land collision-type intermediate-acid magma, which rose and invaded to form the main body of Type-S granite. According to the occurrence of Type-S granite in batholith, there should be two types: high intrusion and para autochthone, both of which shall have the same mechanism and homology.

It shall be pointed out that "satellite-shaped" rock mass was outcropped in Mengsong and Bulangshan at the southern end of Lincang-Menghai Granitic Batholith, of which the lithology is quite special and dominated by muscovite granite and two-mica granite; and the lithology of each rock mass is quite different. The lithology is dominated by muscovite granite, two-mica granite, and mixed with monzonitic granite locally. Biotite is represented by siderophyllite. Being rich in Si and Al but poor in Fe and Mg, muscovite belongs to phengite. Besides zircon and apatite, the accessory minerals include scheelite and cassiterite. This type of rock mass is closely related to tungsten and tin mineralization, and the tin ore deposit in Mengsong is related to this kind of granite. This kind of rock mass belongs to alkaline rock, accounting for 35.7%, which is the magma product of post-orogenic spreading.

Southern Lancang River Melange Zone

This zone is distributed along the Lancang River Valley. According to the observation of the outcropped tectonic-stratum (about 4 km) in the west of Reshuitang, Simao-Lancang Highway, by taking a sequence of dark gray thin-layer shale and gray medium-grained-coarse-grained turbidity greywacke as matrix, the zone is mixed with limestone and siliceous rock lenses, and also contains basic volcanic rocks and ultrabasic rocks locally, among which limestone rock contain Maokou fossils. There are also gray thick-layer medium-grained-coarse-grained lithic sandstones, gray thin-layer fine-grained lithic greywacke, gray thin-layer fine-grained quartz sandstone, and dark gray,

greyish-green thin-layer shale, all of which have been sheared significantly, so it is difficult to restore their original sequence. However, there are still tectonic stratigraphic sequences with weak strain, so some scholars think that this sequence of strata is mainly composed of abyssal turbidite sandstone, and also mixed with clastic sandstone related to contour current and argillaceous rocks in distal-basins. Turbidite sandstone is closely related to volcanic island arc, and the rock debris content is more than 50%.

The abyssal sediments between Lincang Magmatic Arc and volcanic arc on its east side are incomplete due to tectonic damage, and partly covered due to the overlapping of Triassic volcanic arcs. As a result, it is difficult to restore the basin prototype of this abyssal basin distributed along the present Southern Lancang River Fracture Zone. Some scholars believe that it is a fore-arc-basin sediment (Nan Jinhua, doctoral dissertation 1993), while we think it may be a back-arc-basin based on the regional analysis. The reasons are as follows:

(1) In the western margin of Pu'er Block, Daxinshan Formation composed of Late Paleozoic strata is a tectonic stratum, in which the LRE distribution pattern of Late Permian basalt is flat type, with the depletion of incompatible elements such as Rb, Ba, U and Th, etc. and the enrichment of elements such as Ti, Co and Ni, etc., showing the characteristics of oceanic crust tholeiite. This fact indicates that the eastward subduction of Tethyan Ocean in Changning-Menglian has formed a tectonic process of back-arc ocean basin spreading on the east side of Lincang Magmatic Arc.

(2) Basic and ultrabasic rock mass intruded in groups and zones along Southern Lancang River Zone from Chahe, Waili, Banpo in the west of Jinggu County, as well as Nanlian Mountain, Manshuai, Manhuai, Manshan, etc. in the south of Jinghong County. In the field of Waili, significantly foliated dunite (serpentinite) is sandwiched with bedded gabbro (cumulate), and basic and ultrabasic rock tectonics intrude into Late Paleozoic strata, mainly Daxinshan Formation. Because the ophiolite in the inter-arc deep basin was significantly decomposed during the orogenic process, it was destroyed to be incomplete and some tectonics intruded into the volcanic zone in Yun Country-Jinghong Arc.

(3) Nanguang Village which is located in the southeast suburb of Jinghong and near Lancang River is the coarse clastic rocks of Nanguang Formation of Middle-Upper Devonian strata, which was once considered as the symbol of Caledonian orogeny, but it was actually a sequence of Late Devonian abyssal submarine fan glutenite after field research by, which marked the initial spreading of Southern Lancang River

back-arc-basin. Pillow lava and Early Permian radiolarian siliceous rocks (*Follicucullus* sp., *Pseudoalbaillella* sp.,) were found in Jinghongxiaojie (Fig. 2.27). The geochemistry study of rocks shows that an ocean ridge volcanic zone of tholeiite series existed along the Daxing Mountain-Xiaojie line, of which the pattern diagram of rare earth elements and trace elements (Fig. 2.28) is consistent with that of N-type slowly spreading ocean ridge basalts (Shen 2002). According to the analysis of the distribution of Daxinshan Formation starting from the west piedmont of Wuliang Mountain and running through Anle, Minle, Yongping, Mengyang and then extending into Padang where it extends into Myanmar, Daxinshan Formation in Reshuitang (P_1) should be an interarc and abyssal sedimentary basin between Lincang Magmatic Arc and volcanic arc on the east side. Geochemistry study on trace elements shows that the primitive sedimentary characteristics of basalt in Reshuitang and volcanic rocks in intra-oceanic arc (Shen 2002) are distal abyssal turbidite mixed with basic lava, which represents the significant post-arc spreading occurred in the early stage of Early Permian. Volcanic rocks and continental arc volcanic rocks in Daxing Mountain-Reshuitang Intra-oceanic Arc show different geochemistry characteristics (Table 2.2).

The characteristics of geophysical field and the evidence of field geological observation show that a large-scale thrust nappe ductile shear zone is found along the east side of Lincang-Menghai Granite Batholith Zone, which shows the thrust of granite from west to east. The geological interpretation of gravity data inversion shows that granite is a rootless nappe. The same understanding has been obtained based on the recent magnetotelluric sounding data. A ductile shear zone is found between granite and Daxinshan

Formation at the fault on the east side of Lincang Granitic Mass along the river in Qianliu and Mangpa, Lancang County. On the one hand, due to the significant reworking and compression of the later tectonic orogeny, Lancang Fracture Zone shows significant thrust nappe characteristics from west to east, that is, a large mylonite zone was formed along the fracture zone, indicating that it has the nature of a ductile shear zone formed by deep tectonic orogeny and makes Huafeng rock mass in Lincang be a rootless nappe. On the other hand, a strike-slip fault with a curved vertical section is found along Lancang River Valley, which was considered by Wang to be the product of intense intracontinental reworking and deformation in Himalayan.

Yunxian-Jinghong Volcanic Arc

The volcanic arc tectonic rock zone is located between the Lancang River Junction Zone and Jiufang Fracture, and the oldest exposed stratum in the zone is the Middle-Upper Devonian, which is distributed in Jinghong-Nanguang. According to plant fossils and pyroclastic rocks, the Nanguang Formation in Upper Devonian was considered as continental sediment. The on-the-spot investigation reveals that there are mainly a sequence of fluvial sand body sandstone, submarine fan conglomerate, fine clastic turbidite, siliceous turbidite and siliceous rocks, with well-developed Bouma Sequence. It is equivalent to the sediment of low-stand system tract and the condensation section in the maximum flooding period, reflecting that Nanguang Formation is a sedimentary environment from slope to basin margin. There are carbonized plant imprints in the sandstone and volcanic breccias in the breccias, reflecting an ancient land uplift with volcanic activity to the east, and Lamping-Pu'er Basin may have an Early Paleozoic folded basement beneath it as in Qamdo Basin. Carboniferous-Permian is an island arc volcanic-sedimentary formation composed of sand-slate, marl, limestone, pyroclastic rock, basalt, andesite and

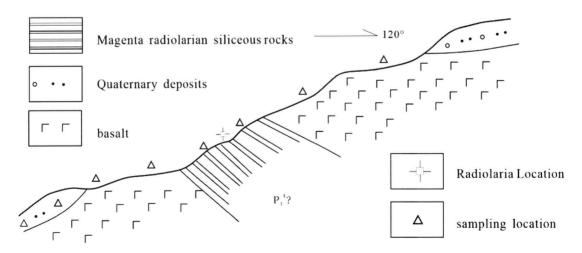

Fig. 2.27 Profile of the place 500 m away from Xiaojie, Henan and Dongzhai in Northwest direction (Shen 2002)

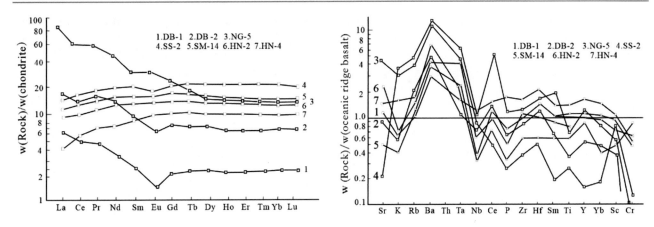

Fig. 2.28 Pattern diagram of rare earth elements and trace elements in rift-ocean ridge volcanic rocks of Southern Lancang River zone (Shen 2002)

Table 2.2 Comparison of geochemistry characteristics of three types of arc volcanic rocks in Southern Lancang River zone

Environment	Intra-oceanic arc volcanic rocks (P_1^1, middle-south member)	Continental arc volcanic rocks (P, south member)	Intracontinental arc volcanic rocks (T2-T3, northern member)
Series	Low-potassium tholeiite-medium-potassium calc-alkaline rocks	Low-medium potassium calc-alkaline rocks	Shoshonite-high-potassium calc-alkaline rocks
Rock types and distribution ages	Q-TH-BA-A-D, $P_{1\text{-}2}$, daxing mountain	A P1, Simao-Lancang BA-A-λ P_1, Bangsha A-D P2, Jingha	K-TB-B-HK-λ, T_3, Xiaodixing SHO-La, T_3, Xiaodingxi HK-λ, T_2, Xiaodingxi SHO-La-Tra, T_3, Wenyu HK-λ, T_2, Minle
(w(La)/w(Yb)) NREE type	1.91–5.13, weak enrichment	2.02–2.62, weak enrichment	5.32–15.30, medium enrichment
Pearce curve	Depletion of Nb, Hf, Ti, Cr, k, Rb Enrichment in Th, Ba, Ta	Depletion of Nb, Ti, Cr Enrichment in Rb, Ba, Th	Weak depletion of Nb, P, Ti, Cr Enrichment in K, Rb, Ba, Th
w(Ti)/w(V)	20 ~ 40	20 ~ 40	20 ~ 65
w(TiO₂)/w(P₂O₅)	6–2, low contents of Ti and P	6–2, low contents of Ti and P	6–2, medium contents of Ti and P
w(Ti)/w(Zr)	< 60, low contents of Ti and Zr	< 60, low contents of Ti and Zr	< 60, medium–low content of Ti, high content of Zr
w(Ta)/w(Yb), w(K)/w(Yb)	< 1.5, relatively low	< 1.5, relatively low	< 1.5, relatively low
w(Nb)/w(Y)	< 0.67	< 0.67	0.67 ~ 1.17
w(TiO₂)/%	Low content of TiO₂, < 1.3	Medium and low content of TiO₂, 3–1.3	Medium and low content of TiO₂, < 3
Associated rock	Graywacke and mudstone interlayer, pyroclastic rock	Siltstone, Siliceous siltstone and mudstone	Siltstone, mudstone interlayer, Pyroclastic rocks

Mo et al. (1993), revised and supplemented

rhyolite. In Bangsha, Jinghong, a sequence of volcanic-sedimentary rock series of continental margin arc was originally in the Middle Triassic, and the Late Permian radiolarian Neoalbaillella ornithoformis assemblage was found in its siliceous rocks. There are Carboniferous-Permian island-like isolated carbonate plateau and deep-water trough of intra-arc rift in the area around Ganlanba, which has typical island arc geomorphological characteristics.

The Middle and Upper Triassic stratum is a sequence of volcanic-sedimentary rock series dominated by volcanic

rocks, with the lower volcanic rocks being acidic and the upper volcanic rocks being lagged arc volcanic rocks. This volcanic-sedimentary rock zone spreads from Xiaodiding and Minle to Nanguang in the south. The Middle Triassic Manghuai Formation and Upper Triassic stratum are the Xiaodixing Formation and Manghuihe Formation. According to the research of Zhang the volcanic eruption in this sequence of strata can be divided into three cycles: The first cycle (Manghuai Formation) formed a sequence of high-potassium rhyolitic volcanic rocks; the second cycle

(Manghuihe Formation) is caused by "bimodal" volcanic eruption, from rhyolitic volcanic rock → subalkaline basalt → medium-long basalt → potassium coarse basalt rhyolite; the geochemical element gradients and trace element diagrams of the first and second cycles show the tectonic environment of continental margin volcanic arc; the "bimodal" volcanic rocks in the third cycle show the back-arc rift environment after hard collision. In this volcanic-magmatic arc, besides the volcanic rocks mentioned above, there are some mafic–ultramafic rocks and diorite developed along the volcanic arc. Its rock assemblage and occurrence characteristics belong to Alaska mafic–ultramafic rocks and peridotite-diorite mafic–ultramafic rocks. The former is mainly distributed in Banpo, Jinggu, while the latter is mainly distributed in Jinghong-Damenglong. The lithochemistry characteristics are similar to those of the same type of California peridotite-diorite. Its general characteristics are similar to those of calc-alkalic basaltic-andesite in the orogenic belt and mafic rocks in some Alaska rocks. The formation of mafic–ultramafic and peridotite-diorite rock bodies in this area is directly or indirectly related to the source areas of magmatic activities formed by volcanic-magmatic arcs in this area. Combined with their close symbiosis in space, it shows that they should be the products of magmatic intrusion in island arcs/orogenic belts.

The tectonic deformation of this volcanic-magmatic arc is relatively significant, generally characterized by eastward thrust fractures, and the folds extending linearly in the same direction can be found. Near Lancang River Fracture, the Jurassic-Cretaceous strata have undergone local metamorphism and deformation to form slate, which formed primary bedding which is significantly replaced by axial cleavage and were mainly related to the late action of the Lancang River Fracture. The Late Paleozoic-Triassic strata have ununiform metamorphism and deformation, and the strata close to the Lancang River Fracture have deepened metamorphism and increased deformation, some of which can reach slate and phyllite, and the highest grade of metamorphism can reach schist with low greenschist facies. It can be seen from the highways in East Jinghong, the part with significant metamorphism and deformation is often closely related to thrust nappe, and a series of imbricate thrust nappe tectonics and synclinal inverted folds with westward reversal and eastward axial tilt are formed, which indicates the kinematic characteristics of nappe from east to west, which may be related to the superimposed transformation of Jinghong left strike-slip fracture in NWW direction.

2.2.7.2 Evolution of Tenasserim Arc-Basin System

It is difficult to understand the tectonic evolution of Tenasserim Arc-Basin System only by a simple open-close evolution. If the Bangong Lake-Nujiang River-Changning-Menglian Ocean Basin is only considered to be formed from the early period of Mesozoic to Paleozoic, it is difficult to properly explain the geological processes in each period in this zone. In our model, the Bangong Lake-Shuanghu Lake-Nujiang River-Changning-Menglian Mage-suture Zone is defined as the northern boundary of Gondwana. As the relic of the continuous evolution and destruction of Proto-Tethys to Paleo-Tethys, Qamdo-Lanping-Pu'er Block was the southwestern margin of Yangtze Landmass in the Early Paleozoic. Jitang Group (including Xixi Group), Chongshan Group, Lancang Group, etc. are all formed by the subduction of the Proto-Tethyan Ocean in the western margin of Yangtze Landmass to the accretionary wedge. In essence, these metamorphic bodies include the products of the convergence and collision of Neoproterozoic Rodinia supercontinent, which can disintegrate island arc volcanic rock-pyroclastic rock assemblage, Early Paleozoic Suyi-Nanlang blueschist assemblage, and residual arc volcanic rock assemblage such as Youxi Group.

Baoshan Block on the west side of Changning-Menglian Junction Zone has been transformed into plateau evolution since Late Cambrian, while the sediment of Gengma-Cangyuan passive continental margin in Early Paleozoic has been developed on the east edge of Baoshan Block. From Late Cambrian (Manggao Rock Formation) to Silurian (Mengdingjie) group complex, flysch sediment occurred. There are typical turbidites in Mengdingjie Group Complex, which reflects the ocean in the east of Baoshan Block. This ocean is the Proto-Tethyan Ocean in the Early Paleozoic. Therefore, when analyzing and studying the geological records of the Proto-Tethyan Ocean in the northern member, the researchers should not only seek from the composition of Bangong Lake-Dingqing-Bitu Ophiolite Melange Zone, but also identify and understand the ocean from the basement, overlying strata, characteristics of MABT, similarities and differences of crustal lithosphere structure and the evolution of ocean-land transition. The consistency of Paleozoic plateau sediment in Gangdise and Himalayan periods reflects that there is a wide passive continental margin sedimentation in the northern part of India Continent, which is quite different from the active continental margin in the northern part of Bangong Lake-Nujiang during Paleozoic. Therefore, Gangdise Block, Tengchong-Baoshan Block (Danbang Microcontinent) and India are part of Gondwana Continental Group, while Qamdo-Lanping-Pu'er Block (Indo-China Microcontinent) and Yangtze Continent are part of Pan-Huaxia Continental Group.

The Tethyan Ocean is located between Gondwana Continental Group and Pan-Huaxia Continental Group since Phanerozoic and is an inherited ocean. The development conforms to the life cycle law of ocean evolution history of more than 600 Ma, rather than the closure of Proto-Tethys and the reopening of Paleo-Tethys.

The subduction of Tethys Ocean to the northeast in the early Late Paleozoic formed Tenasserim Frontal Arc. A series of stable blocks and island arcs behind the frontal arc and arcs to the ocean basin formed the MABT. The restored Paleo-Tethyan Ocean Basin of Dingqing-Zhayu-Bitu-Changning-Menglian includes ocean ridges, abyssal basins, ocean islands, intra-sea arcs and intra-oceanic arcs. The sea mountain carbonate plateau is a sign of the shrinking of the ocean basin in Tethyan Ocean during ocean-land transition. The "bimodal" back-arc rift in Early Carboniferous in North Lancang River to the back-arc ocean basin, and the back-arc ocean basin in South Lancang River have only a few tens of millions of years of life. From Late Permian to Triassic, the back-arc-basin shrank and subducted in both directions, and magmatic arcs appeared on both sides, which led to the formation of continental margin volcanic arcs on the east side. The arc-arc collision in Late Triassic and basin-mountain transition in Jurassic-Cretaceous have been transferred to the intraplate evolution.

2.2.8 Space–Time Structure and Evolution of Baoshan Block (Northern End of Danbang Micro-Landmass)

Baoshan Block is located between Changning-Menglian Junction Zone and Nujiang River Fracture Zone (south member). According to the exposure of strata and sequence composition, it is generally a stable plateau.

At the end of Precambrian and the beginning of Paleozoic, there were significant tectonic thermal events in Baoshan Block: granite in Laojiezi, Ximeng, with an Rb–Sr isochron age of 687 Ma, two-mica granite in Zhibenshan, Luxi, with an Rb–Sr isochron age of 645 Ma (Zhang et al. 1990), and granite in Pinghe, with an Rb–Sr isochron age of 529.9 Ma, which may represent the product of Pan-Africa events. The strata of this landmass are well developed. The oldest stratum is outcropped in the western part of the landmass, namely Gongyang River Group (Z–$\in_2$), which is composed of bathyal and neritic sandstone, mudstone (shale) with a small amount of silty siliceous rock and thin limestone, which are slightly metamorphic. It was caused by rift development. It locally conformably contacts the Upper Cambrian stratum, and most part is overlapped by Lower Ordovician and Middle Permian strata. The Lower Paleozoic strata in the landmass is well developed, and it is the continuous sediment. It includes Hetaoping Formation ($\in_1$), Shahechang Formation ($\in_2$), Baoshan Formation ($\in_3$), Laojianshan Formation (O_1), Shidian Formation (O_{1-2}), Pupiao Formation (O_{2-3}), Renhe Bridge Formation (O_3–S_1) and Lichaiba Formation (S_{2-3}) from bottom to top, and is composed of neritic-tidal sandstone, shale, limestone with a

total thickness of 5000 m. Graptolite shale can be found in Renhe Bridge Formation. It has abundant paleobios and the mixed biological characteristics of North China and South China.

The Upper Paleozoic stratum is dominated by stable plateau carbonate rocks, which are deposited continuously with the underlying Silurian stratum, with a total thickness of about 3000 m, including Xiangyangsi Formation (D_1), Heyuanzhai Formation (D_2), Dazhaimen Formation (D_3), Xiangshan Formation and Pumenqian Formation (C_1), Dingjiazhai Formation and Woniusi Formation (P_1), Bingma Formation (P_2) and Shazipo Formation (P_3). During the late period of Early Permian (Woniusi Formation), there were continental margin marine basic volcanic rocks, the thickest of which was up to 700 m. There is a parallel unconformity between the Lower Permian stratum and the Lower Carboniferous stratum, and the Upper Carboniferous stratum is in lacuna. The lower part of Lower Permian-Middle Permian (Dingjiazhai Formation-Bingma Formation) has Gondwana mudstone sediment and cold-water animal molecules represented by Stepanoviella and Eurydesma, indicating its significant Gondwana affiliated characteristics.

The Triassic stratum is composed of Hewanjie Formation (T_{1-2}) and Nanshuba Formation (T_3) basically inherits the carbonate environment of the Late Paleozoic plateau, but the basin has become closed and dominated by dolomite. In recent years, a large number of Early Triassic conodonts have been found in the lower part of Hewanjie Formation, which is deposited continuously with Permian stratum. At the end of the Late Triassic, some intermediate-basic and intermediate-acid volcanic rocks were active. The Mengjia Formation in the lower part of Middle Jurassic stratum is composed of red conglomerate, sandstone and mudstone, containing gypsum-salt, locally mixed with basalt, and unconformity with Triassic stratum and its lower strata. The Liuwan Formation and Longhai Formation on Mengjia Formation are neritic limestone, sandstone and shale (mudstone). The marine environment ended after the Late Jurassic and was followed by significant intracontinental orogenic activity, which accumulated Paleogene red molasse coarse clastic rocks. In the Neogene, small fault basins developed along various fractures, forming limnetic-alluvial-lacustrine clastic rocks with lignite beds.

Baoshan Block can be further divided into three tertiary tectonic units (Li et al. 2002).

(1) Shuizhai-Muchang Foreland Depression Zone (P_2-T_3). The Late Permian–Triassic strata are mainly developed in this zone, and its lithofacies change from east to west, from continental to marine, and the Upper Triassic stratum overlaps eastward and unconformities on the underlying strata of different ages. This zone is a

foreland depression formed on the front margin of Baoshan Block on the west side of Gengma-Cangyuan passive margin thrust fold zone after Danbang Micro-landmass subducted eastward and Tethyan Ocean closed.

(2) Baoshan-Shidian Uplift Zone This zone is characterized by the lacuna of Late Permian stratum in Baoshan-Zhenkang Zone, and is an uplift zone formed in Mesozoic.

(3) Liuku-Mengjia Mesozoic Foreland Depression Zone. This zone was developed on the basis of another depression on the west side of Triassic Shuizhai-Muchang Foreland Depression Zone. In the Early Jurassic, Dangbang (Baoshan) Micro-landmass was obliquely subducted to the west, the Nujiang Ocean was closed, and Gaoligong Mountain was pushed to the east to form the frontal depression zone. This zone is composed of Middle Jurassic continental molasses, marine clastic rocks and carbonate rocks, and some submarine volcanoes erupted.

This landmass (including the whole Dangbang Micro-landmass) belonged to Gondwana Continental Group in the Paleozoic, but it is still inconclusive whether it is attached to Australia or India in terms of tectonic attributes. On the whole, this zone seemed to be connected with an adjacent continent but not closed in Paleozoic. According to the Cambrian-Neogene outcrop and the research results of paleogeography characteristics of the sedimentary basins by Luo, it can be roughly divided into 3 stages: ① Cambrian-Silurian: This zone was mainly composed of stable neritic clastic rocks mixed with carbonate rocks; ② Devonian-Early Permian: This zone was mainly composed of stable neritic clastic rocks and carbonate rocks, and only basic volcanic rocks appeared in Carboniferous-Permian, which generally showed deceptive conformity with the previous strata; ③ Mesozoic: The sedimentary layer was superimposed and unconformity on the underlying strata of different ages, and this zone was composed of a sequence of clastic rocks and carbonate rocks mixed with intermediate-basic and intermediate-acid volcanic rocks, with magnesium carbonate rocks at the bottom and red molasse gravel at the top.

2.2.8.1 Formation Stage of Stable Plateau (C-S)

At the end of Precambrian and the early and middle Cambrian, a sequence of flysch submarine fan turbidite with a thickness of several kilometers was formed in the continental marginal sea. There were no volcanic rocks, and the stratum was slightly metamorphic, with no large fossils. Siliceous rocks and a small amount of limestone as well as spongy spicules and micro-paleobotany can be found. Sediments became thinner from southwest to northeast, forming the base of carbonate plateau and siliceous clastic rock pad, and changed into neritic-shelf sediments in Late Cambrian-Early Ordovician, mainly siltstone, slate, marl and limestone, with more purple shale in the upper part, rich in trilobites, brachiopods, graptolites and other biological fossils. In Middle Ordovician, the neritic-shelf yellow mudstone and marl alternately deposited, and lacustrine argillaceous dolomite sediments occurred in Lux. In Late Ordovician, there was mainly a regressive sequence, characterized by tidal flat mottled aleuropelitic sediments, which later transformed into estuarine graptolite shale sediments. The actinoceras cephalopoda in Renhe Bridge, Baoshan was *Pararmocers* cf., which was common in Nyalam, South Tibet.

The Ordovician stratum in Gengma-Cangyuan-Ximeng on the east side of Baoshan Block is a plateau margin to continental slope, which is composed of sericite slate, lithic sandstone and siliceous rocks with thin limestone.

The three series of Silurian are complete, the black shale developed in Lower Silurian, and a sequence of continental margin deep-water reticulate limestone with graptolite shale developed in Middle-Late Silurian, which reflects the sedimentary environment from the outer shelf to plateau margin slope.

2.2.8.2 Formation Stage of Basin Series (D-P1)

The sediments in Early Devonian were mainly composed of sand, argillaceous carbonate rocks and bioclastic limestone, which belonged to neritic plateau basin facies, and contain fossils of thin shelled tabasheer and graptolites and brachiopod with thin shell and fine stria. They were anoxic, hydrostatic low energy organisms in a weak reducing environment. Middle Devonian sediments were mainly littoral gravel and sandy argillaceous carbonate rocks. The upper part formed a stable neritic carbonate sediment, while the water body on the east side became deeper, and the sediment became muddy or marly carbonate rocks. The salinity and temperature of the sea water were normal, and the benthos were flourishing, and some reefs were formed. Late Devonian sediments were mainly composed of microcrystalline limestone mixed with siliceous rocks, argillaceous limestone and black shale, which belonged to neritic basin facies, with fossils of conodonts and brachiopod with fine stria.

Early Carboniferous carbonate plateau facies is distributed in NS direction and can be divided into two sub-facies zones according to sedimentary characteristics: The first one is an open plateau sub-facies zone, which is distributed in Baoshan, Shidian and other places. Its sediments are composed of osseous pelsparite, flint-banded strip pelsparite, bioclastic limestone and oolitic limestone, etc., which are rich in marine benthic fossils, forming biological beaches or reefs locally; the second one is plateau slope or basin sub-facies zone, which is distributed in Qingshuigou and Guanpo, Baoshan. Its sediments are composed of dark gray medium-thin pelsparite, argillaceous pelsparite and slump breccia limestone. The slump breccia is of different sizes, irregular shapes and disorderly arrangement, supported by miscellaneous matrixes. The breccia is composed of biosparite and bioclastic limestone, and the matrix is gray to gray-black micritic, with slumping deformation bedding and horizontal bedding developed. There are many pyrite crystals and siderite nodules in the rocks, and there are few biological fossils, mainly benthic and planktonic biota, which represent the sediments in the stagnant environment of deep water. Middle Carboniferous regional uplift sedimentary can be found locally. The Late Carboniferous stratum was influenced by Gondwana Continental Glaciation, and the continental margin glaciomarine debris flow sediments were developed in this zone, which were composed of fine conglomerate, gravel-bearing coarse sandstone and gravel-bearing mud shale. The gravel components were granite, limestone, gneiss, sandstone, etc., and the biological assemblage from cold water to warm water developed. The main biota assemblages in the early period were Eurydesma sp., Stepanoviella sp., Wikingia sp., Schizodus sp., Trigonotreta sp., Marginifera sp. and Lytvolasma. Eurydesma sp. and Stepanoviella sp. are the representatives of typical cold-water biota assemblage, which not only can be found in this zone, but also were widely distributed in Gondwana, the Gangdise, Himalayas, Pakista Salt Ridge and other places, and coexisted with moraine or ice water sediments. The sediments associated with this assemblage in this zone were also ice-water debris flow conglomerate, so this cold-water biological assemblage was closely related to the glacial climate and glaciation in Gondwana. The main biota assemblages at the end of Late Carboniferous were Triticites, Qua-sifusulina, Fenstella, Squamularia, Neospirifer and Marginifera, which have warm habits.

In the Early Carboniferous, Baoshan Biogeographic Region was a mixed biota with environmental characteristics at that time, and a typical cold-water biota assemblage was developed in Late Carboniferous, so Baoshan biota was a part of Gondwana.

The Early Permian stratum was characterized by the developed littoral sediments on the uplifted denuded basement at the end of Carboniferous. It was composed of iron-bearing aluminum sand shale, with basal conglomerate and animal and plant fossils. The Late Permian and Middle-Late Permian were characterized by the developed plateau subfacies, which were composed of limestone, dolomitic limestone and bioclastic limestone, with oolitic limestone and breccia limestone locally and benthic fauna.

2.2.8.3 Formation Stage of Marginal Foreland Basin (P2-T3)

The Tethyan Ocean Basin in Changning-Menglian subducted at the end of Early Permian, and then destructed due to the arc-land collision in Middle Triassic, which can be found on the passive margin of Gengma-Cangyuan in the west of the zone, forming the Shuichang-Muzhai Margin Foreland Basin and Baoshan-Shidian Foreland Uplift and their sedimentary records. In the early period of Late Permian, Gengma-Cangyuan Marginal Foreland Basin was characterized by coal-bearing sand shale sediments, and composed of littoral-neritic molasse. In the late period, it was carbonate sediment. The Baoshan-Shidian Foreland Uplift in the west of the basin was largely devoid of Late Permian and Triassic sediments.

The Triassic stratum was deposited in Shuichang-Muzhai Marginal Foreland Basin. The Hewanjie Formation in the Early and Middle Triassic was mainly a sequence of shallow clastic rocks, dolomitic limestone and limestone interbedded, rich in low salinity bivalves and gastropods fossils, 300–900 m thick, and composed of the supratidal and lacustrine sediment, which was in deceptive conformity contact with the underlying lower Permian stratum. According to recent research, the lower part of this formation contains Early Triassic conodont fossils.

The Late Triassic Nanshuba Formation, consisting of yellow-green silty shale, mudstone mixed with marl, has bivalves Halobia pluriradiata and H.yunnanensis assemblage and ammonite Tropites assemblage, with a thickness of 800–1400 m, and was composed of neritic clastic rock molasse.

2.2.9 Space–Time Structure and Evolution of Boxoila Ling-Gaoligong Arc-Basin System

The Baxoila Ling-Gaoligong Arc-Basin System, that is, the part where Gangdise MABT spreads eastward to the south and bends (bulges from the north to the east), is mainly composed of Baxoila Ling-Gaoligong Frontal Arc on the west side of the Nujiang Ophiolite Melange Zone.

2.2.9.1 Formation Period of Nujiang Ocean Basin and Development Period of Gaoligong Mountain Magmatic Arc

Because of the significant eastward thrust of Gaoligong Mountain Magmatic Arc Orogenic Belt in Late Mesozoic and Early Cenozoic, the Nujiang Ocean Basin overlapped the ophiolite melange zone in the middle part of Nujiang River, and covered the volcanic-magmatic arc of Biluo Snow Mountain and Baoshan Block in the south member, forming a strike-slip oblique thrust ductile shear zone composed of mylonite. Many geological records are denuded and destroyed, so it is difficult to determine the period of formation. The northern member is Dingqing-Zhayu-Bitu Junction Zone, while in the south member, only the remnants of ophiolite melange zone are preserved in Santaishan, Luxi on the side of the northeast Longling-Ruili Right-handed Strike-Slip Fracture Zone. It can be seen that ultramafic rocks intruded in Triassic-Early Jurassic deep-water turbidite and had the characteristics of ophiolite melange. There are significant differences in stratigraphic records, sedimentation, magmatic activity and metamorphic deformation between Tengchong Arc-basin System and Baoshan Block in Late Paleozoic and Triassic, which reflects that the both split from the ocean. Therefore, it is speculated that the ocean basin was finally destroyed in the Middle Jurassic.

Boxoila Ling-Gaoligong Frontal Arc was a relic of continental margin arc in the northern margin of India in Carboniferous. It was mainly composed of Late Triassic-Cretaceous granitic batholith, namely Type I granite, with features of island arc, and ages of mainly 227–210 Ma and 129 Ma. According to the biotite granodiorite in Changma, Yongsong, the Rb–Sr isochron age is (195.3 ± 7.0) Ma. Therefore, the Late Triassic stratum was formed by subduction arc, and Jiali-Parlung Zangbo Inter-arc-Basin and Yarlung Zangbo Back-arc Ocean Basin developed in the south and west of the island arc.

2.2.9.2 Spatial Pattern of Boxoila Ling-Gaoligong Arc-Basin System

From east to west, Gaoligong Overthrust Zone (Santaishan ophiolite melange zone), Boxoila Ling-Gaoligong Magmatic Arc and Bomi Parlung Zangbo Arc-arc Collision Zone are developed in the arc system.

Gaoligong Overthrust Zone (Santaishan Ophiolite Melange Zone)

Gaoligong Overthrust Zone spreads from Gongshan to Longling for more than 350 km in NS direction, and its southern member is restricted by Longling-Ruili NE strike-slip fracture zone, which is generally characterized by dipping westward of Gaoligong Group tectonic foliation and dipping eastward of a series of imbricate thrust nappe tectonics. The original system is the sedimentary cover on Baoshan Block, including Gongyang River Group and the overlapping. The whole-rock Sm–Nd isochron age of the amphibolite in nappe zone is (194.2 ± 20) Ma, the K–Ar age of mylonite biotite in strike-slip fracture is 203 Ma, and the molasse in Middle Jurassic foreland basin contains ultramafic rocks, all indicate that the thrust started at the end of Early Jurassic.

Near the 80 km monument of Luxi-Ruili Highway, along Santaishan and Nongbing in Luxi, ultramafic rocks, siliceous rocks and limestone can be seen in rock blocks, mixed with rocks with the matrix of Late Triassic-Early Jurassic turbidite. The folded axial plane and permeable foliation turned eastward indicate that the melange has also been significantly deformed, forming the imbricated thrust layer.

According to the research by Wang, the Lower Carboniferous marble limestone in the west of Gaoligong Mountain and near Tuantian in the west of Longchuan River, is similar in lithology and biological characteristics to the Lower Carboniferous stratum in Baoshan Block, but obviously different from the Lower Carboniferous stratum in Tengchong Landmass, which may be the tectonic window exposed by Gaoligong giant nappe tectonics. It can be inferred that most of the foreland basins on the west side of Baoshan Block are pressed under the metamorphic rock series of Gaoligong Mountain, while the Nujiang River Junction Zone should be in the west of Tuantian, which was overthrust eastward by at least 20 km in Paleogene.

Boxoila Ling-Gaoligong Magmatic Arc

The magmatic arc is based on the continental crust. Except for the Precambrian Gaoligong Group and Guqin Group, the Lower Ordovician limestone under the overlying strata is only exposed in Guqin in the northern member, and Late Paleozoic strata constitute the surrounding rock invaded by magmatic rocks.

Yingjiang Area in the southern member of Devonian stratum was a fluvial-littoral sandy sediment in Early Devonian, with slow ascending and descending, a small sediment range and a thickness of 62 m. Besides marine organisms, there were terrestrial organisms such as heterostraci, charophytes and ostracods. In Middle Devonian and Late Devonian, the subsidence and transgression expanded, and aleuropelitic clastic rocks and carbonate rocks were deposited. The northern area of Basu-Laigu was the coastal detrital sediment, basically forming a stable carbonate plateau. The semi-pelagic slope zone was developed on the west side of Nujiang River in the north member. Sediments were mainly composed of terrigenous sand, argillaceous strip and layered limestone and argillaceous limestone, and a large number of slumping tectonics and flexible folding layers were developed, reflecting the deepening of seawater from south to north. Nujiang River was the oceanward side.

On the northeast side of Bomi-Tengchong Ancient Land, significant transgression occurred since Early Carboniferous, and on the basis of carbonate plateau, the stratum developed into semi-pelagic to abyssal sediments. From southwest to northeast, the lithofacies zone can be divided into littoral-neritic zone in the outer margin of Chayu-Tengchong Ancient Land (the sediments are composed of sandstone, siltstone, argillaceous ribbon limestone and bioclastic limestone); semi-pelagic zone in Songzong, Laigu, Ranwu and Ridong (slump breccia limestone, terrigenous and volcanic fine turbidite and intermediate-basic volcanic rock sedimentary assemblage are developed). This zone can be divided into two sub-facies zones: the upper slope sub-facies zone and the lower slope sub-facies zone. The former zone is developed in Ranwu and other places, and the sediments are composed of limestone with eye ball-shaped siliceous structure, nodular limestone, fine lamellar limestone, calcareous siltstone, shale, and a large number of slump breccia limestone, with common horizontal bedding, convolute bedding, large cutting bedding, slumping and buckling deformation bedding, etc. The latter zone is developed on the east side of upper slope collapse breccia, and can be found in Zhongba, Ranwu and other places. The sediment is a sequence of fine terrigenous and volcanic turbidite sedimentary assemblage, and composed of fine sandstone, siltstone and siliceous mudstone, with Bouma Sequence, and generally layers a, c and e can be seen. There are a large number of volcanic materials, and the thickness of the low-density turbidite series is more than 1000 m, including basalt, andesite basalt and volcanic breccia. According to the petrochemistry analysis, the volcanic rocks of Early Carboniferous Nuocuo Formation in Bomi-Parlung Zangbo are formed in the continental margin rift, which may represent an arc-basin.

Significant climatic and environmental changes occurred in this zone during Middle-Late Carboniferous to Early Permian, and littoral-neritic gravel sandstone and sand-slate sediments developed in this zone. As in Baoshan, glacial marine sediments can be found in this zone.

The fluvial-littoral zone in Bomi, Ranwu and Tengchongin on the south side: The sediments are composed of grayish green gravel sandstone, feldspathic quartz sandstone, gravel slate, siltstone and shale. Sedimentary facies sequence of barrier island sand bank-tidal flat-estuary sand bank developed from bottom to top. From southwest to northeast, the neritic zone is distributed in Basu-Ridong, and the sediments are composed of gravel slate, siltstone, shale and marl. Gravel components of gravel sandstone and gravel slate in this zone include limestone, sandstone, vein quartz, granite, gneiss, etc., and all of them are near-shore glacier debris flows.

The Paleo-Tethyan Ocean Basin represented by Zhayu-Bitu Nujiang River Zone is located in the further northeast direction. Boxoila Ling-Gaoligong Magmatic Arc is dominated by Bomi-Tengchong Granite Zone, and the northern member of which is longer than 100 km is mainly composed of Zhaxi, Bomi, Zhongba and Dedanla Batholith. This arc is mainly composed of granodiorite, quartz diorite and monzonitic granite, invaded in Carboniferous-Permian strata, and distributed in the north of Parlung Zangbo-Guqin Fracture Zone (the arc-arc collision junction zone can be found in Parlung Zangbo) and the south of Bangong Lake-Nujiang River Zone.

Zhaxize batholith was formed in 108–127 Ma, and Dedanla batholith was formed in 120–128 Ma. The granites in later period are S-shaped and formed in 78–85 Ma. According to the petrochemical composition and digital diagram of the main rock mass, the rocks are mainly the subduction Type I granite and mainly formed in the Early Cretaceous. The southern member is located in eastern Tengchong and intrudes into Gaoligong Group along the granite batholith of Gaoligong Mountain. The U–Pb age of zircon in Naiwang rock mass is 249–258 Ma, and the whole-rock Rb–Sr isochron age of Menglian rock mass is (354.2 ± 154.2) Ma, which was caused by Carboniferous-Permian magma emplacement. The rock assemblage of Late Paleozoic granite is mainly monzonitic granite, followed by granodiorite. The assemblage belongs to the calc-alkaline rock according to the lithochemistry. During the field investigation in Tengchong-Lianghe in 1986, the researchers found rhyolite, crystal tuff and volcanic breccia in Menghong Group of Early Carboniferous, which constituted the host rock of the tin deposit on Siguangping stratum, and a large area of intermediate-basic and intermediate-acidic volcanic lava flow and eruption in Early Carboniferous in Ranwu in the north, which reflected that the Early Carboniferous-Permian Tethyan Ocean had subducted to the southwest to form continental magmatic arc. The Mesozoic rock masses in the west of the watershed of Gaoligong Mountain are widely distributed in the form of rock stocks and small batholith, mainly formed in Middle Jurassic-Late Jurassic (140–190 Ma) and Late Cretaceous (70–85 Ma). The west part was formed later than the east part.

The Middle-Late Jurassic granite was caused by collision orogeny, and the Late Cretaceous granite was caused by the post-orogenic spreading.

Parlung Zangbo Arc-Arc Collision Zone

This zone was found during the 1:250,000 regional survey in Bomi-Medog in recent years. It extends westward and has formed Yongzhu-Namu Lake-Jiali Ophiolite Melange Zone, which is an important tectonic zone between the Gangdise-Motuo-Chayu Magmatic Arc in the south and Boxoila Ling-Gaoligong Magmatic Arc. To the southeast, this zone is covered by the Dulong River-Nmai Hka-Longchuan River thrust nappe fracture zone dipping westward, and is not researched fully. No remnants of ophiolite melange have been found so far.

The ophiolite melange zone is exposed in Chongba Highland, Yabagou and Zhongkang in Bomi, such as olivine pyroxenite, amphibole pyroxenite, gabbro, gabbro-diabase, meta-basalt, quartzite, marble, siliceous rocks and its matrix greenschist mica quartz schist and albite actinolite schist, which are blocks with different sizes and shapes, and are emplaced and wrapped by granites after Cretaceous. The Rb–Sr age of the gabbro-diabase is (215 ± 63) Ma (on a scale of 1:200,000 in Bomi, 1995).

The granites on the north side of this zone are all Type I granites with island arc characteristics, and their Rb–Sr ages are 227–210 Ma and (195.3 ± 7.0) Ma (Changba biotite granodiorite in Yongsong). This zone was formed in Triassic to Early Jurassic. The granitic pluton and the contact zone with surrounding rock are significantly deformed, and the ductile shear zone is developed. The south of this zone is dominated by S-type granite, which was mainly formed in the Cenozoic. The Upper Permian, Triassic and Lower Jurassic strata have not been found in Bomi-Ranwu, and there is angular unconformity of Mariposa Formation of Middle Jurassic on the underlying strata on both sides of the ophiolite melange zone. Therefore, it can be speculated that the residual ophiolite melange in Parlung Zangbo developed on the basis of the inter-arc-basin. The oceanic crust has appeared at least in the Late Triassic, and the arc-arc collision and destruction occurred at the end of Late Triassic and Early Jurassic.

2.3   Geological Evolution of Sanjiang Region and the Adjacent Tethys Multi-Arc-Basin-Terrane (MABT)

2.3.1   Evolution and Tectonic Framework of Sanjiang Multi-Arc-Basin-Terrane (MABT)

To study the tectonic evolution and mineralization of Yidun Island Arc, it is necessary to understand the evolution of the Sanjiang MABT. The evolution process of Sanjiang Tethys tectonic domain includes the formation and evolution of Sanjiang MABT, and this tectonic domain was formed due to the break-up → splicing → splitting of the Pangea. Its evolution includes 6 stages (Li et al. 1999a, b).

2.3.1.1   Evolution of Sanjiang Tethys Tectonic Domain

Destruction Stage of Pangea and Formation Stage of Proto-Tethys

In the Late Neoproterozoic, the global Pangea (Rodinia Supercontinent) was destroyed, forming the Gondwana Continental Group in the south, the Laurasia Continental Group in the north and the Pan-Huaxia Continental Group in the middle. The Proto-Tethyan Ocean lies between both continental groups, and the landmasses of the Pan-Huaxia Continental Group are scattered in the Proto-Tethyan Ocean, as shown in Fig. 2.2.

Formation Stage of Pan-Huaxia Continental Group

At the end of Early Paleozoic, the North China Landmass, Tarim Landmass, Yangtze Landmass, Cathaysia Landmass and the small ocean basins in their intervening areas closed, which made the landmasses of Pan-Huaxia Continental Group once separated combine to form the Pan-Huaxia Continent.

Formation and Evolution of Paleo-Tethys

In Late Paleozoic, Qiangbei, Zhongza, Qamdo-Pu'er and other blocks split off from the western edge of Pan-Huaxia Continent, forming the Jinsha River-Ailaoshan Ocean, Lancang River and Ganzi-Litang Ocean of the Paleo-Tethys, and then developed into the four ocean basins of Sanjiang Paleo-Tethys together with the continued Changning-Menglian Ocean, and forming the tectonic framework of MABT with the blocks, island arcs and ancient volcanic island chains. In Zhenyuan (Laowangzhai) Gold Deposit and its south area, Devonian carbonaceous, argillaceous and radiolarian siliceous rocks are developed, which indicates that the Jinsha River-Ailaoshan Ocean began to rift in the early period of Devonian.

Ocean-Land Transition of Paleo-Tethys

From the end of Late Paleozoic to the beginning of Mesozoic, the oceanic crust of ocean basins of Pan-Huaxia Continent dived either in two directions or in single direction, which made Changning-Menglian Ocean, Lancang Ocean, Jinsha River-Ailaoshan Ocean and Ganzi-Litang Ocean destructed synchronously at the end of Late Triassic, thus the Paleo-Tethyan Ocean has gradually closed from west to east, and the intracontinental convergence and orogeny of Sanjiang Region started. Subsequently, the inherited evolution of Bangong Lake-Nujiang Ocean in Late Triassic occurred, and the Neo-Tethyan Ocean in Yarlung Zangbo River was formed and ran through the whole Gondwana and Eurasia.

Closure of Neo-Tethyan Ocean and Intracontinental Convergence

With the complete disintegration of Gondwana and the formation and expansion of the Indian Ocean, the Gangdise Block and India Landmass moved northward, and Bangong Lake-Nujiang Ocean and Yarlung Zangbo River-Neo-Tethyan Ocean were completely closed at the end of Early Cretaceous and Late Cretaceous-Eocene respectively. Thus, the intracontinental convergence of Sanjiang Region and Qinghai-Tibet Plateau started.

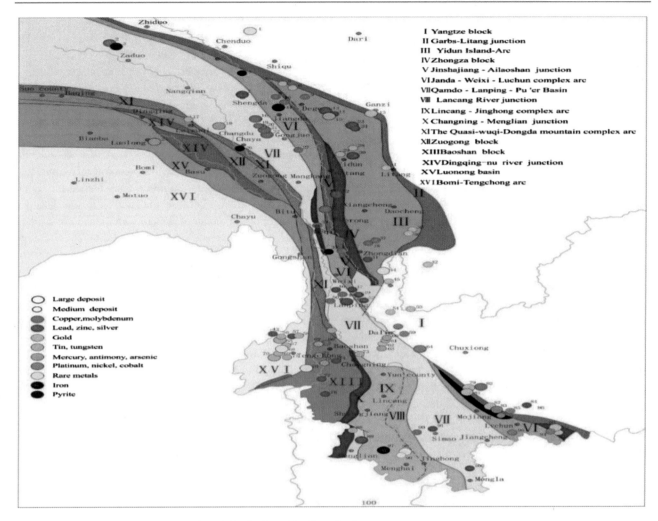

Fig. 2.29 Geotectonic framework and mineral distribution of Sanjiang Region

Transformation and Orogeny of Cenozoic Sanjiang Intracontinental Tectonics

Due to the oblique collision between Indian Plate and Eurasian Plate in Sanjiang Region and the tectonic balance and stress release, large strike-slip, shear and nappe occurred in Sanjiang Region, resulting in a series of tectonic–magmatic-metamorphic zones, and the intracontinental tectonic transformation. The orogeny of the Sanjiang Region fully started since Cenozoic.

2.3.1.2 Structural Pattern of Sanjiang Region

According to the study of ophiolite melange zone, arc-basin system and its evolution of Sanjiang Region, the slicing inlaying of four major arc-land collision junction zones in Sanjiang Region, namely Ganzi-Litang, Jinsha River-Ailaoshan, Lancang River, and Nujiang River-Changning-Menglian, and their blocks, such as Zhongza-Shangri-La, Qamdo-Lanping, Lincang, Baoshan and Chayu-Gaoligong, can be determined. The geotectonics of Sanjiang Orogenic Belt and the basic pattern of metallogenic zone were formed,

which were constructed at the waist, spread at both ends, and twisted in reverse "S" shape from east to west (Fig. 2.29).

Yidun Arc-Basin System, Zhongza-Shangri-La Block, Jinsha River Arc-Basin System, Ailaoshan Arc-Basin System, Qamdo-Lanping Block, Tenasserim Arc-Basin System, Baoshan Block and Boxoila Ling-Gaoligong Arc-Basin System form the Sanjiang Tethys MABT. Yidun Island Arc is located in the east of Sanjiang MABT and the west side of Ganzi-Litang ophiolite Melange Zone on the southwest margin of Yangtze Landmass.

2.3.2 Geological Evolution of Sanjiang Region and the Adjacent Tethys Multi-Arc-Basin-Terrane (MABT)

Scientists of stratigraphic paleontology, tectonics, petrology, geochemistry, sedimentology, geophysics and other disciplines have successively discovered more than 20 ophiolite melange zones in different periods and collision patterns on

Qinghai-Tibet Plateau (including Sanjiang Region), and their subduction patterns are not always northward. The spreading ocean basin restored by the ophiolite melange zones does not have a specific development trending from north to south; a series of island arcs, continental margin arcs or residual arcs in different periods developed on one or both sides of the adjacent subduction collision zone; the sedimentary basins with different sizes and different tectonic environments can be found. Different tectonic regions have different crustal structures and lithospheric structures, as well as various stratigraphical interfaces at different depths. The new geological events have challenged the Tethys evolution of "conveyor belt" and "accordion" which had the greatest influence in the early period, and most of the basin prototypes represented by more than 20 ophiolite melange zones in Qinghai-Tibet Plateau only have the characteristics of "small ocean basin", back-arc-basin and island arc marginal sea. According to the rock assemblages of different geological units in Qinghai-Tibet Plateau, the space–time structure of arc-basin system and the basic characteristics of the tectonic evolution of MABT in Southeast Asia, the evolution of Sanjiang Tethys MABT can be put into the geological evolution framework of the whole Qinghai-Tibet Plateau and its adjacent areas (Figs. 2.30 and 2.31).

2.3.2.1 Frontal Arc of Pan-Huaxia Continental Group and Early Paleozoic Qinling-Qilian-Kunlun Multi-Arc-Basin-Terrane (MABT)

In recent years, Chinese geologists have made great progress in the research of complex tectonic domain composed of Kunlun Island Arc Orogenic Belt, Sanjiang residual island arcs, blocks and back-arc-basin oceanic crust subduction zone on Qinghai-Tibet Plateau, which confirmed the remains of Paleo-Tethyan Oceanic tectonic-rock assemblage. They also discovered the Early Paleozoic oceanic crust subduction and the corresponding active marginal rock assemblage remnants, which are called the Proto-Tethyan Ocean (Liu et al. 1993). The Kunlun Volcanic-Magmatic Island Arc Orogenic Belt on the north side of Proto-Tethyan Ocean is the Early Paleozoic frontal arc of Pan-Huaxia Continent.

In the Kudi ophiolite in West Kunlun, the pillow lava is composed of oceanic tholeiite, with the Nd age of 600–900 Ma. The age of acid magmatic rocks invaded in pillow lava and volcanic rocks is 458–517 Ma (including the age of U–Pb, Rb–Sr, and $^{40}Ar/^{39}Ar$). The Precambrian metamorphic complex on the south side of the ophiolite melange zone is the basement and intermediate-acid intrusive rocks of different ages (mainly the granodiorite, diorite and granitoid rocks, with the age of 540–400 Ma and 260–200 Ma), which may be island arc thrust schists napped southward. The island arc magmatic zone extends eastward to East Kunlun. On the Qinghai-Tibet Highway to the south of

Golmud, Late Devonian continental volcanic lava can be found. The plant fossils can be found in clastic rocks. The North Kunlun Island Arc Magmatic Zone extends over 1000 km from east to west and remains intact, while the Nachitai Group to the south of the Kunzhong Fracture is called the Early Paleozoic eugeosyncline sediment or classified as Ordovician–Silurian turbidite. Some scholars found stromatolites in the marble of the opening of Wanbaogou, and named it Wanbaogou Group separately from Nachitai Group, but in essence, it is still a Neoproterozoic melange zone containing ophiolite. Some scholars named this group "Wanbaogou Group" because the marble in this group is 5 km long and hundreds of meters wide, and most of them are lens blocks in a green schist series of different sizes. Just like Nachitai Group, Wanbaogou Group is a rock assemblage mainly composed of pillow lavas, volcanic rocks and ophiolites. According to the author's observation, the sedimentary rocks containing Ordovician–Silurian fossils are actually blocks in the melange. There are Carboniferous-Permian fossils in Xiaonanchuan, and Xiaonanchuan Group (Bureau of Geological Exploration & Development of Qinghai Province, 1991) was also separated from Nachitai Group. A typical distal turbidite outcrop was found along the road where the east and west beaches meet to the north, indicating that it was composed of the mylonite with significant tectonic shear. From this point of view, the so-called Wanbaogou Group, Nachitai Group and Xiaonanchuan Group on the south side of Kunlun Magmatic Arc do not belong to Smith strata except for some specific slices (blocks) which maintain sequences, but are the non-Smith subduction complex zone which continuously subducts northward from Proto-Tethyan Ocean to Paleo-Tethyan Ocean (possibly since Ordovician). This subduction zone extends eastward with the frontal arc and bends significantly northward in Kuhai. The destruction of the Paleo-Tethyan Ocean Arc-Basin System can be found in the Triassic volcanic arc and fore-arc subduction zone in Burhan Budai-Ela Mountain.

On the north side of Paleozoic frontal arc in Kunlun, the geological history in Early Paleozoic in Tarim, Qaidam, Qimantag, Altyn Tagh and Qilian Mountains mainly includes back-arc sea floor spreading, back-arc-basin shrinkage, arc-arc collision and arc-land collision. Most of this zone has been transformed into land in Devonian, forming a part of the southwest margin of the North China Landmass of the Pan-Huaxia Continental Group. There are Carboniferous relic back-arc-basins only in Zongwulong Mountain, and Tarim and Qaidam are the largest ones. There are always different opinions on the formation environment and tectonic properties of marine volcanic rocks and ophiolite in Early Paleozoic in the Altyn Tagh-Qilian Mountain. Many scholars have clearly demonstrated the ocean ridge basalt in Qilian Mountain, but there are different opinions on

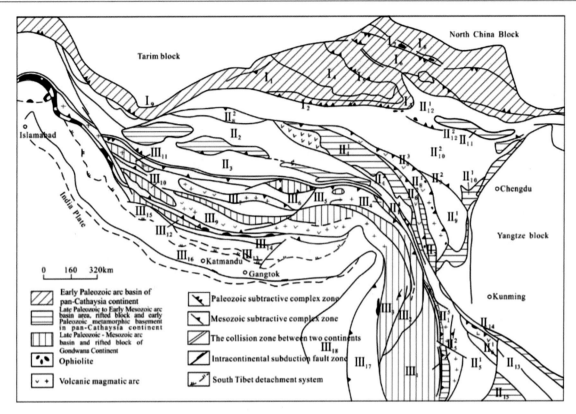

Fig. 2.30 Distribution of archipelagic arc-basin tectonics in Qinghai-Tibet Plateau and its Adjacent areas. I—Pan-Cathaysian continent early Paleozoic arc-basin zone: I_1—Kunlun frontal arc of Pan-Cathaysian continent, I_2—Paleozoic subduction complex zone at the south margin of Kunlun Frontal arc, I_3—Late Paleozoic-Triassic volcanic-magmatic arc and fore-arc accretionary wedge of Burhan Budai-Ela mountain, I_4—Qaidam late Paleozoic residual back-arc-basin, I_5—Ordovician subduction zone at the north margin of Qaidam, I_6—Central Qilian-Datongshan Mesozoic residual Island arc, I_7—North Qilian back-arc ocean basin subduction zone, I_8—Corridor Island chain zone, I_9—Tarim late Paleozoic residual back-arc-basin; II—Pan-Cathaysian continent late Paleozoic-early Mesozoic arc-basin zone: II_1—Late Paleozoic Island chain frontal arc (Neoproterozoic-early Paleozoic accretionary wedge metamorphic basement of Pan-Cathaysian continent), II_1—Lumajiangdong Cuo deformed metamorphic unit (P, volcanic island chain), II_1—Central Qiangtang Island chain, II_1—Tanggula deformed metamorphic unit, II_1—Leiwuqi Island arc, II_1—Lincang Island arc (with accretionary wedge of Lancang group in the east side), II_2—Northern Qiangtan late triassic back-arc-basin, II_3—South Qiangtang foreland basin (J), II_4—Permian–Triassic composite back-arc-basin (with Kaixinling-Zadoi Island arc at the west side), II_4—Qamdo back-arc foreland basin (J-K), II_5—Lanping-Pu'er Late Paleozoic-triassic back-arc-basin, II_5—South Lancang river late Paleozoic-triassic volcanic-magmatic arc, II_6—Mojiang-Lvchun river permian–triassic volcanic-magmatic arc, II_6—Jiangda-Deqin Permian–Triassic volcanic-magmatic arc, II_7—Zhongza Paleozoic plateau (with Jinsha river junction zone at the west side), II_8—Yidun late Triassic Island arc, II_8—Samatuojiari late Triassic volcanic arc and Hoh Xil subduction complex in the north side, II_8—Ganzi-Litang junction zone, II_9—Yajiang late Triassic residual basin, II_9—Xianshuihe Melange zone, II_{10}—Early Paleozoic coastal mountain chain marginal basin at the west margin of Yangtze, II_{10}—Bayankala late Paleozoic-middle triassic back-arc-basin (T_2–T_3 indicate foreland basins); II_{11}—Xiqing mountain late Paleozoic Plateau; II_{12}—Gonghe triassic residual basin, II_{12}—Animaqin snow mountain Melange zone, II_{13}—Gemia early Paleozoic accretionary wedge, II_{14}—Ailaoshan junction zone, II_{15}—Indosinian microblock; III —Late Paleozoic–Mesozoic arc-basin zone at the north margin of Gondwana: III_1—Sibumasu microblock (with Changning-Menglian junction zone at the east side), III_2—Gaoligong mountain late Paleozoic frontal arc, III_3—Lhasa-Bomi-Chayu Mesozoic-Cenozoic volcanic-magmatic arc, III_4—Jiayu bridge late Paleozoic deformed metamorphic unit (with Jiayu Bridge-Zhayu-Bitu junction zone at the east side), III_5—Naqu Jurassic back-arc-basin, III_6—Nierong Paleozoic deformed metamorphic unit (with Anduo-Dingqing junction zone at the north side), III_7—Zenong-Wenbu early cretaceous volcanic-magmatic arc (with Guomangco-Namucuo Melange zone at the north side), III_8—Tsochen-Nyenchen Tanglha early Permian-Mesozoic Island chain, III_9—Gangdise Cretaceous-Paleogene volcanic-magmatic arc, III_{10}—Kongbogangri Cretaceous-Paleogene volcanic-magmatic arc, III_{11}—Anglonggangri Jurassic-Cretaceous magmatic arc (with Pangong Lake-Dongqiao junction zone at the north side, and Shiquan river Melange zone at the south side), III_{12}—Zhongba late Paleozoic Plateau (with Pulan Melange zone at the south side, and Gongzhucuo Melange zone at the north side), III_{13}—Laguigangri Metamorphic core complex zone, III_{14}—Continental margin splitting accretionary wedge (with Yarlung Zangbo river junction zone at the north side), III_{15}—High himalayan overthrust zone, III_{16}—Low himalayan overthrust zone, III_{17}—Burma central lowland volcanic arc, III_{18}—Najia-Arakan front accretionary wedge

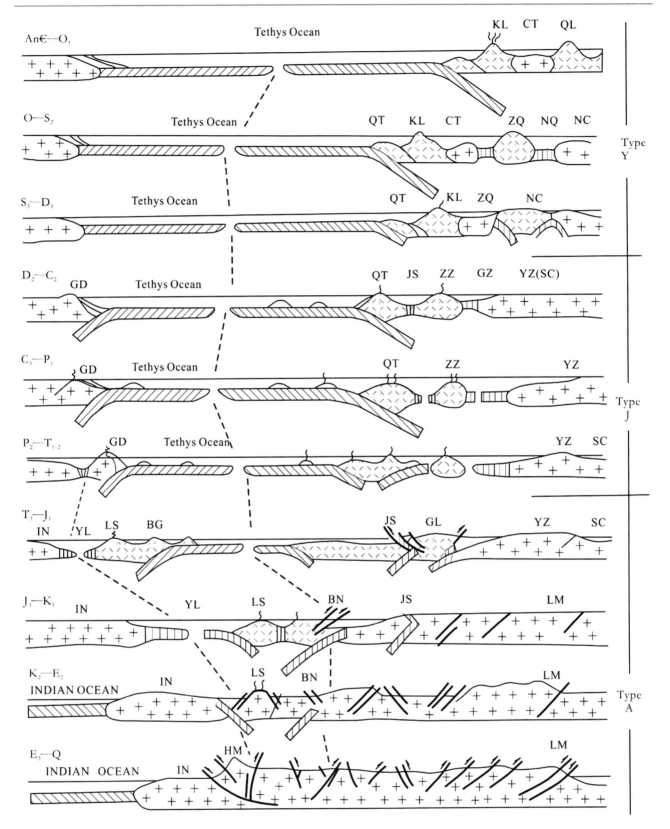

Fig. 2.31 Schematic diagram of tectonics evolution of MABT in Qinghai-Tibet Plateau and its Adjacent areas. KL—Kunlun; CT—Qaidam; QL—Qilian mountain; QT—Qiangtang; ZQ—Central Qilian; NQ—North Qilian; NC—North China; SC—South China; GD—Gangdise; JS—Jinsha river; YZ—Yangtze; ZZ—Zhongza; GL—Ganzi-Litang; LM—Longmen mountain; LS—Lhasa; HM—Himalaya; IN—India; BG—Bangor-Jiali zone; BN—Bangong Lake-Nujiang river; YL—Yarlung Zangbo; Type Y—Indonesia and Southeast Asia split type; Type J—Kuril-Japan Islands split type; Type A—Andes split type

subduction to the north or south. Thought that the North Qilian Mountain was a typical trench-arc-basin system in the Paleozoic active continental margin, and the subduction zone was composed of ophiolite and melange (including high-pressure and low-temperature blueschist and eclogite). Lai identified volcanic rock assemblages with different tectonic backgrounds, such as oceanic ridge, oceanic island (seamount) and island arc, in Qilian Mountain (including the northern margin of Qaidam) by geochemistry methods, and considered that there were mainly three oceanic ridge (oceanic island) volcanic zones and associated arc volcanic rocks. The assemblages were separated by Proterozoic crystalline basement, and constituted three independent Ordovician ocean basins in North Qilian, among which Yushigou-Dakecha Ocean Basin spread in Early Ordovician, while Sunan-Yongdeng Ocean Basin and Zhangye-Jingtai Ocean Basin spread mainly in Middle-Late Ordovician, and the widths of the three ocean basins were estimated to be 2400 km, 600 km and 640 km respectively.

The basic volcanic rocks of Xitie Mountain-Lvliang Mountain-Saishiteng Mountain in the northern margin of Qaidam, which were formed almost at the same time as the North Qilian, feature rock assemblages with high TiO_2 content and low K_2O content, and are similar to ocean ridges, while the intermediate-acid volcanic rocks mainly show the evolution trend of island arc calc-alkaline volcanic rocks, and the ocean basin in the northern margin of Qaidam is 1000 km.

Three typical ocean floor spreading basins in Qilian Mountain, including Hongliugou, Altyn Tagh-Lapeiquan Sea Floor Spreading Basin, Apa-Mangya Sea Floor Spreading Basin, the ocean basin at the northern margin of Qaidam and Qimantag Rift Basin, are all a series of back-arc-basins behind Kunlun Frontal Arc (Fig. 2.29). The tectonic paleogeographic pattern in Early Paleozoic (especially in the $O-S_2$ period) can be compared with Indonesia Island Arc and the archipelagic arc-basin pattern in the north of the arc. The formation of these basins was restricted by the two-way subduction of Proto-Tethyan Ocean and Paleo-Asian Ocean, similar to that of the MABT in Southeast Asia controlled by the two-way subduction of the Indian Ocean and Pacific Ocean. Tuolainanshan-Datongshan Micro-landmass in Central Qilian is similar to the Kalimantan Island Landmass in Southeast Asia. The volcanic rocks in different tectonic settings in Qilian Mountain are closely associated in space. The arc volcanic rocks with different maturity are produced in the same tectonic zone, and the melange zone composed of arc-basin subduction complex and ophiolite tectonic slice is the tectonic contact boundary, which are the main signs to identify the arc-arc collision in Qilian Mountain. In a word, in

Qinling-Qilian-Kunlun Mountains, the ocean basins are mostly developed in Late Cambrian-Ordovician, and they are few small ocean basins and back-arc ocean basins. The volcanic-magmatic arcs were developed in Late Ordovician–Silurian, and the configuration and development of arc-basin systems are similar to those in Southeast Asia.

2.3.2.2 Late Paleozoic Qiangtang-Sanjiang Multi-Arc-Basin-Terrane (MABT) on the Southwest Margin of Pan-Huaxia Continental Group

The evolution of Tethyan Ocean and Pan-Huaxia Continent from Late Paleozoic to Mesozoic was concentrated in Qiangtang-Sanjiang Region. The available data show that the Kunlun Island Arc and Longmen Mountain-Kangdian are the Early Paleozoic coastal mountains in the southwest of Pan-Huaxia Continent. There are two types of basements in this mountain, the land side is composed of the pre-Sinian crystalline hard basement, and the outer side is mainly composed of the metamorphic soft basement of accretionary wedge in Neoproterozoic-Early Paleozoic Continental Margin. The soft basement is covered with Late Paleozoic overlying strata, which is in extensional or angular unconformity. In the western margin of the Yangtze Landmass, the Ordovician–Silurian stratum is a sequence of turbidite dominated by clastic rocks. Lower Ordovician pillow basalts can be found in Muli, and the Ordovician–Silurian stratum in Diancang Mountain-Ailaoshan to the south is composed of continental margin turbidites. Along the southeast margin of Yangtze Landmass, the Sinian-Ordovician stratum is also composed of siliceous stucco turbidite on the continental slope. Significantly, on both sides of Jinsha River, that is, in Haitong-Qingnidong of Qamdo Block on the west side, the Devonian stratum is in unconformity contact with the underlying Ordovician flysch turbidite. In Yidun-Baiyu of Zhongza Residual Island Arc on the east side, the plateau-type Late Paleozoic overlying strata and the Early Paleozoic strata with underlying metabasite basalt, volcanic breccia, trachyte and intermediate-acid volcanic rocks also show different deformation and metamorphism. A sequence of calc-alkaline island arc volcanic rocks and Type I tonalite are developed in the Lower Paleozoic (possibly including Precambrian) Boluo Group on the east side of Qamdo Block. In the ophiolite melange zone of Gemia, Vietnam (northeast of Indosinian Block), the Early Paleozoic stratum is composed of tuffaceous green schist, which was covered by Devonian stratigraphic unconformity. However, Kaixinling-Wuli in the northwest extension of Qamdo Block is an island arc system based on continental crust, although only the rock assemblage of seamount-ocean island-island arc of Early Permian is exposed.

In particular, in the west area of Pu'er-Lanping-Qamdo-Kaixinling, a sequence of metamorphic rocks of Lower Paleozoic Lancang Group, Chongshan Group, Precambrian Jitang Group and Lower Paleozoic Youxi Group are also exposed. The $^{40}Ar/^{39}Ar$ age of blueschist in Neoproterozoic-Lower Paleozoic metamorphic basic volcanic rocks of Lancang Group is 410 Ma, and the age of Lincang batholith on the east side of Lancang Group is 433–422 Ma (Zhang et al. 1990). Jitang Group is composed of intermediate-acid volcanic rocks mixed with fine debris and limestone, while Youxi Group is composed of metamorphic glutenite, sandstone and volcanic rocks with low greenschist facies, including chlorite albite schist, two-mica quartz schist and chlorite quartz schist (Rb–Sr isochron metamorphic age is 371 Ma $\pm$ 50 Ma) of basalt and dacite, showing the geochemical characteristics of island arc environment in Early Paleozoic. The deformed metamorphic basement in North Qiangtang is called Amugang Group, and the horizon of the lower gneiss is equivalent to that of Jitang Group. The middle and upper green schist, mica quartz schist, siliceous rock and volcanic rock members can all be compared with those of Youxi Group. The overlying strata are all stable cap rocks of Late Paleozoic, and the biological features are mainly warm-water organisms.

The above data show that ① The overlying strata of the stripped block group of North Qiangtang-Kaixinling-Qamdo-Lanping-Pu'er are the stable Devonian-Carboniferous stratum, and its underlying sedimentary strata are accretionary wedge-type greenschist metamorphic complex; ② Metamorphic complexes of Amugang Group in Qiangtang, Youxi Group in the west of Qamdo, Chongshan Group in the west of Lanping and Lancang Group in the west of Pu'er are all Precambrian-Early Paleozoic rock units, mainly the volcanic-magmatic arc formed by the subduction of Tethyan Ocean to the northeast (present position) and the remnants of accretionary wedge under the island arc. The Early Ordovician-Middle Ordovician passive marginal flysch sediments seen in Qingnidong-Haitong are the residual sediments after the expansion of Early Paleozoic back-arc-basin. ③ This continental crust strip may have started in Early Devonian and split off from the Early Paleozoic coastal mountains in the southwest of Pan-Huaxia Continent in the form of Japan-Ryukyu Islands back-arc expansion; ④ Following the idea of subduction and "splitting", the history of archipelagic arc orogeny of back-arc expansion, arc-arc collision, and arc-land collision from the south of Kunlun and the west margin of Yangtze to the north of Tibet - Sanjiang Region in Late Paleozoic-Triassic can be found, rather than the so-called "opening-closing" evolution of the subduction and destruction of Proto-Tethyan Ocean and the opening of Proto-Tethyan Ocean. ⑤ Tenasserim Frontal Arc (including Norther Qiangtang Residual Arc, Jitang Group Residual Arc, Chongshan Residual Arc and Lancang Residual Arc) is the frontal arc of the archipelagic pattern on the southwestern margin of Pan-Huaxia Continent in Late Paleozoic, and the southwest side of the frontal arc is the Tethyan Ocean during the shrinking.

There are many arguments about Paleo-Tethys. Some scholars proposed that Late Paleozoic Tethys was developed on the basis of the expansion of the closed back-arc-basin of Proto-Tethyan Ocean; some scholars believed that the Paleo-Tethys was a new ocean basin formed by the drag and spreading of the subduction plate (i.e., the passive continental margin side), featuring the pattern of archipelagic ocean, which was composed of Yangtze affiliated block and Gondwana affiliated block group and the ocean basin between them; some believed that the Paleo-Tethyan Jinsha River-Ailaoshan Ocean might be the result of stretching and splitting on the basis of the closed residual sea or foreland depression of Proto-Tethys, while Paleo-Tethyan Lancang Ocean was formed by the spreading on the basis of Changning-Menglian Back-arc-Basin formed by the subduction of Proto-Tethyan Ocean. Based on years of research, the author believes that Late Paleozoic Paleo-Tethys is the inheritance and development of Proto-Tethyan Ocean, or it can be regarded as a residual ocean. The Late Paleozoic-Triassic island arc, back-arc-basin, margin magmatic arc and inter-arc-basin indicate that the lithosphere of Proto-Tethyan Ocean began to shrink since Devonian.

The ophiolite melange zones in Changning-Menglian, Western Yunnan and Bitu-Zhayu, Eastern Tibet, located between the Indosinian-Qamdo Block in Yangtze affiliated landmass and Sibumasu-Baoshan Block in Gondwana affiliated landmass, is a relic of the Paleo-Tethyan Ocean, which is composed of siliceous distal turbidite and radiolarian siliceous rocks in Devonian-Permian. The Late Paleozoic Paleo-Tethyan Ocean Crust subducted eastward, forming Lincang-Menghai Magmatic Arc, South Lancang River Arc-Basin, Late Paleozoic-Triassic volcanic-magmatic arc on the east side of Lancang River, Pu'er Late Paleozoic-Triassic Back-arc-Basin (converted into back-arc foreland basin from Late Triassic to Cretaceous), Mojiang-Lvchun Permian–Triassic Volcanic-Magmatic Arc and Ailaoshan Back-arc Spreading Ocean Basin in the southern member (Figs. 2.32 and 2.33). It should be noted that in addition to the spreading ridge tholeiite, a small amount of high alkalinity andesite basalt, Early Paleozoic subduction complex (ultramafic rock with a Rb–Sr age of 418 Ma) and residual blocks of remnant arcs can be found in the ophiolite melange zones caused by Late Paleozoic oceanic crust subduction in Ailaoshan. Late Paleozoic Tethyan Ocean subducted eastward to form Kaixinling-Zadoi Permian Volcanic Arc, Qamdo-Markam Permian–Triassic Back-arc-Basin (transformed into back-arc foreland basin in Jurassic-Cretaceous), Jiangda-Weixi Late Permian-Early and Middle Triassic Volcanic-Magmatic Arc,

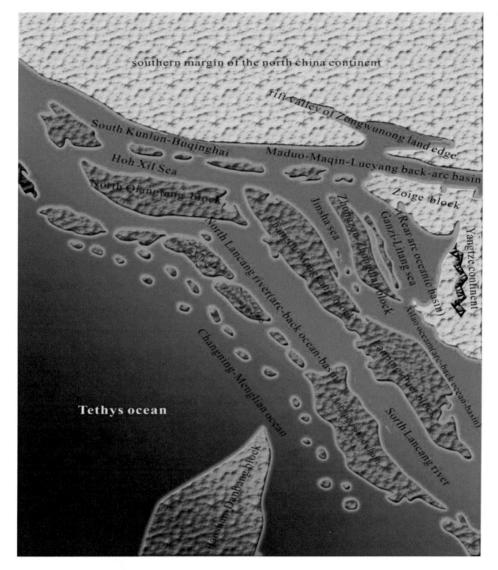

Fig. 2.32 Schematic diagram of Sanjiang carboniferous-permian MABT paleotectonic plane

Fig. 2.33 Schematic diagram of Sanjiang carboniferous-permian MABT paleotectonic profile

Jinsha River Back-arc Spreading Ocean Basin, Baiyu-Shangri-La Residual Island Arc Block and Late Triassic Volcanic Arc, Keludong-Xiangcheng Late Triassic Inter-arc-Basin, Que'er Mountain Late Triassic Magmatic Arc, Ganzi-Litang Late Permian-Early and Middle Triassic Subduction Margin Rift Ocean Basin, Yajiang Residual Basin and South Qinling Back-arc Ocean Basin.

The latest geological survey in Qiangtang, North Tibet shows that ophiolite melange with blueschist and greenschist as matrix has been found along Gangmacuo, Guoganjianian Mountain, Mayigangri, Jiaomuri, Qiagangcuo and Shuanghu Lake, with an extension of 350 km from east to west. The $^{40}Ar/^{39}Ar$ age of the blueschist is 222.5 Ma ± 3.7 Ma (Li et al. 1995), and the Middle Triassic radiolarian siliceous rocks covered by Late Triassic strata unconformity can be found in the melange zone. These records show that there is also an MABT based on a residual arc in the north side of the Paleo-Tethyan Ocean.

Arc-arc collision and arc-land convergence of East Tethyan Pan-Huaxia Continent mainly occurred in the Triassic. The oblique wedging of the Yangtze Landmass to the west was caused by the arc-arc collision and arc-land collision between Danbang Block (Hanbao Block) of Gondwana affiliated landmass and the split frontal arc of Yangtze Landmass as well as Indosinian, Qamdo and Zhongza blocks, including the shrinking and subduction of the back-arc-basin of the expanding ocean basin. The subduction polarity was mostly opposite to that of Paleo-Tethyan Ocean. Jinsha River experienced forward (westward) subduction collision $(P_1-P_2) \rightarrow$ left oblique subduction collision (T_{1-2}) and subsequent collision. This process is also manifested in the fact that in the western margin of Yangtze Landmass, Ailaoshan Back-arc Ocean Basin was destroyed at the end of the Middle Triassic, and the ocean basin developed northward in East Sichuan and East Tibet was destroyed in the Late Triassic. Correspondingly, in the northern margin of Yangtze Landmass, there is an oblique continuous collision in Middle Triassic-Late Triassic-the end of Late Triassic from West Qinling to East Kunlun and Hoh Xil. The eastern margin of Bayankala Late Paleozoic Back-arc-Basin showed the characteristics of a foreland basin in Latin (T_2)-Late Triassic (Xikang Group), which proved that the Yangtze Ancient Land was caused by westward, oblique and two-way subduction, which resulted in the closure of Paleo-Tethyan Bayankala Ocean Basin.

In a word, in Qiangtang-Sanjiang Region, the ocean basins developed mainly in Carboniferous-Early Permian, and are not large, with a width of only thousand kilometers. The volcanic arcs developed in Late Permian-Early and Middle Triassic, and Tenasserim frontal arc is similar to Kuril-Japan-Ryukyu Islands volcanic arc.

2.3.2.3 Gondwana Frontal Arc and Mesozoic Gangdise-Himalayan Multi-Arc-Basin-Terrane (MABT)

In recent ten years, many scholars have carried out much fruitful research on Jiayu Bridge "metamorphic complex" and its south extension in Nujiang River Junction Zone. According to the 1:200,000 regional geological mapping, it is considered that the Bitu Junction Zone in the south is connected with Changning-Menglian Junction Zone through the western slope of Meri Snow Mountain, while the main part of Jiayu Bridge Metamorphic Complex is considered as the northeast continental margin arc of Gondwana. When Xu et al. started their international expedition to Qinghai-Tibet, they considered that the Gangdise-Boxoila Ling-Gaoligong Mountain should be Gondwana Late Paleozoic–Mesozoic

frontal arc, and that Nierong Uplift and Jiayu Bridge Metamorphic Complex are the remnants of the frontal arc. The Late Paleozoic–Mesozoic Tibet Islands are located in the south of this forward arc. During the field investigation, many renowned scholars, such as Xu Jinghua, agreed to see the Bangong Lake-Nujiang River Suture Zone as the northern boundary of Gondwana. The recent research on the differences of geological and geophysical characteristics between the north and south sides of Bangong Lake-Nujiang River Suture Zone further confirms the rationality of the northern boundary of Gondwana in the Bangong Lake-Nujiang Suture Zone.

In Bomi, Songzong, Laigu and Ridong of Gangdise, gravity flow slump breccia limestone, volcanic fine turbidite and intermediate-basic andesite basalt of Nuocuo Formation in Early Carboniferous, and Laigu Formation, dacite, rhyolite and pyroclastic rocks in Middle Carboniferous can be found. Given a large number of Late Carboniferous intermediate-basic volcanic rocks found in Angjie Formation in Cuoqin, this volcanic-sedimentary rock series with volcanic arc characteristics is located intermittently in Gangdise-Nyenchen Tanglha Block (Pan et al. 2004), which may indicate that the North Gondwana transformed into an active continental margin since Early Carboniferous. In Leqingla on the northern margin of Linzhou Basalt in Gangdise, the Middle Permian basalt in Luobadui Formation shows the geochemical characteristics of island arc volcanic rocks (Fig. 2.34); although the main body of Gangdise-Nyanchen Thanglha Mountain Chain lacks Triassic sedimentary records, Middle Triassic arc volcanic rocks and Late Triassic island arc granite have been found in Quesang Temple and Menba in Lhasa. On the west side of Nujiang in Bitu, East Tibet, Jurassic continental magmatic arc in Sanmen Village, volcanic rocks of the Late Triassic Xieba Arc and Late Triassic flysch sediments in Quehala Formation in Gula Fore-arc-Basin can be found; in Dazi, Lhasa, a sequence of low-grade metamorphic calc-alkaline arc volcanic rocks and pyroclastic rocks (Yeba Volcanic Arc) developed in Early-Middle Jurassic in Yeba Formation. The volcanic rocks have the geochemical characteristics of arc volcanic rocks (Fig. 2.34), spreading about 300 km from east to west, with a residual width of only 30 km; in the further south area, the calc-alkaline arc volcanic rocks of Sangri Group (J_3-K_1) on the north side of Yarlung Zangbo Suture Zone (Sangri Volcanic Arc) spread more than 400 km from east to west, and the residual width is only 20–40 km. Due to Shiquan River Ophiolite Melange Zone between Anglonggangri Accretionary Arc (J_3-K_1) and South Gangdise found and explored in Gangdise, and the Guomacuo-Namu Lake-Jiali-Bomi Ophiolite Melange Zone between Guopucuo-Wenbu Volcanic-Magmatic Arc (J_3-K_1) and Bangor-Sangxiong-Boxoila Ling Magmatic Arc (J_3-K_1) discovered in Gangdise Zone in recent years, it can be found

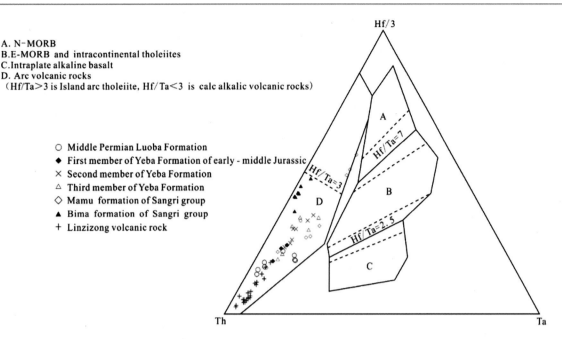

A. N-MORB
B.E-MORB and intracontinental tholeiites
C.Intraplate alkaline basalt
D. Arc volcanic rocks
(Hf/Ta>3 is Island arc tholeiite, Hf/Ta<3 is calc alkalic volcanic rocks)

○ Middle Permian Luoba Formation
◆ First member of Yeba Formation of early - middle Jurassic
× Second member of Yeba Formation
△ Third member of Yeba Formation
◇ Mamu formation of Sangri group
▲ Bima formation of Sangri group
+ Linzizong volcanic rock

Fig. 2.34 Permian-Eocene arc volcanic rocks in Gangdise (Wood 1980)

that Gangdise is not only a continental volcanic arc formed in Late Cretaceous-Paleogene, but also a volcanic arc formed in Middle-Late Carboniferous, Permian, Triassic and Jurassic. Therefore, the new island arc volcanic rocks in different periods in Gangdise are not only caused by the northward subduction of Yarlung Zangbo Oceanic Crust, but also related to the southward subduction of Bangong Lake-Nujiang River Oceanic Crust (Tethyan Oceanic Crust), which represents the northern boundary of Gondwana in earlier periods (Middle-Late Carboniferous, Permian, Triassic and Jurassic). The identification of these arc-arc collision zones and volcanic arcs in different periods in Gangdise indicates that Gangdise-Nyanchen Thanglha is not a simple terrane, but a Gangdise archipelagic arc and its back-arc (inter-arc) basin.

The ophiolite melange zone between Gangdise Mountains and Himalayas is regarded as the Yarlung Zangbo River Suture Zone, and both are classified into the same one. In the investigation in 1994, it was found that this suture zone diverged in the western part of Saga, and Devonian-Middle Triassic Zhongba Neritic Plateau was mixed between two ophiolite melange zones. Gansser (1974) believed that Pulan Ophiolite Zone in the south was napped over the ophiolite from Gongzhucuo Ophiolite on the north side of the plateau to the south during the northward subduction of the Indian Plate. Some researchers also think that the Paleozoic stratum is the residual of giant nappe. These explanations are unsubstantiated because these zones cross a branch of suture. If we study this paleogeographic pattern from the archipelagic arc, we may find that the Zhongba Paleozoic Plateau is the

detachment block that separates Pulan and Gongzhucuo back-arc-basins. What's more, Pulan and Gongzhucuo ophiolite melange is caused by the back-arc-basin shrinkage, arc-arc or arc-land collision at the end of Mesozoic. The ophiolite of Yarlung Zangbo River in Mesozoic is the best-preserved and most complete "trinity" assemblage of ophiolite in Qinghai-Tibet Plateau and even in the Chinese mainland, but its thickness is smaller than that of the ophiolite in West Tethyan Orogenic Belt and some major ocean basins, and it is presented by geological and geochemical characteristics of small ocean basins called by Xiao.

In a word, in the south of Bangong Lake-Nujiang River Suture Zone, the ocean basins developed in Late Triassic-Early Cretaceous, and they are all small ocean basins and back-arc ocean basins. Volcanic arcs developed in the Middle and Late Carboniferous-Eocene. The Gangdise Zone should be Middle-Late Carboniferous Andes active continental margin on the south side of Tethyan Ocean represented by Bangong Lake-Nujiang River Suture Zone. The small ocean basin with ophiolite assemblage interbedded is a series of "entangled" back-arc or inter-arc-basins induced by the southward subduction of Tethyan Ocean.

References

Cong BL, Wu GY, Zhang Q, Zhang RY, Zhai MG, Zhao DS, Zhang WH (1993) The tectonic evolution of western Yunnan Paleo-Tethys structural rocks, China. Sci China (d Ser) 23:1201–1207 (in Chinese with English abstract)

Hsü KJ (1994) An archipelago model of orogenesis. Geol Today 12:290–293

Huang JQ, Chen BW (1987) The evolution of the Tethys in China and adjacent regions. Geological Publishing House, Beijing, pp 1–109 (in Chinese with English abstract)

Li XZ, Xu XS, Pan GT (1995) Evolution of the Pan-Cathaysian landmass group and Eastern Tethyan tectonic domain. Sed Geol Tethyan Geol 4:1–13 (in Chinese with English abstract)

Li JL, Sun S, Hao J, Chen HH, Hou QL, Xiao WJ (1999a) On the classification of collision orogenic belts. Chin JGeol 34:129–138 (in Chinese with English abstract)

Li XZ, Du DX, Wang YZ (1999b) The basin range transition and mineralization: examples from the Qamdo-Pu'er Basin and Jinshajiang-Ailaoshan orogenic belt in southwestern China. Tethyan Geol 22:1–16 (in Chinese with English abstract)

Li DM, Wang LQ, Xu TR, Diao ZZ, Chen KX, Lu YF, Wei JQ, Zhou ZX (2002) The Cu-Au metallogenesis and exploration in Jinshajiang tectonic belt. Geological Publishing House, Beijing, pp 1–259 (in Chinesewith English abstract)

Liu BP, Feng QL, Fang NQ (1993) Tectonic and paleogeographic frameworks of Paleo-Tethys poly-island-ocean in Changning-Menglian and Lancangjiang Belts, Southwest Yunnan, China. Earth Sci-J China Univ Geosci 18:529–539 (in Chinese with English abstract)

Liu BP, Feng QL, Fang NQ (1991). Tectonic and Paleogeographic frameworks of the poly-island Paleo-Tethys ocean in western Yunnan: symposium on tectonic evolution and mineralization of the Tethys in western China. Chengdu University of Electronic Science and Technology, pp 1–212 (in Chinese)

Mo XX, Lu FX, Shen SY, Zhu QW, Hou ZQ, Yang KH (1993) Sanjiang Tethyan volcanism and related mineralization. Geological Publishing House, Beijing, pp 1–267 (in Chinese with English abstract)

Pan GT, Wang LQ, Yin FG, Zhu DC, Geng QR, Liao ZL (2004) Charming of landing of plate tectonics on the continent as viewed from the study of the MABT. Geol Bull China 23:933–939 (in Chinese with English abstract)

Pan GT, Wang LQ, Li RS, Yuan SH, Ji WH, Yin FG, Zhang WP, Wang BD (2012) Tectonic evolution of the Qinghai-Tibet Plateau. J Asian Earth Sci 53:3–14

Pan GT, Chen ZL, Li XZ, Yan YJ (1997) Tectonic evolution of the East Tethys geology. Geological Publishing House, Beijing, pp 1–218 (in Chinese with English abstract)

Pan GT, Hou ZQ, Xu Q, Wang LQ, Du DX, Li DM, Wang MJ, Mo XX, Li XZ, Jiang XS (2003) Archipelagic orogenesis, metallogenic systems and assessment of the mineral resources along the Nujiang-Lancangjiang-Jinshajiang area in southwestern China. Geological Publishing House, Beijing, pp 1–420 (in Chinese with English abstract)

Shen GF (2002) Weathering crust of Baihuanao granite: a potential superlarge-scale Rb, Cs, Y, Sc, quartz and albite ore deposit. Bull Miner, Petrol Geochem 21:182–184 (in Chinese with English abstract)

Wood DA (1980) The application of a Thsingle bondHfsingle bondTa diagram to problems of tectonomagmatic classification and to establishing the nature of crustal contamination of basaltic lavas of the British Tertiary volcanic province. Earth Planet Sci Lett 50:11–30

Xu ZQ, Hou LW, Wang ZX (1992) Orogenic processes of the Songpan-Ganzi orogenic belt of China. Geological Publishing House, Beijing, pp 1–190 (in Chinese)

Zhang YQ, Xie YW, Wang JW (1990) Rb and Sr isotopic studies of granitoids in Tri-river region. Geochimica 4:318–326 (in Chinese with English abstract)

Formation and Evolution of Sanjiang Collision Orogenic Belt

3.1 Definition and Classification of Collision Orogenic Belt

3.1.1 Definition of Collision

In terms of collision, Suess (1875) divided orogenic belts into Pacific type and Tethyan type according to collision types of orogenic belts and took Alpine and Himalayan orogenic belts as typical examples of collision orogeny. Sengor (1992) divided collision orogenic belts into Alps type, Himalayan type and Altai type according to the collisional type and different internal tectonics of the orogenic belt. Based on the various tectonic units involved in the collision, Li et al. (1999) pointed out that collision may occur in various geological bodies (such as land-land, land-frontal arc, land-residual arc, land-accretionary arc, arc-arc, land-arc-land), which is the most complete description of collision behavior between collision objects and various geological bodies so far.

In terms of collision modes, Ren et al. (1999) called the collision between micro-landmasses soft collision and the collision between macro-landmasses hard collision based on the scale and size of landmasses and considered that the landmasses were not connected yet and were in a state of "linked but not joined" after the soft collision; different macro-landmasses were finally joined into a whole and entered a unified dynamic evolution model only after the hard collision. In terms of collision stage, Harris et al. (1988) divided the collision into three stages: syn-collision, late-collision or post-collision and rear-collision; Liégeois et al. (1998) divided the background related to the collision into three tectonic environments: collision, post-collision or rear-collision and intraplate collision; Deng et al. (2002) divided the collision into 3 stages: continental collision between continents, intracontinental collision and post-orogenic collapse.

We believe that the initial collision between the Indian continent and Eurasia was the first collision of moving arc-arc and moving arcs-lands, and the subsequent continuous land-land collision had experienced a long development process, reflecting the interaction of the mesosphere in the earth system. Therefore, the comprehensive convergence collision between Bangong Lake-Nujiang River Junction Zone and Yarlung Zangbo Junction Zone, including Gangdise MABT, can be called the main collision zone of Qinghai-Tibet Plateau. After the comprehensive convergence collision, the sedimentary environment will change radically (e.g., passive margin basin will be transformed into foreland basin, and back-arc basin will be transformed into back-arc foreland basin), and extensive and intense magmatic activity will occur along the frontal arc in front of the collision zone. Post-collision is marked by the emergence of various significant intracontinental deformations such as molasse, two-mica granite and foreland thrust zone.

3.1.2 Classification of Collision Orogenic Belts

As the main orogenic belts throughout the world were formed at different ages during the evolution of the earth and distributed in different tectonic parts of the global tectonics, their internal tectonics, shapes and formation mechanisms are complex and diverse, even those in different parts of the same orogenic belt are different. Therefore, there are various types of orogenic belts. Different scholars have formulated different classification standards and principles according to time or space, tectonics or composition and formation mechanism from different perspectives, so different classification schemes are obtained (Miyashiro et al. 1984; Sengor 1992; Xu et al. 1992; Li et al. 1999).

The key to the study on collision orogenic belt is to determine the temporal and spatial structure, composition, collision process and nature of collision zone. Based on the discovery and determination of more than 20 collisional melange zones in Qinghai-Tibet Plateau, they are generally divided into three types: arc-arc collision zone, arc-land

W. Li et al., *Metallogenic Theory and Exploration Technology of Multi-Arc-Basin-Terrane Collision Orogeny in "Sanjiang" Region, Southwest China*, The China Geological Survey Series, https://doi.org/10.1007/978-981-99-3652-6_3

collision zone and land-land collision zone. In a sense, the type of collision orogenic belt is closely related to the nature of collision zone in the determination. The orogenic belts are divided into three types according to the collision unit, temporal and spatial structure and assemblage relationship of orogenic belts (Table 3.1): circumoceanic continental margin arc (ocean-arc collision) orogenic belts, collision orogenic belts and continental margin orogenic belts. Collision orogenic belts are divided into land-land collision orogenic belt, arc-land collision orogenic belt and arc-arc collision orogenic belt. Arc-land collision orogenic belts are generally divided into three types based on the different properties of the basement (or substratum) of the island arc: magmatic arc with the continental crust (metamorphic basement or old stratum) as the basement, accretionary arc of volcanic magma with the subduction complex as the basement and volcanic arc with the oceanic crust (i.e., frontal intra-oceanic arc) as the basement. After further study on fine temporal and spatial structure and composition, it can be divided into three types: residual arc-land collision type, accretionary arc-land collision type and frontal arc-land collision type.

Examples of collision orogenic belt classification: ① circumoceanic continental margin arc orogenic belt. This kind of orogenic belt was formed by subduction of oceanic plate before the closure of the ocean basin. It mainly refers to modern Cordilleras and Andes. After the closure of the ocean basin, it can be transformed into a land-land collision

orogenic belt or a land-arc collision orogenic belt. ② Land-land collision orogenic belt. This is an orogenic belt formed by collision between two circumoceanic continental plates or blocks, and a mountain system formed by the collision between the passive margin of one continent and the active margin (i.e., the continental margin arc) of the other continent. Although this collision orogeny will be reflected in the arc-shaped mountain chain around the ocean margin formed in the early stage, it has long been combined on the continent as the main body of the arc-shaped mountain chain around the ocean margin, and together with the passive continental margin fold-thrust nappe orogeny, it forms the land-land collision orogenic belt. For example, Gangdise Mountain of the continental margin arc type and the Himalayas of the passive continental margin fold-overthrust nappe type together form a land-land collision composite orogenic belt. ③ Land-arc collisional mountain system. This is an orogenic belt formed by the collision between a mountain chain in the island arc zone and a microcontinent. There is a marginal ocean basin between the island chain and microcontinent, and the core of its mountain system is a back-arc basin subduction zone. For example, Taiwan Province coastal mountain orogenic belt, North Qilian Mountain and Middle Qilian Mountain of Qilian Mountain belong to this kind of arc-land collision orogenic belt. ④ Arc-arc collision orogenic belt. This refers to the orogenic belt formed by the collision between the

Table 3.1 Classification framework of collision orogenic belt types	Orogenic belt type			Example
	Circumoceanic continental margin arc orogenic belt			Andes
	Collision orogenic belt	Land-land collision orogenic belt		Himalayas, Ailaoshan
		Arc-land collision orogenic belt	Residual arc-land collision orogenic belt	Northern Qilian Mountains, Taniantaweng Mountain
			Accretionary acr-land collision orogenic belt	Gangdise Mountain, Kunlun Mountain
			Frontal arc-land collision orogenic belt	Taiwan Coast Mountains, Boshula Ridge, Gaoligong Mountain
		Arc-arc collision orogenic belt		Western Junggar and Shaluli Mountain Maluku arc-arc collision orogenic belt
	Continental margin orogenic belt			Longmen Mountain-Jinping Mountain (western margin of Yangtze Landmass)

island arc mountain chain and the island arc mountain chain due to the closing of the ocean basin and the two arcs of the two-way subduction. These types of mountain systems will be formed after some marginal seas in the western Pacific Ocean disappear. Such mountain systems will be formed by the collision between Mariana Arc and Ryukyu Arc after the Philippine Sea is closed. Maluku Sea is an example of arc-arc collision orogeny in action in modern times. Arc-arc collision orogenic belt is formed by the collision between Cretaceous Gangdise Arc and Bange-Bengcuo Accretionary Arc.

3.2 Types and Spatio-temporal Structure of Orogenic Belt in Sanjiang Area

Our years of research revealed that the spatial and temporal structure of Hengduan Mountain Orogenic Belt is mainly as follows: ① It was based on the evolution of MABT on the northwest side of Paleo-Tethyan Ocean from Late Paleozoic to Triassic, and it had experienced subduction orogeny, arc-arc or arc-land collision orogeny and intracontinental subduction in the middle and late stage of Mesozoic to intracontinental strike-slip contraction orogeny in Cenozoic since Triassic; ② from the perspective of three-dimensional geometry, the four backbone collision junction zones and the landmasss or volcanic arcs sandwiched between them show a reverse s-shaped tectonic framework with the waist in the middle being contracted and two members being scattered from north to south; ③ the sectional structure shows the Lancang River Junction Zone and Taniantaweng Island Arc Orogenic Belt, Jinsha River Junction Zone and Ningjing Mountain-Ailaoshan Arc-land Collision Orogenic Belt are of fan-shaped mountain system, while Qamdo-Markam and Lanping-Pu'er Depression Zones in the southern member between two recoil fan-shaped mountain systems of Jinsha River and Lancang River form a hedged strike-slip fan-shaped mountain system (Fig. 3.1). Although the spatial and temporal structure and evolution process of "Hengduan Mountain" Orogenic Belt are very complicated, the geometry of the orogenic belt located at present is concise and clear.

The Sanjiang area in southwest China has experienced the subduction and closure, collision and orogeny of Proto-Tethys, Paleo-Tethys and Meso-Tethys, especially the Yanshanian-Himalayan orogeny stage due to general intra-continental convergence (or over-collision), which is a complex orogenic belt with various orogenic types. According to the above-mentioned classification scheme of orogenic belt types, the classification of orogenic belt types and its temporal and spatial structure of Hengduan Mountain area in Sanjiang are described as follows from east to west.

3.2.1 Continental Margin Orogenic Belt of Bayankala (Longmen Mountain and Jinping Mountain) in the Western Margin of Yangtze Landmass

The mountain system was transformed from Late Triassic passive continental margin system on the western margin of Yangtze Landmass to the fold-overthrust nappe-type mountain chain of foreland basin. The whole land-arc collision orogenic belt was formed the period from the end of Triassic to Cretaceous, which was further developed in Paleogene, as evidenced by the development of a large number of Yanshanian and Himalayan crustal melting granites and the formation of Western Sichuan Foreland Basin. The overall shape is an asymmetric mountain zone with a noncoaxial shear and eastward thrust nappe stack.

3.2.2 Shaluli Mountain Arc-Arc Collision Orogenic Belt

The main body of this orogenic belt is the island arc orogenic belt, which is a composite orogenic belt formed by the westward subduction, destruction and closing of Ganzi-Litang Back-arc Ocean Basin at the end of Late Triassic, subduction and shrinkage of the continental margin on the western margin of Yangtze Landmass and Changtai-Xiangcheng intra-arc rift basin and collision between Que'er Mountain-Haizi Mountain Magmatic Arc and Yidun Volcanic Arc Mountain Chain and between Que'er Mountain-Haizi Mountain Magmatic Arc and Zhongza-Shangri-La Block. As rock strata and island arc mountain chain on the west side of Zhongza-Shangri-La Block generally thrust eastward, the orogenic type is an asymmetric mountain zone with noncoaxial shear. However, due to the westward thrust of Haba Snow Mountain and Yulong Snow Mountain on the east side and the eastward thrust of Shangri-La Block, the hedged mountain chain type is formed in its southern member (Fig. 3.2). During the Late Cretaceous-Eocene, a sequence of transitional alkaline granites invaded from the post-orogenic belt to the nonorogenic belt.

3.2.3 Ningjing Mountain-Ailaoshan Land-Land and Land-Arc Collision Orogenic Belt

This is a complex orogenic belt, showing different types in different zones. In the middle-south member of Jinsha River, namely Zhubalong-Pantiange area, due to the development of an Early Permian intra-oceanic arc, a Late Permian continental margin arc is developed in Jiangda-Weixi on the

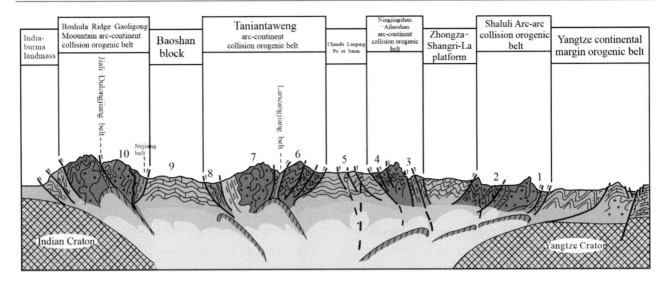

Fig. 3.1 Schematic Diagram of Tectonic Framework and Regional Mineralization of Sanjiang and "Hengduan Mountain" Orogenic Belt. ① Ganzi-Litang Au and Cu Metallogenic Zone; ② Dege-Xiangcheng Cu, Pb, Zn, Ag, Au Polymetallic Metallogenic Zone; ③ Jinsha River-Ailaoshan Au, Cu, Pt, Pd Metallogenic Zone (including Pb, Zn, Cu metallogenic zone on the margin of Zhongza-Shangri-La Block); ④ Jiangda-Weixi-Lvchun Fe, Cu, Ag, Pb and Zn Polymetallic Metallogenic Zone; ⑤ Qamdo-Pu'er Cu, Ag, Pb and Zn Polymetallic Metallogenic Zone ⑥ Zadoi-Jinggu-Jinghong Cu, Pb and Zn Polymetallic Metallogenic Metallogenic Zone; ⑦ Riwoqê-Lincang-Menghai Sn, Fe, Pb, Zn Polymetallic Metallogenic Metallogenic Zone; ⑧ Changning-Menglian Pb, Zn, Ag and Cu Polymetallic Metallogenic Zone; ⑨ Baoshan-Zhenkang Hg, Pb, Zn Rare Metal Metallogenic Zone; ⑩ Tengchong-Lianghe Sn and W Rare Metal Metallogenic Zone

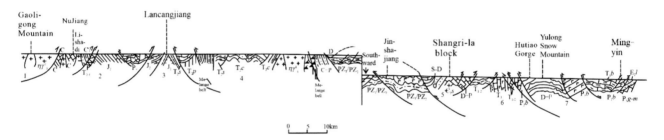

Fig. 3.2 Schematic Diagram of Lishadi-Mingyingou Tectonic section. 1—Gaoligong Mountain Thrust Schist; 2—Biluo Snow Mountain-Chongshan Thrust Schist; 3—Lanping-Pu'er Back-thrust Spreading Foreland Basin; 4—Thrust Schist in Jinsha River Tectonic Zone; 5—Thrust Schist of Zhongza-Shangri-La Block; 6—Yidun-Xiaqiaotou Back-arc Basin Orogenic Belt; 7—Haba Snow Mountain-Yulong Snow Mountain Thrust Schist in the Thrust Schist on the Western Margin of Yangtze Landmass

west side. Therefore, the land-arc collision orogenic belt formed by the collision of Zhongza-Shangri-La Landmass and Zhubalong-Pantiange intra-oceanic arc and the land-arc collision mountain belt formed by the collision of intra-oceanic arc and continental margin arc in the west side were formed after the closure of the ocean basin in the middle-south member. In addition, a mountain chain formed after the closure of Late Triassic rift basin was developed in the Xuzhong-Luchun-Cuiyibi area, which forms the Yunling Mountain Range together with the continental margin arc mountain chain. Therefore, the middle-south member of Jinsha River is a composite orogenic belt.

No intra-oceanic arc was found in the northern member of Jinsha River Zone, and Zhongza-Shangri-La Landmass directly collided with Qamdo Landmass, forming Ningjing Mountain land-land collision orogenic belt. The mountain

zone was formed earlier than the middle-south member; for Early Triassic Pushuiqiao Formation had unconformably overlaid on Jiangda collision crustal melting granite body.

Like the northern member of Jinsha River Zone, no intra-oceanic arc was found in Ailaoshan Zone, which forms Ailaoshan land-land collision orogenic belt. That said, the passive margin overthrust nappe-type mountain chain of Yangtze Continent and Taizhong-Lixianjiang continental margin arc-type mountain chain constitute the land-land collision orogenic belt.

The outstanding features of Ningjing-Ailaoshan Orogenic Belt are as follows: ① The crust on the subduction plate, melange in the junction zone and the volcanic-sedimentary strata on the volcanic arc overthrusts backward in the direction of the superimposed plate in turn, forming a wedge-shaped mountain chain with subduction in the lower part and obduction in the upper part. For example, in the south-central member south of Batang in Jinsha River Zone, Cambrian-Ordovician strata and Devonian-Carboniferous strata on the subducted Zhongza-Shangri-La Landmass thrust westward on the passive margin rock stratum of Jinsha River Zone, or even on the melange zone in the junction zone, showing reverse polarity orogeny (Figs. 3.3 and 3.4). The strata in Permian–Triassic arc volcanic zone in the west thrust westward on Mesozoic red beds in Qamdo-Pu'er Basin (Figs. 3.2 and 3.5). In Southern Diancang Mountain-Ailaoshan Zone, Cangshan Group and Ailaoshan Group in the crystalline basement of Yangtze Landmass thrust westward (or southwestward) on the melange (Fig. 3.6) and volcanic arc zone of the junction zone and cover the junction zone and volcanic arc zone in Yangbi-Midu area. However, in Xiongsong of Eastern Tibet and Tuoding-Tacheng of Western Yunnan, the folds and thrusts on the passive continental margin, which thrusted eastward, were formed in the early subduction and collision stage, were not reworked due to the later deformation but preserved and can still be found at the lower part and the frontal margin the east nappe. ② A sequence of calc-alkaline volcanic rock series was developed, and intermediate-acid magmatic rocks were invaded in the post-orogenic stage of Late Triassic.

The main mountain ranges of this collision orogenic belt include Aila Mountain, Chali Snow Mountain, Yunling Mountain and Xuelong Mountain of Eastern Tibet and Western Yunnan in its west, Baiyu-Batang-Derong (such as Gajin Snow Mountain) of Western Sichuan and Western Yunnan in its east, the snow mountains on the west side of Shangri-La and Diancang Mountain and Ailaoshan in its south.

The Jinsha River-Ailaoshan Ocean subducted westward in the Late Permian and closed in the Early and Middle Triassic, and the volcanic arc mountain chain on the continental margin was formed in the period from Late Permian to Triassic. Collisional orogeny occurred in Mesozoic. The thrust zone composed of crystalline basement rock series in Diancang Mountain and Ailaoshan was mainly formed in Paleogene-Neogene; that is, the primary peaks of the two mountains were formed in Paleogene-Neogene.

3.2.4 Taniantaweng Residual Arc-Land Collision Orogenic Belt

The orogenic belt was formed by the eastward subduction of back-arc oceanic crust and closure of ocean basin of Paleo-Tethys Lancang River, continental-arc collision between Qamdo-Pu'er Block and Dongda Mountain-Lincang Magmatic Arc. This orogenic belt was not only

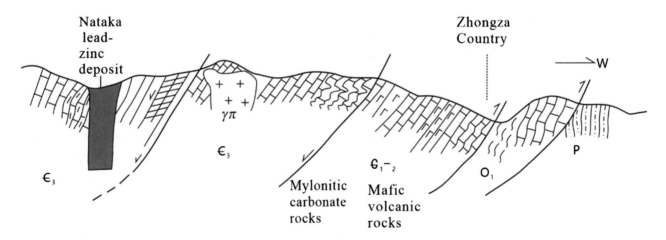

Fig. 3.3 Lower Paleozoic Strata on Najiao Zhongza Block Najiao in Zhongza Township, Batang County, thrusting over Permian Stratum in Jinsha River Tectonics

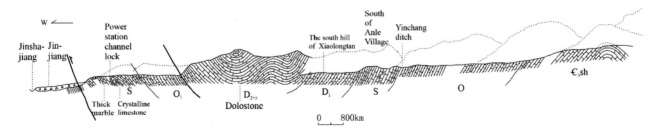

Fig. 3.4 Section of Yinchanggou-Jinsha River Route

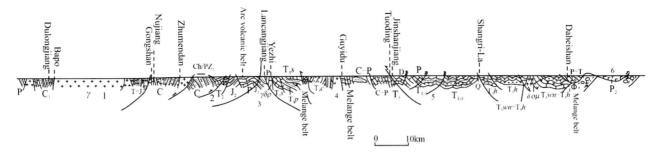

Fig. 3.5 Schematic Diagram of Bapo-Luoji Tectonic section. 1—Gaoligong Mountain Thrust Schist; 2—Biluo Snow Mountain-Chongshan Thrust Schist; 3—Lanping-Pu'er Back-thrust Spreading Foreland Basin; 4—Thrust Schist in Jinsha River Tectonic Zone; 5—Thrust Schist of Zhongza-Shangri-La Block; 6—Haba Snow Mountain-Yulong Snow Mountain Thrust Schist in the Thrust Schist on the Western Margin of Yangtze Landmass

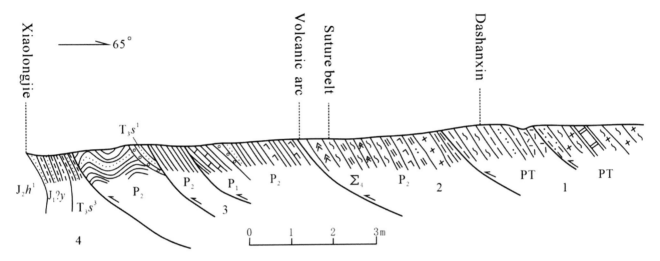

Fig. 3.6 Mountain Core Tectonic Section of Ailaoshan, Xiaolongjie Village, Jingdong County (Based on Regional Geological Survey Report on a scale of 1:200,000 in the eastern part of Wanjing, 2004, revised). 1—Diancang Mountain-Ailaoshan Thrust Schist; 2—Ailaoshan Junction Zone and the Thrust Schist of the Passive Margin Rock Strata on the East Side; 3—Thrust Schist in Taizhong-Lixianjiang Arc Volcanic Zone; 4—Lanping-Pu'er Back-Thrust Spreading Foreland Basin Zone

formed at the same time as Jinsha River-Ailaoshan Orogenic Belt, but also had the same tectonic deformation and mountain chain shape (wedge-shaped), while they had opposite direction. That said, the upper crust of the subduction plate on the west side reversely obducted eastward on the Qamdo-Pu'er Block of the superimposed plate.

A back-thrust spreading foreland basin is also formed at the front margin of the thrust zone, that is, Qamdo-Pu'er back-thrust spreading foreland basin shared with Jinsha River-Ailaoshan Orogenic Belt (Figs. 3.2, 3.5, 3.7, 3.8 and 3.9). At the same time, their magmatic activities are similar. A sequence of calc-alkaline volcanic rock series was developed outside the original volcanic arc zone in the post-orogenic stage and high-potassium shoshonite was found.

It is worth mentioning that an abyssal sedimentary basin was developed in the period from Carboniferous to Permian in Xiaohei River in the west of Yunxian-Puer-Zhendong of Yunnan. The southern extension member of the basin is likely to be connected with the small ocean basin represented by the ophiolite zone in the Nan River Area in the north of Thailand. If so, this small ocean basin may have been a back-arc basin formed by the eastward subduction of the Lancang River Ocean in Permian. Therefore, the Southern Lancang River Zone may have been a land-arc collision orogenic belt, while the Northern Lancang River-Ulaan-Uul Lake zone is a land-land collision orogenic belt. This orogenic belt experienced orogeny again in the Yanshanian-Himalayan super-collision stage. The main mountains include Ulaan-Uul Mountain, Taniantaweng Mountain and part of Nu Mountain.

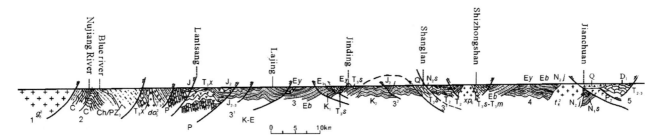

Fig. 3.7 Schematic Diagram of Bijiang-Jianchuan Tectonic section. 1—Gaoligong Mountain Thrust Schist; 2—Biluo Snow Mountain-Chongshan Thrust Schist; 3—Lanping-Pu'er Back-thrust Spreading Foreland Basin; 31—Thrust Schist in the Western Part of Lanping-Pu'er Basin; 32—Huachang Mountain Thrust Schist in the East of Lanping-Pu-er Basin; 33—Tongdian-Madeng Thrust Schist; 4—Thrust Schist in Jinsha River Tectonic Zone; 5—Diancang Mountain-Ailaoshan Thrust Schist

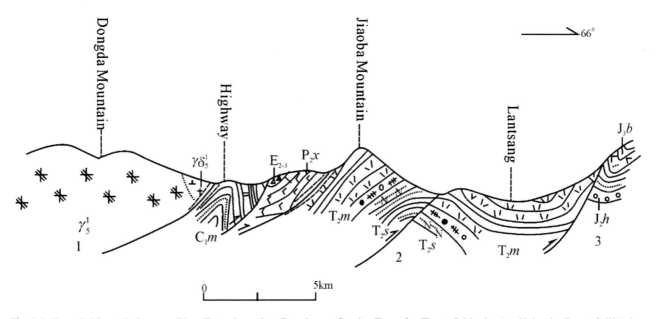

Fig. 3.8 Dongda Mountain-Lancang River Tectonic section (Based on Regional Geological Survey Report on a scale of 1:200,000 in Yanjing and Markam, 2004, revised). 1—Thrust Schist in Dongda Mountain Granite Zone; 2—Thrust Schist in Arc Volcanic Zone (C-T2); 3—Qamdo Basin Zone in the Northern Member of Qamdo-Pu'er Back-thrust Spreading Foreland Basin

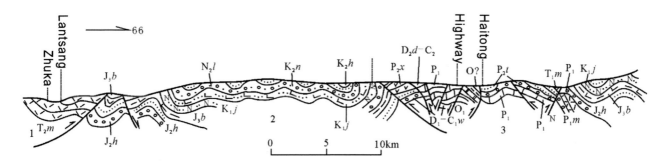

Fig. 3.9 Tectonic Section of Zhuka-Haitong (Based on Regional Geological Survey Report on a scale of 1:200,000 in Yanjing and Markam, 2004, revised). 1—Thrust Schist of Arc Volcanic Zone in Lancang River Tectonic Zone; 2—Qamdo Back-thrust Spreading Foreland Basin Zone; 3—Thrust Schist of Arc Volcanic Zone in Jinshan River Tectonic Zone

3.2.5 Meri Snow Mountain-Biluo Snow Mountain Arc-Land Collision Orogenic Belt

The orogenic belt was formed by the eastward subduction of the oceanic crust in Mali-Bitu and Changning-Menglian Zone in its south as well as the closure and collision of the ocean basin, that is, the collision between Jiayuqiao residual island arc and Zuogong-Baoshan Block. Due to large-scale eastward thrust and coverage of Gaoligong Mountain, there are different opinions on whether there are ocean basins between Bitu Ophiolitic Melange Zone and Changning-Menglian Zone, but the sea areas are connected between them, that is, there is an abyssal trough between them. The time of oceanic crust subduction and collision orogeny is roughly the same as that of Ulaan-Uul Lake-Lancang River Zone, namely the period from Late Permian to Triassic, and the subduction and collision orogeny was intensified in the Paleogene.

This collision orogenic belt is connected to Chongshan Metamorphic Zone in its south, and Chongshan Metamorphic Zone is characterized by a significant strike-slip ductile shear zone, and the melange zone which may be a tectonic pattern of the composition of Changning-Menglian Zone and the subzone of Lancang River Zone. Baoshan Block on the west side of Changning-Menglian Zone collided with Lincang Island Arc on the east side, forming a noncoaxial asymmetric fold-overthrust nappe orogenic belt on the eastern passive margin of Baoshan Block. Its basic characteristics are similar to those of the passive continental margin orogenic belt on the west side of Yangtze Landmass. Its foreland zone is composed of Shuizhai-Muchang Mesozoic Foreland Basin, while Ximeng Metamorphic Core Complex is outcropped in its rear margin due to spreading and splitting (Liu et al. 1993). The mountains formed include Laobei Mountain, Bangma Mountain and Ximeng Mountain. Due to the eastward thrust nappe in the east, a fan-shaped pattern will be formed when jointing with the mountains in the Lincang-Menghai area in the east. This fan-shaped pattern can also be found in the Wayao-Caojian-Liuku area at the north end of Baoshan Block and Chongshan area in the east. Its magmatic activity is characterized by the development of a sequence of Cretaceous boron-rich granite zone in the passive margin orogenic belt on the east side of Baoshan Landmass, which may be mainly formed in period from Triassic to Paleogene, and the mountain chain shape is generally asymmetric due to noncoaxial extrusion, that is, the super-posed mountain system with eastward thrust nappe is formed.

3.2.6 Boshula Ridge-Gaoligong Mountain Frontal Arc-Land Collision Orogenic Belt

Dingqing-Basu-Santaishan Ocean Basin (i.e., Bangong Lake-Nujiang River Ocean) subducted and closed toward the south and west, causing the Gangdise island arc to collide with Zuogong-Baoshan Landmass and then forming an arc-land collision mountain system. The member in this area refers to the area east and south of Dingqing. The famous Tanggula Mountain and Boshula Ridge are formed in the north member, and Gaoligong Mountain Range is formed in the south member, which were mainly formed in the Late Cretaceous-Paleogene.

The shape of the mountain chain varies in different areas. As mentioned above, Gaoligong Mountain is an asymmetric mountain chain, and Liuku-Mengga foreland basin was formed in its foreland-Baoshan Landmass (Liu et al. 1993), and Zuogong Late Triassic Foreland Basin was formed in Jitang Landmass, showing a passive continental margin fold-overthrust nappe orogenic belt. In the Basu area, the upper crust of the subduction plate on the north side reversely obducts southward to form a wedge shape on the superimposed plate, and Boshula Ridge and Gaoligong Mountain on the south and west sides are the frontal arcs on the south side of Tethyan Ocean, in which a sequence of tin-bearing crust-melting granite is developed to constitute the main tin ore zone in western Yunnan. Calc-alkaline volcanic activity occurred in the Tengchong area during the Neogene post-orogenic stage.

3.3 Sanjiang and "Hengduan Mountain" Orogenic Process and Dynamics

Due to the spreading of the Indian Ocean, the Indian plate continuously pushed toward the north, thus the collisional deformation pattern of Qinghai-Tibet Plateau since 50–60 Ma varies in terms of material movement states, kinematics and dynamics characteristics in different time periods, different parts of the crust and different levels of three-dimensional transformation process of lithosphere. Qinghai-Tibet Plateau is dominated by intracontinental strike-slip orogenic belt of Cenozoic Hengduan Mountain in the southeast part, borders the "Namjagbarwa Mountain Tectonic Junction Zone" at the northeast end of Indian plate in the west, connects to the western margin of Yangtze Landmass in the east and finally rotates clockwise in the north–south direction to form a strike-slip and is longitudinally faulted in the east–west Tethyan orogenic belt. The topography, tectonic deformation pattern and geodynamics are obviously different from those in the interior, the west and the north of the plateau. It is of special significance to study the collisional deformation, kinematics, collision dynamic mechanism of Qinghai-Tibet Plateau on its lateral collision effect and spatial and temporal difference, continental collision process, crustal spreading process and the coupling relationship between the deep development process and the dynamic change of the crust surface.

3.3.1 Global Plate Tectonic Setting of "Hengduan Mountain" Orogenic Process

From the perspective of global tectonics, the large-scale Alps-Himalayan Tethyan orogenic belt runs from east to west and then bends eastward and connects Cenozoic MABT in South Asia through Sanjiang Hengduan Mountain-indosinian peninsula arc. By taking Hengduan Mountain-Taima intracontinental north–south orogenic belt as the hub, completely different orogenic processes are shown in both the east and west sides. The southeast side is characterized by the evolution of the Cenozoic MABT in South Asia, which is controlled by the northward subduction of the Indian Ocean. Indonesia frontal arc is distributed in an arc shape and near E-W direction. Subduction, volcanic arc uplift, back-arc spreading, micro-landmass splitting, back-arc and foreland thrust, strike-slip, arc-land and arc-arc collision and other geological events occurred in Cenozoic.

The west side of Hengduan Mountain and the northern margin of India Landmass connecting Indian Ocean created the Cenozoic Himalayan orogenic belt after Gangdise arc-arc collision orogeny in the early stage of Late Cretaceous, subsequent collision with the Asian Continent and continuous intracontinental subduction. As a north–south transform fault, Ninety East Ridge of the Indian Ocean may play a role in regulating the different subduction collisions of the east and west sides, while it is characterized by different types of intracontinental deformation in the part extending into the interior of the continent, such as Hengduan Mountain strike-slip transition, rotational extrusion and extensional detachment, which are widely deformed in the north–south direction. The continuous northward extrusion of Indian Plate caused significant shortening and thickening of Himalayan-Gangdise crust; in addition, the passive impedance of lithosphere of Yangtze Landmass caused the dynamic imbalance between them, forming special strike-slip transformation form of Hengduan Mountain intracontinental orogeny. This transformation and convergence strain includes the westward overthrust nappe of India-Myanmar Naga Mountain in the west of Hengduan Mountain in Sanjiang, which extends to Longmen Mountain-Jinping Mountain area in the east.

3.3.2 Basic Characteristics of "Hengduan Mountain" Orogeny

"Hengduan Mountain" Orogenic Belt is one of the most complex orogenic belts in the world. Since the 1990s, many achievements have been obtained in the study of strike-slip transformation deformation of Hengduan Mountain zone caused by the collision between India Plate and Asiatic plate at home and abroad (1995). Hengduan Mountain Zone is characterized by tectonic patterns due to the oblique collision between the Indian Plate and Yangtze Landmass, such as significant thrust, overthrust nappe and strike-slip rotation; in addition, the associated pull-apart basin tectonics was formed in this zone.

3.3.2.1 Thrust Tectonics

After the collision between continents, the tectonic pattern was superimposed and Sanjiang Hengduan Mountain Zone was dominated by an asymmetric strike-slip thrust tectonic pattern with Qamdo-Markam-Lanping-Pu'er Basin as the central axis under the action of continuous shortening and extrusion of the crust (Fig. 3.8). According to the main marginal oblique overthrust zone and its fold-thrust sheet, the characteristics of both sides of the thrust tectonic are briefly described as follows.

Thrust Sheet with Westward Overthrust Nappe in the East

Haba Snow Mountain-Diancang Mountain-Ailaoshan Thrust Sheet
Cangshan Group and Ailaoshan Group in Yangtze Basement and Paleozoic strata overlaid on it thrust westward and southwest on Mesozoic and Cenozoic strata in Lanping-Pu'er Basin. This overthrust event mainly occurred at the end of Paleogene, showing that Yangtze Triassic limestone overthrust to the southwest on Paleogene red bed in Jinding Lead–Zinc Deposit. This event also caused the tectonic stratigraphic units in Yidun Island Arc Zone, especially in the southern end of Ganzi-Litang Junction Zone, to be covered and pinched out.

The eastern part of Lanping Basin is located in Weixi-Lanping-Yunlong, and Mesozoic strata in the basin overthrust westward on Paleogene red beds. Li et al. (1999) divided it into two thrust nappe schists: Huachang Mountain thrust schist and Tongdian-Madeng (and its southern area) thrust schist. ① Huachang Mountain thrust schist system was formed by Huachang Mountain thrust fault thrusting

Upper Triassic limestone on the east side westward on Paleogene and Neogene strata. In the east and northwest of Hexi Township, Lanping County, limestone (T_3) of Sanhedong Formation covers Yunlong Formation of Paleogene and Neogene and Cretaceous strata. The northward extension of Changshan Fault is cut by the NW-trending Weixi-Qiaohou fault or covered under the thrust schist of the volcanic-sedimentary rock zone on the west side of Jinsha River tectonic zone and connects with Bijiang River fault to the south. ② Tongdian-Madeng Thrust Schist. This thrust schist is on the west side of Tongdian-Madeng and its southern area and is mainly composed of rock strata of Upper Triassic Sanhedong Formation and Maichuqing Formation, which thrust westward on Middle Jurassic to Cretaceous strata. A sequence of detached blocks mainly composed of Upper Triassic strata is developed in Tongdian-Longtang-Wenshuimiao-Jinding Deposit, which is covered on Jurassic, Cretaceous, Paleogene and Neogene strata. Drilling in Jinding Lead–Zinc Deposit revealed that the stratigraphic sequence in the detached block composed of Mesozoic strata are inverted, indicating that they were once an inverted limb of an inverted or flat fold and suggesting that there are folds and thrusts inverted from east to west. The thrust zone is also cut by Weixi-Qiaohou Fault to the north.

Zhongza-Shangri-La Thrust Sheet

The sedimentary facies and biological features of Paleozoic strata of Zhongza-Shangri-La Landmass are similar to those of Yangtze Landmass, so it is a micro-landmass split from Yangtze Plate. In the Triassic, the back-arc oceanic crust of Jinsha River subducted westward and collided with the Qamdo Landmass. Influenced by the collision between Indian and the Asian continents, the Paleogene stratum is dominated by a large-scale thrust sheet that thrusts westward from Zhongza to Shangri-La.

The frontal fracture zone of the western thrust starts from Dongpu, Dege Country in the north, passes through Batang, Zhongza, Derong, Riyu, Nixi, Shangri-La City and Tuoding to Shigu in the south, generally extending over 600 km from north to south in the east boundary zone of Jinsha River Ophiolitic Melange Zone. The thrust zone thrusts westward on Jinsha River Ophiolitic Melange Zone, in which detached block in nappe and slip nappe composed of Paleozoic limestone from Zhongza-Shangri-La Landmass can be found everywhere. Bengzha Village, Batang Country, was once considered as the unconformity between the Upper Permian stratum and the Lower Permian stratum. It is judged that the slide nappe overlaid on Jinsha River Melange Zone based on our observations on sections of three routes. Inclined overturned folds are developed in the nappe zone, and the

significant tectonic mylonitization, flow cleavage and dynamic metamorphism can be found in the boundary thrust fault.

Obducted Sheet in Jinsha River Melange Zone

Obducted sheets in Jinsha River Melange Zone significantly thrust westward on Permian–Triassic Jiangda-Weixi Arc Volcanic Zone on the west side, and its frontal inclined fault is Aila Mountain-Xiquhe Bridge-Baima Snow Mountain-Gongnong fault. This sheet includes intra-oceanic arc remnants and back-arc basin subtraction complex assemblage (Fig. 3.10). A series of thrust faults and secondary shear zones can be found in this zone. Melange contacts with Paleogene red bed thrust fault in Baimang Snow Mountain and other places.

Thrust Sheet in Arc Volcanic Zone in Jiangda-Weixi

The thrust sheet of the arc volcanic zone in Jiangda-Weixi is mainly composed of Permian–Triassic volcanic-sedimentary rock series, under which Paleozoic Ordovician, Devonian, Carboniferous and Precambrian metamorphic rocks are developed and thrust westward on Mesozoic strata in Qamdo-Lanping. It can also be divided into two sub-thrust sheets: one is Jiangda-Mangcuo-Deqin arc volcanic rock thrust sheet in the east, of which the frontal tectonics is Zixiasi-Deqin Thrust Fault and the arc volcanic rock thrust on Paleogene Gongjue red bed; the other is Qingnidong-Haitong Foreland Thrust Zone in the west.

On the basis of Early Qingnidong-Haitong Thrust Zone, the frontal area of the thrust zone in Paleogene had been represented by Ziwei-Xiangdui frontal thrust fault system through the study of regional geological mapping and comprehensive mapping with a scale of 1:200,000 in Qamdo area in the eastern part of Qamdo Back-arc Foreland Basin (as shown in Fig. 3.11). Only in the west of Gongjue Basin, regional oblique thrust zone with an extension member of more than 50 km from northeast to southwest includes: Chesuo-Canggu Oblique Thrust Fracture Zone; thermal oblique thrust fault; Jueyong-Longda Oblique Thrust Fault; Shaxie River Oblique Thrust Fault; Ziduo Oblique Thrust Fracture Zone; and its thermal oblique thrust fault; Babeng Oblique Thrust Fault; Kangba-Jinda Oblique Thrust Fault; Mangzong-Zongbu Oblique Thrust Fault; Juelong Oblique Thrust Fault; Zalong Oblique Thrust Fault; Deri-Gangda Oblique Thrust Fault; Tuoba Oblique Thrust Fault; Lado Oblique Thrust Fault; Wengda Oblique Thrust Fault; Xiama Oblique Thrust Fault; Dama (Lajila) Oblique Thrust Fault; and Ziwei-Xiangdui Frontal Oblique Thrust Fault.

A sequence of strike-slip oblique thrust faults is superimposed to the southwest in a gentle arc shape, and Paleogene anatectic hypabyssal porphyries are developed along the anticline of the oblique fold on the hanging side of the

Fig. 3.10 Section of Geological Structure of Intra-oceanic Arc and Back-arc Basin in Zhubalong-Xiquhe (Based on Pan et al. 1996). 1—basalt; 2—arenopelitic slate; 3—gabbro-diabase; 4—argillaceous rock; 5—basalt; 6—andesite; 7—basaltic andesite and andesite; 8—basalt breccia; 9—Quaternary stratum

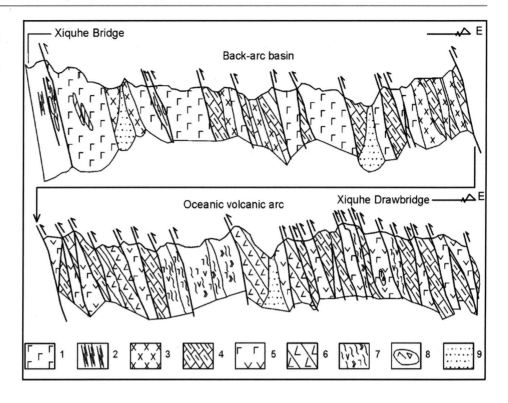

oblique thrust fault. Interfaces from Mesozoic strata in Qamdo Basin conformably overlaid on continuous marine to continental sedimentary stratum, so its tectonic deformation occurred in the Paleogene. It is closely related to the significant subduction and wedging of Indian Plate to Eurasia and the extrusion tectonic system of Qinghai-Tibet Plateau. Due to its special tectonics, the frontal faults of each thrust sheet are characterized by strike-slip and oblique thrust.

Thrust Sheet with Eastward Overthrust Nappe in the West

Dongda Mountain Thrust Sheet

In Riwoqê-Jitang-Dengba area on the west side of Qamdo Basin, the metamorphic rock series of Jitang Group (An∈) and Dongda Mountain Granite Zone on the west side thrust eastward on Carboniferous-Permian–Triassic strata in the arc volcanic zone. The arc volcanic zone thrusts eastward on Jurassic-Cretaceous strata in Qamdo Basin, with Lancang River Fault as its frontal fault. Accordingly, a sequence of schist which thrusts from west to east is also developed in the basin. From Rongxubing Station to the east of Zhuka, Markam Country, through Dengba, Dongda Mountain granite overthrusts eastward on Carboniferous-Permian strata, which overthrusts eastward on Upper Triassic and Paleogene red beds, and Middle and Upper Triassic strata overthrusts eastward on Jurassic-Cretaceous strata. In Zhuka-Lawu area, folds of Jurassic strata are characterized by inclined folds overturned eastward due to eastward thrust.

Thrust Sheet of Jiayuqiao Metamorphic Terrane

Its eastern margin is the Mali-Bangda-Chawalong Frontal Thrust Fault, and Paleozoic volcanic-sedimentary rock series in Xiyuqiao thrusts eastward on the littoral-neritic and marine-continental coal measure strata of the Late Triassic marginal sea and thrusts on Upper Paleogene red bed in Mali. The southwest side of the thrust sheet is developed with Luolong-Basu Thrust Fault, and Jiayuqiao metamorphic rock series is characterized by fan-shaped extrusion and uplift.

Chongshan-Lincang Thrust Sheet

In the northern member-Biluo Snow Mountain-Chongshan area, the metamorphic rocks of Chongshan Group (An∈-PZ) thrust eastward on Carboniferous-Permian strata in the arc volcanic zone. Similarly, Carboniferous-Permian strata in Lanping Basin in its front margin from west to east thrust eastward on Manghuai Formation and Xiaodixing Formation of Late Triassic. The latter thrust on Jurassic-Cretaceous strata, which thrust on Paleogene and Neogene strata, forming many detached blocks (Fig. 3.12). Finally, in Baiyangping-Yingpan-Yunlong area, Daqing Mountain-Beimang Mountain Gault thrusting eastward stands opposite Huachang Mountain-Bijiang River Fault thrusting westward on the east side of the basin. These two faults are close at hand, forming a thrust belt with Paleogene and Neogene basins as its axis.

In the Yun Country-Jinghong area in the southern member, the metamorphic rocks of Lincang Granite Zone

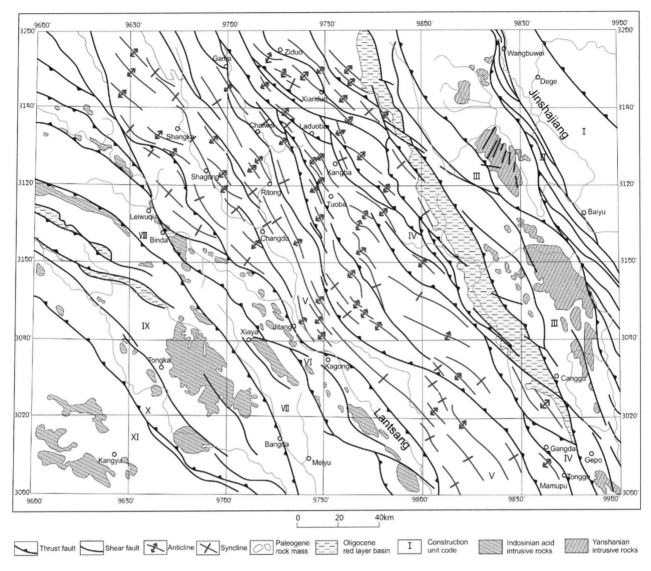

Fig. 3.11 Deformation of Cenozoic Strike-Slip-Thrust Tectonics in Eastern Tibet (Based on geological survey mapping data on a scale of 1:200,000, 2009, map generalization)

Fig. 3.12 Section of Qianzhuhe-Wenshuimiao Nappe Tectonics in the North of Lanping

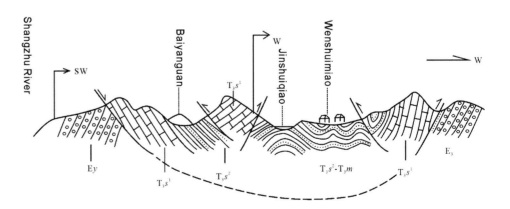

and Lancang Group (An∈) thrust westward on Permian–Triassic arc volcanic zone, and a sequence of thrust sheets is also formed in Lincang Granite Zone. The thrust zone and

detached block formed by Upper Triassic NE-treading thrust nappe is developed on the north bank of Lancang River between Yun County and Weishan in Lanping-Pu'er Basin.

In Wuliang Mountain area, Wuliang Mountain Group at the basement of the basin is also involved in the thrust nappe tectonics; that is, it thrusts northeastward on Mesozoic strata.

Gaoligong Mountain Thrust Sheet

Its frontal fault is Gaoligong Mountain Thrust Fracture Zone (including Nujiang River Fault, Lushui Fault and Longling-Ruishanyu Fault). The southern member of the fault overthrusts the metamorphic rock zone (An∈) in Gaoligong Mountain Group and the granite zone of Gaoligong Mountain on Paleozoic stratum on the west side of Baoshan Block to the east.

Gaoligong Mountain Fracture Zone is the dividing boundary between Baoshan Block (the northern extension member of Shan Block) and Gangdise continental crust island arc zone in the division of Sanjiang tectonic unit in Southwest China. In the Cenozoic intracontinental deformation, it also acts as the eastern end of continental crust block and significantly thrusts the frontal area of nappe eastward, showing as a large-scale ductile nappe fracture zone. Among them, the ductile shear zone with a width of 3–4 km and the mylonite zone with a width of 1 km on the west side are found, with a dip angle of 30–60. Closed synclinal folds that overturn eastward are often found along Gong Mountain-Fugong-Longling-Luxi area. The NW-treading extension member of the northern member of the fault is connected with Jiali strike-slip fault and may also be connected with the fracture zone in Zhaxi, Basu Country, which is limited by the low degree of study and has not been verified on the spot. The middle-south member of fracture zone extends in near N-S direction, slightly bulges to the east in an arc shape, turns to the SW direction at the place near Longling and is cut by the NS-treading dextral fracture zone.

The development of horizontal lineations on the foliation of the ductile shear zone along Gaoligong Mountain Watershed and a large number of noncoaxial rotating tectonics indicate the dextral shear of the fault, with the horizontal displacement of the shear zone above 50 km. The samples of mylonite from the ductile shear zone of Baoshan-Tengchong Highway collected by Zhong Dalai et al. are used to test $^{40}Ar/^{36}Ar$ ages of biotite and muscovite, which are 14.4 Ma, 15.0 Ma, 11.66 Ma and 23.8 Ma, respectively. These deformation ages indicate that Gaoligong strike-slip deformation peak is about 15 Ma in the Miocene and 23.8 Ma in the thrust peak shortened due to collision. Zhong proposed that the forward collision between two blocks shall be shortened to a certain stage. In order to adjust the deformation space, the oblique and tangential convergence collisions should be shortened gradually, and the secondary faults in the block are detached by rotation, being characterized by strike-slip tectonics. It is worth paying attention to the viewpoint that Shan Block slides to the south and east. Since Miocene, Pliocene–Quaternary fluviatile-lacustrine sediments in the extensional basins of Tengchong-Lianghe River-Longchuan-Longchuan River area in Southwestern Yunnan and the basalt interlayer and high-potassium calc-alkaline volcanic groups are the products of strike-slip pull-apart of crustal blocks.

3.3.2.2 Large-Scale Sinistral or Dextral Strike-Slip Tectonics

Deformation of India-Myanmar Mountains

Xilong protrusion is wedged obliquely from east to north on the west side of the "wasp waist" tectonic knot in Hengduan Mountain. Motuo NE-trending sinistral strike-slip fracture zone is found on the north side, and Daoji dextral strike-slip fracture zone is generally used as the southern sliding interface on the southwest side. In the Indo-Burma Mountains, Paleogene gully slope sedimentary rock series in the western depression overthrust westward, and in the Mugu belt of Myanmar, there are granite, gneiss and migmatite in the middle Oligocene–Miocene, which indicates that a thermal event occurred in the middle Paleogene and Neogene, which may be caused by the westward thrust along the Lochte thrust fault, and the Lochte thrust fault is northwest to Yarlung Zangbo River. Along the Sagaing-Namming fault in central Myanmar, there has been a right-handed translation of about 430 km since Miocene. It is pointed out that the basis of this fault displacement is as follows: ① the northern end of the eastern belt of the Indo-Myanmar mountains (in Naga Mountain) and the junction zone of Dagong-Myitkyina (north of Mandalay) were connected in Mesozoic, but now they are right-handed; ② Mayedeng complex on the west side of the fault belongs to the faulted part of the Precambrian wiped valley belt on the east side; ③ the schist and quartzite of Jiesha-Ganshan Mountain were originally connected with the Mish metamorphic rock body and were staggered. Nanming fault extends from the north into Assam, also called Miyou Fault, and then intermittently extends westward into Himalayan Boundary Fault. Quaternary basalts were found in three sites on the east side of Nanming-Sagaing Fault. Pliocene–Quaternary volcanic-magmatic activities and porphyry copper mineralization occurred in Bopa and Dongtonglong in the south of Naga Mountains. There is a strong pleistoseismic zone in the northern tectonic knot of Myanmar, and the focal depth tends to increase eastward. The above statement fully demonstrates that the tectonic knots are still in the convergence effect today. Indo-Myanmar mountain belt is strongly curved, and the right rotation of the Burma Sage fault for 13 Ma leads to the expansion of the Andaman Sea for 435 km, which may be related to the clockwise rotation of

South Asia to the south of the tectonic knot relative to the Indian plate, in addition to the oblique wedging of the west uplift and the westward pushing of Yangtze Landmass.

X-Type Strike-Slip Fault System

In the wedge of Assam in the northeast corner of India plate to the northeast, the Sanjiang Hengduan Mountain belt shows a large-scale left-handed or right-handed strike-slip structure in addition to the overlapping of crustal blocks caused by hedging and recoil in different parts. For example, Sagaing-Nanming dextral strike-slip fracture on the west side of Sanjiang Hengduan Mountain as mentioned above, with a strike-slip displacement of 430 km. At the same time, the strike-slip fault system with X-type distribution is formed on the two sides of Qamdo-Lanping block in Hengduan Mountain, Sanjiang, which regulates the strain components of the Changdu-Lanping block that are squeezed and contracted and extruded northward and southward, respectively. And the displacement of Jinping displacement body (similar tectonic stratigraphic unit to Haidong of Dali) in Ailaoshan belt, which is left-handed strike-slip by 350 km, reflects the west-to-left movement of the Yangtze Landmass.

In the eastern part of Qamdo block, Jurassic-Cretaceous strata formed a series of axial or NW-trending folds and faults, and the Eocene Gongjue strike-slip pull-apart basin formed, reflecting the right strike-slip characteristics of Zigasi-Deqin fault. The left strike-slip fault and Nangqian strike-slip pull-apart basin are mainly developed on the west side, which is characterized by the left strike-slip. Both of them reflect that Qamdo Block was split to the north due to extrusion (Li et al. 1999).

In Weixi-Qiaohou-Ailaoshan Fracture Zone on the northeast edge of Lanping-Pu'er Block, the left strike-slip fault formed a series of folds in Jurassic-Cretaceous in Lanping and Nanjian areas, which were obviously the product of this left strike-slip movement. The magnetic fabric analysis of 8 mylonite samples taken near the 95 km highway monument of Enle-Shuitang Highway shows that the easy (main) magnetization direction $D = 325°$ and the extrusion pressure direction is $58.2°$. According to the obtained anisotropy parameters of magnetic susceptibility E (flatness) $= 1.06°$, T (shape factor) $= 0.33$, P (anisotropy) $= 1.25$, the geometrical shape of the magnetic susceptibility value ellipsoid is a squashed ellipsoid. The easy magnetization direction is nearly horizontal, which is very close to the stretching lineation direction of $327°$. Magnetic fabric research also reflects the characteristics of strike-slip deformation. The K–Ar age of the whole felsic mylonite is (18.7 ± 1.9) Ma, which is roughly similar to the $^{40}Ar/^{39}Ar$

age (20 Ma) of biotite, potash feldspar and amphibole provided by Wu et al. (1989).

On the western edge of Pu'er block, a mylonite belt with a width of tens of meters was found at the contact between Lincang granite in Xiaodixing, Yun County and Mesozoic Xiaodixing Formation, showing the right strike-slip characteristics, which corresponds to the southeast slip of Pu'er block. The interior of the block is mainly characterized by the formation of a series of Paleogene and Neogene strike-slip pull-apart basins, such as Yunlong Basin, Zhenyuan Basin, Jiangcheng-Mengla Basin, Weishan Basin and Jingdong Basin. In the Paleogene and Neogene of Zhenyuan Basin, there are folds with axial near east–west and slightly northward inclination, which indicates that strike-slip is still going on after the deposition of Paleogene and Neogene. Paleomagnetic data show that Zhenyuan was in the Cretaceous period about 3 north of today, which further proves that Pu'er block was pushed away from southeast. The mineralization of Jinding Lead–Zinc Deposit in Lanping is closely related to the geological background of this extrusion-detachment and induced surge of ore-forming materials.

Strike-Slip Tectonics on the East Side of Hengduan Mountain

The eastern edge of Hengduan Mountain is adjacent to the Longmen Mountain-Jinping Mountain Overthrust Zone, connecting with the West Qinling Mountains in the north, curving through Muli in the south and connecting with the Sanjiang belt on the west side of Yulong Snow Mountain. With a total length of nearly 1000 km, it is composed of a series of imbricate thrust sheets, nappes and metamorphic base blocks (Yu 1996; Xu et al. 1992). Longmen Mountain-Jinping overthrust orogenic belt was a continuous complex tectonic belt in the middle and late Mesozoic. Because of the displacement of thrust fault and large-scale strike-slip, two adjacent tectonic units rotate relatively and the development of Daxiangling structural knot, the continuity between them becomes blurred. Among them, the oblique cutting of Xianshuihe strike-slip fracture zone is the main reason that the eastern edge of the plateau is divided into two zones.

As shown in the tectonic map, affected by Xianshui River Fault Zone, the sinistral shear dislocation occurred between Longmen Mountain Thrust Zone and Jinping Mountain Thrust Zone; however, by which we cannot judge that the thrust zone in the eastern plateau was formed first and then was dislocated and split by Xianshuihe Strike-Slip Fault. Both of them were formed at the same time along with the strong uplift of the Qinghai-Tibet Plateau in Cenozoic. They are the Bayankala trough-shaped tectonic unit in the east of the Qinghai-Tibet Plateau and the Sichuan-Yunnan rhombic fault block on the southwest side of the Qinghai-Tibet Plateau, which both slide to the south and converge with Yangtze Block to the west. On the one hand, a huge Longmen Mountain-Jinping Mountain overthrust belt is produced. On the other hand, the sliding rate of Sichuan-Yunnan fault block to the southeast is higher than that of Bayan Har area to the east, resulting in the left-handed sliding nature of its boundary fault. At the same time, the emplacement of Zheduo Mountain-Gongga Mountain syntectonic granite occurred, with the main age of 10–15 Ma (Yu 1996). A series of alkali-rich hypabyssal granite porphyries, monzonitic porphyries and syenite porphyries of Himalayan crust-mantle mixed source are known in Muli-Yanyuan and Ninglang-Binchuan areas where the main boundary fault extends to the south. Their isotopic ages are 35–65 Ma, and rock formations and beds intrude into the Upper Triassic and Paleogene strata, respectively, and some intrude directly into the boundary fracture zone. This kind of magmatic rock formed at the same time as nappe tectonic belt may be closely related to the dynamic process of "up-thrusting and down-thrusting wedge" in which the lower crust in the eastern part of the plateau wedged into the Yangtze continental crust and the upper mantle and the middle and upper crust overthrusted onto the Yangtze Landmass (Yu 1996).

Like the Himalayan orogenic belt, Longmen Mountain-Jinping Mountain overthrusts eastward, while a large-scale extensional detachment takes place at the rear edge (western edge). The extensional structure is characterized by the formation of Wenchuan-Maowen ductile shear zone, the bending of intestinal strata with gravity detachment and the development of large recumbent folds. And more than 20 dome-shaped deformations and metamorphism with different sizes have been divided into metamorphic core complex, magmatic core complex and structural dome according to their structure, composition and mechanism. The fission ages of apatite obtained from Pengguan metamorphic complex and Baoxing metamorphic complex in Longmen Mountain are 4.3–18.2 Ma. The $^{40}Ar/^{39}Ar$ sealing temperature age of amphibole and biotite at the upper limit of fast cooling time of Danba gneiss dome is 20 Ma (Xu et al. 1999). The ages of amphiboles around Xuelongbao Metamorphic Core Complex range from 25 to 30 Ma (Xu et al. 1992, 1999), for example, the age of uplift of the Qinghai-Tibet Plateau, exhumation inversion of the metamorphic complex and formation of Longmen Mountain-Jinping Mountain Thrust Zone is between Oligocene to Miocene, which is completely consistent with the syntectonic magmatic activity, the detached block overlying on the omnidirectional anticlinal limb with Paleogene red bed as the core before the formation of Longmen Mountain and geological records of no orogenic unconformity in strata after the formation of Phanerozoic

strata in the middle-southern member of Longmen Mountain. However, the shortening range of the earth's crust is very small, the height difference between the mountain range uplift and the Sichuan Basin is over 4000 m, and the landscape of Hengduan Mountain falls slowly from north to south. The research on the coupling relationship between these geomorphic surface processes and deep tectonic processes will be an important direction for further research in the future.

3.3.2.3 Extensional Detachment Tectonics

It refers to the detachment fault formed by detachment lithosphere extension, including detachment fault in the process of metamorphic core complex formation and anti-slip normal fault caused by gravity potential at the rear edge of large-scale overthrust nappe structural belt. Anti-slip normal fault was often accompanied by the metamorphic core complex. The former is the cause of the latter, while the latter is the consequence of the former. The large extensional detachment faults or anti-slip normal faults in Sanjiang area are mainly formed at the rear edge of large overthrust nappe, such as the anti-slip normal faults and basin-range structures in Tengchong-Yingjiang area of the rear edge of Gaoligong Mountain thrust belt, the anti-slip normal faults at the rear edge of Lincang-Lancang nappe, the extensional detachment faults and basin-range structures in the eastern edge of Ximeng Group metamorphic core complex and Diancang Mountain metamorphic core complex and its detachment faults. This paper focuses on the Ximeng metamorphic core complex.

Ximeng Metamorphic Core Complex is located in the backland area of Gengmadashan fold-thrust zone on the eastern margin of Baoshan Block. Ximeng Group is outcropped in dome shape, ranging from Laojiezi Formation and Pake Formation in the core to Wangya Formation and Yungou Formation in the margin. Its metamorphic grade has changed from amphibolite facies to greenschist facies, and the sericite chlorite zone of greenschist facies is outcropped in the outermost margin and has been subjected to retrogressive metamorphism. There is a detachment fault between the dactylic erosive prism or mica plagioclase gneiss in Laojiezi and the mylonite marble in Pake Formation, and its tension lineation is southeast or east–west, and Ximeng tin deposit is located in this detachment fracture zone. Pake Formation was in contact with Wangya Formation by normal fault, and a sequence of normal faults tilting eastward or southeastward was developed to the east from the interior of the core complex. The larger faults include Wanggong Fault, Kuanghai Fault, Dabannang Fault and Kelai Fault. According to the research of Li et al. (1999), a series of faults with different sizes is also developed in Mengjiao, Cangyuan and the Ordovician-Carboniferous in Taierbu, Lancang.

On the north side of Laochang Mining Area, the basalt was reduced to only a few meters thick due to the stripping fault between the carbonate rocks in Middle and Upper Carboniferous strata and the basalt of lower Carboniferous strata and formed a traction fold together with the underlying Devonian sandstone. The bottom of carbonate rock is obviously broken, and this stripping fault extends southeastward into Laochang mining area, where some ore bodies of Laochang lead–zinc mine are located in the fracture zone between basalt and carbonate rock. Carbonate rocks have been mylonite in the waste slag of the tunnel, and sheath folds can be seen in large mylonite blocks. We measured the fluid inclusion temperature of carbonate mylonite at 350–370 °C, the pressure of 1600×10^5–1400×10^5 Pa and the ore-forming temperature of Laochang lead–zinc mine at 200–520 °C, which are close to each other. The measured temperature and pressure of the fluid inclusions in felsic mylonite in Laojiezi Formation of Ximeng Group are 460–520 °C and 2750×10^5 Pa, respectively, which is obviously much higher than that of carbonate mylonite formation in peripheral Laochang area, indicating that the temperature and pressure decrease outward from the core complex and reflecting the different formation depths or levels of mylonite (Li et al. 1999). The formation of metamorphic core complex was accompanied by acid magma intrusion, resulting in the formation of large and small Sama rock bodies and outcropping of concealed granite porphyries and quartz granitic thin vein containing disseminated chalcopyrite by drilling holes in Laochang lead–zinc mine area. The age of the large and small Samas and granite porphyries is about 50 Ma. The initial strontium isotope ratio of granite porphyry is 0.7113, belonging to crustal melting granite. The general buried depth of rock mass is large in southeast and small in northwest which is consistent with the western uplift of the metamorphic core complex.

At the same time, the Jurassic extensional faulted basins in the north–south direction are developed on the east and west sides of the Ximeng metamorphic core complex, forming a basin-range tectonic framework with alternating basins and mountains. In the middle basin, there is a sequence of red molasse formations from Middle Jurassic to Lower Cretaceous, and there are coal-bearing formations in Paleogene and Neogene inter-mountain basins in some places. According to the development period of basin-ridge structure, the extensional detachment fault started in Jurassic during the multi-island arc-land collision orogeny, and the emplacement of many Yanshanian crust-melting granite bodies in this area may be related to this action. This effect lasted until Paleogene and eventually led to the structural stripping of Ximeng Group. The absence of crust in this belt may also be the result of this detachment fault. The formation of Ximeng metamorphic core complex was caused by the tensile collapse of the rear edge of nappe belt, that is, the

east side of Gengma Mountain. The delamination and detachment of the crust caused by this tensile collapse may further promote the formation and development of the back-edge thrust fault of Lincang-Jinghong Island Arc Zone and Lancang River thrust fault.

Ximeng metamorphic core complex is on the same tectonic belt as the metamorphic core complex on the eastern edge of Shan block discovered in 1997 on the west side of Chiang Mai, Thailand, in the south. There is a large area of granite basement in the metamorphic core complex in Chiang Mai area. From this point of view, there is probably a large granite base lurking under the Ximeng metamorphic core complex, and the large and small Sama rock bodies, the hidden rock bodies in Laochang lead–zinc mine area and the hidden rock bodies implied by Xinchang skarn-type lead–zinc mine are just the rock branches protruding upward from their rock bases. Therefore, the intrusive activities and stripping faults of the latent bedrock and its branches accompanying the formation of metamorphic core complex play an extremely important role in the formation of tin, copper, lead and zinc minerals in Ximeng area. The buried granite (porphyry) rocks in the deep part of the factory should pay attention to looking for copper, molybdenum and other deposits.

3.3.3 Stress Field and Kinematic Model of Intracontinental Deformation After Hengduan Mountain Collision

3.3.3.1 Division of Deformation Zones

The analysis of Cenozoic stress field and kinematics is based on the field data in eastern Sichuan, Tibet and Western Yunnan (Fig. 3.13). The deformation in the late Cenozoic can be divided into four zones, which are separated by weak deformation zones and have obviously different forms of kinematics and stress fields (A, B, C and D in Fig. 3.13). ① The Pailong-Basu-Nujiang River dextral fault system in the northeast of Nanjiabawa structural junction, the principal compressive stress δ_1 is distributed radially around this structural junction. ② The middle part of Jinsha River-Ganzi conjugate fault system is located between Nanjiabawa structural junction and Gongga structural junction at the western end of Longmen Mountain. The principal compressive stress $\delta 1$ is east-northeast and the minimum compressive stress $\delta 3$ is northwest-southeast. ③ Xianshuihe sinistral fracture zone is on the west side of Longmen Mountain. It is a strike-slip fault extending from Tibet to Yangtze, passing through the end of Longmen Mountain fault and connecting with Xiaojiang fault system. $\delta 1$ and $\delta 3$ are roughly east and north, respectively. From the strike-slip fault of Xianshuihe in the west to the nappe fracture zone of Longmen Mountain Zone in the east, the stress field also

Fig. 3.13 Features of modern deformation in Eastern Tibet-Western Yunnan (Based on data provided by Pan et al. 1996)

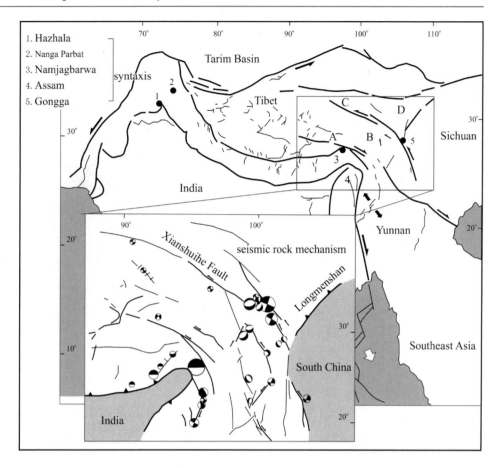

changes clockwise from west to east. ④ Longmen Mountain Nappe Fold Zone, δ1 and δ3 directions are NW and NE, respectively.

In the central part of eastern Tibet, western Sichuan and northern Yunnan, the most obvious feature of neotectonic activity is the NW–SE stretching from eastern Tibet to western Sichuan. (Batang-Litang Fracture Zone, Xianshui River Fracture Zone) is characterized by east–west stretching and north–south nearly horizontal and vertical shortening in the south, resulting in the north–south belt of Sichuan and Yunnan and the north–south strike-slip pull-apart basin in western Yunnan.

3.3.3.2 Stress Field and Kinematics Model

According to the development characteristics of strike-slip fault system and the stress field data obtained from the analysis of epicentral mechanical mechanism and geometric and kinematic data of large-scale faults in the area (Fig. 3.14a), we will discuss two late Cenozoic deformation modes: ① Elastic mode (Fig. 3.14b). Between the flank of Nanjiabawa and Gongga structural junction at the western end of Longmen Mountain, the track of principal stress is similar to the stress field distribution of the compressed rigid

plate, and there is an expansion along the unsupported edge of this rigid plate. ② Subsequent rheological model, the kinematics of the southeast Qinghai-Tibet Plateau fault can be explained by convergent streamline. This linear convergence toward the neck between the two tectonic nodes (Fig. 3.14c) reflects that the crustal material flowing out of central Tibet proliferates toward the neck between the Nanjiabawa tectonic node and Gongga Mountain tectonic node and decelerates and radially expands toward the Indosinian block on the south side. This stress field and streamline shows that the material is rotating clockwise from central Tibet to the southeast (Indo-China), but the moving speed is inconsistent, with a decreasing trend from west to east.

3.3.3.3 Regulation of Strike-Slip Transformation in Paleogene and Neogene

Jiali-Pailong fault and Jinsha River right-lateral strike-slip fault have been active since 20 Ma. According to the formation of syntectonic magmatic rocks, the neotectonic activity of Xianshuihe sinistral fracture zone can be traced back to Miocene at least, and they adjust the lateral shortening of the blocks on both sides. Paleogene and Neogene

Fig. 3.14 Main active faults in Eastern Qinghai-Tibet Plateau and their dynamic analysis (Based on data provided by Pan et al. 1996). **a** Main active fault types, recent strain field and 3 tracks in the central and eastern plateau, western Sichuan and Yunnan (1 > 2 > 3 principal strain); **b** elastic mode; **c** modern deformation mode

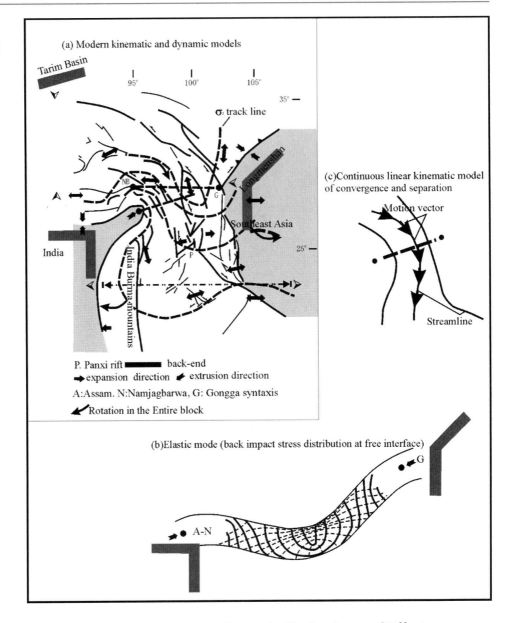

orogenies in Hengduan Mountain between Bangong Lake-Nujiang River Suture Zone and Ganzi-Litang Fracture Zone at 32° N in Eastern Tibet are dominated by dextral compression and twist, accompanied by nappe shortening. The orogenic belt presents a series of spatio-temporal structural patterns with dextral thrust or recoil fault structures. The Paleogene and Neogene kinematic forms of several laterally shortened blocks bounded by conjugate strike-slip faults indicate that apart from the near-vertical stretching, there is also the adjustment effect of strike-slip transformation. In India-Myanmar Mountains adjacent to Hengduan Mountain, Naga Mountain in the west of Shijian Fault overthrusts westward, while Gaoligong Mountain in the east of Sagaing Fault overthrusts eastward, indicating that the dextral strike-slip of Sagaing Fault has a shift and adjustment effect along the strike.

3.3.4 Dynamic Mechanism and Effect of Intracontinental Deformation After Hengduan Mountain Collision

Once the lateral extrusion model of plateau crust was proposed, it has been supported and demonstrated by many scholars, with considerable influence. With a great deal of research on crustal geology and deep geology, especially the GPS survey, paleomagnetism survey, VLBI survey and the appearance of surface geology and deep geology, many new understandings have been obtained (Harrison et al. 1992; Molnar et al. 1993; Li et al. 1995; Zhong and Ding 1996; Lv et al. 1996). Generally speaking, the characteristics of plateau uplift, such as segmentation, multi-stage and multi-factors, and deep dynamic action (gravity equilibrium, mantle plume, large-scale partial melting rheology of lower crust, plate bottom cushion, etc.) are emphasized. Typically, Owens and Zandt (1997) put forward a new understanding on the basis of a large number of geophysical achievements. They believed that the uplift of the plateau was related to these joint influences, such opposite subduction of South India Plate and Eurasian Plate in the north, and the large-scale melting and rising of the lower crust in the plateau and the underpinning of the plate bed. These understandings have had a great impact on the early representative model, the theory of double crust caused by the simple intracontinental subduction of the Indian plate and the theory of horizontal shortening and vertical thickening of the crust caused by the simple extrusion. However, it is worth noting that the above-mentioned model is mainly based on the study of the north–south large section inside the Qinghai-Tibet Plateau, emphasizing the north–south dynamic action between India and Europe and Asia, while little consideration is given to the Cenozoic geological records and deep geological data analysis in the eastern part of the Plateau. Many geological and geophysical data show that the collision deformation and uplift mechanism of the plateau must take into account the northwest subduction or intracontinental obstruction of the east Yangtze Landmass.

3.3.4.1 Deformation Mechanism in the Eastern Part of Qinghai-Tibet Plateau

In recent years, surveys from GPS, paleomagnetism and VLBI have reached similar conclusions. Since the Pleistocene, the movement rate of the Indian plate to Eurasia has been 50 mm/a, only a small part of which is absorbed by compression shortening and most of which is absorbed by the shear component dominated by strike-slip. If the simulation experiments done by Tapponnier and Molnar are correct, as they put forward, the Qinghai-Tibet Plateau will "escape" eastward along a large strike-slip fault due to a large-scale oblique collision, which will inevitably lead to a large-scale eastward migration of Yangtze Landmass, and the uplift of the Qinghai-Tibet Plateau will lose important boundary conditions. The VLBI data show that the migration rate of the Yangtze Landmass to the southeast is only about 8 mm/a. Therefore, Molnar held that the penetration of the Indian plate into Europe and Asia is mostly regulated by the thickening of the Asian crust, and the eastward "escape" is only a small effect. However, he neglected the large-scale intracontinental subduction along the Longmen Mountain-Jinping thrust nappe orogenic belt on the northwest margin of the Yangtze plate. It may be that this subduction prevented the eastward "escape" of the Qinghai-Tibet Plateau material and formed a huge vortex structure in Hengduan Mountain orogenic belt.

The Muli-Yanyuan Cenozoic nappe in Jinping Mountain, southwest China, moved at least about 150 km to Yangtze Block. In the late stage of Longmen Mountain-Jinpingshan orogenic belt, Xianshuihe-Xiaojiang fault, which turned from north to south to northwest, split into two parts. In the early stage (20 Ma), it was a large left-handed ductile shear zone with a translation distance of 80–100 km (Xu et al. 1992). A lot of data show that Xianshuihe and Xiaojiang faults are ancient faults, especially Xiaojiang fault's activity time can be traced back to Paleozoic. Before the Cenozoic, they were not connected with each other. Since the Cenozoic, they have been connected into one, which has transformed into a sinistral strike-slip extensional property. Four north–south sinistral strike-slip faults in central Yunnan have shifted 45 km, 55 km, 57 km and 60 km from east to west, respectively, which play an important role in controlling crustal deformation and stress distribution in this area. This situation is related to the short-range effect of the Yangtze plate on the intracontinental collision of Qinghai-Tibet Plateau.

3.3.4.2 Formation of Yunnan-Tibet Vortex Tectonics Due to Oblique Collision of Indian Continent

Recent research data show that the collision between India and Eurasia is not a simple forward collision but an oblique collision from west to east, and the oblique collision may be the most obvious in the eastern part of the plateau. According to Koons's research on the Southern Alps orogenic belt in New Zealand, the frontal collision forms a wide and slow peak-valley mountain range, while the oblique collision forms a narrow and steep peak-valley mountain range. The formation of the steep Hengduan Mountains and the Sanjiang valley in the eastern part of the Qinghai-Tibet Plateau is the important lateral effect of this collision. Several important tectonic thermal events of oblique collision were recorded. Three thermal events of 25–18 Ma, 13–7 Ma and 3 Ma–modern times have been recorded in the mineral fission track ages of granite exposed at different altitudes in several granite belts in Demula-Chayu, east of Gangdise. The peak activity times of Gongrigabu strike-slip fault in Red River in western Yunnan and eastern Tibet are 23 Ma and 24 Ma, respectively. The dextral shear of Sagaing Fault in Myanmar for 13 Ma led to the opening of Andaman Sea for 435 km, and the dextral displacement of Gaoligong fault occurred for about 100 km at 12.7 Ma. The estimated slip distance along Jiali-Pailong dextral strike-slip fault was over 200 km since Miocene (20 Ma), and the strike-slip speed of Quaternary stratum is between 10 and 2 mm/a.

Modern crustal movement in Eastern Tibet-Western Yunnan is controlled by vortex tectonic. From 1991 to 1997, we carried out a GPS survey to obtain the velocity vector direction of the modern crustal movement from eastern Tibet to western Yunnan: eastern Tibet turned east, then turned to nearly north–south direction in northwest Yunnan, and then turned to southwest direction. If Basu, Shangri-La, Lijiang, Lanping and Tengchong are considered as the inner trace of the eddy, Yajiang, Xichang and Chuxiong are considered as the outer trace of the eddy, and Kunming, Tonghai and Gejiu are considered as the periphery of the eddy, the speed of each trace is relatively close, namely the speed of the inner trace ranges from 8.38 to 17.49 mm/a, the speed of the outer trace ranges from 7.7 to 8.61 mm/a, and the peripheral speed ranges from 1.57 to 5.64 mm/a. Due to the difference of rotation speed between the inner and outer circles, there will be dextral strike-slip (such as Jiali-Pailong, Gaoligong Mountain and Jinsha River Zone), conjugate shear (Sichuan-Yunnan block) and sinistral strike-slip (Xianshuihe-Xiaojiang sinistral strike-slip shear zone) from the core to the outer circle, respectively. The core of the vortex structure is the Nanjiabawa structural knot.

Based on the explanation of focal mechanism of natural earthquakes, we analyzed the kinematics and deformation stress field of the East Himalayan tectonic junction and its adjacent areas and concluded that the movement speed of western Sichuan, eastern Tibet and Yunnan was right-handed shear and rotated clockwise relative to that of southern China, and its rotation rate could reach 1.7°/Ma. We also concluded that the Himalayan wedged northward at a speed of $(38 \pm 12 \text{ mm})$/a and its P axis distributed around the tectonic junction in a fan shape.

The tomography study on the natural earthquakes reveals that the subducted plate of Namcha Barwa-Assam is inserted in the "horn" shape in a NNE direction, forming a high-speed anomalous body in the lower crust and the upper mantle and a NNE-trending "horn" shape within the depth of 90–300 km, which extends to the north or affects the north of 32° N. That is, the plastic vortex of material occurs between the deep high-speed body and the low-speed body, resulting in a new deep process. When the deep vortex is not in harmony with the upper crustal structure, it will be detached from the upper crustal structure, resulting in the disharmony of structure and landform and promoting the segmentation of active structures. Under the push of deep plastic rheology and the transmission of tectonic stress, corresponding geological structures will be produced, resulting in the coupling of the deep process and the upper crustal structure. Driven by the plastic rheology of deep material, the corresponding convoluted structure will be produced in the upper crust. The Yunnan-Tibet vortex is the material flow around this structural junction driven by the down-inserted subduction plate of Nanga Bawa-Assam, from the high-speed (high-density) area to the low-speed (low-density) area.

3.3.4.3 Intracontinental Subduction and Blocking of Yangtze Plate and Comprehensive Effect of the Oblique Collision of Indian Plate

The blocking effect of the Yangtze plate and the oblique collision of the Indian plate have obvious short-range effects on the eastern Qinghai-Tibet Plateau. Traces can be found from magmatic activity, mineralization, tectonic deformation and sedimentary records in this area.

Magmatic Activity and Mineralization

Magmatic activity is an important link to understanding the abyssal dynamic mechanism. Cenozoic magmatic activities in Qinghai-Tibet Plateau are mainly distributed in the west and southeast, corresponding to the two low-speed mantle zones in Qinghai-Tibet Plateau. The western part is mainly volcanic, with little intrusive activity; in the east, on the contrary, intrusion is the main activity. Volcanic rocks in the west are studied in detail, with obvious time–space zoning from south to north, which can be divided into three active periods: Eocene–Oligocene are mainly concentrated in Gangdise belt, and only Nangqian and Basu are distributed in the east; the Miocene was concentrated in a large area of northern Tibet, and only Jianchuan had a small amount of activities in the east; The Pliocene–Pleistocene was concentrated in Qiangtang-Kunlun and Yumen, Qilian, with only a small amount of activities in Tengchong on the south side in the east. In recent years, it has been found that Yulong, Markam and Xiaojiang fault in Xianshui River in Sanjiang area have a large number of small intrusive rock masses, which were formed in 10–50 Ma, which are consistent with the distribution of Cenozoic strike-slip pull-apart basins, accompanied by large-scale mineralization and have an inevitable relationship with the unique stress release of the intracontinental collision of Qinghai-Tibet Plateau in the eastern arc structural belt. Further research also shows that these intrusions in the eastern margin of the Qinghai-Tibet Plateau can be divided into three categories: the first category is alkali-rich porphyry of crust-mantle mixed source, which was formed in the compression environment of 35–65 Ma; the second type is mantle-derived potassium lamprophyre, which was formed in the local extensional environment during 30–40 Ma. The third type is crust-derived granite, which was formed in the compression strike-slip environment in 10–15 Ma period. Tint granite in Zheduo Mountain-Gongga Mountain (Xu et al. 1992), Chayu-Bowo Himalayan Zone may be the product of partial melting of crust caused by ductile shear.

In a word, the relationship and difference between the eastern and western magmatic activities in time and space, the change of activity nature, provenance type and the coupling relationship between the main fracture stress states in time and space all reflect the characteristics of multi-block, multi-stage and multi-factor control of intra-continental collision in Qinghai-Tibet Plateau, which is also an important part of studying the interaction and dynamic mechanism between the eastern plateau and the Yangtze plate in different stages of Cenozoic.

Tectonic Deformation

Tectonic deformation is the most direct dynamic manifestation of continental collision. The eastern part of Qinghai-Tibet Plateau is adjacent to the famous eastern tectonic knot in the west, the eastern boundary of Yunnan-Tibet orogeny in the east and located in the eastern arc tectonic belt of Qinghai-Tibet Plateau. It is an important member for studying the east–west tectonic transformation and the release of continental lateral collision stress to the east. The data show that the core, margin and interior and east part of Qinghai-Tibet Plateau are dominated by the north–south extrusion, the thrust nappe and the east–west strike-slip, respectively. A series of east–west depressed basins and extensional basins were formed by compression and local extensional action in the interior of the Qinghai-Tibet Plateau and the central and western regions, but they were all transformed into strike-slip pull-apart basins arranged in a geese-like manner from the east to the arc. Many data also prove that the fault system in the eastern plateau may have undergone different nature changes since Cenozoic. With Xianshuihe-Xiaojiang fault as the boundary, the NW-trending fault (such as Red River fault) on the west side has been left-handed since about 20 Ma and right-handed since Quaternary (Royden et al. 1997); in addition, the NW-trending fault of the fault to the east of Xianshui River shows that it may be right in the early stage and left since Quaternary (Molaner et al. l996). The whole active fault in the east has been restricted by clockwise vortex structure since Quaternary. Earthquakes along the Jianchuan and Lijiang-Dali lines often take place in the changing zones of crustal movement velocity. Measurements show that the movement velocities of Shangri-La, Lijiang, Lanping and Xiaguan stations relative to Chengdu railway station are 8.6 mm/a, 17.0 mm/a, 12.9 mm/a and 7.2 mm/a, respectively, and their orientations are 176°, 223°, 218° and 216°, respectively. In addition, the Xianshuihe-Xiaojiang fault cut the Longmen Mountain-Qinghe Cenozoic thrust fracture zone on the western boundary of Yangtze Block at about 20 Ma, which was just coupled with the east–west lateral deformation of this fault and the internal deformation of the Cenozoic nappe structure in Longmen Mountain experienced several important changes (Xu et al. 1992). On the one hand, these problems illustrate the characteristics of tectonic deformation in the eastern part of the

plateau, and on the other hand, they also reflect the obvious short-range effect of continental collision on the eastern part of the Qinghai-Tibet Plateau. From this, we know more about the role of the intracontinental subduction of the Yangtze Plate to the northwest on the uplift of the plateau.

Sedimention Records

Sedimention record is an important part of understanding geodynamics. There are a series of different types of Cenozoic sedimentary basins in the eastern Qinghai-Tibet Plateau and western Yunnan, among which the foreland basin in the front of Youmenshan-Qinghe nappe and the strike-slip pull-apart basin in the north–south direction of eastern Tibet-western Yunnan are the most important ones. According to Wang Guozhi's research on sedimentary sequence interface, sedimentation rate and sedimentation flux of Cenozoic Longchuan Basin, Gengma Basin, Baoshan Basin and Yinggehai Basin at the outlet of Red River in western Yunnan, it is found that there is an excellent correspondence with the uplift history of Qinghai-Tibet Plateau. The uplift history can be divided into five stages: the initial stage of 23–19 Ma, the rapid uplift stage of 16.2–11 Ma, the denudation and leveling stage of 11–5.3 Ma, the sharp uplift stage of 5.3–1.6 Ma and the acceleration stage of denudation and uplift of 1.6–0 Ma. This understanding is similar to the conclusion obtained by Ding et al. (1995) on the fission track of granite in Bowo, Chayu. The research on the filling sequence of sedimentary basins in eastern Tibet-western Sichuan region, especially in the western region where the mechanical properties are controlled by Xianshuihe-Xiaojiang fault, and the response of plateau uplift is rarely involved, but this problem is very important for studying the outward expansion and deformation history of Qinghai-Tibet Plateau crust.

It is worth noting that the data of petrology, geological thermometer and geological chronology of the core complex in Nanga Bawa show that the average stripping rate is (4.5 ± 1.1) mm/a since 10 Ma, and at least 14 km of material has been denuded since 1 Ma. Strengthening the study of the coupling relationship between the deep process and the crust surface action is a very important subject for the collision deformation and dynamic analysis of the Qinghai-Tibet Plateau.

To sum up, the crustal deformation in Hengduan Mountain and its adjacent areas is influenced by the intracontinental collision of India plate, Eurasian plate and Yangtze Landmass (even the influence of the Pacific plate), with a very complicated evolution process. Hengduan Mountain is the best place to study the oblique collision between India and Eurasia and the blocking effect of Yangtze Plate.

References

Burchfiel BC, Chen ZL, Liu YP, Royden LH (1995) Tectonics of the Longmen Shan and adjacent regions, Central China. Int Geol Rev 37:661–735

Deng JF, Mo XX, Xiao QH, Wu ZX, Zhao HL, Luo ZH, Su SG, Wang Y, Liu C, Zhao GC, Qiu RZ (2002) Sequence of geological events and pTt paths of orogenic processes. Acta Petrologica Et Mineralogica 21:336–342 (in Chinese with English abstract)

Ding L, Zhong DL, Pan YS, Huang X, Wang QL (1995) Fission track evidence for the Neocene rapid uplifting of the eastern Himalayan syntax is. Chin Sci Bull 40:1497–1500 (in Chinese with English abstract)

Harris NBW, Xu RH, Lewis CL, Jin CW (1988) Plutonic rocks of the 1985 Tibet Geotraverse: Lhasa to Golmud. Philos Trans R Soc Lond A327:145–168

Harrison TM, Copeland P, Kidd W, Yin AN (1992) Raising Tibet. Science 255:1663–1670

Li JL, Sun S, Hao J, Chen HH, Hou QL, Xiao WJ (1999) On the classification of collision orogenic belts. Chin J Geol 34:129–138 (in Chinese with English abstract)

Li XZ, Xu XS, Pan GT (1995) Evolution of the Pan-Cathaysian landmass group and Eastern Tethyan tectonic domain. Sediment Geol Tethyan Geol 4:1–13 (in Chinese with English abstract)

Liégeois JP, Navez J, Hertogen J, Black R (1998) Contrasting origin of post-collisional high-K calc-alkaline and shoshonitic versus alkaline and peralkaline granitoids. The use of sliding normalization. Lithos 45:1–28

Liu ZQ, Li XZ, Ye QT et al (1993) Dividing of tectono-magmatic belts and distribution of the ore deposits in Sanjiang region. Geological Publishing House, Beijing, pp 1–246 (in Chinese)

Lv QT, Ma KY, Jiang M (1996) Seismic anisotropy beneath southern Tibet. Earthq Sci 9:279–287 (in Chinese with English abstract)

Miyashiro A, Aki K, Şengör AMC (1984) Orogeny. Wiley, Chichester, p 242

Molnar P, England P, Martinod J (1993) Mantle dynamics, uplift of the Tibetan Plateau, and the Indian Monsoon. Rev Geophys 31(4):357–396

Owens TJ, Zandt G (1997) Implications of crustal property variations for models of Tibetan plateau evolution. Nature 387:37–43

Pan GT, Chen ZL, Li XZ, Xu QH, Jiang XS (1996) Models for the evolution of the polyarc-basin systems in eastern Tethys. Sediment Facies Palaeogeogr 2:52–65 (in Chinese with English abstract)

Ren JS, Niu BG, Liu ZG (1999) Soft collision, superposition orogeny and polycyclic suturing. Earth Sci Front 6:85–93 (in Chinese with English abstract)

Sengör AMC (1992) Plate tectonics and orogenic research after 25 years: a Tethyan perspective. Earth Sci Rev 27:1–201

Suess E (1875) Die Entstehung der Alpen. Wien (Braumüller), p 168

Wu HW, Zhang LS, Ji SC (1989) The Red River-Ailaoshan fault zone—a Himalayan large sinistral strike-slip intracontinental shear zone. Chin J Geol 1:1–8 (in Chinese with English abstract)

Xu ZQ, Hou LW, Wang ZX (1992) Orogenic processes of Songpan-Garze Orogenic Belt of China. Geological Publishing House, Beijing, 190 pp (in Chinese)

Xu ZQ, Yang JS, Jiang M, Li HB (1999) The deep subduction of the continent and the uplift of orogenic belts around the Tibet Plateau. Earth Sci Front 6:139–151 (in Chinese with English abstract)

Yu RL (1996) A Cenozoic continental transform orogenic belt in southwest China. Acta Geologica Sichuan 1:1–5 (in Chinese with English abstract)

Zhong DL, Ding L (1996) Rising process of the Qinghai-Zizang (Tibet) Plateau and its mechanism. Sci China Ser D 4:369–379 (in Chinese with English abstract)

Mineralization and Metallogenic System in Sanjiang Region

As an important tectonic unit of Eastern Tethyan tectonic domain, Sanjiang orogenic belt started from the expansion, subduction, subtraction and closure of the Paleozoic Paleo-Tethys Ocean, went through the opening and closure of the Neo-Tethys Ocean and the subduction and collision process of the Indian continent and ended in the Himalayan comprehensive intracontinental convergence and uplift orogeny, forming a giant composite orogenic belt composed of several island arc collision orogenic belts and several stable blocks. With the wide range of orogeny, huge uplift amplitude, strong magmatic activity and intense mineralization, it ranks first in the global orogenic belt. The Sanjiang orogenic belt is also an important part of Tethys metallogenic domain, which is comparable to the Andean metallogenic belt in South America for its huge metal reserves, concentrated and large ore deposits, diverse ore deposit types, complex mineralization and huge prospective scale. There are not only a number of world-famous large and super-large ore deposits, such as Yulong copper deposit, Gacun polymetallic deposit, Jinding lead–zinc deposit, Ailaoshan gold deposit and Tengchong tin deposit, but also a number of new large and super-large ore deposits, such as Pulang copper deposit, Beiya gold deposit, Baiyangping silver polymetallic deposit, Xiasai silver deposit, Gala gold deposit, Yangla copper deposit, Dapingzhang copper polymetallic deposit, Nongduke gold-silver deposit, Duocaima lead–zinc deposit, Duri lead–zinc deposit, Hetaoping copper polymetallic deposit and Luziyuan lead–zinc polymetallic deposit.

The anatomy and metallogenic regularity of typical deposits in the Sanjiang orogenic belt have undergone several rounds of scientific and technological research, and important progress has been made. There are also multi-perspective studies on the summary of the metallogenic regularity of Sanjiang. Ye (1993) summarized the regional metallogenic regularity of the Sanjiang orogenic belt from the concept of metallogenic series and identified 19 metallogenic series of ore deposits; Li et al. (1999) extended the concept of metallogenic series, emphasizing that it is strictly controlled by the process of mountain-to-basin, basin-to-mountain and mountain-controlled basin in orogenic belt, and divided the Sanjiang mineralization into three structural metallogenic series and ten metallogenic series; Hou and Li (1999) systematically analyzed the mineralization of the Sanjiang orogenic belt from a new perspective of plume tectonics and proposed that the formation and evolution of the Sanjiang orogenic belt were restricted by plume tectonics, and the mineralization was composed of hot plume metallogenic giant system and cold plume metallogenic giant system. Based on the data obtained from multiple rounds of scientific and technological research, this chapter will explain the metallogenic characteristics of the Sanjiang orogenic belt and the temporal and spatial distribution of ore deposits under the guidance of the concepts of archipelagic arc-basin metallogenic theory and intracontinental tectonic transition metallogenic theory.

4.1 Metallogenic Event of Archipelagic Arc-Basin-Block System

It has been mentioned that the Sanjiang Tethys giant orogenic belt is an archipelagic arc-basin-(continent) block system formed by the shrinking and subduction of the Tethys Ocean in different periods of Phanerozoic and has been assembled through the combination of archipelagic arc collision orogeny. Along with the archipelagic arc-basin-block phylogeny from the Paleozoic to Triassic, multi-episodic metallogenic event sequence occurred, showing regular metallogenic spatial structure characteristics.

4.1.1 Paleozoic Metallogenic Events

The Paleozoic metallogenic events mainly occurred in two peak periods, namely, the intersection of early Paleozoic $\in$/O and $\in$/P. The former one developed in the interior and

W. Li et al., *Metallogenic Theory and Exploration Technology of Multi-Arc-Basin-Terrane Collision Orogeny in "Sanjiang" Region, Southwest China*, The China Geological Survey Series, https://doi.org/10.1007/978-981-99-3652-6_4

edge of the stable landmass at the initial stage of the formation of the MABT, while the latter one developed in the continental margin rift-ocean basin and the volcanic-magmatic arc formed by the subduction and closure of the MABT at its peak.

4.1.1.1 Early Paleozoic €/O Metallogenic Period

The mineralization mainly develops on the passive margins of the Zhongza block and Baoshan block on the east and west sides of the Qamdo-Pu'er block, mainly forming exhalative sedimentary lead–zinc deposits, which occurred in the Early Paleozoic marine carbonate rocks. The metallogenic age ranges from 426 to 655 Ma (Liu et al. 1993a, b). Representative ore deposits include Luziyuan lead–zinc deposit (€₃), Mengxing lead–zinc deposit (O₁), Hetaoping lead–zinc deposit on Baoshan Block and Najiao System lead–zinc deposit on Zhongza Block. Most of these deposits have undergone the transformation of tectonic–magmatic metallogenic events in the later period. The ore deposits in this metallogenic period are mainly lead–zinc mineralization and the superposition of copper and iron mineralization in the later period, so the ore deposits are large in scale. In recent years, new ore deposits have been discovered continuously in this area, showing great potential.

4.1.1.2 Late Paleozoic C/P Metallogenic Period

The mineralization mainly develops in three important metallogenic environments. The first is the Changning-Menglian ocean basin environment, in which Tongchangjie copper deposit related to Permian oceanic ridge basalt series (with Cyprus-type ore deposit characteristics) and Laochang Pb–Zn–Ag polymetallic deposit related to alkaline intermediate-basic volcanic rock are produced, showing the characteristics of volcanic-associated massive sulfide deposit (Yang et al. 1992). The second is the Late Permian volcanic arc environment generated by the eastward subduction of the Lancang River ocean basin, mainly producing massive sulfide copper deposits related to the Late Permian marine intermediate-acid volcanic rock series and volcanic-sedimentary iron deposits related to marine basic volcanic rock series. The former is represented by the Sandashan copper deposit, and the latter is represented by the Manyang Fe deposit. The third is the intra-oceanic arc environment produced by the westward subduction of Jinsha River ocean basin, in which volcanic-associated massive sulfide deposit related to Permian arc volcanic rock series is produced, represented by Yangla copper deposit. In general, the metallogenic period is dominated by Cu, Pb and Zn mineralization, followed by Fe mineralization. Volcanogenic massive sulfide deposits are the main types of deposits, which are large in overall scale and rich in prospects. The basic-ultrabasic rock masses of Jinggu Banpo, Jinghong Nanlin Mountain, etc., recently measured in the Lancang

River volcanic-magmatic belt are formed in the middle and late Permian (258–292 Ma, Lehman 2007), which may also be related to the basic eruption of the mantle series at the same time as the Emeishan basalt eruption. Pyrite, pyrrhotite and other sulfide mineralization are widely developed in the rock mass, which is an important target area for finding copper-nickel-platinum-palladium deposits in Yunnan except the Jinsha River-Ailaoshan belt.

4.1.2 Late Triassic Metallogenic Events

The Late Triassic became one of the most important metallogenic periods in the Sanjiang orogenic belt. The deposits were mainly formed in the closed period of the development of MABT. During this period, some arc-basin systems continued to develop (such as the Yidun arc), and most arc-basin systems entered the post-collision and extension stage. Therefore, there are at least three types of metallogenic environments in the Late Triassic: arc-basin environment, superimposed volcano-rift basin environment, and rift basin environment inside the block.

4.1.2.1 Metallogenic Events in the Yidun Island Arc Orogenic Belt

The mineralization of the Yidun Island Arc Orogenic Belt occurred mainly with the Indosinian subduction orogeny, and the deposits were mainly produced in the volcanic-magmatic arc and back-arc extension basins of the island arc orogenic belt, respectively forming two important metallogenic belts with different deposit types and metal associations (Fig. 4.1).

The copper polymetallic metallogenesis related to arc volcanic rocks develops along the volcanic-magmatic arcs, starting from Zengke in the north and reaching Shangri-La in the south, and is divided into two metallogenic sub-belts in the north and south. The deposits in the north sub-belt are concentrated in the Changtai Arc and are mainly volcanic-associated massive sulfide deposit (VMS) related to submarine volcanic exhalation. The ore deposits in south sub-belt are concentrated in Shangri-La arc, mainly include porphyry and skarn deposits.

At present, a super-large ore deposit (Gacun), a medium-sized ore deposit (Gayiqiong) and a series of small ore deposits and ore occurrences have been discovered in the sub-belt of massive sulfide deposits.

They occurred in the Late Triassic intra-arc rift belt of Changtai arc, and their metallogenic tectonic environment is similar to that of the Okinawa Trough Back-arc-Basin (Letouzey and Kimura 1986) and Miocene back-arc-basin of Japan (Cathles et al. 1983). Among the intra-arc rift belts, the most typical rift basins are Changtai and Zengke fault basins, where typical bimodal rock assemblages and limited

Fig. 4.1 Distribution of mineral deposits in the Yidun island arc orogenic belt

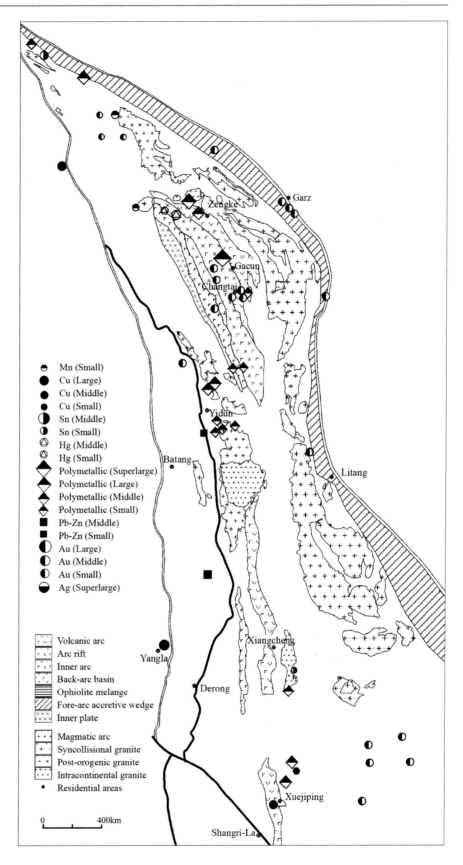

basin facies deposits are developed, with a water depth of about 800–1200 m (Hou et al. 2001a, b). Almost all massive sulfide deposits occur in confined or depressed basins within fault basins (Hou and Mo 1990; Xu an Fu 1993; Hou et al. 1995). The Re–Os age of the sulfide in the Gacun deposit is (217 ± 12) Ma (Hou et al. 2003a, b), and the K–Ar age of the altered surrounding rock of the Gayiqiong deposit is 210–221 Ma (Hou et al. 1995), that is, the mineralization time of the two deposits is basically the same.

In the porphyry-type polymetallic sub-belt, a super-large porphyry-type deposit, a large-scale porphyry-type deposit, a large-scale skarn polymetallic deposit and a series of small and medium-sized deposits have been discovered. The volcanic rocks in the belt belong to calc-alkalic basaltic andesite andesite-andesite-dacite series, the intrusive rocks are mainly a series of ultrahypabyssal porphyry and porphyrite homologous to volcanic rocks and the rock assemblage is diorite porphyrite-quartz dioritic porphyrite-monzonite porphyry-quartz monzonite porphyry-granite porphyry. Porphyry (porphyrite) occurs in groups of small stocks, bosses and dikes, controlled by NNW-striking fracture zone, intrudes into volcanic-sedimentary rock series and forms east–west belt. The diagenetic ages of the western belt are 235 Ma (Tan et al. 1991) and 249 Ma (Zeng et al. 2003, 2004); the diagenetic ages of the eastern belt range from 214 to 216 Ma (Zeng et al. 2000, 2003). The metallogenic age of the western belt is 224.6 Ma (Tan 1985), and the metallogenic age of the eastern belt is 213–216 Ma (Li et al. 2007). There are Xuejiping large porphyry copper deposit and Chundu porphyry copper deposit in the west, Pulang super-large porphyry copper deposit and Hongshan large porphyry-skarn copper deposit in the east. The deposit is located in a relatively concentrated porphyry (porphyrite) group complex and where several groups of structures are interlaced.

The epithermal Au–Ag–Hg metallogenic belt related to volcanic rocks is mainly developed in the back-arc-basin of Changtai arc. The mineralization is related to the "bimodal" (shoshonite-rhyolite) volcanic activity in the late Triassic back arc expansion period. The ore-bearing rock series is a high-potassium rhyolitic volcanic series, with an Rb–Sr isochron age of 213 Ma (Hu et al. 1992). The representative deposits are Kongma temple large mercury deposit and Nongduke medium gold-silver deposit. The deposit of Kongma temple is located at the northern end of volcanic zone in back-arc-basin. The volcanic rock belongs to the Miange Formation in Upper Triassic (T_3m). The deposit is produced in the ultra-acidic rhyolite in the middle section of Miange Formation, with its mineralization controlled by shear fracture zone in near north–south direction. The surrounding rock alteration is dominated by strong silicification, followed by sericitization and clayization. The ore body consists of altered rhyolitic clastic rocks in the fracture zone. The mineralized rock is rhyolite with breccia-like and porous structure and strong silicification-sericitization. Its occurrence is consistent with the surrounding rock. The ore belt is 8–27 km long and consists of more than ten lenticular ore bodies. The metal minerals are mainly cinnabar, with a small amount of pyrite, galena, sphalerite, orpiment and livingstoneite. The Nongduke deposit is located in the middle of the volcanic rock belt in the back-arc-basin (Qu et al. 2001). The mineralized acid volcanic rock series of the Miange Formation in Upper Triassic is rich in tuff and is characterized by high silicon content. In addition to native gold, there are antimonite, ramdohrite, silver-tenantite and so on. The ore-forming process includes two stages: the pre-enrichment of ore-forming elements in volcanic rocks and post-enrichment of magmatic hydrothermal reformation (Qu et al. 2001).

4.1.2.2 Metallogenic Events of Jinsha River Arc-Basin System

In the Jiangda-Weixi continental margin arc, a volcano-rift basin was formed by the late Triassic collision and extension, which was superimposed on the Permian epicontinental arc terrain. The metallogenic events represented by the VMS deposit were mainly developed in the superimposed volcano-rift basin on the continental margin arc, becoming an important polymetallic ore belt (Fig. 4.2).

In the volcano-rift basin in the southern segment of the continental marginal arc, the filled volcano sequence of sedimentary rock is composed of at least ten volcano-sedimentary cycles. The lower cycle is dominated by the basalt series, with thick basalt alternating with thin sand-slate and siliceous rocks; the middle cycle is composed of basalt in the lower part, calcareous siltstone in the middle part and limestone and thick rhyolite in the upper part; the upper cycle consists of siliceous slate and turbidite in the lower part and rhyolite series in the upper part. The volcanic activity shows a typical "bimodal" feature. At the top, it is a sequence of littoral-neritic clastic rock formations with molasse properties, intercalated with neutral-intermediate-acid volcanic rocks and pyroclastic rocks. In the Luchun deposit, its ore-bearing rock series can be divided into four volcanic-sedimentary rhythm units, with its lithofacies characteristics reflecting the development process of the extensional rift basin from tension fracture to fault depression to atrophy, as well as the evolution process of paleosedimentary environment of basin water body from shallow to deep and then to shallow again. In the "bimodal" volcanic rocks, subabyssal facies tuffaceous turbidite and sandy argillaceous flysch formations, there is volcanic-associated massive sulfide deposit. The Rb–Sr age of the ore-bearing rhyolite is 224–238.9 Ma (Wang et al. 1999), and the Rb–Sr age of the regional rhyolite with equivalent horizon is (235 ± 7) Ma (Mu et al. 2000). VMS deposits are controlled by horizon strictly, occur mostly in layered or stratoid form

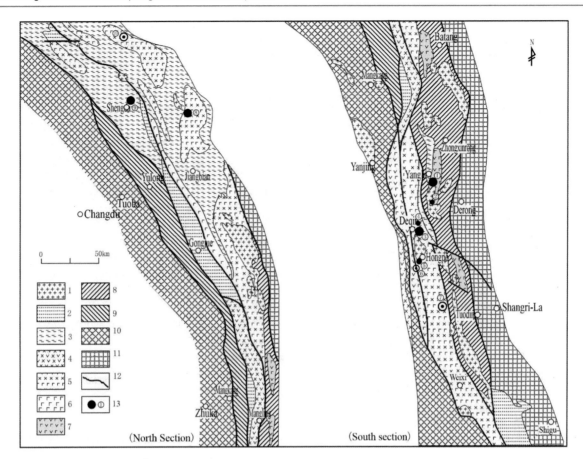

Fig. 4.2 Schematic diagram of Jinsha River Arc-Basin System structure-magma-mineralization. 1—granite (γ4-5); 2—sandstone sedimentary area (J-K); 3—clastic rock sedimentary area (T_3); 4—intermediate-acid volcanic rock belt in continental marginal arc (T_{1-2}); 5—Jiangda-Deqin-Weixi volcanic rock belt of superimposed rift basin (T_2-T_3); 6—Adenge-Nanzuo continental marginal arc intermediate-basic volcanic rock belt (P_1-P_2); 7—Jubalong-Benzilan basic volcanic rock belt in the intra-oceanic arc (P_1^2-P_2); 8—Deqin-Shimianchang tectonic melange belt; 9—Jinsha River tectonic melange belt; 10—Qingnidong-Haitong thrust zone; 11—Qamdo-Lanping landmass; 12—Zhongza-Shangri-La landmass; 13—Ore deposit (occurrence) and its number

and coexist closely with lamellar siliceous rocks and ribbon limestone. Luchun copper polymetallic deposit is the most typical deposit (ore deposit ③ in Fig. 5.2). However, there is no essential difference between the submarine hydrothermal fluid activity and sulfide sedimentary environment and the intra-arc rift environment produced by black ore-type minerals. Both types of rift basins provide the following important mineralization conditions, including magma system driving convective circulation of submarine hydrothermal fluid, volcanic rock series producing ore-forming materials, fault system transporting fluid migration and drainage and fault depression environment accumulating fluid and ore-forming materials. The upper cycle is composed of clastic rocks, intermediate-acid volcanic rocks and gypsum-salt formation with littoral-neritic molasse, and the sedimentary facies characteristics reflect the evolution history of rift basin with gradual shrinkage and shallow-water body. Among them, the Chugezha silver-rich siderite deposit is produced in the intermediate-acid volcanic rock series.

The ore bodies are disseminated, vein, reticulated and lenticular, which are strictly formed in stratiform-stratiform-like siderite beds and closely associated with barite siliceous rocks.

The sequence of sedimentary assemblages in the upper superimposed basin of the northern segment of the Jiangda-Weixi continental margin arc is generally consistent with that of the southern segment. The basin is filled with Upper Triassic sediments, with a sequence of red siliceous conglomerate molasses generally developed at its bottom, reflecting that the basin was formed in the extensional environment after the collision orogeny. The lower sequence is composed of fluvial-lacustrine facies clastic rocks and neritic limestone, which evolved into flysch turbidite formation composed of deep-water facies sand-slate series. Pillow-shaped basic lava and its clastic rocks are widely developed in some sections. Granite porphyry is commonly emplaced near the crater, showing a similar "bimodal" combination feature, which can be stratigraphically compared with the "bimodal" volcano-sedimentary sequence in the lower part of the Luchun-Hongpo basin; in this horizon, VMS deposit (Deposit ⑧ in Fig. 4.2) is typical of Zuna lead–zinc deposit. The ore-bearing rock series consists of carbonate rock, sandstone interbedded with barite layer and siliceous rock/hematite/ barite interlayers with stripes or bands, showing the geological features of exhalative-sedimentary mineralization. The upper sequence consists of littoral-neritic red-gray clastic rock and thin-layered limestone, with intermediate-basic and intermediate-acid volcanic rocks widely distributed. The volcanic rock assemblage is mainly andesite-dacite-rhyolite of potassium calc-alkaline series, which can be compared with the ore-bearing horizon of the gold-bearing and silver-rich siderite deposit in Chugezha. In this horizon, there are VMS deposits represented by Zhaokalong and Dingqin-nong silver polymetallic deposits. The ore-bearing rock series in Zhaokalong area is composed of thick stratiform micro-crystalline limestone in the lower part, middle gray intermediate-acid pyroclastic rocks and gray sand-slate, upper dark purple neutral volcanic rocks and medium-thin sand-stone and sand-slate. The direct host rocks mainly consist of clastic rock series composed of sandstone, siltstone, slate and tuff and exhalative sedimentary rock series composed of dolomite and siderite deposit. The metal sulfides are mainly veined, reticulated, densely disseminated and strictly pro-duced in the siderite deposit. The ore-bearing rock series in the Dingqinnong mining area consists of lower gray intermediate-acid volcanic clastic rocks and strongly altered siliceous layers, middle gray-white giant thick marble and upper acidic volcanic tuff. The ore-hosting rocks are mainly strong silicified rocks and acidic volcanic rock, and the ore body is produced between the middle and lower parts in layered and stratoid form and folded in the same shape as the surrounding rock strata.

4.1.2.3   Metallogenic Events of the Southern Lancang River Arc-Basin System

In bimodal assemblage rock series of shoshonite-rhyolite in the superimposed basin of Zuogong-Jinghong continental margin arc, a large number of ore occurrences and miner-alization sites have been discovered, especially Dapingzhang copper deposit, which was thought to have formed in Car-boniferous. The recent ore-bearing rock series and metallo-genic age dating results show that it may have formed earlier in the middle and late Silurian, which is a typical VMS deposit. See Fig. 4.3 for the distribution of main deposit deposits in this belt.

4.1.2.4   Lanping Basin

The late Triassic intracontinental rift evolution stage is one of the important metallogenic periods in Lanping Basin. Rift-rift basins and mineralization are constrained by the large-scale delamination of the lithosphere in the Late

Triassic. At this stage, the lower crust was greatly delami-nated, the hot mantle material upwelled in large quantities and the large basin extended, forming a central axis fault and a series of small contemporaneous faults. Driven by the upwelling source, the hydrothermal fluid migrated vertically along the contemporaneous fault, and intense hydrothermal activity occurred along the syngenetic fault zone, forming hydrothermal sedimentary siliceous rocks and hydrothermal sedimentary polymetallic ore bodies or mineralized bodies, which are scattered in the quantity of up to more than 100. Among them, there are two large-scale silver deposits and three medium-sized copper–silver polymetallic deposits, represented, respectively, by Heishan-Huishan silver lead–zinc deposit, Yanzidong silver–copper–lead–zinc deposit, Xiawu District-Dongzhiyan silver-copper deposit and Dongzhiyan-Hexi strontium deposit. The deposit is pro-duced in the Upper Triassic Sanhedong Formation, and the ore-bearing rock series is mainly composed of siliceous rock, fine-grained layered crystalline dolomite and dolomitic limestone, showing the characteristics of hydrothermal sed-imentation. Most of the ore bodies are layered-lens-like, controlled by specific stratigraphic horizons. The ore is in the shape of ribbon, breccia, mass and disseminated and is mainly composed of galena, sphalerite, chalcopyrite and freibergite. The ore-forming fluids mainly migrate vertically along contemporaneous faults, with a large content of Cl-, F-, Pb, Zn, Sb, Cu, Ag and other ore-forming material. Although these medium–low-temperature hydrothermal sedimentary polymetallic deposits have been strongly reformed in the Himalayan period, there are still a lot of hydrothermal sedimentary characteristics (Fig. 4.4).

In conclusion, with the development of the Sanjiang Tethys archipelagic arc-basin system from infancy, maturity to closure, three different episodes of metallogenic event have developed, which mainly occur in the inner margin of stable landmass, MABT and post-collision extensional rift basin. There are several important metallogenic belts and ore concentration areas in space, forming a metallogenic lineage dominated by base metals. The main metallogenic types consists of VMS-type, porphyry-type, exhalative-sedimentary-type and tectonic-hydrothermal deposits, and the metallogenic metal assemblage includes Pb–Zn, Pb–Zn–Cu–Ag, Cu and Sr-Ba assemblage, from simple to complex with the evolution of archipelagic arc-basin system.

4.2   Metallogenic Event of Intracontinental Conversion Orogeny

The intracontinental orogenic metallogenic event is the most significant metal metallogenic event in the Sanjiang orogenic belt and is typically represented by the Himalayan metallo-genic event. The development of this metallogenic period is

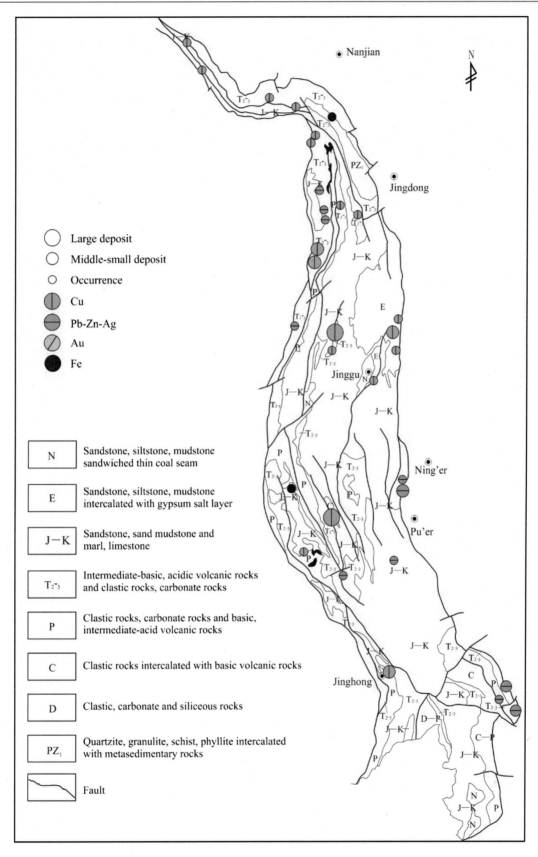

Fig. 4.3 Main deposits' (points) distribution map in the southern section of Jinggu-Jinghong metallogenic belt

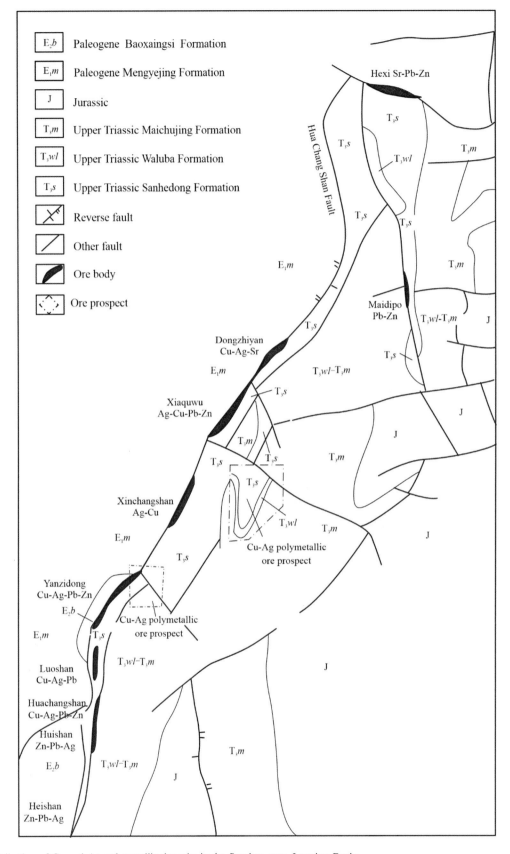

Fig. 4.4 Distribution of Cu and Ag polymetallic deposits in the Sanshan area, Lanping Basin

closely related to the strike-slip extension, nappe shear and associated porphyry system and basin fluid system produced in the Sanjiang orogenic belt by the collisional uplift of the Qinghai-Tibet Plateau. Many large and super-large deposits, such as Yulong porphyry copper deposit, Jinding lead–zinc deposit and Ailaoshan gold deposit, were formed during this metallogenic period.

4.2.1 Porphyry Copper–Gold Metallogenic Events

The porphyry copper–gold metallogenic events are mainly developed in the eastern margin of the Qinghai-Tibet Plateau formed by the collision of the India-Asian continent. Tectonically, this area is a tectonic transfer zone absorbing and adjusting the collision stress and strain of Indo-Asian continent and has successively experienced Paleozoic Paleo-Tethys orogeny and Himalayan large-scale intracontinental deformation. Its Paleozoic orogeny is mainly manifested in the subduction and reduction of the Paleo-Tethys ocean basin of the Jinsha River and the development of the Jiangda-Weixi arc. The Cenozoic deformation is mainly manifested as Eocene–Oligocene (40–24 Ma) transitional compression-torsional deformation, Early-Middle Miocene (24–17 Ma) transitional tension-torsional deformation and east–west extension since Neogene, with a series of strike-slip fault combinations in different directions developed successively. Among them, the western assemblage includes Jiali and Gaoligong Mountain strike-slip faults, developing around the eastern tectonic knot; the central assemblage includes the Batang-Lijiang fault in the northern section and the Ailaoshan-Red River fault in the southern section. The former is distributed in SN direction with right strike-slip, and the latter is extended in NW direction with left strike-slip. The two constitute the boundary fault zone between the Yangtze landmass on the east side and the Qiangtang terrane on the west side; the eastern assemblage includes the Longmen Mountain thrust zone and the Xianshuihe and Xiaojiang strike-slip faults. A series of derivative extensional basins are developed along the strike-slip fault, such as Gongjue, Jianchuan and Dali basins, among which Cenozoic alkali-rich intrusive rocks and potash volcanic rocks are developed, forming the famous Jinsha River-Red River alkali-rich magmatic rock belt. In this huge magma belt, there are two ore-bearing alkali-rich porphyry belts that are eye-catching. One is the Jiangda-Heqing-Dali alkali-rich porphyry belt, distributed along the suture between the two continental blocks, with the isotopic ages ranging from 48 to 27 Ma; another is the Shangri-La-Yanyuan-Yaoan alkali-rich porphyry belt, produced in the western margin of the Yangtze, with its isotopic age ranging from 48 to 31 Ma.

In the Jinsha River-Red River alkali-rich magmatic rock belt, there are two important porphyries copper–gold belt (Fig. 4.5) formed with the large-scale copper–gold mineralization occurred along with porphyry emplacement. Figure 4.5 shows the spatial distribution of important porphyry deposits in the tectonic transfer zone on the eastern margin of the plateau. It is roughly bounded by the Jinsha River ancient suture zone and is divided into two metallogenic belts in the east and west. The west belt starts from Jiangda in the north and reaches Xiangyun in the south, with a length of 750 km, including the Yulong porphyry Cu belt in the north; the east belt starts from Shangri-La in the north and stretches from the middle-south section to Beiya porphyry Au (Cu) deposit and Machangqing Cu–Mo–Au deposit in the south section, with a length of about 300 km. Generally, the western belt is produced on the suture zone between the Yangtze landmass and the Qiangtang terrane, controlled by a large-scale strike-slip fault zone; the eastern belt is produced on the western margin of the Yangtze block, with its northern section controlled by strike-slip faults, and its southern section is restricted by the inherited faults following the Panxi rift period. Porphyry metallogenic belts appearing in "pairs" have the following similar characteristics:

(1) Metallogenic age: the metallogenic age of the porphyry deposit can be accurately dated by using the Re-Os of molybdenite, or it can be estimated indirectly based on the crystallization age of the ore-bearing porphyry. Generally speaking, porphyry mineralization usually occurs 1–3 Ma before the last intrusion of ore-bearing porphyry. In the western belt, the Re-Os age of molybdenite in the Yulong porphyry belt is 35.6–35.8 Ma. The metallogenic age of the Machangqing porphyry copper-molybdenum deposit has not been directly determined, but it is estimated to be around 36.0 Ma according to the Rb–Sr isochron age of the ore-bearing porphyry. The mineralization history and mineralization type of the Beiya porphyry deposit are relatively complex. According to the K–Ar age of the Beiya syenite porphyry (48 Ma), its main metallogenic age is estimated to be (45 ± 2) Ma. In the eastern belt, the $^{40}Ar/^{39}Ar$ plateau ages of Cu-bearing monzoporphyry amphibole in Xifanping is 47.52 Ma, with the isochron age of 46.8 Ma, and the estimated metallogenic age of 44 ± 2 Ma. The K–Ar age of the Yao'an syenite porphyry varies from 31 to 50 Ma, and its metallogenic age of the deposit is estimated to be 34 to 47 Ma. The above data are somewhat speculative, but the metallogenic ages of the two belts are consistent in sequence.

(2) Ore-bearing porphyry. Similarities are shown as follows: ① The ore-bearing porphyry assemblages in the

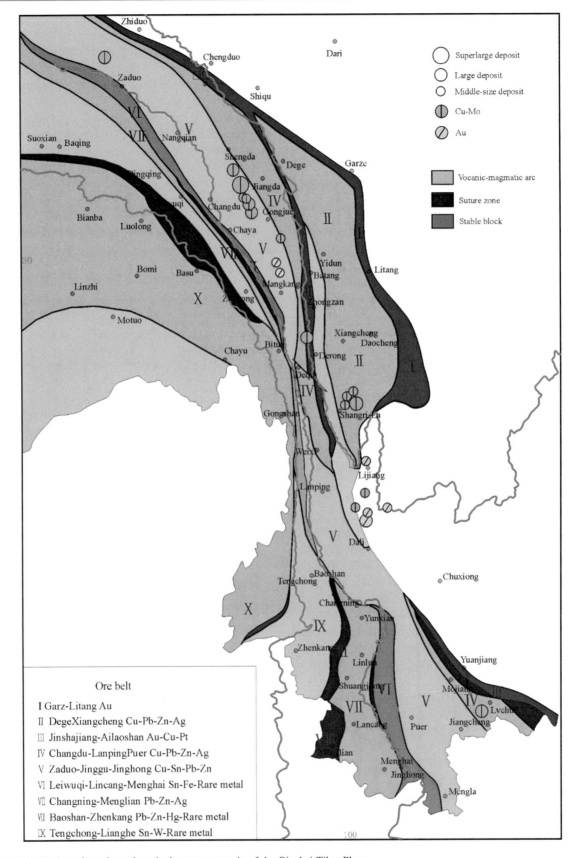

Fig. 4.5 Distribution of porphyry deposits in eastern margin of the Qinghai-Tibet Plateau

eastern and western belts are monzonitic granite porphyry, monzonitic porphyry and a small amount of syenite porphyry. From north to south in space, it changes from monzonitic granite porphyry to syenite porphyry. ② Ore-bearing porphyry bodies are mostly produced as small rock strains, and most of them are complex rock bodies with multiple intrusions. ③ In the complex rock bodies, the mineralization is mostly closely related to the meta-acidic porphyry intruded in the middle and late stages.

(3) Surrounding rock alteration. Surrounding rock alteration is mainly centered on rock mass, developing in an annular shape. Silicification nuclei are frequently found in mineralized rock mass, which are potassium-silicate lithification zone, quartz sericitization zone and propylitization zone in turn outward, and skarn lithification zone, marble lithification zone and hornstone zone are developed in outer contact zone.

(4) Metallogenic characteristics. Similarities are shown as follows: ① Although the production environments of the eastern Tibet porphyry deposits and the Pacific Rim porphyry deposits are different, the mineralization characteristics are basically consistent; ② similar mineralization assemblages appear in the eastern and western belts. The western ore belt is Cu, Cu–Mo and Au–Pb—Zn assemblages, and the eastern belt is Cu, Cu–Au and Au–Pb–Ag. ③ The mineralization types are similar, such as disseminated mineralization of veinlets in the porphyry body, sulfide-rich plate-like bodies in the contact zone and layer-like, lens-like and vein-like bodies in the surrounding rocks.

4.2.2 Metallogenic Events of Gold Deposits in Shear Zones

The shear zone-type gold mineralization events related to large-scale sinistral strike-slip and nappe shear have formed the famous Ailaoshan large-scale gold belt, which occurs in the strongly sheared Ailaoshan ophiolite melange belt (Hu et al. 1995). The metallogenic belt is 120 km long and 500–5000 m wide, consisting of four large Au deposits (Laowangzhai, Donggualin, Bifishan, Shangzhai, Jinchang, Daping), eight medium-sized gold deposits and over 30 small and ore occurrences. The main part is distributed along the Red River strike-slip fault zone and occurs in the ophiolitic melange tectonic slices near the three faults and distributed in the right-line oblique row (Fig. 4.6).

Tectonically, Ailaoshan fault and the Jiujia-Mojiang fault control the distribution of the Ailaoshan gold belt, the intersection of NW-trending brittle shear zone and near EW-trending thrust fault zone, the distribution of gold field or gold deposit. Brittle-ductile shear zones of different lithological layers control the formation of single deposits or ore bodies (Hu et al. 1995). Horizontally, the gold belt is generally controlled by the Upper Paleozoic tectonic-stratigraphic unit. The gold deposits are developed in tectonic slices composed of basic volcanic tuff-sedimentary tuff-clastic rock-crystalline limestone and radiolarian siliceous rock of Upper Paleozoic. Its mineralization intensity is positively correlated with the development intensity of Carboniferous basic-ultrabasic rocks and Yanshanian-Himalayan lamprophyre and granodiorite porphyry, reflecting that the mineralization is closely related to Ailaoshan oceanic crust material (ore source) and late magmatic activity (heat source) (Huang et al. 1997). According to the output characteristics and ore types of gold deposits in the Ailaoshan gold belt, the deposits are mainly of tectonic altered rock type or shear zone type (Hu et al. 1995). At present, such deposits are widely accepted abroad as orogenic gold deposits. It can be divided into three types according to the gold-bearing formations and ore types: Laowangzhai type, Jinchang type and Kudumu type. Laowangzhai-(Donggualin) type ore deposit directly occurs in basic lava, breccia, breccia lava and sedimentary tuff, quartz complex sandstone and sericite with strong pyritization, dolomitization and sericitization in Lower Carboniferous. Kudumu ore deposit occurs in the bedding shear zone of pyritization and sericitization tuff and basic lava of Middle Carboniferous. The Jinchang deposit mainly occurs in the outer contact zone of ultrabasic rock mass, forming Au ore body of strong silicification and carbonation ultrabasic rock type and Au ore body of metasomatic siliceous rock type. These ore bodies are mostly vein-like, lens-like and layer-like. The veins fill the fracture zone to form gold-bearing quartz veins and lens bodies; hydrothermal metasomatism in the contact zone of mafic ultrabasic rocks forms the layered and lenticular gold-bearing quartzite ore bodies (Hu et al. 1995). The altered mineral dating results of the ore-bearing host rocks and altered surrounding rocks indirectly indicate that the metallogenic age of the gold belt varies from 180 to 28 Ma, but the main metallogenic period is the Himalayan period, with an estimated age of 35–43 Ma (Huang et al. 1997).

4.2.3 Metallogenic Events of Tectonic-Fluid Polymetallics

Structured with strong intracontinental convergence and compression during the Himalayan period, thrust nappe, detachment gliding nappe and strike-slip-shear are the main tectonic forms of the crust surface in the Sanjiang area. The mountain systems on both sides of the Qamdo and Lanping foreland basins respectively shift toward the basins on a

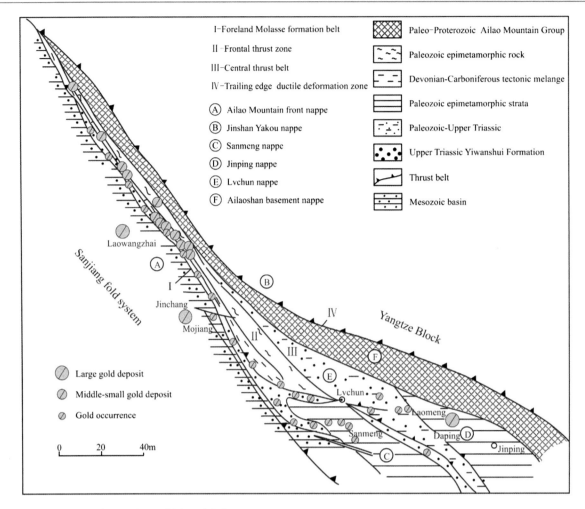

Fig. 4.6 Geological map of Ailaoshan gold deposit belt

large scale, and the medium and shallow crustal fluids controlled thereby migrate and converge from the orogenic belt to the basin, discharge and unload along the nappe belts or internal faults on both sides of the basin, resulting in metallogenic event of massive drainage-metal accumulation. The trans-unit tectonism in the process of intracontinental convergence, large-scale and multi-source fluid convergence and migration, as well as extensive material source fusion and massive metal accumulation have formed a medium-low-temperature hydrothermal polymetallic metallogenic belt with huge extension scale and a number of promising deposits.

4.2.3.1 Thrust Nappe Structure-Fluid Metallogenic Event

Regional fluid flow caused by tectonic compression during intracontinental convergence is the most important metallogenic factor. Driven by regional tectonic stress, massive fluids are discharged from the middle and upper crustal basins and orogenic belts and greatly converge and mix with

metamorphic hydrothermal fluids from the deep crust, magmatic hydrothermal fluids, tectonic hydrothermal fluids formed by intracontinental subduction and even mantle fluids and possibly heated atmospheric precipitation. Under the tectonic compression, these fluids from different sources can migrate and move on a large scale across tectonic units, extract and transport ore-forming material from wider provenance areas and form a huge metallogenic-fluid system. The fluid-metallogenic system formed mainly acts in the middle upper part of the crust. It is a medium-low-temperature hydrothermal fluid, with fluid action appearing in scale, and controlled by regional fracture. The tectonic channel for fluid reservoir discharge is a favorable place for mineralization. Alteration and discoloration, mineralization and geochemical anomalies of hydrothermal ore-forming elements related to hydrothermal activities are widely developed in the Middle and Upper Triassic, Jurassic-Cretaceous, Paleogene-Neogene red beds in the thrust belts on both sides of the Qamdo Basin, which well records and reflects such tectono-fluidic interaction and its spatiotemporal scale. The North Lancang River belt in the west of the basin is the largest thrust nappe belt developed along the Qamdo-Markam foreland basin and the Leiwuqi-Dongda Mountain island arc orogenic belt in the west of the Qamdo block. In the process of intracontinental subduction of the former to the latter or thrust nappe of the latter to the front, on the one hand, a large amount of tectonic hydrothermal fluid may be formed in the subduction zone and rise along the subduction surface; on the other hand, the large-scale eastward tectonic nappe has strongly squeezed the Qamdo-Markam basin, resulting in a large amount of drainage. Then, it mixes with the tectonic hydrothermal fluid from the subduction zone, rises along the thrust front, releases and unloads and causes extensive medium-low-temperature hydrothermal alteration and mineralization in the region (Fig. 4.7). The main types of alteration include silicification, argillization, carbonation, baritization, etc., accompanied by asphalt and nappe faults along Huoyexiong, Erluo Bridge and other fronts. Cold and hot springs are developed in modern times, and the (medium) low-temperature hot brine fluid is very active. The metallogenic belt and geochemical anomaly belt controlled by this run through the north and south, with mineralized sites and geochemical anomalies scattered all over the place, forming a number of potential ore deposits including Oluoqiao large arsenic deposit, Zhaofayong lead–zinc deposit, Xigang lead–zinc deposit and Laruoma lead–zinc deposit, indicating a promising prospecting. The ore-forming element assemblages are mainly (copper), lead, zinc, silver, arsenic, antimony, mercury, etc., the typical medium and low temperature assemblage. The secondary fault structures in the thrust front belt are the main ore-bearing structures, and the ore-forming hydrothermal fluids are mainly formed by filling and metasomatism. Though the medium-low-temperature metallogenic belt is formed in the Qamdo Basin, the metallogenic-fluid system is the result of the joint action of two adjacent tectonic units, and the ore-forming fluids and ore-forming materials have multiple sources. The same fluid-metallogenic system also appears in the thrust belt on the east side of the basin. During the further pushing process of the Jinsha River orogenic belt during the Himalayan period, a group of faults represented by the Tuoba fault may have been active again, and a westward thrust occurs, causing the drainage of ore-forming fluid in the basin on the western front belt, forming a medium-low-temperature hydrothermal deposit represented by the Duri large lead-silver deposit. The scale of the eastern thrust belt is far less than that of the northern Lancang River belt in the west, so the scale of the metallogenic belt formed is also small. However, the eastern thrust belt is adjacent to the Yulong Himalayan porphyry belt.

In addition to originating from the basin system itself, the ore-forming fluid and ore-forming material are likely to be joined by deep magmatic fluid or metamorphic fluid, which still have the conditions for forming large ores locally. The same orogenic belt-foreland basin thermal and dynamic fluid metallogenic system is also fully displayed in the Lanping Basin. The most prominent event in the intracontinental convergence process of the Lanping Basin is the large-scale nappe expansion of the orogenic belts on both sides of the basin, eventually forming a tectonic pattern of two large-scale thrust nappe tectonic belts symmetrical to the central axis of the basin. Along with the development and formation of hedging tectonic belts and strike-slip activities, the activity and migration of fluids in orogenic belts and basins were triggered, which generated a significant impact

Fig. 4.7 Relationship between the Jiqu Basin and the Machala uplift nappe structure in the North Lancang River belt and the mineralization of medium- and low-temperature hydrothermal fluids. 1—Upper Triassic—Jurassic; 2—Upper Paleozoic Carboniferous—Permian; 3—Hot brine; 4—Stratabound reformed copper-silver deposit; 5—Filling metasomatic polymetallic deposit

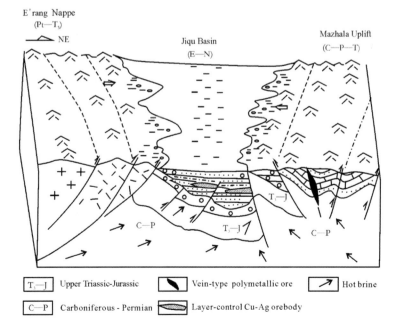

on the mineralization and metallogenic enrichment of the Cenozoic basin, and the basin entered a new era of large-scale mineralization since the Paleocene. In the Cretaceous period of unidirectional thrust nappe from the northeast to the southwest, the fluids mainly migrated and flowed laterally to the basin on a large scale. In the late Eocene–Oligocene, under the conditions of NE–SW offset compression and dextral strike slip, the fluid had undergone lateral-to-vertical migration-convection, and the overpressure fluid system in the deep basin mainly rose along the central axis fault zone with strike-slip properties in the basin and mixed with the basin fluid in the shallow red beds, providing a large amount of ore-forming materials. Then, it is metallized in the dome trap structure formed by extrusion thrust and nappe, forming the famous Jinding lead–zinc deposit and the Sanshan-Baiyangping large-super-large silver-copper polymetallic metallogenic prospect. Compared with the Qamdo Basin in the north, the fluid system of the intracontinental convergence basin in the Lanping Basin is mainly discharged through the central axis fault of the basin and the thrust fault on the east side of the basin. It is discharged through the front fault of the thrust belt on both sides of the basin.

4.2.3.2 Metallogenic Events of Strike-Slip Pull-Apart Basin

In the process of intracontinental convergence in the Cenozoic Sanjiang area, a series of Paleogene and Neogene tectonic basins of different sizes were formed, which were juxtaposed with nappe structure, strike-slip structure and gliding nappe structure, forming an important component of intracontinental convergence. The Cenozoic basins are mainly dominated by the regional deep and large faults in the foreland basins of Qamdo and Lanping and between them and the orogenic belts on both sides and are the strike-slip pull-apart products of these faults. The eastern Tibet area in the north is controlled by the dextral activity of the Chesuoxiang fault that divides the Qamdo Basin and the Jinsha River orogenic belt. The Paleogene Gongjue Basin formed by it extends for hundreds of kilometers from north to west, with a sedimentary thickness of more than 4000 m, and locally contains calc-alkaline intermediate base-intermediate-acid volcanic rocks. Controlled by the secondary fault on the east side of the North Lancang River giant thrust fault zone, the Nangqian Paleogene Basin and the Jiqu Paleogene-Neogene Basin were formed in the central and western parts of the basin, respectively. The Nangqian Paleogene Basin developed alkaline trachyandesite-trachyte volcanic rocks. The Cenozoic is related to the sinistral activity along the Red River fault. On both sides of the Lanping Mesozoic basin, the Qingkou-Mishajing-Qiaohou fault and the Bijiang River fault on the central axis of the basin are controlled to form strike-slip basins. The Lanping-Yunlong basin controlled by the latter is the largest, with a north–south length of 140 km and a east–west width of 5–10 km. These basins mentioned above are characterized by extensional or compressive strike-slip. Due to their connection with deep and large faults, some of them contain magma eruptions and become favorable places for deep fluid to rise and unload. Taking the famous Jinding lead–zinc deposit in Lanping-Yunlong basin as an example, the mineralization is mainly related to the discharge of deep fluid rising in the basin during strike-slip along Bijiang River fault (Fig. 4.8). In the Jiqu Basin of Qamdo, the Paleogene-Neogene red molasse clastic rocks have both sandstone-type copper mineralization and hydrothermal-type copper-silver-mercury polymetallic mineralization formed along basin-controlled faults, as well as discoloration and alteration of rocks, indicating that there is also the mechanism of deep fluid discharging to the basin along the fault structure.

4.2.3.3 Extensional Detachment Structure-Fluid Metallogenic Event

During the process of intracontinental convergence, the old orogenic belt further undergoes unidirectional or bidirectional outward thrust nappe, and extensional detachment structures appear to varying degrees at the trailing edge of the orogenic belt to control the activities from the orogenic belt and atmospheric precipitation. The basic model is that the fluid extracts minerals from the rocks of the orogenic belt with rich provenance and converges into the detachment and gliding nappe structures in the extensional environment. During the eastward thrust nappe of Jinsha River orogenic belt in the Himalayan period, the detachment and stripping structures occur along different tectonic layers in the extension and tension of the western rear edge, resulting in the filling and metasomatism of ore-forming hydrothermal fluids, forming the gold polymetallic deposits represented by Azhong (Fig. 4.9).

In the south, the Tuoding copper deposit, the Najiao system, the Gangriluo and other lead–zinc deposits and even the Xiasai super-large silver polymetallic deposits all have the obvious characteristics of such tectonic control or transformation. In the Lancang River orogenic belt, due to its large-scale thrust nappe to the Qamdo-Pu'er block, an extensional stripping zone developed in the Baoshan block at its trailing edge, forming the Shuangjiang extension basin, the Ximeng metamorphic core complex, and the Mengsheng extensional basin, accompanied by crustal molten granite and granite porphyry and related tin and lead–zinc deposits, such as Aying tin ore, Xinchang lead–zinc ore, and were simultaneously controlled by detachment faults (Li et al. 1999).

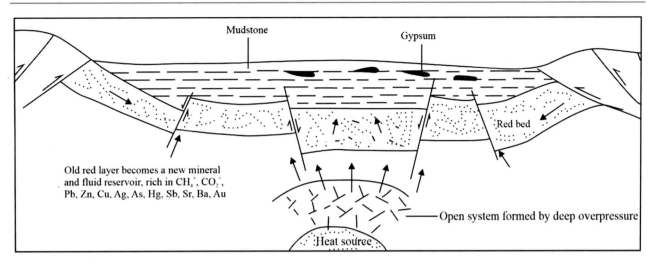

Fig. 4.8 Distribution map of fluid in the evolutionary stage of extensional strike-slip in the Lanping Basin

Fig. 4.9 Relationship between himalayan detachment structure and gold (silver) polymetallic mineralization in Jinsha River Orogenic Belt. 1—The metallogenic pre-enrichment layer related to Indosinian volcanism; 2—The deposit transformed and enriched by the Himalayan detachment structure

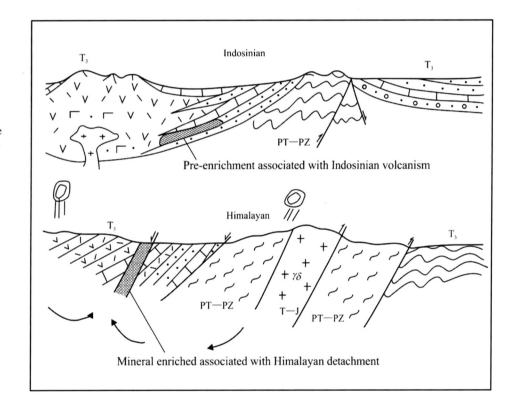

4.2.4   Metallogenic Events of Syn-Collision Granite Tin

The syn-collisional granite tin metallogenic event mainly occurred in Tengchong-Bomi granite belt, which constitutes an important tin and rare mineral and rare earth ore concentration area. Tectonically, the Tengchong-Bomi granite belt is produced on the Lhasa terrane between the Pangong Lake-Salween River suture (BNS) and the Yarlung Zangbo River suture (IYS) and belongs to the southeastern extension of the Gangdise granite base. Due to the strong wedging of

the Indian continent to the northeast and the subsequent development of the Nanga Bawa tectonic knot, the Gangdise granite base dominated by the Yanshanian granite takes a strong turn, and the western section is distributed in the near EW direction and is separated from the Tethys Himalaya in the south by the IYS. The eastern section is generally distributed in an arc in the near SN direction, limited by the BNS in the east and separated by the Naga Hills suture zone (where IYS is located in Myanmar) in the west. The strike-slip fault system or large shear zone that regulates the strong collision strain between India and the Asian continent is distributed around the tectonic knot. The northwest section is the Jiali fault zone, and the southeast section is the Gaoligong fault zone (Yin and Harrison 2000), especially the large-scale strike-slip-shear action of the latter, which fundamentally controls the formation and development of the Tengchong-Bomi granite belt. The Chayu-Tengchong granite belt can be roughly divided into two sets of granites based on the time limit of the Indo-Asian continental collision: Yanshanian granites before collision and collision-period Cenozoic granites. The Yanshanian granites before collision are related to the subduction of the Neo-Tethys oceanic plate, with the emplacement age and rock assemblages consistent with those of the Gangdise granites in the western segment; the collision-period Cenozoic granites and Yanshanian granites are coherently distributed, but the early potassium-rich granites are mainly concentrated in the Tengchong-Lianghe area, while the late two-mica granites are concentrated in the Chayu-Bomi area (Lu et al. 1993). The exposed area of granite in Tengchong area accounts for more than 50% of the whole area, forming three granite belts of different ages, different genetic types, interdependent in time and space and distributed in parallel. From east to west, they are the Early Cretaceous-Late Jurassic Donghe granite belt, late Cretaceous Palaeoyong granite belt and Paleogene Binlang River granite belt. The Binlang River granite belt mainly exposes complex rock bodies such as Huashui, Xintang, Lailishan, Xinqi and Bingwai. The Xintang complex rock bodies are produced in the form of a rock base, with an area of more than 36 km^2; the main body of the Lailishan complex rock mass is produced by rock strains, with an area of more than 7 km^2; the plane distribution of the Xinqi rock mass is elliptical, with the rock plant area of about 40 km^2. These Himalayan rock masses are clearly controlled by a series of strike-slip faults. Isotopic dating data show that (Chen et al. 1991; Lu et al. 1993) the Cenozoic granites during the Tengchong collision period have multi-stage emplacement characteristics. The Rb–Sr isochron age and K–Ar age of the earliest emplaced monzonitic granite limit the magma crystallization age to 58–60 Ma. The Rb–Sr isochron age and K–Ar age of the subsequently emplaced moyite are concentrated between 51 and 54 Ma, which limits the crystallization age of the second phase emplaced magma to about 52.5 ± 1.5 Ma. The Rb–Sr isochron age of the neoporphyritic granite or granite porphyry produced by the rock strain has a narrow range of 52.4–52.5 Ma. Although there are no direct dating data for muscovite granite and muscovite albite granite intruding into moyite in the form of bedrock intrusion, it is speculated that their formation age should be less than 51 Ma.

Rock geochemical studies show that the Al saturation index (ASI) of monzogranite and moyite is close to 1, which belongs to meta-aluminate to peraluminous granite. The ASI of muscovite (albite) granite varies from 1.02 to 2.63 and belongs to peraluminous to strong peraluminous granite. The $w(K_2O)/w(Na_2O)$ ratio of these granites is greater than 1, and the $w(K_2O)/w(Na_2O)$ ratio and SiO_2 content increase in turn. The content of trace elements in the former is relatively high in Sr, Ba and low in Rb, while the latter is obviously high in Rb and abnormally low in Sr, Ba. The muscovite granite is characterized by an abnormally high Y, and the muscovite albite granite is characterized by an abnormally high Rb; the former has a right-dipping LREE enrichment type, with obvious negative Eu anomaly; the latter has a "swallow-type" REE distribution patterns, suggesting that different types have different magmatic source rocks or melting mechanisms. The analysis of the tectonic environment shows that although the Tengchong Cenozoic granites were produced in the continental collision zone, the monzonitic and moyite at 51–60 Ma were formed in the stress relaxation stage after the strong collision between the Indian and Asian continents, while the formation of muscovite granite was related to the activity of Gaoligong strike-slip fault zone.

Tengchong tin and rare mineral and rare earth mineralization are related to moyite and muscovite albite granite, respectively. The former is represented by the large-scale tin deposit in Lailishan, and the latter is characterized by the rare ore and rare earth ore of Baihuanao. Together with various tin mineralizations in the Yanshanian period before the collision in the area, they constitute part of the 800 km-long tin belt in Southeast Asia. The ore-bearing moyite is characterized by high K, F, S and high initial ratio of $^{87}Sr/^{87}Sr$ (0.7124 to 0.7138), with its $w(Sn)$ varying from 150 to 200 µg/g, $w(Mg)/w(Ti)$ varying from 1.5 to 3.0 and $w(Zr)/w(Sn)$ varying from 10 to 416, showing the typical geochemical characteristics of Sn-bearing granites (Lehman et al. 1990). The tin ore body mainly occurs in the contact zone between the granite body and the surrounding rock and in the fracture zone of the surrounding rock. Tin ore is mainly of massive sulfide type, composed of cassiterite, a large amount of pyrite, pyrrhotite and a small amount of sphalerite and galena, with the grade of tin ore varying from 0.63 to 1.58% (Liu et al. 1993a, b). The Baihuanao muscovite albite granite is characterized by rich LILE (K, Rb, Cs and Li) and REE, and the mineralized granite has undergone a complex process from REE mineralized granite to Sn-W granite through Nb–Ta mineralized granite (Liu et al. 1993a, b). Disseminated and greisen type ores are the main types of mineralization, mainly produced in the inner and outer contact zones of granite. The mineralized metals are mainly of Cs, Li, Ta, Sc, Y, Sn and W, of which Rb resources account for about 1/3 of the global total Rb resources.

In conclusion, intracontinental orogeny and mineralization are mainly related to the large-scale collision process of Indo-Asian continent, and the main metallogenic events are related to four major geological processes: large-scale strike-slip (fault, pull-apart), large-scale strike-slip-shear action, large-scale nappe slippage and syn-collision magmatism. Four major metallogenic belts or ore concentration areas are mainly formed, namely porphyry Cu-Au ore belt, ductile shear (orogenic) Au ore belt, composite basin polymetallic ore concentration area and collision granite Sn ore concentration area.

4.3 Archipelagic Arc-Basin System and Collisional Orogenic Metallogenic System

The Sanjiang orogenic belt is an important tectonic-metallogenic unit of the East Tethys metallogenic domain. The metallogenic characteristics of the Paleozoic early Mesozoic and Tethys ocean continent transition process are attributed to the tectonic background of the archipelagic arc-basin system. After the opening and closing of the Neo-Tethys Ocean and the subduction and collision of the Indian continent, it ended in the Himalayan full intracontinental convergence and uplift orogeny, forming various metallogenic systems with specific metallogenic functions coupled by various elements controlling mineralization in a specific geological space–time structure. This section will describe the Sanjiang archipelagic arc-basin system and the collisional orogenic metallogenic system based on the data obtained from multiple rounds of scientific and technological research and from the ideological point of view of the metallogenic system.

4.3.1 Metallogenic System Types in the Sanjiang Orogenic Belt

4.3.1.1 Concept and Connotation of Metallogenic System and Metallogenic Series

Metallogenic series and metallogenic system are common modes of thinking and conceptual models adopted by the current ore deposit academia to study regional mineralization and explore the temporal and spatial evolution and distribution laws of ore deposits. The concepts of metallogenic system and metallogenic series are guided by the theory of metallogenic system, emphasizing the internal connection and overall function of metallogenic processes and environmental elements, and revealing the temporal and spatial evolution and distribution laws of metallogenic processes and their products. The metallogenic series refers to a group of ore deposit-type assemblages that are closely related in time, space and genesis, formed under the dominant metallogenic process in a certain geological period and geological environment (Chen 1999). However, since the

formation of ore deposits often goes through a complex and long-term process of accumulation, transportation and deposition of ore-forming materials and is often subjected to the superimposition and transformation of the later geological process, resulting in repositioning, re-enrichment or redestruction, it is difficult to accurately demarcate the ore-forming age or grasp the internal relationship between the deposit and the environment. Therefore, the determination and division of the metallogenic series often have its limitations. Additionally, due to the characteristics of multi-source composite genesis of many deposits, it is difficult to identify their exact geological mineralization or determine their exact metallogenic series; some important metallogenic processes, such as crustal (submarine or intracontinental) hydrothermal mineralization, are still difficult to be classified into the existing metallogenic series. The metallogenic system is composed of all the geological elements that control the formation, change and preservation of deposits, the metallogenic process and the series of deposits and abnormal mineralization series formed in a certain geological time and space domain. It is a natural system with metallogenic function (Zhai 1998), emphasizing the organic connection between mineralization and environmental factors, processes and laws of material accumulation and dispersion, and regarding mineralization as the related product of coupling of various elements in the system. Therefore, the metallogenic system not only studies the mineralization products of deposits in the temporal and spatial structure of the geological environment, but also studies the four elements system of material, energy, time and mechanism that control the metallogenic process, so as to reveal and identify the mechanism and internal relations of various mineralizations on the whole. Although the thought of metallogenic system is in the process of research and exploration, its concept, connotation and extension are still vague and need to be determined, and some issues (such as the boundary and scale of metallogenic system, classification and naming) have not yet reached a consensus and need to be studied in depth. Nevertheless, the thinking mode and scientific connotation of the metallogenic system should undoubtedly become an important guiding ideology for the study of regional mineralization and the formation law of deposits.

4.3.1.2 Metallogenic System Types

Boundary Scale and Classification Level of Metallogenic System

The boundary scales of metallogenic systems are generally defined according to the temporal and spatial scales of their respective research objects, ranging from the global scale to a single deposit. No consensus has been reached so far. We believe that the global tectonic evolution usually goes through the evolution and transformation of the oceanic-continental tectonic system. Different stages of the oceanic-continental tectonic system evolution usually have different tectonic backgrounds, shaping different tectonic–magmatic–sedimentary formations, resulting in different metallogenic environments, and constraining different metallogenic systems. Based on this, we have determined four metallogenic giant systems according to the tectonic evolution and the development characteristics of the Sanjiang giant orogenic belt: continental marginal splitting metallogenic giant system, continental marginal convergent metallogenic giant system, intracontinental convergent metallogenic giant system and intracontinental rift metallogenetic giant system. The boundaries of metallogenic giant systems are usually discrete continental margins, island arcs or collision orogenic belts and intracontinental orogens or rifts. One or several metallogenic systems can be developed in the metallogenic giant system. The orogenic belt usually experiences the subduction orogenic stage and the collisional orogenic stage, in which the collisional orogenic stage also goes through the process of syn-collision orogeny, root demolition and subsidence and post-orogenic extension (Dong 1999), developing the subduction orogenic metallogenic system, the collisional orogenic metallogenic system and the post-orogenic extensional metallogenic system. The boundaries of metallogenic systems are defined as secondary tectonic units of large tectonic belts. In the metallogenic system, according to the dominant factors (such as stress, magma, fluid) that constrain the ore-forming material convergence and metallogenic process, metallogenic product assemblage and possible ore deposit genesis types, metallogenic subsystems can be further determined, such as arc magma-hydrothermal ore-forming subsystem, intra-arc rift hydrothermal fluid ore-forming subsystem, back-arc volcano-hydrothermal metallogenic subsystem.

Metallogenic System Classification

By comparing the evolution characteristics of the ocean-continental tectonic system, the types of metallogenic environments, the main controlling factors and processes of the system, the mineralization assemblages and the genetic types of the deposits, four metallogenic giant systems and 11 metallogenic systems are preliminarily divided.

(1) Continental margin splitting metallogenic giant system
   • Rift metallogenic system.
   • Ocean basin metallogenic system.
(2) Continental marginal convergent metallogenic giant system
   • Subduction orogenic metallogenic system, arc magma-hydrothermal metallogenic subsystem; intra-arc rift hydrothermal fluid metallogenic subsystem, back-arc volcanic hydrothermal metallogenic subsystem.

• Collisional orogenic metallogenic system.
• Syn-collisional magmatic metallogenic subsystem, syn-collisional volcanic metallogenic subsystem.
• Post-orogenic extensional metallogenic system.
• Magmatic hydrothermal metallogenic subsystem, tectonic-fluid metallogenic subsystem.

(3) Intracontinental convergent metallogenic giant system
   • Intracontinental magmatic metallogenic system, intracontinental crustal magmatic metallogenic subsystem, intracontinental mantle-derived magmatic metallogenic subsystem.
   • Tectonic dynamic fluid metallogenic system.
   • Thrust nappe tectono-fluid metallogenic subsystem, shear detachment tectono-fluid metallogenic subsystem.
   • Foreland basin fluid metallogenic system.
   • Metallogenic system of strike-slip pull-apart basin.

(4) Intracontinental rift metallogenetic giant system
   • Fault basin metallogenic system.
   • Extensional stripping metallogenic system.

4.3.2 The Continental Margin Splitting Metallogenic Giant System

In the Sanjiang giant orogenic belt, the continental margin splitting metallogenic giant system is mainly developed in the Yunxian-Jinghong rift volcanic zone and the Changning-Menglian rift-oceanic basin volcanic rock belt. The former develops the rift metallogenic system, while the latter develops the post-rift-ocean basin metallogenic system. There are other continental margin splitting metallogenic systems, such as the Jinsha River-Ailaoshan continental margin splitting metallogenic system.

4.3.2.1 The Background of Metallogenic System

The Proto-Tethys Ocean may have been formed in the Neoproterozoic. From the Sinian to the early Paleozoic, the oceanic crust plate began to subduct eastward, forming the Jitang-Jinghong ancient island chain and the Lancang River back-arc-basin. Entering the middle and late Paleozoic, it transformed from contraction and extrusion to extension and expansion. In addition, possibly due to the demolition and subsidence of the mountain roots of the orogenic belt and its induced mantle uplift and thinning of the continental crust, it rifted in the Yun County-Jinghong area on the east side of the Jitang-Jinghong ancient island chain and gradually expanded into a back-arc ocean basin, which lasted until the Middle and Late Triassic, and finally closed in the later period of the Late Triassic. The Yun County-Jinghong continental margin rift-type volcanic rock belt was formed due to the intense volcanic activity in rifts and ocean basins.

Yun County-Jinghong Rift Volcanic Rock Belt

As stated in the geophysical data, the gravitational field is similar to the characteristics of the Oslo rift zone, which is bounded by the Lancang River fault in the west and the Mengla fault in the east (Wang et al. 2001). Volcanic rocks can be divided into two periods: Late Paleozoic and Mesozoic.

Late Paleozoic volcanic rocks are mainly Carboniferous and Permian volcanic rocks. Carboniferous volcanic rocks refer to a set of lava assemblages dominated by quartz keratophyre and secondary spilites, with pyroclastic rocks developing here. Lava includes spilite, keratophyre, dacite and quartz keratophyre. As described in petrochemical, rare earth and trace element analysis, the primitive magma was mainly formed in the lower crust-upper mantle. According to the characteristics of the Dapingzhang geological section, the composition and strength of the magma and the types of volcanic eruptions, Yang (2001) divided it into an eruption cycle and three subcycles from early to late: ① The first subcycle is dominated by the overflow of sodium-rich lava on the seabed. It started from the quartz keratophyre magma eruption, and then the composition of the magma changed to kratophyre. At the same time, the eruption of basic spilite magma appeared in a short period of time, and finally, it ended with submarine volcanic deposits, forming massive sulfide ore bodies. ② The second subcycle is dominated by magma eruption, which has a certain explosive effect. In the early stage, rhyolite magma and intermediate-basic magma erupted mainly; in the middle stage, the eruption activity weakened, accompanied by eruption in the intermittent period, forming volcanic breccia and tuff; in the later stage, the eruption and outbreak activity weakened, forming the tuff and sedimentary pyroclastic rock far away from the crater. ③ The volcanic activity of the third subcycle comes to an end. In the early stage, it was neutral magma eruption, and after a short pause, it turned to be dominated by acid magma eruption; after that, it was mainly an eruption to form volcanic breccia; finally, it ended with the formation of tuff, tuffaceous-siliceous rock and normal sedimentary rock.

Permian volcanic rocks. The Early Permian is dominated by tuff, intercalated with basic lava; the Late Permian is composed of dacite, breccia lava, breccia tuff, tuff, tuff lava and other intermediate-acid rocks.

Triassic volcanic rocks mainly include Middle and Late Triassic volcanic rocks. The volcanic activity in this period shows a decreasing trend from north to south. From the Middle Triassic to the early period of Late Triassic and to the late period of Late Triassic, the eruption environment showed a change of marine facies—land–ocean interaction—terrestrial facies (Regional Geology of Yunnan Province 1990). The Middle Triassic volcanic rocks are distributed in the Yun County-Lincang Bangdong area in the north and

Jinghong-Mengla area in the south. The northern volcanic rocks are a set of high-potassium rhyolite volcanic rocks, which are well-developed, with Jingdong Minle copper deposits produced here. The Upper Triassic volcanic rocks are all over the region, and the lower segment is dominated by neutral-basic volcanic rocks; the upper segment in the northern Yun County is potassium-rich trachyte basalt and potassium-rich rhyolite volcanic rock, forming a "bimodal" volcanic rock assemblages. The volcanic rocks in the central and southern parts do not show the "bimodal" feature and are trachyandesite-dacite-rhyolite volcanic rock assemblages. Special mention should be made of the volcanic rocks in the Sandashan mining area in Jinghong. When the Yunnan regional survey team made a Jinghong amplitude of 1:200,000 in 1979, the strata in the mining area were changed to the middle and lower sections of the Upper Triassic Xiaodingxi Formation (the No. 16 team of the original Yunnan Geological Bureau classified it as the Lower Paleozoic; Sanjiang Regional Mineral Records defined it as the Upper Permian). The Wenyu, Guanfang and Sandashan copper deposits occur in the Upper Triassic Xiaodingxi Formation (Yang 2001).

The research on the Lancang River belt is still at a low level, but the copper deposits discovered are all produced in the "bimodal" assemblage of volcanic rock series, revealing that the extension and expansion of the Lancang River belt, the "bimodal" volcanic activity and the formation of copper deposits provide superior metallogenic environment and conditions.

Changning-Menglian Rift-Ocean Basin Volcanic Rock Belt

There are two magma evolution series in the belt: tholeiitic magma series and alkaline magma series. Rock types of the tholeiitic magma series include tholeiitic basalt, basaltic andesite and its pyroclastic rocks; rock types of the alkaline magma series include alkaline picrite basalt, alkaline olivine basalt, trachybasalt, trachybasal andesite and its pyroclastics rocks. These two series obviously belong to two types of parent magma with different properties.

Spatially, the alkaline basalt is the most widely distributed, while the tholeiitic basalt is limited; in terms of time, alkaline basalt is distributed in the early and middle Carboniferous, and tholeiitic basalt is distributed in the late Carboniferous (Laochang area). According to the rock assemblage, lithochemistry and geochemistry characteristics of volcanic rocks, most of them belong to continental rift-type volcanic rocks, with a few of them typical of ocean ridge basalt (Tongchangjie tholeiitic series volcanic rocks).

The above-mentioned types of volcanic rocks represent different stages of rift development: alkaline basalts are developed in the early stage, reflecting continental rift stage, with Laochang (black ore type) polymetallic deposits produced; ocean ridge basalt of new oceanic crust is developed in the late stage, and ophiolite is developed in some areas (Mo et al. 1990), reflecting the development stage of small ocean basin, with Tongchangjie (Cyprus type) copper-zinc deposit produced.

We cooperated with Professor Lehman of Germany to conduct research on the Lancang River volcanic belt. Sporadic molybdenite was found in Dapingzhang copper ore, and two Re-Os isotope samples obtained well matched isochrons with ages of 428.8 ± 6.1 Ma and 442.4 ± 5.6 Ma, respectively. Therefore, the occurrence and development of the Lancang River ocean basin in the southern section of the Sanjiang seem to have new implications for the evolution from Proto-Tethys to Paleo-Tethys. We believe that the Lancang River Ocean has not been completely closed since it opened in the Early Paleozoic (i.e., the Proto-Tethys stage), and the Lancang River volcanic rock belt shows multiple periods of expulsive sedimentary mineralization from the late Early Paleozoic to the early period of the Early Mesozoic, such as Dapingzhang copper polymetallic deposit (late Silurian), Minle copper deposit (middle and late Triassic) and Sandashan copper deposit (late Triassic).

In the Late Permian, there may be basic-ultrabasic magmatic intrusion and eruption, Jinggu Banpo, Jinghong Damenglong and Nanlianshan basic (local ultrabasic) intrusions in this area, with good conditions for searching copper–nickel sulfide deposits.

4.3.2.2   Metallogenic System Types

The continental margin splitting metallogenic giant system is related to the tectonic–magmatic–sedimentary environment in different stages of the self-rifting-rift-ocean basin due to the extension and expansion of the continental margin crust, which correspondingly constitutes a rift metallogenic system and a rift-ocean basin metallogenic system of a continental margin splitting metallogenic giant system.

Rift Metallogenic System

The rift metallogenic system is developed in the Yun County-Jinghong rift belt. It is generally believed that during the Silurian-Carboniferous period, the volcanic eruption metallogenic subsystem was developed in the middle segment of the rift (Dapingzhang) in the eruption cycle of sodium-rich quartz keratophyre-keratophyre-spilite-pyroclastic rock deposits, forming Dapingzhang VMS copper deposit and the same type and ore-forming anomaly. However, from the analysis of the Re-Os age of the molybdenite in the Dapingzhang copper polymetallic deposit, the time of the Dapingzhang submarine volcanic exhalation and mineralization may be as early as the Silurian. The types of metallogenic systems in the Triassic were different in different stages of

rifting evolution and in different regions. The metallogenic subsystems are further divided according to the factors that restrict the mineralization and the difference of ore deposits. In the Middle Triassic, epivolcanic dacite porphyries related to volcanic rocks developed, appearing the magmatic hydrothermal metallogenic subsystem. For deposits such as Minle copper deposit, some people regard it as a new type, that is, ash-flow type copper ore, which is closely related to the sodium-rich welded tuff flow (Yang 2001). In the Late Triassic, a volcanic hydrothermal metallogenic subsystem was developed in the basaltic andesite series in the northern section of the rift, forming the hydrothermal copper deposits represented by Guanfang and Wenyu; the volcanic-sedimentary metallogenic subsystem was developed in the neutral or moderately acidic tuff in the remote crater of the southern segment of the rift valley, forming the Sandashan CVHMS (Cyprus type) copper deposit.

The above-mentioned deposits related to Carboniferous and Triassic rift volcanism constitute the South Lancang River copper belt.

Ocean Basin Metallogenic System

The rift-ocean basin metallogenic system was developed in the early Carboniferous Changning-Menglian rift-ocean basin volcanic rock belt. Different stages of rift-ocean basin evolution have different types of metallogenic systems. In the early stage, the volcanic activity in the rift development stage formed rift basalts with a high degree of alkalinity, with the hydrothermal fluid metallogenic subsystem developed and forming the KVHMS deposit represented by Laochang; in the late stage, the volcanic activity in the ocean basin stage reduced the alkalinity of volcanic rocks, forming ocean ridge basalts, developing a volcanic-sedimentary metallogenic subsystem, as well as the CVHMS deposits represented by Tongchangjie.

4.3.2.3   The Spatiotemporal and Chemical Structure of the Metallogenic System

Rift Metallogenic System

The rift metallogenic system developed in the County-Jinghong rift belt is bounded by the Lancang River fault in the west and limited by the Mengla fault in the east. The system originated in the Silurian period (442.4 ± 5.6 Ma), and later in the Late Permian and the Middle and Late Triassic, large-scale volcanic-magmatic activities occurred again in this area. Although the mineralization is generally controlled by the rift zone, the ore deposits and ore fields are mainly distributed along the volcanic apparatus of Jiufang fault zone and volcanic depressions, and the types and chemical structures of the deposits in different sections of the rift are also different. Taking the Dapingzhang VMS deposit

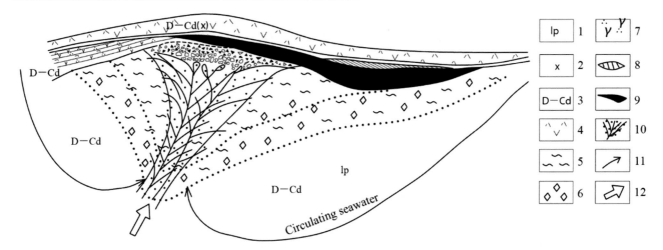

Fig. 4.10 Metallogenic model of the dapingzhang copper polymetallic deposit. 1—rhyolite porphyry; 2—dacite; 3—Silurian-Carboniferous Daaozi Formation; 4—dacite; 5—sericitization and silicification alteration zone; 6—carbonation alteration zone; 7—tuffaceous-siliceous rock; 8—barite; 9—massive ore body (Cu, Pb, Zn, Au, Ag polymetallic ore body dominated by sphalerite); 10—reticulated vein—disseminated ore body; 11—movement direction of infiltration circulating seawater; 12—movement direction of magmatic water

of the submarine volcanic eruption metallogenic system in the central and southern section as an example, the mineralized metals mainly contain Cu–Pb–Zn, associated with Au–Ag. The metal elements of the deposit show vertical zoning, that is, the upper basin facies massive ore bodies are Cu–Pb–Zn, associated with Au–Ag assemblages; for the lower pipeline facies veinlet disseminated ore body, the metal element is mainly Cu (Fig. 4.10); for the Minle porphyry copper deposit in the magmatic hydrothermal metallogenic subsystem in the middle and northern section, the metal element is mainly copper; for the Guanfang and Wenyu hydrothermal copper deposits in the volcanic hydrothermal metallogenic subsystem in the north section, the mineralized metal elements are mainly Cu-Pb; for the Sandashan VMS deposit (Cyprus type) of the volcanic-sedimentary metallogenic subsystem in the south, the mineralized metal element is mainly Cu (Fig. 4.11).

Ocean Basin Metallogenic System

The ocean basin metallogenic system is developed in the rift-ocean basin metallogenic system of the Changning-Menglian belt, bounded by the Kejie-Menglian fault in the west and limited by the Changning-Lancang fault in the east. The system was formed in the Early Carboniferous. Although the volcanic activity is generally controlled by rift-ocean basins, the deposits are obviously dominated by volcanic apparatus and fault structures. However, different stages of rift-ocean basin evolution have different types and chemical structures of metallogenic systems. In the early stage, the type of metallogenic system was the hydrothermal fluid metallogenic subsystem (Laochang), with the mineralized metal elements dominated by Ag–Pb–Zn, associated

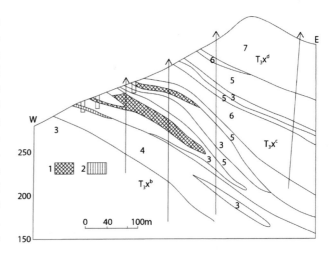

Fig. 4.11 Sandashan copper deposit exploration line profile (in the scale of 1:200,000 Jinghong amplitude). 1—dense copper-bearing pyrite; 2—iron cap; 3—carbonaceous sericite schist; 4—off-white massive metamorphic volcanic tuff; 5—off-white sheet metamorphic volcanic tuff; 6—dark gray sheet metamorphic volcano tuff; 7—bean-green massive metamorphic volcanic tuff

with S and Cu. A single ore body showed the characteristics of "black ore" in the upper part and "yellow ore" in the lower part, that is, the upper part was a massive silver–lead–zinc ore body, and the lower part was a copper-bearing pyrite ore body. Fine vein disseminated structure appeared in the lower part of the copper ore body, and there might be concealed porphyry in the deep part; in the late stage, the type of metallogenic system was the volcanic-sedimentary metallogenic subsystem (Tongchangjie), with the mineralized metal elements dominated by Cu–Zn assemblages.

4.3.2.4 Main Controlling Factors of Metallogenic System

The Main Controlling Factors of the Rift Metallogenic System

Magmatic Hydrothermal Metallogenic Subsystem
The system is represented by Minle copper deposit, and the main controlling factors include rock mass, rock mass emplacement strata and hydrothermal alteration.

(1) Rock mass. The ore-forming rock mass is gray-purple dacite porphyry, with a mottled, rhyolitic, massive structure. The phenocrysts mainly include plagioclase and potassium feldspar, followed by biotite, amphibole (pyroxene) and accessory minerals such as apatite and magnetite. The matrix is felsic, with felsic and sub-graphic texture (Li 1996). The rhyolitic structure of the rock is developed, the xenoliths often have a tendency of directional arrangement, and the dark minerals (biotite) often appear dark edges and other phenomena, which belong to the volcanic-subvolcanic facies rocks.
(2) The stratum where the rock mass is emplaced. The main host stratum of the ore deposit is the Middle Triassic.
(3) Hydrothermal alteration. Alterations mainly include potassium silicification, mudstone, propylitization, kaolin-sericitization, chlorite-epidedization (Ren 2000), etc., indicating that gas–liquid activities are superimposed in the later stage of magmatic activity.

Volcanic Hydrothermal Fluid Metallogenic Subsystem
The system is represented by the Dapingzhang copper deposit, and the main controlling factors include volcanic apparatus, hydrothermal alteration and regional fault structures.

(1) Volcanic apparatus. Occurred in the volcanic dome structure, the deposit is controlled by the volcanic exhalative sedimentary rock system composed of spilite-keratophyre and other open tectonic systems such as various primary structures, secondary structures and cryptoexplosive breccia pipes of the intrusive body. Volcanic massive sulfide deposits are controlled by paleo-sea-floor topography and damaged by later nappe structures. The ore bodies at the volcanic eruption center and the eruption pipeline are thick (Yang 2001).
(2) Hydrothermal alteration. The hydrothermal alteration is mainly developed in the veinlet disseminated ore bodies, with weak hydrothermal alteration of the massive ore bodies. The main types of hydrothermal alteration include silicification, chloritization, pyritization, sericitization, carbonation and baritization. The mineralized-altered rock pipe is formed with the cryptoexplosive breccia belt controlled by NW-trending fault as the center. The hydrothermal alteration of the rock pipe is characterized by lateral zoning, in the order from the center to the outside as follows: pyritization-chloritization-silicification zone → sericitization-silicification zone → sericitization-carbonation zone. Silicification, chloritization and pyritization are closely related to mineralization.
(3) Regional fault structure. The NW-trending regional Jiufang and Liziqing faults control the distribution of the ore-bearing spilite keratophyric volcanic series rocks. The volcanic eruption takes Jiufang fault zone on the west side as the channel and erupts from west to east in a fissure style. The Jiufang fault plays a role in controlling the mineralization.

The volcanic hydrothermal metallogenic subsystem developed in the northern section of the rift is represented by the Guanfang copper deposit. The ore body occurs in the middle-basic volcanic lava of the Upper Triassic Xiaodingxi Formation, with its mineralization controlled by the combination of lithology, structural fissures and hydrothermal alteration.

Volcanic-Sedimentary Metallogenic Subsystem
The volcanic-sedimentary metallogenic subsystem is developed in the southern section of the rift, represented by the Sandashan copper deposit. The ore body occurs between the neutral or moderately acidic tuff (roof) and the carbonaceous sericite schist (floor) in the far crater of the Upper Triassic Xiaodingxi Formation. The ore body tends to pinch out with the pinching out of carbonaceous sericite schist, and there is a phenomenon that pyrite is more enriched with the increase of the carbonization degree of sericite schist. The roof and floor surrounding rocks undergo intense hydrothermal alteration, including pyritization, silicification, sericitization, chloritization and carbonation. The formation of copper-bearing pyrite is that the sulfur-bearing basic volcanic eruptive material supplements the iron and copper elements in seawater with the participation of organic matter. In a strong reducing environment, they combine and precipitate with each other. After hydrothermal alteration, copper, cobalt and other elements are further enriched and mineralized.

Main Controlling Factors of Ocean Basin Metallogenic System

In the Early Carboniferous, a fissure-type intermediate-basic volcanic eruption occurred in the Changning-Menglian rift-ocean basin. In the Lower Carboniferous, the northern segment is called the Pingzhang Formation, and the lower part is an alkalescent series of sodium-alkaline basalt-latite; the upper part is carbonate rock, with the thickness of 105–595 m. The volcano-sedimentary metallogenic subsystem is also developed. In the Lower Carboniferous, the southern segment is called the Yiliu Formation, which is almost entirely composed of volcanic rocks. The lower part is the alkalescent series sodium–potassium-alkaline basalt-latite; the upper part is dominated by pyroclastic rocks, with the thickness ranging from 530 to 1159 m. For the potassium-alkaline basalt-latite, the hydrothermal-fluid metallogenetic subsystem is developed.

Main Controlling Factors of Volcanic-Sedimentary Metallogenic Subsystem

The system is represented by the Tongchangjie Cu–Zn deposit. The metallogenic subsystem is mainly controlled by horizon and lithology. The host stratum is a marine sodium-alkaline basaltic volcanic-sedimentary rock series in the lower part of the Lower Carboniferous Pingzhang Formation (1/200,000 regional geological survey division), and the copper-bearing pyrite bodies are mainly produced in the lower part of the ore-bearing rock series. The volcanic-sedimentary rocks are dominated by basic tuffs and sedimentary tuffs, intercalated with unstable sodium basic lava, jasper rock, carbon-bearing tuff or tuffaceous mudstone. The ore bodies are layered, lentil-like and lenticular (Fig. 4.12), folded synchronously with the strata, indicating that the ore bodies are controlled by horizons and lithology. In addition, small vein body and high-grade tectonic metamorphic hydrothermal copper-bearing pyrite veins occasionally penetrate along the fault. The ore-forming material originates from the sodium-based volcanic gas–liquid and is adsorbed and deposited underwater by tuffaceous, argillaceous and organic matter to form CVHMS (Cyprus-type) deposits.

Main Controlling Factors of Hydrothermal Fluid Metallogenic Subsystem

The system is represented by Laochang Pb, Zn, Ag deposits. The main controlling factors include host strata and lithology, volcanic apparatus and fault structures and hydrothermal alteration.

(1) Host strata and lithology. The host strata in the mining area include the marine volcanic-sedimentary rocks of the Lower Carboniferous Yiliu Formation and the Middle and Upper Carboniferous carbonate rocks. The volcanic-sedimentary rock series of the Yiliu Formation belongs to the calc-alkaline tholeiitic basalt series and the alkaline basalt series (dominant), which can be divided into three major volcanic eruption-sedimentary cycles with eight layers (Fig. 4.13). Each cycle begins

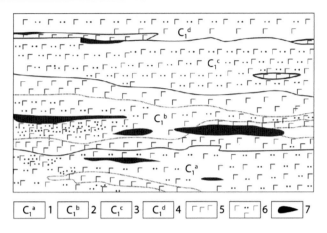

Fig. 4.12 Schematic diagram of ore-bearing layers and their lithofacies change characteristics of Tongchangjie copper deposit in Yun County (Adapted from the Tongchangjie report). 1—The first ore-bearing layer of the Lower Carboniferous Pingzhang Formation; 2—The second ore-bearing layer of the Lower Carboniferous Pingzhang Formation; 3—The third ore-bearing layer of the Lower Carboniferous Pingzhang Formation; 4—The fourth ore-bearing layer of the Lower Carboniferous Pingzhang Formation; 5—Basalt; 6—Basaltic tuff, sedimentary tuff intercalated with tuffaceous sedimentary rock; 7—Ore body

with the overflow of basalt magma and ends with the eruption of andesite magma. The ore bodies mainly occur in the lower volcanic cycles composed of trachybasalt, trachyandesite lava, breccia tuff, sedimentary tuff, exhalite, tuffaceous siltstone, limestone, etc. The ore bodies are developed in the upper part of the rhythm, accompanied by sulfide siliceous exhalite (Yang 1989). In the Middle and Upper Carboniferous, there are dolomite and limestone intercalated with dolomitic limestone, with vein-like ore bodies.

(2) Volcanic apparatus and fault structure. The ore body is obviously controlled by volcanic apparatus and fault structures. The volcanic apparatus consists of two volcanic domes and three volcanic depressions (Liu et al. 2000), which in order from west to east includes the Xiangshan volcanic depression (corresponding to the Xiangshan syncline), the Qinglongqing volcanic dome (corresponding to the Qinglongqing anticline), Xiongshishan volcanic depression (corresponding to the Xiongshishan syncline), Laochang dome (corresponding to the Laochang anticline), Shuishishan volcanic depression (corresponding to the Shuishishan anticline) (Fig. 4.14). Among them, the Laochang volcanic dome is distributed along the east side of the F1 fault in a nearly north–south direction, with a relatively large scale, and the Qinglongqing volcanic dome is distributed along the north side of the F4 fault in a northwest direction. The main ore bodies are distributed

Erathem	System	Series	Code	Histogram 1:5000	Thickness m	Lithologic description
Cenozoic	Quaternary		Q		0~90	Residual slope deposit, alluvial layer: gravel, sand mud lead, slag and other composition. Part of the area is caving limestone.
Paleozoic	Permian	Lower	P_1^2		50~140	Biolimestone: grey massive micrite with a medium-thick layered limestone on top containing Neoschwagerina.
			P_1^1		200~280	Massive limestone: gray massive dolomitic limestone sandwiched with limestone, with local tortoise breccia limestone. Lead-bearing limonite veins can be seen in limestone fissures. Containing fossils such as Nankinella sp..
	Carboniferous	Middle-upper	C_{2+3}^3		50	Dark gray medium-thick layered and massive coral limestone, marked by corals. Contains Triticites, Schwagerina sp. and other fossils.
			C_{2+3}^{1+2}		310~430	The upper part is gray-white micrite, with occasional argillaceous bands; the middle part is gray massive mesocrystalline to coarse-crystalline dolomite with micrite and oolitic limestone; the lower part is dark gray medium-thick layered micrite, with interlayers intercalated siliceous strips and shale. There are lead-bearing limonite veins in this layer, the lower part is No. III ore body, and the limestone at the bottom are mineralized. The limestone contains Psedoschuagerina sp, Eostelleua sp. and other fossils.
		Lower	C_1^8		0~150	The upper part is gray-white tuff, the purplish red and shale parts are light green tuff, basaltic volcanic breccia tuff; the lower part is yellow-green, purplish red sand shale, with black shale and lenticular limestone.
			$C_1^7 b$		55~160	The upper part is gray-green massive basalt and sloping peridotite, with a small amount of basalt tuff, and the lower part is gray-green basalt, fused tuff, and andesite tuff. The horizon and lithology of this layer are stable. The upper part of the tuff is No. II ore body.
			C_1^{6+6}		80~160	The upper part is gray trachy andesite polyclastic tuff intercalated with sedimentary tuff, charcoal siliceous rock, and eruptive breccia; the middle part is variegated trachy basaltic tuff, breccia lava, and the lower part is gray trachy basaltic silicified tuff with strips Silica, lenticular limestone. It is No. I silver-lead-zinc ore body, and it gradually becomes a copper ore body in the south, and most of them have been fully mineralized; some drilling holes still show granite porphyry veins.
			C_1^4		0~120	Gray, gray-green andesite tuff breccia intercalated with tuff, thin limestone.
			$C_1^3 a$		60~130	The upper part is light gray-green andesite agglomerate, and the lower part is purple-gray almond-shaped andesite sandwiched with andesite tuff breccia.
			$C_1^2 b$		50~130	The upper gray-green dense massive basalt, the lower gray-green basaltic lava rhyolite basalt tuff.
			C_1^1		>20	Grey-white andesite tuff, fused breccia.
	Devonian	M-U	D_{2-3}		>70	Gray-green sandstone and gray-black thin siliceous rock.
		Lower	D_1		>330	The upper part is gray-green medium-thick layered fine-grained feldspar quartz sandstone with siliceous rock, and the lower part is homochromatic shale with sandstone.

Fig. 4.13 Comprehensive stratigraphic histogram of the Laochang Ag polymetallic ore district, Lancang

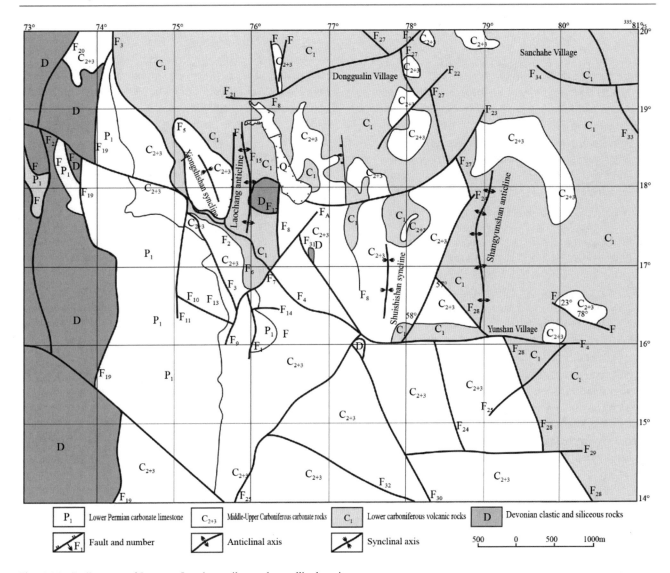

Fig. 4.14 Outline map of Lancang Laochang silver polymetallic deposit

in volcanic depressions, and the volcanic domes are mostly vein-shaped and network-vein-shaped ore bodies. Granitic dykes and skarn copper mineralization have been found recently during deep drilling for construction, and it is speculated that there is a concealed rock mass or porphyry in the lower part.

(3) Hydrothermal alteration. The ore deposit has strong hydrothermal alteration, complex types, multi-phase superposition and obvious zoning characteristics. The main alteration types include iron-manganese carbonatization, propylitization, carbonation, silicification, pyritization and skarnization. It is generally shown that iron-manganese carbonation is developed in the upper ore group (III ore group), propylitization and skarnization are developed in the lower ore group (II and I ore group), and carbonation, silicification and pyritization are all over the ore group.

4.3.3 The Continental Marginal Convergent Metallogenic Giant System

In the Sanjiang giant orogenic belt, the continental marginal convergent metallogenic giant system is mainly developed in four composite orogenic belts with the Qamdo-Pu'er microblock as the central axis of the opposite hedging collage, namely the Yidun Island arc orogenic belt, the Jinsha River orogenic belt, the Lancang River orogenic belt and the Salween River orogenic belt. Along with the formation and evolution of orogenic belts, three important metallogenic systems are usually developed: subduction orogenic metallogenic system, collisional orogenic metallogenic system and post-orogenic extensional metallogenic system. This section mainly systematically expounds the metallogenic systems of the Yidun island arc orogenic belt and the Jinsha River orogenic belt.

4.3.3.1   The Background of Metallogenic System

Yidun Island Arc Collision Orogenic Belt

The Yidun island arc collision orogenic belt is a composite orogenic belt in the Tethys giant orogenic belt, which began during the late Indosinian (Norian-Rhaetian) large-scale subduction orogeny (Hou 1993; Hou et al. 1995), experienced the collisional orogenic process in the Yanshanian period, including arc-continent collision and continental crust shrinkage and thickening, orogenic uplift and extension, and finally suffered from the superimposition and reconstruction of intracontinental convergence and large-scale shearing and translation during the Neo-Tethys period. The westward subduction of the Ganzi-Litang oceanic crust plate resulted in the development of the Yidun remnant arc (237–210 Ma). Given the nature of the thin continental crust underlying the island arc, as well as the inhomogeneity of the subduction angle of the oceanic crust plate, the north–south segment of the remnant arc is shaped. The Changtai arc in the northern section is characterized by tension, and four secondary tectonic units can be identified from east to west, namely outer arc (volcanic-magmatic arc), intra-arc rift, inner arc (residual arc) and back-arc spreading basin. The island arc is characterized by an intra-arc rift basin formed by inter-arc rifting and a back-arc-basin formed by back-arc expansion. The former is marked by the basalt-rhyolite bimodal rock assemblage and the deep-water fault basin (Hou 1993); the latter is characterized by a shoshonite—high-potassium rhyolite bimodal rock assemblage and deep-water black rock series deposits. The Shangri-La arc in the southern segment is characterized by compression, the island arc is dominated by simple volcanic-magmatic arcs and basically no back-arc expansion basins are developed. The volcanic-magmatic arc is characterized by andesite-dacite and super-hypabyssal emplaced intermediate-acid rock mass (Hou et al. 1995), and there is no corresponding volcanic activity in the back-arc area. The arc-continent collision occurred around 210 Ma, which not only led to the uplift of the remnant arc terrane, but induced the continental crust melting to form syn-collision granite, which was superimposed on the volcanic-magmatic arc. At about 189 Ma, the delamination of the lithosphere might have induced asthenosphere upwelling, and the crust began to appear extensional tension fracture, inducing crustal melting and felsic volcanic eruption, and forming the intraplate fissure-type acidic rhyolite series distributed along the high-potassium rhyolite belt in the back-arc-basin. Subsequently, the lithosphere of the orogenic belt was further stretched and even collapsed. At about 80 Ma, there was massive emplacement of A-type granite, forming intermittently distributed A-type granites belts extending for hundreds of kilometers in the NNW direction in the inner (west side) of the orogenic belt (Hou et al. 2001a, b).

Jinsha River Orogenic Belt

The Jinsha River orogenic belt is an important orogenic belt in the Tethys giant orogenic belt, which began in the large-scale subduction orogeny in the early period of Early Permian (Mo et al. 1993; Li et al. 1998; Wang et al. 1999), experienced the collisional orogenic process of the Early and Middle Triassic (including arc-continent collision and continental crust shrinkage and thickening, magmatic activity and orogenic uplift), as well as the post-orogenic extension in the early period of Late Triassic and finally went through the superimposition and reconstruction of intracontinental convergence and large-scale shearing and translation during the Neo-Tethys period. It is a composite orogenic belt formed by successively developed continental marginal arcs and collision orogenic belts affected by the collision and uplift of the Qinghai-Tibet Plateau. Since Devonian, the Jinsha River rift basin was developed between Qamdo landmass on the west side and Zhongza block on the east side. It has expanded strongly in Carboniferous and formed an initial ocean basin and received radiolarian siliceous rock-thick stratiform limestone-black mudstone low-density turbidite series. By the early Permian, the ocean basin expanded strongly and rapidly, forming a mature ocean basin, developing a new ocean crust marked by ophiolite complex and deep-sea sediments represented by siliceous argillaceous-sandy argillaceous flysch formation. At the end of the Early Permian, the oceanic crust plate of Jinsha River may be detached and subducted westward along the intra-oceanic fracture zone, resulting in the development of Zhubalong-Dongzhulin intra-oceanic arc and west canal river-Jiyi single arc back-arc-basin. Among them, a set of basalt-basaltic andesite-andesite island arc volcanic rock assemblages from tholeiite series to calc-alkaline series developed in intra-oceanic arc, while the oceanic crust volcanic rock assemblages marked by diabase sheet wall group and tholeiite and abyssal facies-subabyssal facies deposits represented by siliceous argillaceous-sandy argillaceous flysch formation developed in the back-arc-basin (Mo et al. 1993; Li et al. 1998; Wang et al. 1999). Around the late Permian, the expanding oceanic crust plate of Jinsha River subducted toward the Qamdo block on a large scale, resulting in the development of volcanic along the eastern edge of Qamdo block, forming the Permian continental margin arc of Jiangda-Deqin-Weixi (Mo et al. 1993; Li et al. 1998; Wang et al. 1999). The arc volcanic rocks evolved from tholeiite series through calc-alkali series to shoshonite series from morning till night, which marked the evolution process of volcanic arc's occurrence, development and maturity. Volcanic facies of volcano-sedimentary rock series varied along the continental margin arc, with various kinds of sedimentary and large fluctuation terrain.

Since the early Triassic, the tectonic and sedimentary environment of the Jinsha River arc-basin system and its stable landmasses on the east and west sides have changed dramatically. At the end of the late Permian, the Jinsha River ocean basin subducted and closed and gradually turned into the arc-land collision stage. The main signs are as follows: the Jinsha River Ocean Basin forms a remnant ocean basin; the collision mountain arc is superimposed on the Jiangda-Deqin continental margin arc; the Zhongza landmass on the east side uplifts into land, without Middle Triassic; the P_2/T_1 shows unconformity; the Qamdo landmass becomes a foreland basin, without Middle and Lower Triassic. As a result of arc-land collision and land-land docking, collision or post-collision intermediate-acid volcanic rocks and their associated intrusions developed in Shusong-Tongyou area in the southern segment of the continental margin arc. Collision or post-collision basaltic andesite-andesite-dacite-rhyolite assemblage developed in Jiangda-Xu Zhong area in the northern segment of the continental margin arc, covering the purple-red conglomerate of the early Triassic foothills, and superimposed on the continental margin arc.

In the Late Triassic, crustal extension began to appear in the collision orogenic belt. The mountain root delamination of the orogenic belt and asthenosphere upwelling caused by this may lead to the thinning and strong extension of the continental crust of the orogenic belt, even the collapse of the orogenic belt, thus forming extensional basins and even extensional rift (Wang et al. 1999). In the southern part of Jinsha River orogenic belt, extensional basins were formed by post-collision extension, such as Luchun-Hongpo basin and Reshuitang-Cuiyibi basin, and basalt-rhyolite bimodal rock assemblage, subabyssal facies turbidite and sandy argillaceous flysch were developed. In the northern section of Jinsha River orogenic belt, the post-collision extension was obviously strengthened. Jiangda continental margin arc expanded strongly and formed a volcanic-sedimentary basin of the Late Triassic. Jiangda continental margin arc received a sequence of basic-neutral-acidic calc-alkaline volcanic rocks (Dongka Formation) dominated by marine. On the west side of Jiangda continental margin arc, the crust expanded strongly to form a rift basin, and a sequence of marine pillow basaltic lava and gabbro-diabase wall group developed along the Xiaxiula-Shengda line. By the end of Late Triassic, the extensional basin gradually shrank and destroyed, and a large number of gypsum-salt deposits composed of gypsum, barite and siderite developed.

4.3.3.2 Metallogenic System Types

The metallogenic system in orogenic belts is closely related to the formation and evolution of orogenic belts and the tectonic–magmatic–sedimentary environment in different evolution stages of orogenic belts. With the evolution of orogenic belts from oceanic crust subduction orogeny through arc-continental collision orogeny to post-orogenic crustal extension, due to mineralization, it constitutes a continental margin convergent giant metallogenic system and several important metallogenic systems in orogenic belt.

Subduction Orogenic Metallogenic System

In the Late Triassic ancient island arc belt of Yidun, a subduction orogenic metallogenic system developed. However, the development types of metallogenic systems are different in different sections of island arcs. In the intra-arc rift zone of the Changtai extensional arc in the northern part of the island arc, it developed the hydrothermal fluid metallogenic subsystem and formed large VHMS deposits represented by Gacun deposit and Gayiqiong deposit and a batch of same type deposits, occurrences, mineralized anomalies and ore-induced anomalies, which are strictly confined to the intra-arc rift zone and its internal fault basins and volcanic depressions. In Shangri-La compressive arc in the southern section of the island arc, it developed magmatic hydrothermal metallogenic subsystem and formed porphyry deposits and skarn deposits represented by Xuejiping deposit and Hongshan deposit, which constituted an important copper polymetallic ore concentration area in Shangri-La area. This was closely related to hypabyssal-super-hypabyssal intermediate-acid rock masses and rock stock. In the back-arc-basin of Changtai arc, it developed back-arc volcanic hydrothermal metallogenic subsystem and formed epithermal deposits represented by Kongmasi deposit and Nongduke deposit, which constituted a new silver-gold-mercury metallogenic belt. This is closely related to high-potassium rhyolite. In Jinsha River orogenic belt, the subduction orogenic metallogenic system is developed in early Permian when Jinsha River Ocean Basin subducted westward and intra-oceanic arc and continental margin arc have formed and evolved. In the Zhubalong-Dongzhulin intra-oceanic arc, an arc magma-hydrothermal metallogenic subsystem related to Permian intermediate-basic volcanic rocks (basaltic andesite-andesite) has developed, and VHMS deposit has formed, represented by Yangla copper deposit. In Jiangda-Deqin-Weixi Permian continental margin arc, arc volcanic hydrothermal Cu–Au–Pb–Zn mineralization and corresponding metallogenic subsystem related to arc volcanic rock series have developed, and polymetallic deposits have formed, represented by Nanren Cu–Au–Pb–Zn polymetallic deposit.

Post-orogenic Extensional Metallogenic System

After a short arc-continent collision, Yidun island arc collision orogenic belt immediately entered the post-orogenic lithospheric extension stage. With the linear eruption of acid rhyolitic volcanic rocks and the large-scale intrusion of

A-type granite, the post-orogenic extensional metallogenic system developed. Among them, A-type granite belt controls the development of magmatic hydrothermal metallogenic subsystem, and a tin polymetallic deposit assemblage has formed, represented by Nianlong polymetallic deposit. These deposits constituted an important tin-silver polymetallic ore belt. The extensional zone controls the development of tectonic-fluid metallogenic subsystem, and hydrothermal silver deposits or silver-bearing polymetallic deposits' assemblage has formed, represented by Xiasai deposit. These deposits constituted an important silver polymetallic ore concentration area.

In Jinsha River orogenic belt, the post-collision extensional metallogenic system developed in the late evolution stage of the orogenic belt when mountain root delaminated and crust extended. This is the most important metallogenic system in this orogenic belt. The metallogenic system consists of at least two metallogenic subsystems. One is the hydrothermal fluid metallogenic subsystem. In Deqin (the southern part of the orogenic belt), the submarine sedimentary exhalative VMS deposits, which are related to bimodal rock assemblage and occurred in acid rhyolitic volcanic rock series, were formed. The deposits are concentrated in deep-water volcanic-sedimentary basins caused by extensional fault, represented by Luchun deposit and Hongpo copper polymetallic deposit. In Zhaokalong-Dingqinnong area in the northern part of the orogenic belt, mineralization occurred in shallow-water facies or sea-land intersecting facies intermediate-acid volcanic rock belt, which is formed in the post-collision extensional environment. Then, the exhalative-sedimentary sulfide deposits were formed in shallow-water environment, such as Zhaokalong iron-silver polymetallic deposit and Dingqinnong silver-copper polymetallic deposit. In Xiaxiula-Shengda extensional rift basin in the northern part of the orogenic belt, mineralization is controlled by the activities of contemporaneous fault, rift basins and hydrothermal fluid in the post-collision extensional stage. The hydrothermal sedimentary sulfide deposits with hydrothermal sedimentary rocks as the main host rocks were formed, including Zuna silver-bearing lead–zinc deposit. Although each of the three types of deposits have its own characteristic, they constitute a unified metallogenic subsystem developed in the post-collision extensional environment. Luchun deposit represents the products of deep-water volcanic-sedimentary-hydrothermal mineralization in post-collision extensional environment. Zhaokalong deposit represents the products of volcanic-sedimentary-hydrothermal mineralization in shallow-water environment. Zuna deposit represents the products of hydrothermal sedimentary mineralization far away from volcanic environment in post-collision extensional rift basin. The second is magmatic metallogenic subsystem, which mainly develops in

andesite-diorite porphyrite complex area in Jiangda area of the north-middle section of orogenic belt. Porphyrite iron ore is formed primarily, represented by Jiangda iron ore. This type of deposit may have formed in the early stage of post-collision crustal extension or the transformation stage from continental margin arc to extensional basin.

4.3.3.3 Metallogenic System Structure

The structure of metallogenic system should include temporal, spatial and chemical structure, that is, the temporal framework of metallogenic events in the metallogenic system, the boundary scale of metallogenic system and its mineralization and the metal assemblage, mineralization zoning and distribution of elements anomaly in the metallogenic system.

Subduction Orogenic Metallogenic System

In Yidun ancient island arc zone, the subduction orogenic metallogenic system takes this island arc zone as its boundary, covering secondary tectonic units such as main arc, intra-arc rift and back-arc spreading basin. The main body of the subduction orogenic metallogenic system developed in the Late Triassic, with its age ranged from 237 to 200 Ma. The hydrothermal fluid metallogenic subsystem is limited by the intra-arc rift zone of Changtai arc. Although the mineralization is generally controlled by the intra-arc rift zone, the deposits are mainly concentrated in volcanic depressions and fault basins in the intra-arc rift zone. The metallogenic age ranges from 238 to 210 Ma. Among them, the mineralization age of Gacun VHMS deposit is about 210–221 Ma and that of Gayiqiong VMS deposit is 200 Ma (Hou et al. 1995). Mineralized metal is mainly base metal Zn–Pb–Cu, accompanied by a small amount of precious metal Ag–Au. As far as metallogenic zone is concerned, no obvious metal spatial zoning is found. But single deposit shows clear vertical zoning, which is Cu–Zn → Zn–Pb–Cu → Pb–Zn–Ag → Ag–Pb–Zn → Pb–Ba from bottom to top.

The magmatic hydrothermal metallogenic subsystem is confined to Shangri-La magmatic arc and mainly distributed around some hypabyssal-super-hypabyssal intermediate-acid rock masses and rock stock. Its production is based on ore concentration areas. Mineralization also mainly occurred in Indosinian. Among them, the mineralized porphyry age of Xuejiping copper deposit is 224 Ma, and the metallogenic age of Hongshan copper-bearing polymetallic deposit is 214 Ma (Tan 1985). The deposit types of this subsystem are mainly porphyry and skarn. Among them, ore-bearing porphyry is more neutral than typical porphyry copper deposit, and it is the homologous and heterogeneous product of Shangri-La island arc andesite. Ore-bearing skarn is characterized by stratiform-stratiform-like skarn, which may be the associated product of porphyry copper deposits.

Mineralized metal mainly includes base metal Cu, accompanied by a small amount of Zn–Pb–Fe (Yang et al. 2001). The metal spatial zoning in the ore concentration area is not obvious, but it generally shows the outline of Cu-rich metal zone in the middle section and Pb–Zn–Ag-rich metal zone in the edge.

The back-arc volcanic hydrothermal metallogenic subsystem is restricted to the back-arc-basin. The mineralization takes place in the back-arc acidic rhyolitic volcanic rock area. The epithermal deposit occurs in the high-potassium rhyolitic volcanic rock, forming an intermittently distributed Ag–Au–Hg ore belt. The metal zoning along the ore belt is obvious. Hg mineralization is mainly concentrated in the northern section of the metal zone along the ore belt, represented by Kongmasi large mercury deposit, which occurs in the contact zone between rhyolitic volcanic rocks and overlying limestone. Ag–Au mineralization is mainly concentrated in the middle section of the metal zone along the ore belt, represented by Nongduke medium-sized silver deposit, which occurs in high-potassium rhyolite volcanic rocks and their fracture zones. Au–Ag mineralization is mainly concentrated in the southern section of the metal zone along the ore belt. Because of the extremely low exploration degree, most of the metal zone in the southern section is some ore occurrences, mineralized spots and ore-induced anomalies at present. The post-orogenic extensional metallogenic system is superimposed on the island arc orogenic belt and mainly developed in the A-type granite belt and within 10 km of its two sides. The development age of the metallogenic system is about 138–70 Ma (Qu et al. 2001).

In Jinsha River orogenic belt, the subduction orogenic metallogenic system occurred in its early intra-oceanic arc and continental margin arc, and the main body of the system developed in Permian arc-forming period. In the Permian intra-oceanic arc, although the metallogenic subsystem related to marine intermediate-basic arc volcanic rock series constitutes the main body of Yangla deposit, the initial metal enrichment formed in Permian has suffered from the superposition and transformation of magma-hydrothermal liquid and further metal enrichment in late Indosinian. Therefore, although the main body of metallogenic system occurred in the Permian intra-oceanic arc period, the mineralization may extend to Indosinian and Yanshan. Despite main body of sulfide deposits occurring in specific horizons of arc volcanic rock series, the spatial position of ore-rich bodies is closely related to Indosinian granite porphyry. The mineralized metal assemblage is Cu–Zn, and the regional zoning of mineralized metals has not been found yet. On the Permian continental margin arc, although there are also metallogenic subsystems related to arc volcanic rock series developed, no large-scale VMS deposits have been found, and the mineralized metal assemblage is Pb–Zn.

Post-orogenic Extensional Metallogenic System

The post-orogenic extensional metallogenic system in Yidun collisional orogenic belt takes this collisional orogenic belt as its boundary. The system is mainly confined to the tectonic–magmatic–sedimentary formation formed in late post-orogenic collisional extensional stage. The main body of the system developed in the lithospheric extensional stage and the orogenic belt collapse stage, with the system age ranges from 100 to 70 Ma. Magmatic hydrothermal metallogenic subsystem is mainly confined to A-type granite and its inner and outer contact zones. The major mineralized metals are tin and tin-bearing polymetallic metal, and skarn-type deposit is principal deposit. From Gaogong in the north to Rongyicuo in the south, the emplacement age of ore-bearing rock mass is from 102 to 75 Ma, corresponding to which the main mineralized metals change gradually from tin to silver. The metallogenic subsystem of tectonic fluid is mainly limited within 2–10 km of the periphery of A-type granite belt. The system mineralization is mainly controlled by large fault zone or interbed detachment zone formed in crustal extension stage. The deposit type of the system is mainly fracture zone altered rock, represented by Xiasai deposit and Lianlong deposit. Ag is the main mineralized metal element, followed by Zn–Pb. The spatial zoning of metal seems to be related to the distance from the "heat source" (granite). If the metal zone is close to the rock mass, it is mainly composed of tin-bearing polymetallic ore, with the metal assemblage of Sn + Ag + Pb + Zn. If the metal zone is far away from the rock mass, it is mainly silver-bearing polymetallic ore, with the metal assemblage of Ag + Pb + Zn.

In Jinsha River orogenic belt, the post-collision extensional metallogenic system is distributed as a whole in extensional basins or extensional rift belt formed by mountain root delamination and crustal extension in orogenic belt. The system generally developed in the relatively short geological period of the early period of Late Triassic. Due to the north–south segmentation of collision orogenic belts and the development differences of extensional basins, several different metallogenic subsystems have been formed. The submarine hydrothermal metallogenic system was developed in the deep-water extensional rift basin filled with "bimodal" volcanic rock series in the middle part of the orogenic belt. VMS deposits under deep-water environment were formed, with the mineralized metal assemblage of Cu–Pb–Zn or copper polymetallic type. The hydrothermal-sedimentary metallogenic subsystem was developed in the shallow-water extensional rift basin of intermediate-acid volcanic-sedimentary rock series in the northern part of the orogenic belt. A shallow-water environmental exhalative-sedimentary sulfide deposit was formed, with the mineralized metal assemblage of iron-silver polymetallic type. The magmatic hydrothermal metallogenic subsystem was developed in the weak extensional area between the north and south extensional basins of the orogenic belt, accompanied by a large number of hypabyssal-super-hypabyssal emplacement of intermediate-acid magma. Porphyrite deposits were formed, with the mineralized metal assemblage of iron domination.

4.3.3.4 Main Controlling Factors of Metallogenic System

The main controlling factors of metallogenic system mainly refer to the background, environment, material and the restriction leading to the dispersion-enrichment-deposition of ore-forming materials in the system. The coupling action of various factors in the system leads to the development of metallogenic system and the formation of mineralization assemblage of deposits.

Main Controlling Factors of Subduction Orogenic Metallogenic System

Main Controlling Factors of Subduction Orogenic Metallogenic System in Yidun Orogenic Belt
The main controlling factors of subduction orogenic metallogenic system include island arc tectonic environment, super-hypabyssal emplacement magma or magma chamber, convection hydrothermal fluid, marine felsic rock series supplying ore-forming materials, relatively closed sag basins and caprock deposits.

(1) **Arc magma-hydrothermal metallogenic subsystem**

In arc magma-hydrothermal metallogenic subsystem, relatively compressed magmatic arc environment, relatively closed caprock system and hypabyssal or super-hypabyssal emplacement of intermediate-acid that segregated hydrothermal fluid are the main controlling factors for the development and formation of porphyry copper deposits and skarn copper polymetallic deposits.

The biggest difference between compressional island arc and tensile island arc lies in the stress state of the main arc area and back arc area in the late subduction orogeny. As mentioned above, under the state of tensile stress, volcanic arcs crack to form intra-arc rift, accompanied by "bimodal" volcanic activities and basin fault. Submarine metallogenic hydrothermal systems were developed. In the compressive island arc environment, the volcanic arc is not tensile cracking, but compressive uplift under compressive stress. Under this background, arc magma will be strongly differentiated. Some arc magmas show arc volcanic eruptions, while some arc magmas show hypabyssal or super-hypabyssal emplacement. It eventually forms a series of porphyry, porphyrite, rock stock and rock bodies, constituting volcanic-intrusive

complex with volcanic rocks. Acid porphyry will lay a material base for the formation of porphyry copper deposits, and the emplacement of intermediate-acid rock mass will provide conditions for the formation of skarn-type deposits. In Shangri-La compressive arc, volcanic-intrusion developed on a large scale with magma along the island arc. Among them, volcanic rocks are mainly andesite, which is difficult to form mineralized parent rocks. The super-hypabyssal emplacement of late product of magmatic evolution forms quartz monzonitic porphyry and monzonite granite porphyry, which become important metallogenic parent rocks of porphyry copper deposits and constitute an important basis of arc magma-hydrothermal metallogenic subsystem. For example, Pulang copper deposit and Xuejiping porphyry copper deposit were formed in the arc volcanic-intrusive complex belt and occurred in the inner and outer contact zones of quartz monzonitic porphyry and monzonitic granite porphyry. Diorite porphyrite and/or quartz dioritic porphyrite are formed by hypabyssal emplacement or bedding penetration of arc magma evolution products. Contact metasomatic skarn and/or remote melt-hydrothermal reformed skarn can be produced in the contact zone between them (diorite porphyrite and/or quartz dioritic porphyrite) and limestone, which becomes the main ore-bearing unit of skarn-type copper polymetallic deposits. For example, skarn-type polymetallic deposits in Hongshan, Hongniu and Gaochiping were formed by mineralized skarns in the Upper Triassic sand-slate series in Shangri-La arc wing.

In Shangri-La volcanic-magmatic arc, porphyry-type copper deposits and skarn-type polymetallic deposits constitute a unified arc magmatic hydrothermal metallogenic subsystem. Spatially, skarn-type polymetallic deposits have deep metallogenic position and low horizon and are generally located in the wing of the volcanic magma arc, with relatively low metallogenic temperature. The metallogenic metals are Cu–Pb–Zn type, associated with Mo and Au, etc. Porphyry-type copper deposits have shallow metallogenic position and high horizon and are generally located in the core of volcanic magma arc, with relatively high metallogenic temperature. The metallogenic metal is Cu type. In time, although skarn-type polymetallic deposits and porphyry-type copper deposits were generally formed in the island arc orogenic stage, skarn-type polymetallic deposits may have earlier mineralization, which is roughly in the same period to or later than the arc andesite activity period. Porphyry-type copper deposits are formed relatively late and represented the final stage product of Shangri-La volcanic magma arc. Therefore, the two types of deposits may represent the early and late metallogenic products of arc magma-hydrothermal metallogenic subsystem, respectively, and the formation of the deposits requires a relatively compressive and stable environment to make arc magma fully differentiate and emplacement. A relatively closed caprock is needed to avoid boiling diffusion of ore-forming fluid. A hypabyssal-super-hypabyssal emplacement of intermediate-acid magma is needed to segregate magmatic hydrothermal fluid rich in metal matter in a closed condition.

Main Controlling Factors of Hydrothermal Fluid Metallogenic Subsystem in Intra-arc Rift

In the hydrothermal fluid metallogenic subsystem of intra-arc rift, the key factors for the development of this metallogenic subsystem and VMS deposit are the extension fracture system which promotes the convection of hydrothermal fluid, the deep-water fault basin which prevents the boiling of hydrothermal fluid, the shallow magma chamber which drives the circulation of hydrothermal fluid and the complex rock series composed of extremely thick rhyolitic volcanic rocks and intrusive dikes (metal materials of extremely thick rhyolitic volcanic rocks and intrusive dikes are leached by hot water fluid).

(1) Intra-arc rift and contemporaneous fault. The ancient and modern submarine hydrothermal mineralization studies show that although VMS deposits occur in island arcs, oceanic ridges, intraplate volcanic activity centers and post-collision extensional environments, mineralization occurs in different extension fracture environments. In ancient and modern island arc belts, VMS deposits do not occur on volcanic arcs, but in intra-arc rift or back-arc extensional basins. Mineralization does not develop in the stage of island arc orogeny, but in the stage of island arc cracking. This is true of Japanese black ore, Gacun deposit and Okinawa Trough black chimney sulfide deposit. The reasons are as follows: On the one hand, the extension fracture environment provides an important channel and transport system for long-term stable convection of hydrothermal fluid. On the other hand, the extension fracture environment also provides important conditions for strong water/rock reaction and sufficient leaching of ore-forming materials. In Gacun deposit, a group of contemporaneous basement and basin margin faults, which are distributed in NNW direction and formed in the stage of intra-arc extension fracture, have been identified and constituted the channel of fluid migration. At the same time, the intersection network between them (contemporaneous basement and basin margin faults mentioned above) and the fault system spreading close to EW direction become the discharge vent of hydrothermal fluid on the seabed (Hou et al. 2001).

(2) Fault basins and volcanic depressions. VMS deposits and ore fields are concentrated in the sub-fault basins and volcanic depressions in the intra-arc rift zone. Three deep-water fault basins (800–1200 m) have been identified, namely Zengke Basin, Gacun Basin and Xiangcheng Basin. Zengke Basin is an area with concentration occurrence of three VMS deposits, including Gayiqiong deposit. In Gacun basin, Gacun deposit occurs in volcanic depressions in fault basins. High-precision magnetic survey data also clearly reveal that at least 2–3 similar volcanic lithostatic depressions and large-scale ore-induced anomalies are developed in the southern part of Gacun deposit. In Xiangcheng Basin, VMS deposits and occurrences already known are concentration occurrence in this basin. On the one hand, these fault basins and volcanic depressions provide necessary pressure for preventing hydrothermal fluid boiling. On the other hand, they provide important sedimentary space for hot brine storage and sulfide accumulation.

(3) Bimodal assemblage and felsic rock series. Basalt-rhyolite bimodal rock assemblage is not only the regional volcanic rock symbol of intra-arc rift, but also an important source rock of ore-forming material. In Yidun Island Arc zone, basalt and rhyolite can occur interbedded and coexist in space. Basalt can occur in the lower part of ore-bearing rhyolite series (including Gacun deposit) or cover the top of ore-bearing rhyolite series (including Gayiqiong deposit). The relative scale of basalt and rhyolite series determines the ratio of w (Cu)/w (Pb + Zn) of VMS deposit. The larger the scale of basalt series is, the higher the Cu grade of the deposit will be, and vice versa. Rhyolitic volcanic rock series is not only the direct host surrounding rock of VMS deposit, but also the ore source of metal ore-forming materials dissolved by hydrothermal fluid. Therefore, its development scale directly controls the tonnage of VMS deposits. The larger the acid volcanic rocks are, the larger the deposit occurrence scale will be. As the main ore-bearing rocks, the lithology and lithofacies of rhyolitic volcanic rocks also have restriction on VMS deposits to some extent. It is possibly due to relatively large fluid migration pores, and pyroclastic facies often occur in vein-reticulated sulfide ore belts. Perhaps because rhyolitic magma provides some ore-forming materials, the highly differentiated out-phase of calc-alkaline acidic magma often has a close temporal and spatial relationship with VMS deposits.

(4) Hypabyssal emplacement of magma chambers. The field detection and observation of the submarine metallogenic hydrothermal fluid system show that magma chambers of a certain scale often develop at a depth of 1–3 km below the ancient and modern submarine hydrothermal active areas. In the east of modern submarine active hydrothermal area, Pacific Ocean ultrafast

expanding ridge (Urabe et al. 1995) and Okinawa Trough (Halbash et al. 1993; Hou et al. 1999) are the representatives. In typical ancient submarine hydrothermal area, black deposit in Japan and Gacun deposit (Skinner and Ohmoto 1983; Cathles et al. 1983) are the representatives. Modern magma chamber not only drives the convection of submarine hydrothermal fluid, but also injects some magmatic fluid (gas) and ore-forming materials into the hydrothermal system (Yang and Scott 1996). Ancient magma chamber has now been consolidated into intrusive rocks of different scales. The rock stock and dykes that intrude upward along the magma chamber front fracture zone often penetrate into the overlying volcanic rock series, forming a volcanic-intrusive complex, becoming an important ore-bearing rock series.

(5) Convection Thermal Fluid System In the past, it was generally believed that the convection metallogenic hydrothermal fluid was mainly pore fluid (seawater) which flowed through rocks and had water/rock reaction (Ohmoto 1983). However, increasing evidence showed that the depressurization and degassing of felsic magma could produce magma fluid rich in metal chloride and contribute to the metallogenic hydrothermal fluid system (Urabe 1989–1992; Yang and Scott 1996; Hou et al. 1999). Oxygen isotope evidence and high-salinity fluid inclusion evidence from the Gacun deposit also show that the metal-rich fluid segregated by rhyolitic magma is injected into the submarine hydrothermal fluid system, forming metallogenic fluid rich in ^{18}O and with high salinity (Hou et al. 1999).

Is the convection hydrothermal fluid system single-pass convection or double-diffusive convection? No agreement has been reached yet. Although the former is accepted by most scholars, the latter is supported by some new evidence. In the Gacun mining area, the quantitative calculation of the hydrothermal alteration system, the high-temperature and high-salinity data of the lower stratiform-like epidotization-silicification zone and the material change in the water/rock reaction confirmed that the submarine metallogenic hydrothermal fluid system is a double-diffusive convection system, and the deep hot brine or high-pressure fluid reservoir is formed in the upper part of the fracture zone at the top of felsic magma chamber. As an important heat medium and fluid source, it drives the convection of the upper cold seawater (Hou et al. 1995).

Back-Arc Volcanic Hydrothermal Metallogenic Subsystem

In back-arc volcanic hydrothermal metallogenic subsystem, the back-arc extensional tectonic environment, the shallow magma chamber that drives the fluid circulation, and the complex system formed by the extremely thick rhyolitic volcanic rocks-intrusive dikes (the metal material of the extremely thick rhyolitic volcanic rocks-intrusive dikes was leached by hydrothermal fluid) and the superimposition and transformation of fault tectonics are the key factors for the development of this metallogenic subsystem and volcanic rock-type epithermal Au–Ag-polymetallic deposits. Over the past decade, a number of super-large epithermal Au–Ag polymetallic deposits have been found in volcanic rock areas in the world. These deposits all occurred in the extensional tectonic environment of the back-arc rift in the active continental margin, and the ore-bearing rock series are all acid volcanic rocks formed in the late period of island arc evolution. The bimodal volcanic rock series in the Yidun Island arc back-arc-basin was developed on the continental crust basement by a strong extensional tectonic environment. It has the tectonic conditions for forming volcanic rock-type epithermal Au–Ag polymetallic deposits. At present, along this volcanic rock belt (Miange Formation), besides Nongduke middle-type Au–Ag polymetallic deposit and Kongmasi large mercury deposit, regional geochemical exploration and microwave remote sensing data also show a number of comprehensive mineralization anomalies such as Tage, Darike and Dulonggou, which show a good metallogenic prospect of this volcanic rock belt. Nongke mining area is dominated by Ag and Au, accompanied by As, Sb, Hg and other mineralizing elements. Mineralization is characterized by low-temperature hydrothermal solution. In the Kongsi mining area, a large mercury deposit has formed, and there are also high Ag, Cu, As, Sb and other ore-forming elements. It also shows the characteristics of low-temperature hydrothermal mineralization. Throughout the volcanic zone, the mineralization assemblage is dominated by Ag-Au-Hg, accompanied by metallic elements such as As, Sb, Cu and Zn.

The epithermal mineralization in Yidun Island arc back-arc-basin is related to highly acidic rhyolite rich in volatile matter. These rhyolites occurred in the source region of weakly enriched mantle, and the fluid components from subduction plates participated in the magma formation, accounting for about 50% of the fluid in the magmatism of the main arc zone. During the rising process, magma interacted strongly with crustal rocks before eruption, assimilating a large number of crustal materials. Ore-forming elements Hg, As, Sb may mainly come from these assimilated crustal rocks, and ore-forming elements such as Au and Ag may provide more from mantle source region.

Main Controlling Factors of Subduction Orogenic Metallogenic System in Jinsha River Orogenic Belt

The subduction orogenic metallogenic system of Jinsha River orogenic belt is characterized by submarine hydrothermal fluid mineralization and volcanic hydrothermal mineralization, and the submarine hydrothermal fluid metallogenic subsystem and volcanic hydrothermal metallogenic subsystem related to arc volcanic rocks are correspondingly formed and developed.

The Submarine Hydrothermal Fluid Metallogenic Subsystem in the Oceanic Arc Environment is Located in the Jinsha River Orogenic Belt

The submarine hydrothermal fluid metallogenic subsystem related to intra-oceanic arc volcanic rocks in the subduction orogenic metallogenic system developed in the westward subduction of Jinsha River ocean basin and the formation and evolution of intra-oceanic arc and continental margin arc in the early Permian. The intra-oceanic arc is built on the oceanic crust basement marked by an ophiolite melange belt, in which the basic volcanic rocks are typical oceanic ridge basalts. The zircon U–Pb ages range from 363 to 296 Ma (Chen et al. 1998), which proves that the Jinsha River oceanic basin developed from Carboniferous to Early Permian. The ore-bearing volcanic-sedimentary rock series is basalt-basaltic andesite-hornblende andesite assemblage with a thickness of more than 600 m and deep-water flysch formation. The hornblende of andesite has a K–Ar age of 257–268 Ma (Li et al. 1998), which indicates that the volcanic activity in the oceanic arc occurred from the late period of Early Permian to the late Permian. Intra-oceanic arc environment where submarine hot water system is developed, deep-water fault basin where hydrothermal fluid boiling is prevented, shallow magma chamber which drives fluid circulation, and a large number of intermediate-basic volcanic rocks, in which the metal was leached by hydrothermal fluid, and complex series composed of late intrusive dikes are the key elements for the development of this metallogenic system and Yangla deposit. The deposit was formed in the intra-oceanic arc environment formed during the detachment and subduction of Jinsha River oceanic crust, and volcanic activity provided heat source and related material sources. The ore-forming hydrothermal fluid is mainly the high-salinity fluid formed by the infiltration seawater and the interlayer water of volcanic-sedimentary rock series after fully extracting ore-forming minerals. The metal mainly comes from volcanic rocks, and the fluid is from a mixed source. Mineralization is in the high permeability intra-oceanic arc section, where the submarine hydrothermal solution upwells and accumulates on the spot. Then, a large number of colloidal and cryptocrystalline metal sulfide-bearing sediments and calcium-iron silicate deposits (primary skarn bed) are formed in the high-temperature hot water environment. After the deposit was formed, it experienced many times of tectonic-magmatism and was superimposed and reformed by the late metallogenic hydrothermal activities.

Volcanic Hydrothermal Metallogenic Subsystem in Continental Margin Arc Environment

Volcanic hydrothermal Cu–Au–Pb–Zn mineralization and corresponding metallogenic subsystems related to arc volcanic rock series developed in the Permian continental margin arc of Jinsha River-Deqin-Weixi. It developed in the westward subduction of Jinsha River oceanic basin and the formation and evolution of intra-oceanic arc and continental margin arc in the early Permian, forming polymetallic deposits such as Nanren Pb–Zn deposit and Cu–Au mineralization. The continental margin arc is formed on the continental crust basement marked by the continental margin volcanic arc zone, among which the intermediate-basic and intermediate-acidic volcanic rocks have typical island arc volcanic rock characteristics. The earliest arc volcanic activity occurred in the late period of Early Permian and lasted until the late Permian (Li et al. 1998), and the arc volcanic activity developed tholeiite series → calc-alkaline series → potassium basalt series volcanic rocks from early period till late period, which marked the complete process of island arc generation, development and maturity (Mo et al. 1993), and confirmed that the subduction of Jinsha River oceanic basin was reduced from early Permian to late Permian. The ore-bearing volcanic-sedimentary rock series is basaltic andesite-andesite with a thickness of 200 m and its corresponding subvolcanic rock (diorite porphyrite) assemblage and (tuffaceous) sandstone, siltstone, sandy mudstone, tuffaceous-siliceous rock and thick massive bioclastic limestone formation.

The key factors for the development of this metallogenic system and Nanzuo style deposit are the continental margin arc environment in which intermediate-acid volcanic rocks are developed, the shallow magma chamber that drives the volcanic hydrothermal cycle, a large number of metal material leached by hydrothermal fluid in intermediate-basic-intermediate-acid volcanic rock series and the superimposed transformation of the later fault tectonics. The deposit took shape in the continental margin arc environment formed during the subduction of Jinsha River oceanic crust, and volcanic activity provided heat source and related material sources. The ore-forming hydrothermal fluid is mainly high-salinity fluid formed by infiltration seawater and volcanic hydrothermal fluid, the metal mainly comes from volcanic rocks, and the fluid belongs to mixed sources. Mineralization is in the continental margin arc with high permeability, which is further enriched to form deposits by the superimposition and transformation of late fault tectonics.

Main Controlling Factors of Extensional Metallogenic System After Orogeny

Main Controlling Factors of Post-orogenic Extensional Metallogenic System in Yidun Orogenic Belt

The main controlling factors of post-orogenic extensional metallogenic system include crustal extensional environment, large-scale fracture and decollement structure, A-type granite emplacement and large-scale fluid migration. In different metallogenic subsystems, the types, configurations and coupling modes of these main controlling factors are different.

Magmatic Hydrothermal Metallogenic Subsystem

In the magmatic hydrothermal mineralization subsystem, A-type granite and its segregated magmatic fluid, limestone in granite contact zone and contact metasomatic skarn are the key factors of mineralization.

(1) A-type granite. As a magmatic product in the collapse stage of the orogenic belt, the A-type granitic magma activity started at 100 Ma (Gaogong in the north) and ended at 75 Ma (Batang in the south) and reached the peak of magmatic activity at 80 Ma, forming more than ten rock bodies large and small, constituting the A-type granite belt with NNW direction. It was formed in the extension-collapse stage of the orogenic belt. According to rock assemblage, geochemical characteristics and strontium isotopic composition, granites can be divided into at least two types, namely A1 type and A2 type. The former type is represented by Zhalong, Gaogong, Rongyicuo, with an $^{87}Sr/^{86}Sr$ initial ratio of 0.74407. A1-type granite is a typical crust-derived molten product and formed in the early stage of orogenic belt extension-collapse. A2-type granite, represented by Lianlong rock mass, is small in number, with an $^{87}Sr/^{86}Sr$ initial ratio of 0.7095, which indicates that its magma was originated with the participation of some deep-seated basic materials and was formed in the late stage of the collapse of the orogenic belt (Hou et al. 2001). Nevertheless, both types of granites were originated from the argillaceous rock-rich crustal source area, the source rock of A1-type granite source area is relatively uniform and the composition of A2-type granite source area varies greatly. Both types of A-type granites produce accessory minerals-cassiterite, which reflect that magma is rich in tin. Tourmaline minerals often appear in rocks, suggesting that magma is rich in volatile components. Compared with ordinary granites, granites in this area are significantly enriched in elements Sn, Ag, Bi, Pb and W, with enrichment

coefficients Sn and Ag reaching 8.14 and 4, respectively. In comparison with the tin-bearing granites in the world, the Sn content is fairly, but the $w(Rb)/w(Sr)$ ratio is higher, which reflects the geochemical characteristics of tin-bearing granites in collision orogenic belt. Most rock bodies of A-type granite are tin-silver mineralized in different degrees, which proves that A-type granite is the first main controlling factor for the development of magma-hydrothermal metallogenic subsystem.

(2) Contact metasomatic skarn. Contact metasomatic skarn is usually the main ore-bearing rock unit of tin-silver mineralization, although tin-silver mineralization also occurs in the inner contact zone of granite. According to the contact surrounding rock types, it can be divided into several skarn metasomatic types. In the contact zone between granite and calcareous limestone, diopside skarn and garnet skarn mainly occur, which constitute the most important metasomatism type in the area. In the contact zone between granite and argillaceous limestone, vesuvianite-diopside skarn and melilite skarn occurred. Both types of skarnization were accompanied by important tin-silver mineralization events, forming cassiterite-natural bismuth-polymetallic sulfide assemblages. The hydrothermal fluid after skarnization is transformed to be acidic, which leads to acidic leaching, accompanied by greisenization and sericitization, and corresponding cassiterite-polymetallic sulfide mineralization, superimposed on the mineralized skarn belt.

(3) Tectonic-fluid metallogenic subsystem. In the tectonic-fluid metallogenic subsystem, granite which heats and drives fluid activity, fracture structure and interlayer decollement structure which transport fluid in large scale and structural intersection space suitable for fluid convergence are the main controlling factors of the system.

(3.1) Extensional fracture and decollement structures. Extensional fracture and decollement structures are the main structural forms in the crustal extension and collapse stages of Yidun orogenic belt. The extensional fracture generally strikes NNW, which extends perpendicular to the main extensional stress direction. This extensional fracture may cut down the whole crust and spread along the eastern edge of rigid metamorphic basement blocks. Along the extensional fracture zone, structural breccia, regional linear zonal hydrothermal alteration and stratiform-like and ribbon-shape ore bodies controlled by fracture develop, which proves that extensional fracture zone is not only the main channel for large-scale fluid transportation and migration, but also the ore-holding space for metal deposition of ore-forming fluid. With the strong extension of the earth's crust, a series of decollement structures have been formed. The detachment zone is generally NNW or/and near SN strike and mainly occurs in the extremely thick Upper Triassic sand-slate system. Similarly, these detachment zones are not only the active channel of hydrothermal fluid, but also the ore-holding space of ore bodies.

(3.2) Granite and peripheral rock series. Altered rock-type silver polymetallic deposits, occurrences and ore-induced anomalies in the tectonic-fluid metallogenic subsystem are distributed along the eastern edge of the basement rigid block in the region and spatially occur in NNW extensional fracture and interlayer detachment zones, generally occurring within 1–10 km of the periphery of A-type granite block. In Xiasai mining area, five ore blocks occur in the Upper Triassic sand-slate series within 1.5–5.0 km of the periphery of Rongyicuo granite block. In Nanzhigou, Lianlong mining area, the silver polymetallic mineralization belt occurs in the upper Triassic sand-slate of the periphery of the Lianlong granite block. From the granite body to its periphery, it often evolves from skarnization to propylitization (chloritization + actinolization) to sericitization to silicification, and it is distributed in a linear belt along the NNW direction or/and SN direction fracture zone. This proves that the granite body may act as a "heat engine" to maintain the regional hydrothermal fluid activity, driving the large-scale migration of hydrothermal fluid along the extensional fracture zone, and moving in NNW direction and near S N direction to accumulate and unload at intersection part. The lead isotope data of the Xiasai deposit confirmed that metallic lead originated from volcanic-sedimentary rock series in the upper crust of the orogenic belt, and the sulfur isotope composition characteristics (δ^{34}S average—8.2‰) were far different from that of magma, which confirmed that the sulfur was not provided by granite. The large variation range of δ^{34}S value (−4.9‰ to −10.5‰) indicated that the sulfur in the deposit might be biological sulfur reduced by bacteria, or it may be that the hydrothermal fluid is metasomatism leached from the sedimentary rock series through which it flows, especially the upper Triassic neritic-shelf sand-slate series. The high element abundance of Ag, Pb, Zn, Sb, As, Bi and regional polymetallic register and geochemical anomalies in the Upper Triassic sand-slate series also indicate that the sedimentary rock series in the

upper crust emplaced by A-type granite bodies have important potential to provide a large amount of ore-forming materials.

Main Controlling Factors of Post-orogenic Extensional Metallogenic System in Jinsha River Orogenic Belt

The post-orogenic extensional metallogenic system is constructed on the Jiangda-Deqin-Weixi Hercynian-Indosinian continental margin volcanic arc belt in the eastern margin of Changdu block. It is the product of mountain root delamination and crustal extension in the late evolution of the orogenic belt and is the most important metallogenic system in the Jinsha River orogenic belt. The metallogenic system consists of at least two metallogenic subsystems. One is a hydrothermal fluid metallogenic subsystem, and the other is a magmatic metallogenic subsystem. Among them, the hydrothermal fluid metallogenic subsystem is the most important metallogenic system in this belt, and the coupling of magmatism, sedimentation and fluid action in the superimposed volcano-sedimentary basin formed by the late extension mechanism of continental margin island arc collision orogenic belt is the main control factor of systematic mineralization.

(1) Space–time range and constituent elements of the system. The spatial range of post-orogenic extensional metallogenic system is consistent with the Jiangda-Deqin-Weixi continental margin volcanic arc belt, and it evolves synchronously with the development stage of extensional (T_3) structure in the late orogenic period. From the perspective of the source and evolution of ore-forming materials, the system composition should extend to the lower crust and upper mantle. The main metallogenic elements of the system include magmatic hydrothermal solution related to intermediate-acid intrusion, volcanic-jet hydrothermal solution related to volcanic eruption and underground circulating hot water possibly indirectly heated by deep magma, which are closely related to the formation of porphyrite iron ore, skarn iron-copper ore-magma metallogenic subsystem and jet polymetallic ore-hot water fluid metallogenic subsystem, respectively. Magmatism is responsible for the direct or indirect transport of ore-forming materials and energy to the system, which plays a decisive role in the metallogenic evolution of the system, especially the tectonic–magmatic mineralization in the late Triassic extensional basin stage.

(2) Formation and evolution of the system. Because the host of the system is above the volcanic arc of the continental margin, after the collision and orogeny of the volcanic arc in the early and middle Triassic, the

mountain root delamination and crustal extension in the late stage has resulted in the tensile collapse of the upper continental margin arc, which transformed the volcanic arc from a long-term compressive tectonic system to an extensional system, resulting in strong magmatism and the development of superimposed rift (depression) basins on it. At the initial stage of extensional detachment, it was composed of littoral facies-shelf facies clastic rocks mixed with volcanic rocks formation, and the sediments formed a combination sequence from coarse to fine, and the water body changed from shallow to deep, accompanied by strong homologous magma intrusion in the same period or later. In its early stage, it is neritic-sub-abyssal facies clastic rocks' formation mixed with volcanic rocks. In the middle stage, it entered the peak of extension. The magma stored under the volcanic arc was injected into the basin on a large scale through volcanic eruption or magma intrusion, and the sub-abyssal facies-abyssal flysch and bimodal rock assemblage were formed in a short time. In the late period, it is sub-abyssal facies-neritic clastic rock formation mixed with volcanic rocks. At the end of the period, the basin changed from extensional and rifting to compressional, and the basin gradually shrank and died out, forming clastic rocks with molasse property and basic-intermediate-acid volcanic rocks and volcanic clastic rocks formation in littoral and neritic facies, and a large number of gypsum deposits and purple-red clastic rocks have accumulated. At the same time, the peak of tectonic–magmatic activity is also the time when the metallogenic system matures and enters the main metallogenic evolution stage, forming the hydrothermal fluid metallogenic subsystem. Mineralization is controlled by the syngenetic faults, rift basins and hydrothermal fluid activities in the post-collision extension stage, resulting in sedimentary exhalative deposit in the deep-water environment, such as Zuna silver-bearing lead–zinc deposit in the northern section, Luchun deposit in the middle section and Hongpo copper polymetallic deposit in the southern section of Laojunshan lead–zinc deposit. Mineralization took place in the shallow-water facies or the intermediate-acid volcanic zone in the land-sea interaction facies formed in the post-collision extensional environment, forming the sedimentary exhalative deposits in the shallow-water environment, such as Zhaokalong Fe–Ag polymetallic deposit and Dingqinnong Ag–Cu polymetallic deposit in the north section, Chugezha Fe–Ag polymetallic deposit in the middle section. Magmatic metallogenic subsystem mainly develops in andesite-diorite porphyrite complex area in the northern part of the orogenic belt from Jiaduoling to Jiangda, forming porphyrite-type iron deposits,

represented by Jiangda iron deposit, which may have been formed in the early stage of crustal extension after collision, or in the stage of transition from continental margin arc to extensional basin. At the end of the middle period of the Late Triassic, the extensional activity tended to weaken, the magmatic activity stopped, and the metallogenic system shrank.

(3) Main metallogenic elements

(3.1) Magma properties and ore-controlling effect. Magma provides ore-forming energy and main ore-forming materials for the system. As an important space channel, magma eruption or intrusion apparatus controls the evolution of ore-forming fluid system derived from magma, which is the final positioning space of ore-forming materials-deposits of the system. As the extension and rifting occurred on the "basement" background of volcanic arc, its magmatic rocks generally showed high-potassium calc-alkaline series. The main rock types were basalt, rhyolite and intermediate-acid andesite-rhyolite, with a small amount of basic basaltic andesite or andesitic basalt. The properties of magmatic rocks in the orogenic belt control mineralization obviously, and the minerals related to acidic endmember volcanic rocks (rhyolite) are lead–zinc–copper–silver polymetallic deposits (such as Luchun zinc–copper–lead–silver polymetallic deposit and Laojunshan lead–zinc deposit). Minerals related to intermediate-acid endmember volcanic rocks (dacite rhyolite) are iron-copper-silver polymetallic deposits (such as Zhaokalong iron–silver polymetallic deposit, Dingqinnong silver–copper polymetallic deposit, Chugezha iron–silver polymetallic deposit). Minerals related to basic endmember volcanic rocks (basalt) are Ag–Pb–Zn polymetallic deposits (such as Zuna Ag–Pb–Zn polymetallic deposit). Iron-copper polymetallic deposits (such as Jiangda iron–copper deposit) are related to intermediate (acid) andesite volcanic eruption and diorite intrusion.

(3.2) Ore control by volcanic/intrusion apparatus. The magmatic activity in the Late Triassic volcanic-sedimentary basin was controlled by NW–SN fault structure, and it was a fissure-type multi-center eruption, forming a series of volcanic eruptions of different sizes or intrusion apparatus distributed in belt. On the Zhaokalong-Dingqinnong belt, it is preliminarily recognized that volcanic apparatus may exist in the northwest side of Zhaokalong, Shengjie, Zhenama of northeast of Dingqinnong as well as Gemma and Ba Long between them, which have become important ore-controlling conditions for the formation of volcanic exhalative deposits in the Late Triassic. Jiaduoling area is another big volcanic-intrusive center, which controls the formation of large porphyry iron deposits and skarn deposits in peripheral areas. There is a long-term relationship between the volcanic eruption or intrusion apparatus and the deep magma chamber. During and after the magma activity, it is the channel and spout for the deep ore-forming fluid system to discharge outward. The ore-forming fluid forms a circulation mechanism with this as the center, and ore-forming materials are continuously extracted from the deep and surrounding volcanic rocks, and it was carried to favorable structural parts near the volcanic/intrusion apparatus and unloaded to form ore bodies. Therefore, large volcanic activity centers are often important conditions for the formation of large deposits or ore concentration areas and have the function of accumulating ore-forming materials in the metallogenic system.

(3.3) The ore-controlling function of the basin. Ore-controlling basins are mainly confined depressions between volcanic uplifts and rift in rift basins. The confined depression mainly develops in the marginal zone of volcanic-sedimentary basin, which is located between volcanic highlands at that time. It is closely related to the late volcanic activity and receives volcanic material deposits from volcanic uplift, and a large number of volcanic lava, volcanic breccia/agglomerate and other volcanic eruption rocks are developed. The transition relationship between volcanic rocks and normal sedimentary rocks is crisscrossed vertically and horizontally, and edging carbonate rocks can be developed around the volcanic uplift. A large number of volcanic falling objects are found in edging carbonate rocks. Such confined depression is closely related to volcanic apparatus and formed relatively closed environment in a certain scope, becoming the favorable place for unloading of ore-forming fluid of volcanic apparatus. After entering the basin, the ore-forming fluid is mainly formed minerals through normal sedimentation. The hydrothermal sedimentary rock with its characteristic, namely Sedimentary exhalative (SEDEX) deposits, both Zhaokalong polymetallic deposit and Dingqinnong silver (copper) polymetallic deposit, show this mineralization feature. The rift developed in median axis of volcanic-sedimentary basin (rift basin) mainly grew a set of extremely thick abysmal-bathyal clastic rock,

carbonate rock deposition in its extension process. Constrained by the extension fracture at the same period, it grew a set of volcanic lava with bimodal volcanic rock assemblage as the main part. Mineralization occurred in the main extension period (extension rift period) of the basin, featuring the development of slump breccia, calcite turbidite, silicious turbidite, tuffaceous turbidite and convolution structure. The formation of metallogenic thermal fluid system is related to volcanic-sub-volcanic magmatism. Then, the fluid was discharged to the basin through volcanic channel or syngenetic fracture. The mineralization is realized by hydrothermal deposit, accompanied with a set of hydrothermal sedimentary rock assemblage, namely VHMS or SEDEX deposit, represented by Zuna (sliver)–lead–zinc deposit, Luchun zinc–copper–lead–silver polymetallic deposit, Laojunshan lead–zinc deposit.

The ore controlling of the basin is mainly manifested as the metallogenic-fluid system provides favorable space for discharging and unloading. The coupling of hydrothermal system, discharging channel and relatively closed basin is the key factor determined volcanic eruption deposit and hydrothermal sediment deposit in hydrothermal fluid metallogenic subsystem under extension system in Late Triassic.

4.3.4 Intracontinental Convergent Giant Metallogenic System

4.3.4.1 The Background of Intracontinental Convergent Metallogenic System

The Profound Transformation of Intracontinental Convergence to Physical Structure of Sanjiang Orogenic Belt Is the Most Important Constraint Condition for Mineralization

Since the Late Cretaceous, with the successive closure and died out of Middle-Tethyan Ocean and Neo-Tethyan Ocean, micro and small microblocks have dispersed at the margin of Eurasia and Gondwana has finally welded and combined with the parent landmass through multiple orogenies. Eurasia and Gondwana and Yangtze continent combined to form a new continent. However, the interaction and movement among landmasses did not stop, but entered an overall period of intracontinental convergence and plateau uplift. The Sanjiang area in Southwest China restricted by the Indian plate, Eurasian plate and Yangtze plate constitutes the earth dynamic environment of crust deformation in the area. The extrusion and jacking occurred continuously in Indian plate in NNE is the main power causing convergence and deformation in the area. Influenced by Paleo-Tethys and Meso-Tethys archipelagic arc-basin orogenic system, the physical structure of Cenozoic Sanjiang crust shows extreme nonhomogeneity, block segmentation, soft-hard interbeddings. It went through complex tectonic deformation in intracontinental convergence. It generally showed large-scale shortened horizontal thickness in NE-NEE, forming Hengduan Mountains in nearly NS direction and Sanjiang stream sediment. After strongly extrusion, it formed tectonic wasp-waist near north latitude line 28°. Qamdo block in NW direction and Lanping-Pu'er block have extruded and skidded off toward two sides, respectively, forming huge "X" shape structure knot in Sanjiang area. With the extrusion and shortening of the crust and the skid-off of the block, taking the Qamdo-Lanping block as the central axis, a large-scale tectonic collision took place between the mountain systems on both sides. At the same time, a large-scale strike-slip activity has occurred between the blocks and the orogenic belts on both sides. These huge deformation structures have formed during the intracontinental convergence process not only deeply reformed the upper crust, but also led to the extensive development of thrust nappe, strike-slip pull-apart and extensional gliding nappe structures in the Sanjiang area. The intense intracontinental convergence even affected the lower crust and upper mantle, causing crust-derived and mantle-derived magma activities, which had an extremely important impact on regional mineralization and became an important constraint condition of intracontinental convergent giant metallogenic system.

Intracontinental Convergence Mineralization

Intracontinental convergence is a large-scale transformation, adjustment, recombination and integration of crust tectonic and physical composition after the evolution of Tethys MABT, resulting in widely activization of mantle matter. Two major structural system transformations in Paleo (Meso)-Tethys have laid an important geological structural background and material foundation for intracontinental convergence mineralization. From the successive cracking and stretching to subduction collision of Lancang River, Jinsha River, Ganzi-Litang Ocean Basin and Nujiang Ocean Basin, the tectonic-magmatism in the crust-mantle conversion process has profoundly influenced the physical structure of the Sanjiang crust, including the injection of multiple molten magmas from the upper mantle, lower crust, subduction ocean crust and the crust itself into the upper crust, and the distribution and location of ore-forming materials in the crust due to the fluid action in the magma evolution process, especially in volcanic island arc orogenic belt,

where there are not only deposits formed by metal accumulation, but also caused relatively enrichment of metallogenic elements in stratum rock. Subsequently, the foreland basins (such as Qamdo basin and Lanping basin) developed on the block adjacent to the orogenic belt received a large amount of erosion materials from the volcanic arc orogenic belts on both sides. The materials in the orogenic belt were further differentiated through supergene weathering and sedimentation, and some ore-forming materials were transferred and deposited in the extremely thick stratum sedimentary rocks of the basin. Then, the ore-forming material was distributed again and was enriched in some rocks, such as sandstone-type copper mineralization widely occurring in the Upper Triassic, Jurassic-Cretaceous and Paleogene in Qamdo basin and Lanping basin which is related to this to a certain extent. Taking Lanping Basin as an example, hydrothermal sedimentary mineralization has occurred in the T_3-J_1 intracontinental rift metallogenic stage, followed by sedimentary copper mineralization in the depression basin metallogenic stage (J_{2-3}) and foreland basin metallogenic stage (K), and copper, lead, zinc and other ore-forming materials were initially enriched in the extremely thick Jurassic-Cretaceous red beds. These material conditions influence and restrict the occurrence of Himalayan mineralization to some extent. On the basis of long-term tectonic-metallogenic evolution, sedimentation, accumulation, intracontinental convergence created more favorable conditions and opportunities for mineralization, causing large-scale mitigation, enrichment and mineralization of ore-forming material, forming a batch of large and super-large deposit such as Yulong copper deposit, Jinding lead–zinc deposit, Ailaoshan gold deposit. Some deposits formed before intracontinental convergence were transformed and enriched. It has become a consensus that Cenozoic is the most important metallogenic period of Tethys in Sanjiang.

Himalayan mineralization is not a simple repetition on the basis of early mineralization, but an inherited development and advance of mineralization in a wider three-dimension structure, featuring in metallogenic tectonic environment, metallogenic mechanism, mineralization, intensity and scale, etc. In the process of intracontinental convergence, due to the cross-tectonic unit action of tectonic, magma and fluid, various geochemical domains with constraints on mineralization, which were established based on blocks and sutures (suture zone) in the early days, became open systems. There were extensive material and energy exchanges between blocks and orogenic belts, and ore-forming materials and energy have not only covered the upper crust, but also extended to the lower crust-upper mantle vertically. The multiple sources of ore-forming fluids and ore-forming materials and the complex assemblage of ore-forming elements are important characteristics of Cenozoic intra-continental covergence and mineralization.

Two Tectonic Mechanisms Controlling the Intracontinental Convergent Metallogenic System

In Cenozoic, there were two types of tectonic ore-controlling mechanisms, namely, extension and extrusion. A large-scale and deep-seated extensional tectonic ore-controlling mechanism has mainly occurred in the north–south sector of the giant "X"-shaped structure knot of the Sanjiang River in Cenozoic. This was mainly related to the strike-slip and pull-apart activities in the process of block extrusion and mutual collision and rotation. In the Paleogene Eocene, influenced by the regional extrusion stress field in the northeast, a strong right-hand strike-slip pull-apart has occurred along the Qamdo block in the Jinsha River orogenic belt in the north of Sanjiang River, and salt-bearing pull-apart basins represented by Gongjue and Nangqian were formed on the earth surface. Crustal decompression resulted in partial melting of the lower crust and upper mantle, forming alkaline to alkaline-rich magma, which controlled the mantle-derived magma metallogenic system. In the south of Sanjiang area, the strike-slip pull-apart mechanism of the central axis fault in Lanping-Pu'er basin controls the formation of Lanping super-large lead–zinc deposit. The right strike-slip of Ailaoshan-Honghe Paleo-gene is not only closely related to the formation of Ailaoshan large-scale gold deposit, but also forms alkali-rich porphyry belt at the margin of Yangtze plate on the northeast side. Superficial extensional activities occur in large numbers in the front and rear edges of the strong nappe orogenic belt, represented by gliding nappe faults, whose extensional decollement structure interface is a favorable place for metallogenic hydrothermal activities. It controls the formation of many tectonic altered rock-type polymetallic deposits related to low-temperature hydrothermal solutions in the late orogenic belt.

The Cenozoic "Sanjiang" area tectonic movement is dominated by intracontinental convergence and extrusion, which has the strongest effect on the transformation of the crustal structure. Long-term regional lateral extrusion leads to the vertical superposition and thickening of the crust, and large-scale intracontinental subduction and overthrust nappe are the main ways of crustal thickening and stress absorption. As a result of the convergence and extrusion, some crustal materials melted again in some old orogenic belts, forming S-type granite series, and consituting metallogenic systems related to crust-derived granites, such as W, Sn, Mo and W metallogenic belts, respectively, formed in Leiwuqi-Dongda Mountain in eastern Tibet and two island arc oro-genic belts in Yindun in the late Yanshan period. On both sides of the foreland basin, the convergence has caused the

mountain belts on both sides to advance further to the basin, and a series of low-angle thrust structures are developed in the shallow parts of the crust on both sides of the foreland basin. With the material of the orogenic belt moving to the foreland, the tectonic compression drives the water in the rocks of the orogenic belt and the pressed foreland basin to be discharged. This water is mixed with the meteoric water, magma water from deep sources and metamorphic water, and then, they migrate to the front of nappe structure. A large number of ore-forming materials are extracted from the passing rocks, especially from the early ore-forming pre-rich beds. Particularly, when the sedimentary strata of the foreland basin are rich in gypsum and salt, they are easily dissolved into the fluid, forming a hot brine fluid system for mineralization, which greatly improves the extraction and transportation of minerals in the rocks. Because of the trap effect of overthrust nappe on fluid system, leakage miner-alization and unloading mineralization may occur by the fluid at the front of the system. The middle-low-temperature polymetallic mineralization and extensive geochemical anomalies commonly developed in the east–west thrust fault zones of Qamdo and Lanping-Pu'er basins are mainly con-trolled by this tectonic-metallogenic mechanism.

4.3.4.2 Intracontinental Magmatic Metallogenic System

Intracontinental Crust-Derived Magmatic Metallogenic Subsystem

This metallogenic system is related to the continental crust remelting granite of the orogenic belt in the early stage of the late Yanshanian intracontinental convergence, which mainly occurs in the Leiwuqi-Dongda mountain composite island arc orogenic belt and Yidun island arc orogenic belt to the west of the northern Lancang River, respectively, forming a regional extension Tin-bearing granite belts with a length of 200–300 km, which controls two tin (tungsten) and silver polymetallic metallogenic belts, including Changmaoling-Xiaya-Larong-Suoda and Gaogong-Cuomolong.

Leiwuqi-Dongda mountain composite island arc belt is an island arc orogenic belt formed by the subduction and clo-sure of the North Lancang River at the end of the Late Paleozoic. During the Indosinian, there was a further colli-sion orogeny between the North Lancang River fault and the Qamdo block, which led to a large-scale remelting and transformation of the upper crust material, resulting in a strong continental crust collision magmatism. During the Middle and Late Triassic, high silicon and high-potassium (intermediate) acid andesite-rhyolite volcanic belt in the North Lancang River belt and granite belt in Jitang-Dongda mountain on the west side were formed, respectively. The abundance of mineralization and associated elements such as Sn, W, Ag, Sb, As, Bi, Pb in old metamorphic rocks such as Neijitang Group is several times higher than the average value of the crust, and they are important ore source rocks. In the process of continental crust remelting, the continental crust material has undergone a great degree of transformation and differentiation, and the crustal material has experienced the actions of heating and drainage, magma melting, crystallization differentiation, etc. All kinds of fluids overflew and dispersed outward, carrying the ore-forming materials activated and extracted from the crust to the low-temperature and low-compressive stress parts together. Since the late Triassic, the orogenic belt has been dominated by the overthrust nappe in the northeast direction, and its western rear edge is relatively in an extensional environment. Therefore, the hydrothermal solution tends to migrate and converge to this part, forming the pre-enrichment of minerals. In the middle and late Yanshanian period, with the closure and collision of the Nujiang ocean basin, an intra-continental subduction and collision occurred from west to east, which triggered the melting of the continental crust again. A series of small intrusions were formed along the west side of the Jitang-Dongda mountain plutonic belt in Indosinian. The rock bodies mainly intruded into sandy argillaceous clastic rocks of the Lower Carboniferous and Upper Triassic, with isotopic ages of 87.7–63.7 Ma. The hydrothermal activity after the magmatic period was related to tin (tungsten), (Nb–Ta), heavy rare earth.

Tin (tungsten)-bearing granites are mainly hornblende-biotite monzogranite, biotite (or two-mica) monzogranite, biotite granite and muscovite granite. From the center to the outside, the lithofacies of a single rock mass changes obviously; mineral crystal grains change from coarse to fine; and dark minerals gradually decrease, which is obviously a product of multi-stage full differentiation and evolution. According to the research on this belt of Yong et al. (1991), the lithochemical composition of tin-bearing granite is characterized by high silicon and high alkali, with the average $w(SiO_2)$ of 71.88%, $w(K_2O + Na_2O)$ of 7.95%, and $w(K_2O) > w(Na_2O)$. Low MgO and TiO_2, with average values of 0.64% and 0.36%, indicate that there is a typical cause of continental crust reformation. The total rare earth content of tin-bearing granite is high, and it is of light rare earth enrichment type. $(w(La)/w(Yb))_N$ is 5.15–30.94, and the Eu is strongly deficient (δ EU 0.29–0.55), which indicates that magma is originated from partial melting of the upper crust and experienced good crystallization differentiation. Mineralization is related to the light granite formed by the high separation crystallization of magma. This kind of granite has a gentle rare earth pattern, the total amount of rare earth decreases and the contents of light and heavy rare earths are similar. $(w(La)/w(Yb))_N$ is 1.54–9.9, and the Eu is strongly deficient (δ EU 0.21–0.57). Tin-bearing granite

generally has higher contents of ore-forming or mineralized elements such as Sn, W, Ag, Cu, Mo, B, F, Be, Rb, Bi than similar rocks. These elements may have inherited the early pre-enrichment material source. The above characteristics are similar to those of tin-bearing granites in Southeast Asia, South China and Western Yunnan.

Ore-bearing rock mass is mainly controlled by NW–SE strike fault structure, mostly presenting as small rock mass and rock strain. Mineralization is closely related to post-magmatic period hydrothermal activity, belongs to high-middle temperature hydrothermal filling metasomatism genesis and occurs in the contact zone inside and outside the rock mass. There are three main metallogenic types in Changmaoling-Xiaya-Larong-Suoda belt. The north section is mainly cassiterite (wolframite)-tourmaline-quartz type and greisen type. The former includes Saibeilong-Mabuguo tin deposit and many mineralized occurrences in the Changmaoling area of Leiwuqi County. Mineralization mainly develops in hornfelsic surrounding rock in granite's outer contact zone. Mineralization is controlled by fracture zone and structural fissure, with cassiterite (wolframite)-tourmaline-quartz as the main combination, mainly in single vein and compound vein or reticulated vein. Ore-forming temperature is mainly between 380 and 240 °C, and the ore-forming fluid $\delta^{18}O_{H2O}$ is +6.83‰ ~ +8.56‰, indicating that the ore-forming hydrothermal solution mainly comes from magmatic water. The latter is represented by the Xiaya ore occurrence in Basu County, where mineralization occurs in the contact zone of granite or the altered zone of greisenization altered zone at the top of the granite.

The mineral assemblage is mainly cassiterite-tourmaline-fluorite-quartz, cassiterite-muscovite-quartz and fluorite-topaz-cassiterite-quartz. In the early stage, cassiterite and other minerals are distributed in greisen in a dispersed and stellate manner. In the late stage, vein mineralization is mainly formed, with mineralization temperature ranged from 300 to 220 °C, and the fluid $\delta^{18}O_{H_2O}$ of +7.45‰ (Yong 1991). A series of small acidic rock bodies are developed along the east and west sides of Indosinian-Yanshanian complex granite base in the southern section of Dongda mountain, and the corresponding tin (tungsten) geochemistry or cassiterite (wolframite) heavy sand anomalies are obvious. The main metallogenic types are greisen type or cassiterite-sulfide type, represented by some occurrences such as Larong and Qudeng.

Gaogong-Cuomolong tin-bearing granite belt is located on the inner side (west side) of Yidun Island arc zone, sandwiched between Keludong-Xiangcheng fault and Aila-Riyu fault. Most of the rock bodies are complex, mainly invading the Tumugou Formation of the Upper Triassic, with the isotopic age of the rock bodies of 115.8–76.8 Ma, and the duration of magmatic activity of 8–29 Ma.

The early porphyritic moyite as well as the late porphyritic biotite monzogranite and moyite have the characteristics of crust-derived granite. The content of Sn, Ag, Bi and other ore-forming elements in granite is generally high, and the ore-forming materials are mainly provided by rock mass. Tumugou Formation of Upper Triassic invaded by rock mass also has a high background of ore-forming elements. Controlled by post-magmatic period hydrothermal and circulating meteoric water activities, the mineralization is carried out by filling metasomatism. When the rock mass intrudes into limestone with active chemical properties, it forms skarn-type tin polymetallic deposits after metasomatism, represented by Cuomolong and Lianlong deposits. When there are large structural channels around the rock mass, such as nappe and gliding nappe interface, magmatic fluid and meteoric water are mixed and filled along these channels to form a structurally altered rock deposit, represented by the large silver deposit in Xiasai.

In recent years, a lot of new progress has been made in molybdenum polymetallic prospecting in the southern section of this belt. The Tongchanggou Cu–Mo(W) ore, Xiuwacu Mo–W ore and Donglufang Cu–Mo ore have great resource potential. Our team studied the Cu–Mo(W) deposit in Tongchanggou and obtained that the zircon U–Pb age of mineralized molybdenite granodiorite diorite porphyry (TCG-16) is 79.1.2 ± 1.9 Ma and that of weakly mineralized granodiorite diorite porphyry (TCG-21) is 83.6 ± 1.1 Ma. The formation age of Xiuwacu monzonitic granite porphyry is 83.3 ± 1.7 Ma (MSWD = 2.6). For the mineralization age of molybdenite, according to the mineralization types and the spatial distribution of porphyry, the mineralization types of Cu–Mo–W deposits in late Yanshanian are porphyry-type Mo–Cu–W mineralization → skarn-type Cu–Mo mineralization → hydrothermal-type Cu–Pb–Zn–Ag mineralization from the porphyry to outward direction. The mineralized body is mainly controlled by the structural fissure system in porphyry, and the structural fissure mainly develops at the top or edge of the rock mass, but there is no obvious development in the center of the rock mass. Ore-bearing fractures are mostly distributed in vein and reticulated vein, and the more developed and mineralized the fissures are, the more enriched they are. For example, rich molybdenite mineralization is distributed in the structural fissures developed in the edge of Tongchanggou ore-bearing porphyry, and the ore bodies controlled by the structural fissures are mostly lenticular or stratiform-like. The development of molybdenite in Xiuwacu Cu–Mo deposits is closely related to the initial stage of quartz vein, and molybdenite, chalcopyrite and scheelite are all produced at the edge of wide quartz vein or veinlet-like quartz vein. At the same time, Cu–Mo–W mineralization also indicates the characteristics that porphyry body itself is an ore body.

In a word, the Yanshanian post-collision extension is the main tectonic mechanism of the terrigenous granite metallogenic system, and the magma formed by the remelting of crustal materials is the most important ore-controlling factor of the metallogenic system. Tin (tungsten) and silver polymetallic ore-forming materials mainly come from crust-derived magma. Multi-stage remelting and magma crystallization differentiation lead to mineral migration and enrichment. Hydrothermal driving after magmatic stage leads to the joining of groundwater in different degrees, and some ore-forming materials are brought in by surrounding rocks. Hydrothermal filling metasomatism is the main metallogenic mode.

Intracontinental Mantle-Derived Magmatic Metallogenic Subsystem

The intracontinental mantle-derived magmatic metallogenic system occupies an extremely important position in the intracontinental convergence mineralization of the Sanjiang River. The mineralization is related to the Yulong-Jinsha River-Honghe Cretaceous-Paleogene Eocene porphyry belt, which runs through the north–south tectonic units. The famous porphyry copper (molybdenum) deposit represented by Yulong was formed in the northern Qamdo area, and the porphyry-explosive breccia-type gold-silver (polymetallic) deposit was transited to the south. Extensional tectonic mechanism, deep source magmatism and hydrothermal convection system controlled by hypabyssal porphyry are the three important metallogenic elements of this system.

The Tectonic Environment of Porphyry Formation and Magma Source Area

According to the study of Yulong porphyry belt, the area of Xiariduo-Haitong, where Yulong porphyry belt is located, is the inner side of the ancient continental margin arc formed by the westward subduction of Jinsha River Ocean during Hercynian-Indosinian, rather than the long-term stable intraplate environment. The direct provenance of Himalayan porphyries is the upper mantle or lower crust which was subducted and transformed from the oceanic crust of Jinsha River to the lower part of Qamdo block. The age of 200–240 Ma Sm–Nd isotope model of porphyries (Wang 1995) indicates that the original magma of porphyries left the mantle in the late Indosinian. Therefore, it can be considered that the porphyry was formed in the island arc environment of the ancient continental margin. The magma source area is either the residual ancient oceanic crust of Jinsha River or the island arc mantle modified for it. This porphyry is not significantly different from the porphyry generally formed on the active continental margin of the Mesozoic and Cenozoic convergent plates and the porphyry directly originated from the melting of subducted oceanic crust. In addition, tourmaline-quartz explosive breccia is widely developed in ore-bearing porphyry. The porphyry magma is extremely rich in boron. As the porphyry magma is not contaminated by crustal materials on a large scale, one possible explanation is that boron from the ancient oceanic crust is rich in sediments.

Extensional Tectonics and Deep-Seated Magmatism

The isotopic ages of Himalayan ore-forming porphyry and homologous volcanic rocks in Qamdo area range from 52 to 30 Ma (it is mainly K–Ar age), equivalent to Paleogene Eocene. During the NE jacking of the Indian plate, dextral strike-slip occurred in the Jinsha River orogenic belt, which led to crustal decompression, partial melting of lower crust and upper mantle, forming deep-seated magma upwelling. The coupling between strike-slip pull-apart and magma plays an important role in the formation of porphyry deposits. Chesuo fault, which divides the Qamdo block from the Hercynian-Indosinian Jiangda-Mangling continental margin arc orogenic belt, is a regional super-deep crust fault. The Paleogene strike-slip pull-apart activity is very strong and controls the Gongjue half graben strike-slip extension basin, which stretches for hundreds of kilometers, and is an important rock-guiding structure of deep-seated magma. In the early stage, the right strike-slip-shear was the main activity, and a series of NW–NNW strike folds and compressional-torsional fault structures which presented in a left-hand geese line obliquely crossed with Chesuo fault developed on one side of Xipan fault, which constituted the main rock-bearing structure. Then, the left lateral relaxation accompanied with westward extension activity has occurred, and the secondary structure changed from compressional-torsional to tensional-torsional, thus becoming the most favorable place for the final location of ore-bearing porphyries. The extensional tectonic environment is not only conducive to the rapid and smooth upward migration of magma and ore-forming materials, but also can keep the metallogenic system in contact with the deep-seated magma chamber for a relatively long period of time, so that ore-forming materials can be continuously and fully supplied to the system, which provides an important condition for the system to form metal piles with great commercial value.

Magmatic Mineralization

Magmatic rock, as ore-forming parent rock, not only provides ore-forming fluid and ore-forming materials directly to the system, but also provides essential heat energy for the metallogenic system. It has the function of "heat engine" and drives hydrothermal fluid to circulate, so that mineralization can be fully realized under this mechanism. Porphyry plays a direct and decisive role in mineralization. Himalayan

ore-bearing porphyries are mainly composed of calcium-alkaline to alkaline monzonitic granite porphyries, quartz monzonitic porphyry and orthophyre, which are mainly alkaline ones. These two kinds of porphyry control porphyry copper (molybdenum) deposits and gold-silver polymetallic mineralization series or assemblage, respectively. Porphyry magma formed in the lower crust or upper mantle is rich in ore-forming materials and directly enters the upper crust along active deep-seated faults. The transformation of crust-mantle material is an important constraint for mineralization. Ore-forming materials originating from the lower crust or upper mantle reservoir take magma and its derived geological fluids as carriers. From magma generation to mineralization, the main ore-forming materials basically follow a relatively simple unipolar evolution track of extraction from source area → magma migration → fluid differentiation → cyclic unloading.

The seating environment and output state of magma play an important role in controlling mineralization. Most of the ore-bearing porphyries invaded the anticline dome structure or the intersection of faults with near-surface structural expansion at that time, which was favorable to the accumulation of hydrothermal solution after magmatic stage, and made the ore-forming fluid system have a good ore-gathering function. Most ore-bearing porphyries contain explosive breccia, mainly tourmaline-quartz breccia, the formation of which is related to an extremely boron-rich magma melt mass differentiated from the magma, indicating that the magma has extremely high energy and volatile components during its upwelling. This is conducive to the separation of ore-forming materials from the magma. Porphyry rich in volatile matter can provide abundant ore-forming fluid, and its hydrothermal action is the key factor to form the circulation mechanism of ore-forming fluid centered on porphyry. As the main carrier, magma brings heat and ore-forming minerals from the mantle. Many research showed that the mantle-friendly elements such as Cu, Fe, Pt, Pd, Cr, Ni, Co, Au and Ag in porphyry copper deposits mainly come from the mantle, while the crust-friendly elements such as Mo, Pb, Zn, W, Sn, Sb and Bi mainly come from the crust. They are generally high in the upper crust strata of Qamdo block and are mainly brought in by fluid circulation. Porphyry emplacement is found near the edge of the basin or the fault zone in the basin, which are the active channel of the basin's contemporaneous tectonic displacement fluid. The fluid is abundant, and it is mainly hot brine formed by flowing through gypsum-salt-rich strata such as the Upper Triassic, which has strong ability to extract and transport ore-forming materials. This is in favor of more ore-forming materials to enter the porphyry hydrothermal metallogenic circulation system.

Ore-Gathering Mechanism of Porphyry System. Predecessors Have Established Generally Applicable Genetic Models and Descriptive Models for Porphyry Deposits

It is widely believed that mineralization is related to hydrothermal convection system centered on porphyry, and hydrothermal convection system is the key factor to determine mineralization, alteration structure and the final positioning and shaping of ore bodies. For the porphyry copper deposits and porphyry-explosive breccia gold-silver (polymetallic) deposits, the spatial allocation of mineralization and alteration is summarized in Fig. 4.15. The hydrothermal system of the ore-gathering is composed of three different parts from bottom to top or from inside to outside with porphyry as the center. The depth of the lower part of the system is about 500–1000 m. In and around the porphyry, the hydrothermal solution mainly comes from the magmatic fluid with a temperature between 300 and 500 °C, which is in a relatively closed supercritical temperature state. The fluid is rich in mineralizing elements such as silicon, potassium and boron, fluorine, chlorine and sulfur. In the depth range of about 500 m in the middle of the system, the hydrothermal rising dominated by magmatic water suddenly changed from a closed supercritical state to a depressurized and open environment, which caused the hydrothermal boiling and depressurization at the top of porphyry or cryptoexplosive breccia, and the fluid volume expanded and the density decreased. This might cause pumping and made the peripheral meteoric water with lower temperature and higher density move to the rock mass. They quickly replenished, mixed and neutralized with magmatic water and was quickly heated to move further upward or outward, resulting in the formation of the hydrothermal circulation system. During the repeated circulation of hydrothermal solution, a large amount of quartz, potash feldspar and tourmaline are precipitated in the system, and copper, molybdenum and some minerals such as gold, silver, lead and zinc are mainly deposited in the vicinity of inside and outside of the rock mass of the lower part of the system (such as the contact zone of the rock mass) at first. The main components such as gold, silver, lead and zinc are further migrated outward by the fluid. With the decrease of temperature, these components are mainly deposited in the favorable tectonic channels around the porphyry body. The upper part of the system is located at the place from below the original earth surface to the top of the explosive breccia body, with a depth ranging from 0 to 500 m. It belongs to the discharge outlet of the geothermal pool convection system. The fluid is mainly discharged through the rising steep fault tectonic fissures in the same tectonic period. The rising deep-seated magma water is fully mixed with the surrounding groundwater, and the temperature is further reduced to 50–100 °C. Hot springs or large geothermal springs (fields) were formed on the earth surface. Epithermal mineralization can occur in the upper part of the system. There are still such hot spring activities in Yulong porphyry belt. The sinter contains typical low-temperature hydrothermal elements such as As, Sb, Mn, Hg, (Au), (Ag), forming a mineral assemblage such as natural sulfur, cinnabar, barite, stibnite and male (female) sulfur. From the system structure of the above circulating fluid, the upper part is epithermal mineralization, the middle part is related to the formation of explosive breccia-type gold-silver (polymetallic) deposits and the lower part mainly forms porphyry-type copper (molybdenum) deposits, which may constitute a complete metallogenic assemblage model. To some extent, porphyry copper (molybdenum) deposits seem to be equivalent to the deep system of explosive breccia or epithermal gold-silver (polymetallic) deposits. The fluid from the magma and the groundwater are in a transitional relationship with complete mutual solubility, without any interface. Therefore, the mineralization and alteration from the rock mass to the surrounding rock are also a continuous transitional relationship, and there is no strict boundary between the above-mentioned mineralization. As far as a certain deposit is concerned, due to the difference in the degree of development of the fluid system, as well as the differences in structural elements, surrounding rock conditions, current hydrogeological conditions, fluid temperature and pressure, properties, active stages, ore-forming material supply, etc., the development of its metallogenic assemblages is significantly different from each other, and only a part of them may be relatively developed and valuable deposits are formed.

4.3.4.3 Metallogenic System of Tectonic Dynamic Fluid

The metallogenic system is characterized by a kind of metallogenic mechanism and mineralization, which is a large-scale drainage activity between the orogenic belt and the foreland basin driven by the thermodynamic and tectonic dynamics in the crust during the process of intracontinental convergence. The fluid extracts the ore-forming elements flowing through the stratum rocks and brings them into the overthrust zone or detachment gliding nappe structure at the junction of basin and mountain to unload. Under the tectonic background of intense intracontinental convergence and compression in the late Yanshanian-Himalayan period, thrust nappe, detachment gliding nappe and strike-slip-shear are the most important tectonic forms of the crustal surface in the Sanjiang area. The mountain systems on both sides of the Qamdo and Lanping foreland basins undergo a large-scale hedging to the basin, and the fluids in the middle and shallow layers of the crust controlled by this have migrated and converged from the orogenic belt to the basin and are discharged and unloaded along the nappe belts or internal faults on both sides of the basin, controlling the mid-low-temperature hydrothermal polymetallic metallogenic

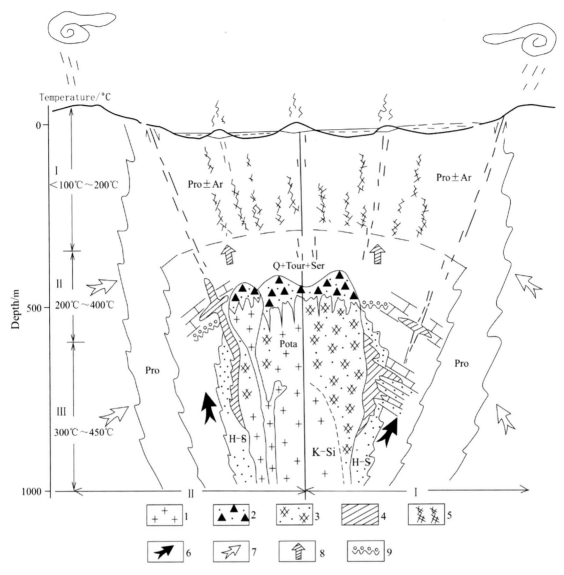

Fig. 4.15 Metallogenic model of porphyry copper deposit and explosive Breccia gold–silver (polymetallic) deposits' fluid system. 1 —porphyry; 2—cryptoexplosive breccia; 3—fine vein disseminated mineralization; 4—hydrothermal contact metasomatism; 5—reticulated vein fracture zone; 6—magmatic hydrothermal migration direction; 7— groundwater migration direction; 8—mixed hydrothermal fluid; 9— boiling surface. pro ± Ar—propylitization (argillization); Q + Tour + Ser—silicification, tourmalinization, sericitization; H–S—hornification, skarnization; Pota—potassization; K–Si—nucleus

belts extending on a large-scale or comprehensive geochemical anomalous zones, and forming a number of promising deposits. Inside the orogenic belt, reverse extensional gliding nappe of different scales often appears on the trailing edge of the intense thrust nappe structure, causing fluids in the orogenic belt to move along these extensional faults for mineralization. To this end, the Tuoding in Jinsha River orogenic belt and medium-low-temperature polymetallic deposits are related to this.

This metallogenic system occupies an extremely important position in Sanjiang Cenozoic mineralization, which truly reflects the conditions and opportunities of trans-unit tectonism in the process of intracontinental convergence,

large-scale and multi-source fluid convergence and migration, as well as extensive material source fusion and massive metal accumulation. The structural elements of metallogenic system mainly include thermodynamic and tectonic dynamic conditions formed by intracontinental convergence, fluid migration and discharge guided by nappe compression or extension strike-slip structural mechanism, abundant provenance bases in stratum rocks of orogenic belts and basins, ability to extract and transport metallogenic elements determined by fluid properties (such as hot brine formed by dissolving gypsum-salts) and ore-forming traps favorable for fluid unloading (such as dome structures, strike-slip basins, extension fractures).

Thrust Nappe Structure-Fluid Metallogenic Subsystem

Regional fluid flow caused by tectonic compression during intracontinental convergence is obviously the most important metallogenic factor of the metallogenic subsystem. In the Sanjiang area, especially in the middle-south section of Sanjiang, during collision, compression and shrinkage, thrust nappe structure-fluid metallogenic events were generally developed, and structure-fluid metallogenic subsystems were formed in many zones. Different orogenic belts have formed different metallogenic subsystems due to their different tectonic evolution backgrounds and differences in the composition of rock series strata. For example, in Ganzi-Litang and Jinsha River-Ailaoshan orogenic belts, accompanied by strike-slip-shear-nappe tectonic action, altered rock gold deposit metallogenic system was formed; on the east side of Qamdo-Lanping-Pu'er basin, Pb, Zn and Ag polymetallic hydrothermal metallogenic subsystems were formed under the participation of mantle fluids, and on the west side, Cu polymetallic hydrothermal metallogenic subsystems were formed. In Pangong Lake-Salween River-Changning-Menglian orogenic belt, Pb, Zn and Ag polymetallic hydrothermal metallogenic subsystems were formed during compression and nappe. The Qamdo-Lanping-Pu'er basin was strongly compressed, and a large amount of water was drained and mixed with the tectonic hydrothermal solution from the subduction zone, rose, released and unloaded along the thrust front belt, causing extensive medium-low-temperature hydrothermal alteration and mineralization in the region (Fig. 4.16), forming a medium-low-temperature hydrothermal mineralization and sedimentary-hydrothermal reworked metallization subsystems.

Extensional Detachment Structure-Fluid Metallogenic Subsystem

During the process of intracontinental convergence, the old orogenic belt further undergoes unidirectional or bidirectional outward thrust nappe, and extensional detachment structures appear to varying degrees at the trailing edge of the orogenic belt to control the activities from the orogenic belt and atmospheric precipitation. The basic model of the metallogenic subsystem is that the fluid extracts minerals from the rocks of the orogenic belt with rich provenance and converges into the detachment and gliding nappe structures in the extensional environment. Many examples of such deposits have been found in the Jinsha River orogenic belt. During the further eastward thrust nappe of the Jinsha River orogenic belt in the Himalayan period, the detachment and stripping structures occur along different tectonic layers in the extension and tension of the western rear edge, resulting in the filling and metasomatism of ore-forming hydrothermal fluids, forming the gold polymetallic deposits represented by Azhong. In the south, the Tuoding copper deposit, the Najiao system, the Gangriluo and other lead–zinc deposits and even the Xiasai super-large silver polymetallic deposits all have the obvious characteristics of such structural ore control or reworked metallization.

Metallogenic System of Strike-Slip Pull-Apart Basin

This metallogenic subsystem refers to the metallogenic mechanism and mineralization that occur in the basin due to deep hydrothermal activity or sedimentary diagenesis, respectively, with the Cenozoic strike-slip basin as the main control element, and the basin is mainly acting as a mineralization carrier or reservoir.

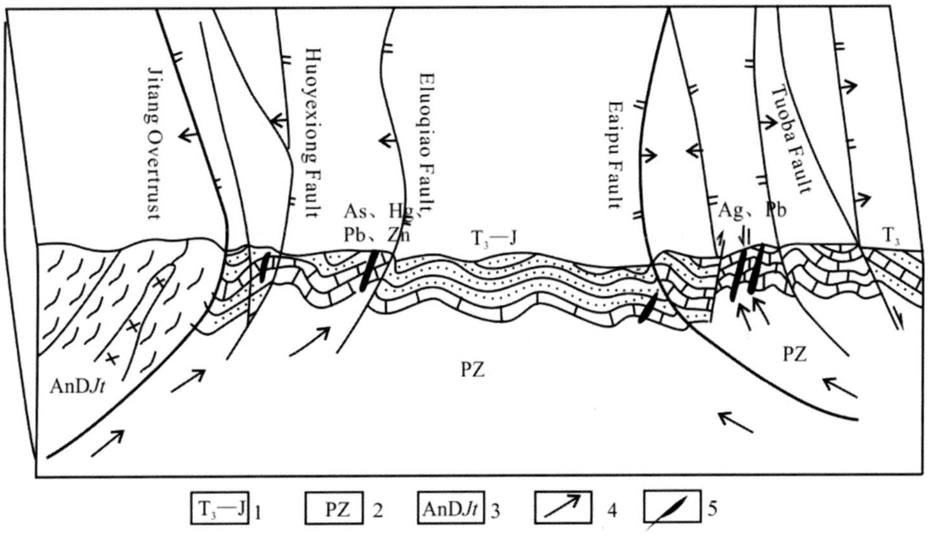

Fig. 4.16 Relationship between the nappe structure in the middle segment of the Qamdo basin and the hot brine mineralization. 1—Upper Triassic-Jurassic; 2—Paleozoic; 3—Pre-Devonian Jitang Group; 4—Hot brine migration direction; 5—Arsenic, silver and lead polymetallic deposit

The strike-slip pull-apart basins in the Sanjiang area are widely developed. For example, a series of pull-apart basins have formed in the Qamdo-Lanping-Pu'er basin. They are the products of the strike-slip pull-apart process of regional deep fractures between the two sides and internal orogenic belts of the basin. The eastern Tibet area in the north is controlled by the dextral activity of the Chesuoxiang fault that divides the Qamdo Basin and the Jinsha River orogenic belt. The Paleogene Gongjue Basin formed by it extends for hundreds of kilometers from north to west. These basins mentioned above are characterized by extensional or compressive strike-slip. Due to their connection with deep and large faults, some of them contain magma eruptions and become favorable places for deep fluid to rise and unload. For example, the mineralization in the famous Jinding lead–zinc deposit located in the Lanping-Yunlong Basin is mainly related to the discharge of deep fluid rise in the basin during the strike-slip process along the Bijiang River fault.

4.4 Analysis of Key Metallogenic Geological Process

The Sanjiang giant metallogenic belt, as an important part of Tethys giant metallogenic domain, one of the three giant metallogenic domains in the world, is famous for its unique forming environment, intense metallic mineralization, complex metallogenic types and huge prospective scale. There are not only a number of world-famous large and super-large ore deposits, such as Yulong copper deposit, Gacun polymetallic deposit, Jinding lead–zinc deposit, Ailaoshan gold deposit and Tengchong tin deposit, but also a number of new large and super-large ore deposits, such as Pulang copper deposit, Beiya gold deposit, Baiyangping silver polymetallic deposit, Xiasai silver deposit, Gala gold deposit, Yangla copper deposit, Dapingzhang copper deposit, Luziyuan lead–zinc polymetallic deposit, Chang'an gold deposit, Hetaoping lead–zinc polymetallic deposit, which have entered a scale development stage. Hou et al. (1999) summarized six characteristics of the mineralization in the Sanjiang area; Li et al. (1999) vividly described it as "a late bloomer surpassing the former, deep-seated source, collision and nappe mineralization". Based on the long-term research on the Sanjiang metallogenic belt, and taking the mineralization geodynamic analysis as the main method, the section determines its main metal metallogenic periods and main metallogenic types, focuses on the dynamic background and metallogenic geological environment of its formation and development and reveals the key geological processes controlling mineralization.

4.4.1 Evolution of Mineralization

With the evolution of the Sanjiang giant orogenic belt experiencing subduction orogeny → collision orogeny → intracontinental orogeny, the evolution of the regional mineralization has the following apparent laws.

4.4.1.1 Evolution of Metallogenic Environment

The metallogenic geological environment of the Sanjiang orogenic belt has undergone three fundamental changes and transformations, corresponding to three important metallogenic peaks. During the Late Paleozoic metallogenic period, the paleo-ocean basins opened in Carboniferous-Early Permian underwent closure and subduction orogeny one after another, resulting in an MABT, which became an important metallogenic background for VMS deposits, porphyry copper deposits and epithermal deposits. During the Late Triassic metallogenic period, lithospheric delamination is occurred in the orogenic belt which had experienced the arc-continent collision in the Early and Middle Triassic, forming the superimposed volcano-rift basin, and developing basin fluid mineralization and submarine deep-water/shallow-water hydrothermal mineralization. In the Himalayan metallogenic period, the collision between the Indian and Asian continents, which started at 60 Ma, caused collision uplift of the Qinghai-Tibet Plateau in the hinterland of Tibet, and large strike-slip faulting in the eastern margin of the plateau, which has not only led to the development of large granite porphyry and alkali-rich porphyry belts becoming the main ore-bearing rock of porphyry Cu–Mo deposits and porphyry Cu–Au deposits, but also formed a series of strike-slip pull-apart basins and a series of large nappe shear zones, which controlled the fluid activity and hydrothermal deposit formation in the strike-slip pull-apart basins, and became an important ore-hosting space and ore-controlling structure for some large deposits (Jinding lead–zinc deposit and Ailaoshan gold deposit).

4.4.1.2 Evolution of Metallogenic Types

Different metallogenic environments are formed in different tectonic evolution stages, accompanied by different metallogenic types. In the Early Paleozoic period, the separation of landmasses and the opening of ocean basins usually occurred only at the boundary of discrete blocks, and hydrothermal sedimentary deposits were developed. In the Late Paleozoic MABT, porphyry copper deposits related to arc magmatic porphyry, VMS deposits and epithermal deposits related to arc volcanic rocks were mainly developed. Three types of submarine hydrothermal mineralization were mainly developed in the volcanic-rift basins formed by extension after collision orogeny in Late Triassic: ① VMS deposits related to marine acid volcanic rocks, mineralization occurred in deep-water and shallow-water environments; ② exhalative-sedimentary deposits related to marine rift basins and contemporaneous faults; ③ hydrothermal sedimentary deposits related to hot brine in silled basins, such as barite and gypsum deposits. In the Himalayan period, mineralization has tended to be complex with various metallogenic types. Among them, at least four types are particularly important: ① porphyry copper–molybdenum and copper–gold deposits closely related to large strike-slip faults; ② hydrothermal deposits related to the dual action of strike-slip pull-apart basin and strike-slip nappe structure; ③ shear zone type deposits related to large strike-slip and nappe shear zone; ④ hydrothermal deposits related to extensional detachment structure or nappe slippage structure.

4.4.1.3 Evolution of Metallogenic Metal Assemblages

The evolution of metallogenic metal assemblages has simple-to-complex and single source-to-multi-source composite features. During the Early Paleozoic metallogenic period, the metallogenic metallic assemblages were mainly Pb–Zn or Fe assemblages. During the Late Paleozoic metallogenic period, the metal assemblages were characterized by the presence of a large amount of Cu and the development of Cu–Zn, Cu–Pb–Zn and Fe–Cu assemblages. During the Late Triassic metallogenic period, the metal assemblages were characterized by the presence of large amounts of Ag, Au and Hg and the development of Zn–Pb–Cu, Cu–Zn–Pb, Zn–Pb–Ag and Ag–Au–Hg assemblages. In the Himalayan period, the metallogenic metal assemblages were extremely complex, which is mainly manifested as follows: ① rare and rare earth elements were enriched and mineralized; ② single elements were mineralized on a large-scale, such as extra-large Au deposit, large Sr deposit and extra-large Cu deposit; ③ multi-source elements coexist, and crust-source elements (Pb, Zn, Ag, Sr, etc.) and mantle source elements (Co, Ni, Au, Cu, etc.) coexist in time and space for enrichment and mineralization.

4.4.1.4 Evolution of Metallogenic Intensity

Although mineralization runs through the evolution of the Sanjiang orogenic belt, large-scale mineralization was mainly occurred in the Late Triassic and Himalayan periods, and the mineralization of many large and super-large deposits in the Sanjiang area was concentrated in the two periods. In the Late Triassic metallogenic period, the scale of submarine hydrothermal mineralization was the largest; in the Himalayan metallogenic period, the scale of hydrothermal fluid mineralization in porphyry copper deposits, shear zone-type gold deposits and strike-slip pull-apart basins was the largest, showing a triple balance of forces.

4.4.2 Analysis of Key Metallogenic Geological Process

As mentioned earlier, although the mineralization in the Sanjiang area runs through the evolution of the giant orogenic belt, the important mineralization was mainly developed in three different evolution stages of the giant orogenic belt and three important metallogenic geological environments resulted therefrom. The large-scale mineralization occurred in ① the arc-basin system related to subduction orogeny, ② the post-collisional orogenic extensional system related to lithospheric delamination (break-off), ③ the large-scale strike-slip fault system related to collisional uplift of Qinghai-Tibet Plateau. This section will conduct a deep analysis of the formation and evolution processes of the three important metallogenic backgrounds and geological environments, as well as the three key metallogenic geological processes.

4.4.2.1 Subduction Orogeny and Arc-Basin System

The evolution of the Paleo-Tethys in the Sanjiang area is a subduction orogeny and a development process of island arc orogenic belt. Since the island arc orogenic belts and their mineralization in the Sanjiang area have been discussed and understood deeply, only their basic characteristics are summarized here.

Before collision orogeny, the Sanjiang giant orogenic belt was formed by the collage of several (island) arc orogenic belts and microblocks sandwiched between them. These arc orogenic belts mainly include Yidun island arc orogenic belt, Jinsha River continental margin arc orogenic belt and Lancang River continental margin arc orogenic belt. Although they have different development histories, they all have formed their own arc-basin systems.

Yidun Island Arc Orogenic Belt

Yidun orogenic belt is a composite orogenic belt in the Sanjiang giant orogenic belt, which started from large-scale subduction orogenesis during the late Indosinian (Rhaetian–Norian) (Hou 1993; Hou et al. 1995), experienced the collisional orogenic process in the Yanshanian period, including arc-continent collision and continental crust shrinkage and thickening, orogenic uplift and extension, and finally suffered from the superimposition and reconstruction of intracontinental convergence and large-scale strike-slip and shearing during the Neo-Tethys period. Its subduction orogeny is originated from the westward subduction of the Ganzi-Litang oceanic crust slab in the early Late Triassic, which has led to the development of Yidun volcanic-magmatic arc (237–210 Ma) and the formation of a complete trench-arc-basin system. Given the nature of the thin continental crust underlying the island arc, as well as the inhomogeneity of the subduction angle of the oceanic crust plate, the north–south segment of the remnant arc is shaped. The Changtai arc in the northern segment is characterized by the development of intra-arc rift basins and active back-arc spreading basins, with extensional arc properties; the Shangri-La arc in the southern segment is characterized by the development of intermediate-acid volcanic-intrusive rock magmatic arcs and inactive back-arc-basins, with compressional arc properties; the Xiangcheng arc in the middle segment lies between extensional arc and compressional, with the development of intra-oceanic arcs marked by high-magnesium andesite (Hou et al. 1995). Four secondary tectonic units can be identified from east to west in the Changtai arc, namely outer arc (volcanic-magmatic arc), intra-arc rift, inner arc (residual arc) and back-arc spreading basin. The intra-arc rift is marked by the development of basalt-rhyolite bimodal rock assemblage and deep-water faulted basin (Hou 1993), while the back-arc spreading basin is characterized by the development of a shoshonite-high-potassium rhyolite bimodal rock assemblage and deep-water black rock series deposits. The Shangri-La arc is mainly a volcanic-magmatic arc, characterized by development of calc-alkaline magma series andesite-dacite and super-hypabyssal emplaced intermediate-acid rock mass (Hou et al. 1995), and there is no corresponding volcanic activity in the back-arc area. The Xiangcheng arc has developed earliest, which is composed of early intra-oceanic arc and late volcanic arc.

The trench-arc-basin system of the Yidun island arc orogenic belt is an important metallogenic environment for the Late Triassic mineralization and development in the Sanjiang area. The intra-arc rift zone in the Changtai arc provides an important ore-accumulating basin environment, magma "heat engine" driving fluid circulation and source rocks supplying ore-forming materials for submarine hydrothermal fluid mineralization. A large number of super-hypabyssal emplacement porphyry systems in the Shangri-La arc control the formation of porphyry copper deposits and skarn polymetallic deposits. The development of epithermal deposits (Ag–Au–Hg) is controlled by the back-arc spreading basin of the Changtai arc and high-potassium rhyolite series.

Jinsha River Orogenic Belt

The Jinsha River orogenic belt was formed and developed from the oceanic crust subduction in Early Permian, experiencing collision orogeny during Early and Middle Triassic and post-orogenic extension during Late Triassic, and suffering superimposed reformation by intracontinental convergence and large-scale shear translation in the Neo-Tethys, which is a composite orogenic belt consisting of

successively developed continental marginal arcs and collision orogenic belts influenced by the collision and uplift of the Qinghai-Tibet Plateau. The oceanic crust remains marked by Carboniferous-Permian ophiolite complex, and the intermittent spreading and melange accumulation of deep-sea sediments along the Jinsha River fault zone represented by siliceous argillaceous-arenaceous flysch construction indicate that the insha River melange zone was once a closed ocean basin. On the west side of the Jinsha River fault zone, the Early Permian Jubalong-Dongzhulin intra-oceanic arc and Xiquhe-Jiyidu back-arc-basin are developed, indicating that the Jinsha River oceanic crust plate might have detached and subducted westward along the intra-oceanic fracture zone in Early Permian, forming a sequence of calc-alkaline magma series intra-oceanic arc volcanic rock assemblages of basalt-basaltic andesite-andesite-dacite. In the eastern margin of the Qamdo block, the Jiangda-Deqin-Weixi Late Permian continental marginal arc is developed, showing that the Jinsha River oceanic crust plate subducted to the Qamdo block on a large scale at the end of Early Permian, and formed arc volcanic rock series evolving from tholeiitic basalt series to shoshonite series through calc-alkaline magma series.

The arc-basin system of the Jinsha River orogenic belt is an important metallogenic environment for the Late Triassic mineralization and development in the Sanjiang area. Although the Ailaoshan gold mineralization belt was mainly formed in Himalayan Period, the host rock-ophiolitic melange undoubtedly provided an important source of ore-forming materials. The Yangla large copper deposit was produced in the Jubalong-Dongzhulin intra-oceanic arc environment and located in calc-alkaline andesitic arc volcanic rock series, indicating that intra-oceanic arc is also an important metallogenic environment for submarine hydrothermal fluid mineralization.

Lancang River Orogenic Belt

The arc-basin system of Lancang River Orogenic Belt is generally similar to that of Jinsha River Orogenic Belt. Although the ancient suture zone of Lancang River may have been covered by a large nappe in the later period, it was subducted eastward at the end of Early Permian, which has been proved by the development of the volcanic arc in Zuogong-Jinghong on its east side in Late Permian (Mo et al. 1993). It is not clear whether a back-arc spreading basin was formed in Lancang River Orogenic Belt in the subduction orogenic stage. However, since the volcanic arc formed in Late Permian was developed in the western margin of Qamdo-Pu'er Block and had characteristics of continental marginal arc, it is estimated that the back-arc-basin would be inactive even if it was developed. At present, only a few small polymetallic ore deposits and numerous ore occurrences have been found in the Lancang River continental marginal arc orogenic belt, but their metallogenic potential is still unclear.

4.4.2.2 Lithospheric Delamination (Detachment) and Post-collisional Orogenic Extensional System

The collisional orogeny was originally formed by continental crust subduction (A-type subduction) and often went through the following processes: shortening of continental crust and thickening of double crusts, mountain root delamination, and asthenosphere upwelling, post-orogenic extension and tectonic regime transformation. The collisional orogeny can be described with the P Tt trajectory, showing extrusion and increasing temperature in early stage, decompression, stretching and increasing temperature in the middle stage and stretching and decreasing temperature in the late stage (Jamieson 1991). In Sanjiang area, several (island) arc orogenic belts formed by oceanic crust subduction were developed and entered into the collision orogenic stage successively in Triassic and went through the following processes of arc-land collision, shortening and thickening of continental crust and lithosphere extension, during which Sanjiang and "countertrust orogenic belt being transversely blocked by mountains" were formed in Yanshanian and large-scale mineralization was formed and developed in Late Triassic.

Collisional Orogeny and Collision Orogenic Belt

(1) Jinsha River arc-land Collision Orogenic Belt. The main body of Jinsha River Orogenic Belt is a composite mountain system formed in Early and Middle Permian by collating Zhubalong-Pantiange intra-oceanic arc in Early Permian and Jiangda-Weixi continental marginal arc in Late Permian. An arc-land collision mountain system was formed in the middle-south member of Jinsha River Zone due to the closure of ocean basin and collision between the intra-oceanic arc and Zhongza Block, and another arc-land collision mountain system was formed due to collision between the intra-oceanic arc and the continental marginal arc on the west side. The collisional volcanic rocks formed in the Early Triassic are intermittently outcropped along the collision mountain system. A land-land collision mountain system was formed in the northern member of Jinsha River due to collision between Zhongza Block and Qamdo Landmass, causing that Pushuiqiao Formation of Lower Triassic unconformably overlays on Jiangda collision crustal melting granite body. Qamdo Block lacks the sediments formed in Early and Middle

Triassic due to the overall uplift. Collision occurred in Early-Middle Triassic not only leads to the continuous westward subduction of the substances in Zhongza Block, but also causes the overthrust of volcanic-sedimentary substances on the subduction plate toward Qamdo Landmass in turn, forming a "snake swallowing frog (smaller on both ends and bigger in the middle)" collision tectonics.

(2) Lancang River Collision Orogenic Belt. The orogenic pattern of Lancang River Collision Orogenic Belt is similar to that of Jinsha River Orogenic Belt. The main body was formed by the collision and collage of Zuogong-Jinghong Continental Marginal Arc formed in Late Permian, Dongda Mountain-Lancang River Magmatic Arc and Qamdo-Pu'er Block in Early and Middle Triassic. With the overall eastward collision and thrust of the orogenic belt, the upper crust of the subduction plate trusted eastward on Qamdo Landmass and the rear margin was spread and split, forming extensional basins and metamorphic core complexes. The collision orogeny in Early and Middle Triassic has not only formed the intrusion of syn-collision granite for constituting the main body of Lancang River granite base, but also developed syn-collisional volcanic rock assemblage for superimposing on Zuogong-Jinghong Volcanic Arc.

(3) Yidun Collision Orogenic Belt. Collision orogeny in Yidun Orogenic Belt has occurred relatively late and mainly occurred in the late stage of Late Triassic (206–200 Ma). During this period, Ganzi-Litang Ocean Basin was closed, Yangtze Landmass collided with Yidun Island Arc, and Zhongza Block thrust eastward as a tectonic slice, forming an asymmetric collisional mountain zone with non-coaxial shear. The typical product of collision is syn-collision granite, which is superimposed in the magmatic arc zone. The remnant arc lacks J-K sediments due to its overall uplift.

Post-collision Extension and Volcanic-Rift Basin

The main body was originally formed by the extension of post-collision orogeny in Late Triassic and was characterized by different tectonic responses and rock records in different collision orogenic belts, forming different metallogenic conditions with different mineralization.

(1) Jinsha River Orogenic Belt. The gravity field of Jinsha River Collision Orogenic Belt changed significantly in Late Triassic after the arc-land collision in Early and Middle Triassic. The significant extension after the collision has not only led to "bimodal" volcanic activity, but also formed a volcanic-rift basin, which was superimposed on Jiangda-Weixi Continental Marginal

Arc formed in Late Permian. At present, three important post-collision extension basins formed in the Late Triassic have been discovered, which are Shengda-Chesuo Volcanic-Sedimentary Basin, Xuzhong-Luchun-Hongpo Volcanic-Sedimentary Basin and Reshuitang-Cuiyibi-Shanglan Volcanic-Sedimentary Basin from north to south. Semi-pelagic volcanic turbidite, tuffaceous turbidite, tuffaceous-siliceous turbidite and arenaceous flysch were developed in the basin. The volcanic rocks in the northern basin are dominated by marine pillow basalt series, accompanied by the development of gabbro-diabase dykes, while the basalt-rhyolite "bimodal" volcanic rocks were developed in the southern basin.

The Rb–Sr isochron age of rhyolite in bimodal magma assemblage is 235–239 Ma (Wang et al. 1999; Mou et al. 2000), indicating that the bimodal rock assemblage was developed in the early stage of Late Triassic. The basalt of bimodal assemblage belongs to tholeiitic basalt, which can be divided into two groups according to the content of TiO_2, namely group with high content of TiO_2 (1.25–1.85%) and group with low content of TiO_2 (0.25–1.08%). The former is represented by basalts in the Jijiading area, which are significantly different from basalts in the island arc or back-arc-basin due to the high content of TiO_2, showing the geochemistry affinity of ocean ridge basalts (Wang et al. 1999). The latter is represented by basalts in Renzhixueshan and Cuiyibi area, showing the geochemistry affinity of basalts in island arc or back-arc-basin. The successive development of these two sequences of basalts may reflect the development process of volcanic-rift basins in the Late Triassic from intense stretching to gradual shrinkage.

Volcanic-sedimentary sequence of volcanic-rift basin: The lower bed is composed of semi-pelagic basaltic volcanic rocks-diabase complex and flysch sediments, the middle bed is composed of abyssal interbedded basalt-rhyolite bimodal assemblage and argillaceous rock series, the upper bed is composed of neritic rhyolitic volcanic rock series and sandy mudstone series and the top bed is littoral-neritic molasse clastic rock series mixed with intermediate-acid pyroclastic rocks, accompanied by gypsum-salt sediments containing barite, gypsum and siderite assemblage, which records the developmental history of volcanic rifted basin, namely, intense stretching and rifting in the early and middle stage and gradual shrinkage and destruction in the late stage. It is estimated that the three volcanic-sedimentary basins are different in the rifting speed and rifting distance. Among them, Xuzhong-Luchun-Jiangpo Basin has the highest rifting speed (0.43 cm/a) and the longest rifting distance (140 km); Shengda-Chesuo Basin has the lowest rifting speed (0.27 cm/a) and the shortest rifting distance (63 km),

while the rifting speed and rifting distance of Reshuitang-Cuiyibi-Shanglan Basin are between the highest rifting speed and the lowest rifting speed and between the longest rifting distance and the shortest rifting distance, respectively.

These post-collision extension basins are important metallogenic basins in copper polymetallic ore deposits in Jinsha River Zone. The hydrothermal activity system with submarine eruption was formed by volcanic activity in the basin; in addition, the "brine pool" under semi-closed and closed conditions was formed in the secondary depression zone of the rift basin to further forming the exhalaive sedimentary VMS deposit by sedimentation. Mineralization was formed in the middle stage of basin extension and splitting, such as, Zuna lead–zinc ore deposit in Chesuo Basin, in which the ore-bearing rock series is composed of carbonate rock-clastic rock-barite formation; Luchun copper-lead–zinc ore deposit in Jijiading Basin, Hongpo copper–gold polymetallic ore deposit and Laojunshan lead–zinc ore deposit in Cuiyibi Basin, in which the ore-bearing rock series is composed of acid volcanic tuff-sedimentary rock-siliceous rock formation. The volcanic-subvolcanic hydrothermal-sedimentary siderite-type gold-silver polymetallic ore deposits related to intermediate-acid volcanic rocks were formed based on extrusion tectonics at the end of the basin development, such as Zhaokalong siderite-type silver-rich polymetallic ore deposit and Dingqinnong copper–gold ore deposit in Chesuo Basin and Chugezha Siderite ore deposit in Cuiyibi Basin, in which the ore-bearing rock series was formed by intermediate-acid pyroclastic rock-sedimentary rock-siderite formation; the gypsum ore deposit was formed in the Yulirenka area of Luchun Basin.

(2) Development history of Lancang River Collision Orogenic Belt is similar to that of Jinsha River Orogenic Belt. Its post-collision extension also occurred in the Late Triassic, and the typical product is the potassic trachybasalt-rhyolite bimodal rock assemblage of Manghuihe Formation, which was overlaid on the shoshonite-latite series of Xiaodixing Formation. This sequence of bimodal rock assemblage is mainly distributed on Zuogong-Jinghong Volcanic Arc formed in Permian in Lancang River Orogenic Belt, but mainly outcropped in the middle-south member. Acid volcanic rock series with a thickness of 3000–4000 m were mainly developed in Northern Lancang River Zone (Eastern Tibet), which are in conformable contact with the underlying strata. The Rb–Sr age is 238.9 Ma. With high content of silicon ($w(SiO_2)$ 67–79%) and high content of alkali ($w(K_2O + Na_2O)$ 6–8.6%), the rocks belong to the high-potassium calc-alkaline series. Its geochemistry characteristics are similar to those of the volcanic rocks of Xiaodixing Formation in the Southern

Lancang River Zone (Western Yunnan), belonging to the collisional volcanic rock series (Mo et al. 1993). The bimodal rock assemblage of Manghuihe Formation is mainly distributed in the area north of Minle in Jinggu and the potassic trachybasalt and high-potassium rhyolite series are outcropped in interbeded way in the Southern Lancang River Zone (Western Yunnan), with seven rhythmic cycles at most. The rock belongs to the high-potassium calc-alkaline series-potassium basalt series. With the decrease of potassium content, the volcanic rocks are developed into a medium-potassium calc-alkaline series in the area south of Jinggu. Potassic volcanic rocks are moderately enriched in LREE, that is, $(w(La)/w(Sm))_N = 5.32–15.13$, relatively scarce in high-field strength element (HFSE), such as Nb, Ta, Zr, Hf, Ti, and relatively enriched in K, Rb, Ba, Th, indicating the geochemistry affinity of volcanic arc shoshonite. In view of the above, Mo et al. (1993) defined it as a lagging arc volcanic rock. However, its temporal and spatial distribution and tectonic environment for outcropping show that this sequence of bimodal volcanic rock series may be originated from the "island-type" mantle source area but was formed in the lithosphere extension stage after collision orogeny and outcropped in the tectonic condition with crust stretching and rifting.

Although the extent of geological study for Lancang River Zone is relatively low, the discovered VMS deposits were outcropped in this sequence of bimodal volcanic rock series, such as Wenyu copper ore deposit in Jingdong occurs in potassic trachybasalt, Minle copper ore deposit in Jinggu occurs in high-potassium rhyolite rock series and the recently discovered Dapingzhang copper deposit also occurs in bimodal rhyolite volcanic rock series, which reveals that the extension of Lancang River Orogenic Belt and "bimodal" volcanic activity provide a favorable metallogenic condition for the VMS formation.

(3) Yidun Collision Orogenic Belt. The post-orogeny extension of Yidun Collision Orogenic Belt occurred later, and the typical magma product is composed of extensional acid volcanic rock series formed in Early Yanshanian and A-type granite formed in Late Yanshanian. Although no volcanic-rift basin superimposed on the collision orogenic belt was developed, two important volcanic-magmatic rock zones were formed. The extensional acid volcanic rock series is closely associated with the high-potassium rhyolite zone in the back-arc area in Yidun Island Arc Zone in space and erupted from Triassic to Jurassic, with Rb–Sr isochron age of (189.2 ± 5) Ma (Hou et al. 2001a, b). Being characterized by high alkali content $(w(K_2O + Na_2O) = 5.58–10.16\%)$ and $w(K_2O)$ $(4.46–8.76\%) > w$

(Na_2O) $(0.12–1.40\%)$, the acid rhyolitic volcanic rocks belong to the shoshonite series. It is rich in K, Rb, Zr, Ta but poor in CaO, Sr and Eu as for its geochemistry characteristics, showing the geochemistry affinity of intraplate volcanic rocks (Qu et al. 2001). Compared with the rhyolite with high-k content (213 Ma) in the back-arc spreading basin (Hou et al. 1995), this sequence of alkaline rhyolite has a typical "swallow-shaped" REE distribution pattern, with the initial strontium ratio of 0.714578, showing that its magma is originated from the typical continental crust melting. Its development indicates a short period of intense collision (206–200 Ma) occurred in Yidun Collision Orogenic Belt, causing the change of regional stress field, that is, the stress field changed from the extrusion system to extension system since (189.2 ± 5) Ma.

A-type granite was outcropped on the west side of the extensional acid volcanic zone in space and then intruded into the sand-slate series formed in the Upper Triassic, constituting an important granite zone in this area. It was formed in Late Yanshanian, with isotope ages of 102–75 Ma and the peak age of magmatic activity of 80 Ma (Hou et al. 2001a, b). The rock assemblage is moyite, granodiorite and monzogranite and mainly composed of potash feldspar (microcline), plagioclase, quartz and lepidomelane, accompanied with accessory mineral assemblage of allanite + apatite + zircon + tourmaline + magnetite + cassiterite.

Rich in alkali $(w (Na_2O + K_2O) = 6.13–8.68\%)$ and AR = 1.9–3.0, the rock belongs to the alkaline series. Being characterized by its relatively high ratio of w $(FeO^*)/w(MgO)$ and $w(Na_2O + K_2O)/w(CaO)$, relatively abundant HFSE (Zr, Hf, Nb, Ta, Ce, Y, etc.) and relatively large ratio variation in trace elements $(w(Rb)/w(Ba)$: 0.52–39.3; $w(Rb)/w(Sr)$: 8.0–39.8), the rock shows the geochemistry affinity of A-type granite and can be compared with A2-type granite (Eby 1992; Hou et al. 2001a, b). With a typical "swallow-shaped" REE distribution pattern and significant negative Eu anomaly, the $^{87}Sr/^{86}Sr$ ratio of granite with $w(La)/w(Yb) = 2.46–5.99$ is 0.7441. These geochemistry characteristics indicate that the granitic magma is originated from the partial melting of typical continental crust characterized by argillaceous rock series.

Therefore, the large-scale development of A-type granite in this area indicates that the lithosphere extension reached its peak at about 80 Ma after collision orogeny. On the one hand, this type of significant extension promotes the significant upwelling and thermal erosion of asthenosphere substances and significantly changes the status of the thermal structure in the lower bed of the orogenic belt, resulting in large-scale melting of the crust and the formation of granite magma; on the other hand, it leads to significant extension of crust and even collapse of the orogenic belt (Hou et al. 2001a, b).

The significant extension and volcanic-magmatic activity of Yidun Collision Orogenic Belt in Yanshanian led to large-scale silver-tin polymetallic mineralization. A series of skarn or granite-type tin polymetallic deposits, occurrences and mineralization points were developed intermittently in the inner and outer contact zone of A-type granite zone, forming an important tin polymetallic mineralization zone. The hydrothermal fluid driven by granite intrusions migrated and converged along the large-scale detachment zone and fault zone formed by extension within 1–10 km around the A-type granite zone, forming a large-scale silver polymetallic ore concentration area represented by Xiasai silver deposit. Although no gold ore deposits for industrial purpose have been found in the acid rhyolite series of 189 Ma at present, the significant whole-rock pyritization and excellent Au geochemistry anomaly reflect that this sequence of alkali-rich rhyolite has a certain metallogenic potential.

Lithospheric Delamination and Plate (Detachment)

The evolution history of Sanjiang giant orogenic belt shows that the regional gravity field experienced a major turning in the Late Triassic, and its tectonic response, rock record and mineralization are quite different from those of MABT. By taking Qamdo-Pu'er Block as the central axis, the arc-land collision has occurred in Early and Middle Triassic and then large-scale extension has occurred in Late Triassic in Jinsha River Orogenic Belt thrusting westward, accompanied by the development of bimodal rock assemblage with basalt-high-potassium rhyolite, forming a series of NW-trending extension basins. The significant extension has also occurred in the Late Triassic in Lancang River Orogenic Belt overthrusting eastward, with the distinctive potassium basaltic magmatic activity and the development of potassic trachybasalt-high-potassium rhyolite bimodal rock assemblage. Although the collision orogeny has occurred late in Yidun Orogenic Belt in the east, it was quickly developed and entered the extension stage at the end of Late Triassic or Early Jurassic. The high-potassium rhyolitic volcanic rocks (189 Ma) were formed in the early stage of extension, and A-type granite zone (102–75 Ma) was formed in the peak stage of extension and even the collapse stage of the orogenic belt. Major turning of regional gravity field and large-scale mineralization have occurred in the Late Triassic. VMS deposit related to hydrothermal activities of seawater, the exhalaive sedimentary copper–niobium polymetallic deposit related to fluid activities in the basin and the silver polymetallic deposit related to large-scale fluid migration-convergence process were mainly developed.

Extension, high-potassium volcanic rocks, bimodal rock assemblages and large-scale mineralization events occurred in Sanjiang Giant Orogenic Belt in Late Triassic are not accidental but are restricted by a unified deep orogeny mechanism. Lithospheric delamination and detachment of subducted continental crust plates may be two possible deep orogeny mechanisms.

High-density mineral assemblages (eclogite facies) were formed in the collision orogenic belt due to crust thickening and rock metamorphism in the lower crust, and the crust and lithosphere were thickened due to shortening of continental crust, resulting in the temperature of mountain root being relatively lower than that of asthenosphere (Dong et al. 1999). This thermal-substance structure will cause potential gravity instability, leading to de-rooting or lower crust delamination. As the direct result of delamination, the hot asthenosphere with low density rises to the crust-mantle boundary and overlays the cold lithospheric mantle, resulting in thickening of lithosphere and rapid heating of the lower crust, so that the crust uplifts rapidly and extends subsequently. The basaltic magma invades the lower crust due to the decompression melting caused by asthenosphere rising, and the lower crust being heated will be further melted to drive granite magma invading into the upper crust. Therefore, large-scale basaltic magma underplating and shoshonite magma exhalation are often considered as petrological evidence of lithospheric delamination (Key 1994; Key and Key 1994; Dong 1999). Sacks and Secor (1990) and Davies and Blanckenburg (1995) considered that the subducted lithosphere plates will be detached under the combined action of the uplift force of the subducted continental crust and the downward force of the subducted oceanic crust, which directly leads to the large-scale upwelling of asthenosphere and partial melting at the bottom of the crust, resulting in large-scale magmatism and the formation of a large amount of syn-collision granite. Obviously, both the delamination of the lower crust and the detachment of the subducted continental crust led to the large-scale upwelling of the asthenosphere and heat the crust, providing necessary thermal melting conditions for large-scale magmatism and a huge "heat source" for regional large-scale fluid activities.

The global seismic tomography data from Fukao (1995) show that the substances in Tethyan subduction plate have been detached, subsided and returned to the mantle on a large scale and are currently located 1200 km underground. The seismic tomography data of Sanjiang area from Zhong Dalai et al. show that the subduction plate was also detached and returned deep to the mantle. The lithosphere extension of Sanjiang Orogenic Belt occurred on a large scale in Late

Triassic, and the significant development of high-potassium volcanic rock, basalt-rhyolite "bimodal" volcanic activity and volcanic-rift basins in the volcanic continental marginal arc indicated that the lithospheric delamination or subduction plate detachment began to develop in Late Triassic, while the large-scale intrusion of A-type granite indicated that the delamination or detachment reached its peak at 80 Ma.

The large-scale upwelling of asthenosphere induced by lithospheric delamination or subduction plate detachment is not only led to mantle-derived and crust-derived magmatic activity, but also brought large amounts of deep ore-forming materials. This is also led to crustal tension and faulting activities and developed important ore-forming basins while creating huge regional thermal anomalies that drove large-scale convective fluid circulations along tension-induced fracture zones as well as long-distance migrations and accumulations, thus inducing large-scale mineralizations in the Late Triassic.

4.4.2.3   Collision Uplifts and Large-Scale Strike-Slip Fault Systems of Qinghai-Tibet Plateau

After experiencing a subduction orogeny, a collision orogeny, and a lithospheric extension successively, the huge Sanjiang Orogenic Belt entered a new stage, i.e., the overall intracontinental convergence orogeny, in the Early Paleogene. With the collision uplifts of Qinghai-Tibet Plateau, large-scale strike-slip extensions and nappe shears successively took place in Sanjiang Orogenic Belt mainly under large-scale strike-slip faulting actions, thus developing a series of strike-slip basins (pull-apart ones), Jinshajiang-Honghe Porphyry Belt and large nappe shear belts therein and controlling the developments of large and huge deposits therein during the Himalayan Period.

Collision of Indian and Asian Continents and Large-Scale Strike-Slip Faulting

Intercontinental convergence might be tolerated through crustal thickening and lateral detachment (McKenzie 1972; Molnar and Tapponnier 1975; Tapponnier and Molnar 1977). The strains of the Qinghai-Tibet Plateau caused by the convergence and collision of India and Asian continents were represented by double crustal thickening and large-scale strike-slip faulting (Tapponnier and Molnar 1976; Peltzer and Taponnier 1988; England and Molnar 1990). The convergence and collision of Indian and Asian continents that started in the Paleocene were manifested as a forward compression from south to north, which led to the occurrence of the Himalayan orogenic belt and caused the crust in the middle of the plateau to shorten dramatically and thicken twofold, and which was accompanied by the generation of the large-scale syn-collision Gangdise Granite Belt about 55 Ma ago. As the Indian continent moved northward in its entirety, its northeast corner wedged forcefully into the east of Tibet and collided with the southwest edge of the Yangtze continent, causing the crust in Sanjiang Orogenic Belt to shorten drastically.

When the Indian continent collided strongly with the Asian continent and continued to move northward in the Eocene, large-scale strike-slip fault systems, such as the nearly EW Sanjiang Fault System (north section) and nearly SN Sagaing Fault Zone (south section), started to develop on the east edge of Qinghai-Tibet Plateau as right lateral strike-slip ones (Fig. 4.17). A series of large-scale NW strike-slip faults were developed in an extensive area east of the plateau, including Kunlun Fault, Xianshuihe Fault, and Honghe Fault successively from north to south (Fig. 4.17; England and Molnar 1990; Taponnier et al. 1990). The land block between the left lateral Kunlun and Xianshuihe faults rotated clockwise (1°–2°/Ma) and shortened (10–20 m/Ma) in the east–west direction under the nearly NW right lateral strike-slip-shear (England and Molnar 1990). The left lateral strike-slip of Xianshuihe Fault created the Zheduoshan-Gonggashan translational granite (10–15 Ma; Luo 1998). The 1000 km-long Ailaoshan-Diancangshan Metamorphic Rock Belt was developed under the left lateral Honghe strike-slip faulting, leading to partial melting of the enriched mantle (Huang and Wang 1996; Zhang and Xie 1997; Zhong et al. 2001) and controlling the development of Jinshajiang-Honghe Alkali-rich Porphyry Belt (41–26 Ma; Zhong et al. 2000). About 23 Ma ago, a large-scale strike-slip movement was occurred along the strike-slip-shear zone, and the Indosinian Block was extruded southward and slipped more than 500 km southeastward relative to South China (Taponnier et al. 1990).

Coupled Strike-Slip and Extension and Generation of Ore-Bearing Porphyry Belts

The convergence and collision of the Indian and Asian continents and the northward thrust of the former put the east part of Tibet in a huge strike-slip-shear system, thus developing a series of small-scale right lateral strike-slip faults, such as Chesuo-Deqin Right Lateral Strike-Slip Fault and Wenquan Right Lateral Strike-Slip Fault, accompanied by a series of en echelon left lateral NW–NNW folds and alternating fault structures (Fig. 4.18). These NNW–NW anticline and fault structures, as important rock-controlling structures, provided ultrashallow emplacement space for Yulong Porphyry Belt. The porphyry magma migrated upward along the Wenquan Right Lateral Strike-Slip Fault and was finally located in a series of dilatation spaces on the west side of the strike-slip fault, thus developing Yulong Ore-bearing Porphyry Belt (33–52 Ma).

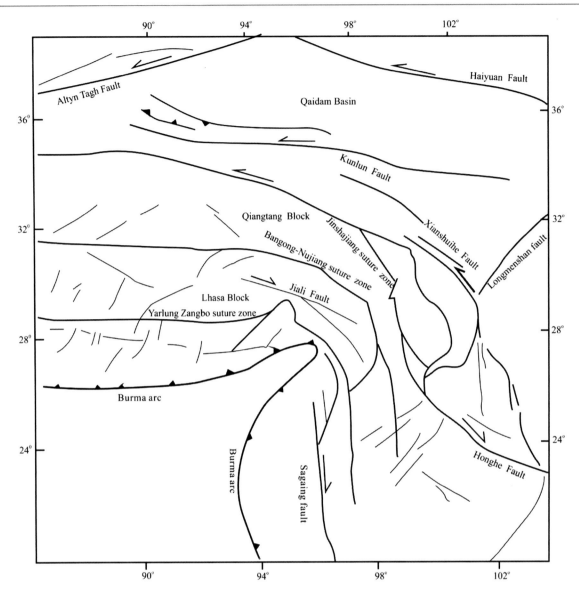

Fig. 4.17 Tectonic framework of Tibet Plateau's East Edge

Coupled Strike-Slip and Torsional Thrust and Generation of Strike-Slip Pull-Apart Basins

Because the Indian continent forcefully wedged northeast and contacted and collided with the Yangtze continent, strong eastward and westward compressive stress components arose and a huge X-shaped structural knot centered on Deqin (Fig. 4.19) was developed in Sanjiang Orogenic Belt. The Pu'er Block at the southern end of the structural knot slipped southward on a large scale along the Honghe Left Lateral Strike-slip Fault about 23 Ma ago (Taponnier et al. 1990); the northern end of the structural knot saw a retreat and slipped northward along the left lateral strike-slip fault.

During the northward slip, Qamdo Block underwent relaxation rebound and strong extension, thus developing a series of strike-slip extension basins, such as Gongjue Basin east of Yulong Porphyry Belt and Nangqian Basin west of Alkali-rich Porphyry Belt (Fig. 4.19). With Chesuo Fault as its eastern margin, Gongjue Basin is a half graben basin. Nangqian Basin is controlled by Tuoba strike-slip fault, with characteristics of strike-slip pull-apart basins. Both of the basins received very thick Paleogene and Neogene fluvial-lacustrine red clastic rock series with latitic-trachytic volcanic rock series sandwiched. It was inferred according to the intrusion ages of these alkaline-slightly alkaline volcanic rocks that the strike-slip extensions and extension pull-apart basins were developed 42–37 Ma ago, when the middle-period magmatic emplacement peak of Yulong Porphyry Belt almost occurred (40 ± 2.3 Ma). The temporal and spatial paragenetic relations between the two K_2O-rich volcanic rock series and the two porphyry rock series show that

Fig. 4.18 Tectonic framework and spatial distribution of Yulong Porphyry Belt on Eastern Edge of Qamdo Block (Wang et al. 2000). I—Volcanic Rock Belt in Gongjue Basin; II—Yulong Granite Porphyry Belt; III—Ritong-Mamupu Syenite Porphyry Belt; IV—Volcanic Rock Belt in Nangqian-Lawu Basin. ① Chesuo-Deqin Fault; ② Wenquan Fault; ③ Tuoba Fault; ④ North Lancang River Fault; 1—asalt; 2—Andesite; 3—Rhyolite; 4—Trachyte; 5—Monzonite granite porphyry; 6—Syenite; 7—Age

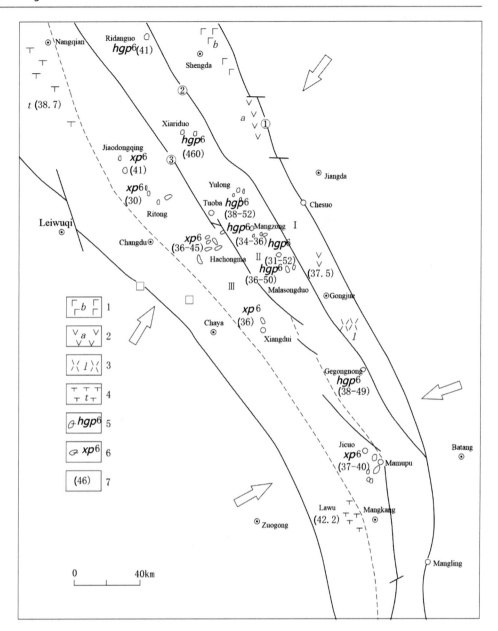

these strike-slip tensile actions might be key to inducing crust/mantle material meltings and controlling magma uprise emplacements.

When Pu'er Block thrusted and slipped southeast, a series of strike-slip pull-apart basins, such as Lanping-Yulong Basin and Jiangcheng-Mengla Basin, were developed as important mineralizing basins. Lanping-Yunlong Basin, which is controlled by the Bijiang Strike-slip Fault and spreads in the north–south direction, might be a pull-apart fault basin developed under both a right lateral strike-slip pull-apart action and a torsional thrust. The basin is filled with very thick continental salt-bearing formations, with slump deposits and fluviolacustrine alluvial fan deposits in its top and alkali-rich porphyry emplacement in its south

extension. In the Jinding Mining Area, there are ore deposits on the side where the Bijiang Strike-slip Fault is located, which are at the bottom of the decollement zone of the thrust nappe's footwall. In Sanshan Mining Area, ore deposits are located exactly in the overthrust zone and its hanging wall's fault zone (Fig. 4.20). Jiangcheng-Mengla Basin, which is composed of a series of fault block uplifts and depressions, might be developed under both a strike-slip pull-apart action and a thrust depression (Liu et al. 1993a, b).

Strike-Slip Faulting and Shearing Napping

With the continuous northward advancement of the Indian continent, a great strike-slip-shear is occurred in the Sanjiang area, thus developing complex nappe structural belts along

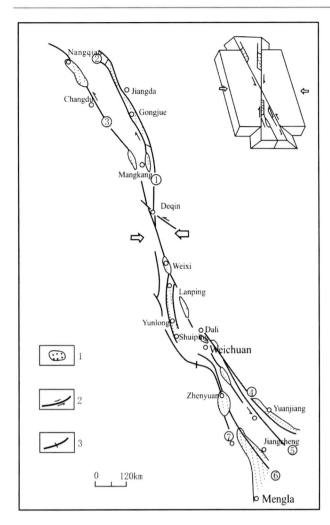

Fig. 4.19 Relaxation rebound and violent extension of Qamdo Block during Northward Slip. 1—Cenozoic strike-slip pull-apart basin; 2—strike-slip fault; 3—anticline Axis. ① Chesuoxiang Fault; ② Qingnidong-Gongjue Fault; ③ Jiezha-Chaya Fault; ④ Honghe Fault; ⑤ Ailaoshan Fault; ⑥ Bijiang River-Zhenyuan Fault; ⑦ Lancangjiang Fault

some large strike-slip faults in addition to composite basins induced by strike-slip pull-apart actions and torsional nappings in Pu'er Microcontinental Block. Ailaoshan Mineralization Belt is a nappe structural belt under strike-slip-shear, which consists of three thrust fault zones, i.e., Honghe Fault Zone, Ailaoshan Fault Zone and Jiujia-Mojiang Fault Zone, with several nappes or tectonic rock slices sandwiched between them, such as Ailaoshan Front Nappe, Yushan Yakou Nappe, Sanmeng Nappe, Lvchun Nappe and Jinping Slip Body. The nappes are bounded by thrust faults, which are distributed in the right lateral en echelon pattern (Fig. 4.6). Nappes are the main ore-bearing rock series of Ailaoshan Gold deposit, and the interlayer slip planes within them provide important ore-bearing space. Research shows that intersections of strike-slip faults might be main migration channels for deep fluids, thrust faults between nappes

might be migration channels for shallow fluids, and interlayer slip zones might provide storage and deposition space for fluids.

4.4.3 Analysis of Important Ore-Forming Environments and Mineralizations

In the huge Sanjiang Orogenic Belt, main mineralizations are mainly occurred in three important environments. One is a trench-arc-basin system formed under subduction orogeny, which mainly involved seabed hydrothermal mineralization and epithermal mineralization; the second is the extension system formed after the collision orogeny, which mainly involved seabed hydrothermal fluid mineralization in deep and/or shallow-water environments and fluid mineralization in extensional fault basins; the third is a system formed by intracontinental convergences and large-scale strike-slip faults caused by collision uplifts of Qinghai-Tibet Plateau, which mainly involved porphyry mineralization, strike-slip pull-apart basin mineralization, and shear nappe fluid mineralization. Here, three important mineralization environments are selected for further elaboration and analysis.

4.4.3.1 Mineralizing Porphyry System

In the huge Sanjiang Orogenic Belt, the intracontinental deep-rooted epithermal porphyries played a decisive role in the Himalayan intracontinental convergence mineralization system. They have unique structural setting, large output scale and huge mineralization potential and are comparable to those of the porphyry copper ore belt in the Andes continental margin arc in these aspects.

Temporal and Spatial Distributions of Ore-Bearing Porphyries

More than 100 Himalayan porphyry bodies have been found at the eastern margin of Qamdo Microcontinental Block in Sanjiang Orogenic Belt, and they are mainly distributed in the microcontinental block's internal uplift area and uplift-depression junctions formed in the Cenozoic Era. On both sides of the uplift area are strike-slip pull-apart basins. On its east side is Gongjue Half Graben Slip Extension Basin and on its west side is Nangqian Strike-Slip Pull-apart Basin. In the basins, there are 4000 m thick gypsum salt-bearing red molasse sediments, accompanied by alkaline volcanic rock series that are sandwiched between Paleogene red beds.

Here, Himalayan porphyries can be divided into at least two belts. One is the eastern belt, i.e., Yulong Ore-bearing Porphyry Belt at the western margin of Gongjue Basin, which extends from Narigongma in Qinghai Province in the north through Ridanguo, Xiariduo, Yulong and Malasongduo into

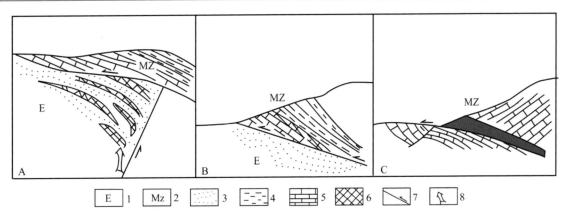

Fig. 4.20 Nappe Thrust structure of Lanping Basin and its deposit-controlling form (Li et al. 1999). 1—Cenozoic; 2—Mesozoic; 3—glutenite; 4 —argillaceous rock; 5—limestone; 6—ore body; 7—nappe structure; 8—ore-forming fluid

Mamupu in the south. It is 200 km long and 15–30 km wide, with more than 20 monzonitic granite porphyry bodies (Fig. 4.18). The other is the western belt, i.e., Alkali-rich porphyry belt consisting mainly of syenite porphyry. As a huge alkali-rich porphyry belt, it extends southward from Ritong and Xiangdui in the north through Mangkang, Shangri-La, Dali, and Jinping into the north part of Vietnam, with a length of more than 1000 km (Zhang et al. 1997). Yulong porphyry belt is characterized by typical porphyry copper/molybdenum deposits (Rui et al. 1984; Ma 1990). In the alkali-rich porphyry belt, there are porphyry-type and cryptoexplosive breccia-type gold-silver polymetallic deposits (Wang et al. 2000). Although the age data of Yulong Porphyry Belt were obtained from different laboratories by using different dating methods, the statistical law based on K–Ar age data clearly shows that the porphyry belt has at least three emplacement age ranges: 48.2–55 Ma, 38–41.5 Ma and 30.9–34.6 Ma, which correspond to three peaks, respectively: 52 ± 2.8 Ma, 40 ± 2.3 Ma, and 33 ± 3.3 Ma, respectively. The whole-rock-mineral Rb–Sr isochron ages of the Yulong monzonite granite porphyry are 52 ± 0.2 Ma and 41 Ma, respectively, and its biotite $^{40}Ar/^{39}Ar$ age is 52.84 ± 1.68 Ma (Ma 1990); the Rb–Sr isochron age of the Duoxiasongduo monzonite granite porphyry is 52 Ma, and its zircon U–Pb age is 41 Ma (Shen 1995). Besides, two zircon U–Pb ages of the Angkenong monzonite granite porphyry are respectively 40.9 Ma and 33.7 Ma (Ma 1990). These age data confirm that Yulong porphyry belt ever saw at least three magma emplacement periods: early, middle and late periods, and the magmatic activity peak periods are 52 Ma, 41 Ma and 33 Ma, respectively.

The magma emplacement ages of alkali-rich porphyry belt are relatively young. The Rb–Sr age of the lamprophyre in the Ailaoshan area is 28.8–49.0 Ma, and its fission track age is 22.7–27.2 Ma (Bi et al. 1996; Zhang and Xie 1997); the age of the alkali-rich porphyry in the Jianchuan area is

26.5–37.6 Ma (Deng et al. 1998) and that of the Jinping area is 27.0–41.0 Ma (Zhong et al. 2000).

The K–Ar age of the neutral volcanic rock in Gongjue Basin on the east side is 37.5 Ma and that of the trachyte in Nangqian Basin on the west side is 38.7–42.4 Ma (Wang et al. 2000), both of which are generally equivalent to the middle-period magmatic intrusion age of Yulong porphyry belt.

Lithogeochemical Characteristics

Monzonite granite porphyry and syenite granite porphyry are the main rock types in Yulong porphyry belt, but alkali feldspar granite porphyry is also contained therein. The rock masses often contain small amounts of dark granodiorite inclusions. The phenocryst assemblage is as follows: calcareous hornblende + magnesian biotite + andesite/ oligoclase + sanidine potassium feldspar + quartz. The accessory mineral assemblage is as follows: magnetite + zircon + apatite + titanolite. The rocks here feature the accessory mineral assemblage characteristics typical of I-type granite and magnetite-bearing granite series. The ALK value range of the rocks in Yulong porphyry belt is 7.46–8.85, and their δ range is 2.10–3.04, with the characteristics of the typical calcic-alkalic series. The A/NKC values of the porphyry rocks are mostly lower than 1.1 and are equivalent to those of I-type granite (Hine et al. 1978; Griffiths 1983). Compared with the copper-bearing porphyritic rocks in Cordillera orogenic belts, the rocks in Yulong Porphyry Belt generally show the chemical characteristics as follows: lower SiO_2 and Fe_2O_3 contents and higher Al_2O_3, MgO, K_2O (K/Na > 1) and $K_2O + Na_2O$ contents. Large-ion lithophile elements (LILE): compared with ocean ridge granite (ORG) and within-plate granite (WPG) (Pearce 1984), Yulong Porphyry Belt is typically rich in K, Rb, Ba and Sr (Fig. 4.21). The Rb, Ba and K_2O content ranges of these rocks are respectively 168–315 μg/g, 95–380 μg/g and 4.15–8.06%, which are respectively

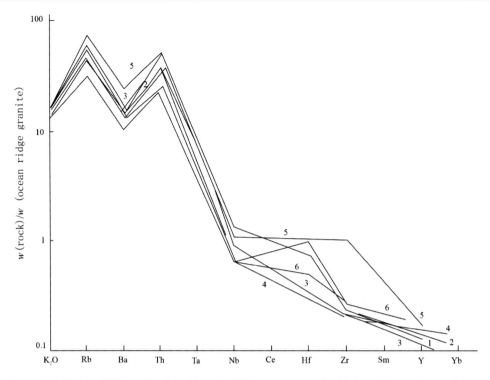

Fig. 4.21 Trace element distribution of Yulong Porphyry Belt. 1—Yulong rock mass; 2—Malasongduo rock mass; 3—Duoxiasongduo rock mass; 4—Mangzong rock mass; 5—Mamupu rock mass; 6—Zhalaga rock mass

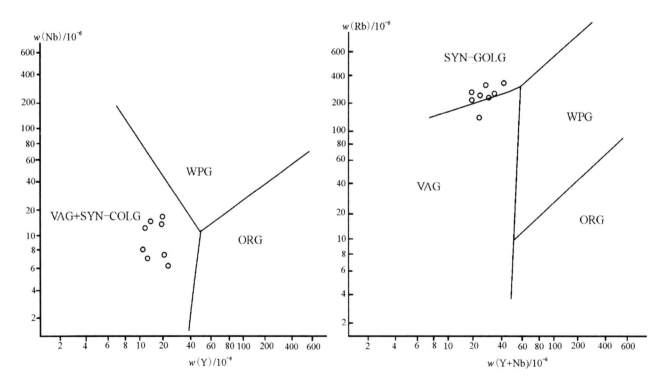

Fig. 4.22 Trace element discrimination diagram for Yulong Porphyry Belt. ORG—Ocean Ridge Granite; VAG—Volcanic Arc Granite; WPG—Within-plate Granite; SYN-COLG—Syn-collision Granite

30–50, 10–30 and 10–20 times those of ORG and are respectively equivalent to those of arc granite and syn-collision granite (Fig. 4.22). These geochemical characteristics indicate that either the porphyry magma source area was strongly enriched with LILEs relative to the ORG-type mantle, or the porphyry magma was contaminated with a large amount of crustal materials during the uprise emplacement process.

High-field strength elements (HFSE): the HFSE (Zr, Hf, Nb, Ta, Ti, U, Th) contents of the rocks in Yulong Porphyry Belt are close to or lower than those of ORG (Fig. 4.22). Their Zr and Hf content ranges are respectively 90–148 μg/g and 5–19 μg/g, slightly lower than those of ORG, respectively, while their Nb and Ta content ranges 6–18 μg/g, which are largely equivalent to those of ORG respectively. Their Ti content range of (0.21–0.52%) is significantly lower than that of ORG, while their Th content range (24–54 μg/g) is significantly higher than that of ORG (Fig. 4.21). The Zr, Hf, Nb and Ta abundance characteristics of Yulong porphyritic rocks show that their magma source area was depleted to different degrees in terms of these elements relative to the ORG mantle, esp. Zr and Hf. Their Ti and Th abundance anomalies relative to ORG might be related to the fluid metasomatism in the magma source area.

All the porphyritic rocks in Yulong Porphyry Belt show the distribution characteristic of strong LREE enrichment (w(Ce)/w(Yb) = 54.1–109.1). Different from crustal remelting granites, they have a distribution pattern with no obvious negative Eu anomaly, indicating that the porphyry magma did not undergo any intensive fractional crystallization. The YbN and LaN ranges of Yulong porphyritic rocks are respectively 3–7 and 60–180, which are respectively lower and higher than those of Jiangda-Weixi Arc's volcanic rock and dioritic porphyrite (3–4 and 30–40). Compared with ORG, Yulong porphyritic rocks are very poor in Yb and Y. The content ranges of the two elements therein are respectively 0.94–1.92 μg/g and 9.33–19.31 μg/g, only 1/100 and 1/10 times those of ORG, respectively. Since Y is a relatively stable trace element that did not participate in mantle metasomatism, its content therein being lower than that of ORG indicates that the porphyritic magma source area had lower HREE, Y, Zr, Hf, Nb and Ta contents relative to the ORG-type mantle, esp. Y, Yb, Zr and Hf. Yulong porphyritic rocks being rich in LREEs reveal that their LREE behavior was similar to their LILE behavior, i.e., either there was a secondary enrichment in the depleted mantle because of mantle metasomatism or the magma was enriched with LREEs and LILEs because of massive crustal material melting.

The ^{87}Sr/^{86}Sr values of Yulong, Duoxiasongduo and Malasongdo porphyry bodies in Yulong Porphyry Belt are respectively 0.70663, 0.70654 and 0.7077 (Ma 1990; Wang et al. 1995), which are equivalent to that (<0.707) of Berrida I-type granite in Australia (Chappel and White 1974) and are within the ^{87}Sr/^{86}Sr range of the typical I-type granites across the world (mostly ranging from 0.706 to 0.708). The εNd values of Yulong and Duoxiasongduo porphyritic rocks in Yulong Porphyry Belt are respectively −3.686798 and −2.828496 (Wang et al. 1995). As shown in Fig. 4.23, they are located near the joint between the depleted mantle and

Fig. 4.23 A εNd(t)-εSr (t) correlation diagram of ore-bearing porphyry of porphyry belt in Eastern Tibet (Wang et al. 1995). I—Australian I-type granite; II—Australian S-type granite; III—I-type granite in eastern Tibet; IV—S-type granite in eastern Tibet; V M-type granite in eastern Tibet; UC—Upper Crust; YC—Young Crust; PC—Paleocrust; 1—Granite in Jiangda rock belt; 2—granite in Yulong rock belt; 3—Dongdashan-Leiwuqi rock belt; 4—Nujiang rock belt; 5—Basu-Chayu rock belt

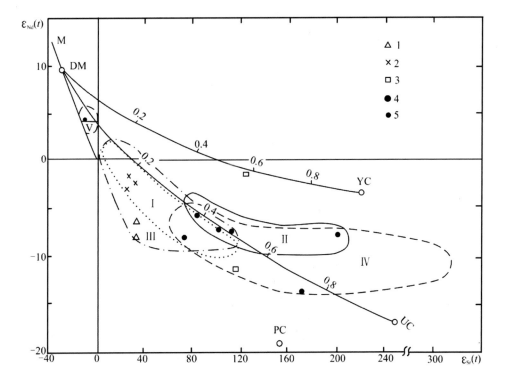

upper crust (Wang et al. 1995) and are within the εNd(t)-εSr (t) range of Australian I-type granite, clearly indicating that the magma forming the porphyritic rocks in Yulong Porphyry Belt consisted of both melted crust and mantle materials, the latter of which prevailed (Faure et al. 1978).

Mineralization Characteristics of Ore-Bearing Porphyries

Yulong porphyry copper deposit belt is the largest porphyry copper deposit belt in China, in which one super-large copper deposit, two large copper deposits and three medium- and small-sized copper deposits have been discovered. All the ore-bearing porphyritic rocks are located in the Triassic volcanic-sedimentary rock series and their mineralizations feature small scale (<1 km^2 outcrop area), shallow emplacement (mostly <1.5 km) and steep inclination. Although they were developed in an intracontinental environment, their mineralization characteristics are very similar to those of the porphyry copper deposits in Andes copper deposit belt. Their mineralizations are summarized as follows.

(1) Mineralization Characteristics. They have three mineralization forms in space: (a) whole or partial mineralization in porphyry; (b) steeply inclined or vertical ore body in contact zone between porphyry and wall rock; (c) stratoid or lenticular ore body in altered wall rock strata. In most cases, the three mineralizations are closely accompanied by each other and show clear mineralization zoning and annular distribution centered on the porphyry mass. Except a small ore-free core in the center, the whole-rock mass has mostly been mineralized to different degrees into Cu/Mo ores that exist in the disseminated or veinlet disseminated pattern; there is a Cu–Fe polymetallic mineralization in the contact zone, in which ore bodies are vein-like or lenticular and ores are massive or veinlet-like; there is a Pb/Zn/Ag/Au mineralization assemblage in the altered wall rock, which contains vein-like, stratoid, and lenticular ores that constitute a mineralization zone of a certain scale.

(2) Wall Rock Alteration with the Ore-Bearing Porphyry as the Center, Hydrothermal alteration zones featuring extensive coverage, high intensity and obvious zoning exist here, including potassium silicification zone, quartz-sericite zone and propylitization zone successively from porphyrite to outskirts. The main part of the potassium silicification zone was developed in the porphyry and is characterized by having large quantities of secondary potassium feldspar and quartz; the quartz-sericite zone was developed around the porphyry, within the contact zone, and above the potassium silicification zone, and is characterized by feldspar sericitization and large quantities of chlorite, epidote,

and tourmaline; propylitization occurs extensively in the strata around the porphyry and around the quartz-sericite zone. Although hydrothermal alterations at medium and low temperatures took place, the characteristics of the original rocks are kept.

(3) Mineralization Stages According to Li (1981) and Rui et al. (1984), there are three mineralization stages here: ① pneumatolytic stage at high temperatures ranging from 400 to 700 °C, at which the porphyry mostly underwent potash alternation and silification and the wall rocks in the contact zone saw skarnization, which was accompanied by Cu-Mo mineralizations; ② high-medium-temperature stage involving temperatures ranging from 200 to 500 °C, at which there occurred sericitization accompanied by Cu–Mo–Fe mineralizations; ③ mesothermal and epithermal stages, at which mineralization temperatures were lower than 230 °C and in the contact zone and peripheral strata there occurred argillation and propylitization accompanied by Au–Ag polymetallic mineralizations.

(4) Mineral Assemblages Mineral assemblage varies with ore body location and ore type. The main metallic mineral assemblage of the veinlet disseminated ore in the porphyry is pyrite + chalcopyrite + molybdenite; the main metallic mineral assemblage of the massive sulfide ore in the contact zone is pyrite + chalcopyrite + magnetite, with small amounts of molybdenite, bismuthite, ilmenite, galena and sphalerite; the main metallic mineral assemblage of the sulfide ore in the altered wall rock is pyrite + chalcopyrite + galena + sphalerite. In addition, small amounts of bornite, chalcopyrite, native gold, gold-silver ore, antimony-gold ore, etc. have been found in the porphyry copper deposits.

Formation Environment and Diagenetic Mode of Porphyry

Generally, each porphyry copper–molybdenum or copper–gold deposit across the world was developed in an active continental margin arc or island arc environment at a Mesozoic or Cenozoic convergence plate boundary, e.g., the porphyry copper belt around the Pacific Ocean, the mineralizing porphyry in which originated from a molten or island arc mantle source area of the subducting oceanic crust and the mineralization involved in which was closely related to subduction orogeny. However, the Himalayan deep-rooted epithermal porphyry and its porphyry-type copper–molybdenum deposits and explosive breccia-type gold–silver deposits in the Sanjiang area were developed in an intracontinental environment, and the diagenesis and mineralization involved are not necessarily related to a plate subduction mechanism or a continental margin arc environment. Although the ore-bearing porphyry occurred in the intracontinental convergence environment, it was related to large-scale strike-slip faults; although it has nothing to do with subduction orogeny, it has geochemical affinity to island arc volcanic rocks. Thus, in order to reasonably explain the development, evolution and dynamic background of the Sanjiang ore-bearing porphyry belt, it is necessary to deeply understand the magmatic source rock characteristics of the ore-bearing porphyry and analyze in detail the strike-slip faults that restricted its distribution.

Large-Scale Strike-Slip Fault System Under Intercontinental Collision-Occurrence Background of Ore-Bearing Porphyries

Researches on the tectonic evolution history of the huge Sanjiang Orogenic Belt show that the Jinsha and Lancang River oceanic crust plates respectively east and west of Qamdo Microblock that was separated from the western margin of Yangtze Continent, ever subducted into each other at the P_1/P_2 intersection, and the orogenic belts on the east and west sides were pushed against each other from the Yanshan Movement Period on, thus causing Qamdo Microblock and its adjacent areas to enter the intracontinental convergence stage in its entirety in Paleogene.

The convergence and collision of the Indian and Asian continents, the northward thrust of the former and the southward movement of Yangtze Continent, which started around 65 Ma ago, caused the eastern Tibet region to be in a huge strike-slip-shear system, thus developing a series of small-scale right lateral strike-slip faults, such as Chesuo-Deqin right lateral strike-slip fault and Wenquan right lateral strike-slip fault. These faults, which were accompanied by a series of en echelon left lateral NW–NNW folds and alternating fault structures, are important parts of the Sanjiang Right Lateral Strike-Slip Fault System. The strikes of these structures intersect with Chesuo Fault and Wenquan Fault at acute angles. These NNW–NW anticline and fault structures, as important rock-controlling structures, have provided ultrashallow emplacement space for Yulong Porphyry Belt. In correspondence with the Gangdise syn-collision granite belt (55–36 Ma) formed by the India-Asia intercontinental collision in Paleocene, the porphyry magma in the eastern Tibet region migrated upward along the Wenquan Right Lateral Strike-Slip Fault and finally settled in a series of dilatation spaces west of the strike-slip fault, thus developing Yulong ore-bearing porphyry belt (52–33 Ma). The two granitic rock belts that occurred at different tectonic positions with different spreading directions were developed in the same period (55–52 Ma), showing that the developments of the porphyry belts were controlled by the right lateral strike-slip faulting derived from the northward convergence and collision of the Indian continent.

Subduction-Related Mixed Crust/Mantle Source—Possible Source of Ore-Bearing Porphyries

In eastern Tibet, the strike-slip faults generated through large-scale strike-slips cut the lithosphere as deep as the upper mantle, thus providing the necessary dynamic conditions for the formation of the ore-bearing porphyry magma. However, the Paleozoic Jinsha River oceanic crust plate subducted to Qamdo Microblock and stayed in the upper mantle for a long time, resulting in crust/mantle material exchange, which may form island arc-type mantle source rocks as magmatic source rocks of ore-bearing porphyries. This conclusion is strongly supported by the following evidence.

(1) Geological Evidence. The oceanic crust fragments represented by ophiolitic melange are distributed in an intermittent way along Jinshajiang-Ailaoshan Fault Zone, indicating that the fault zone was once an ancient suture zone separating Qamdo Microblock on the west side from Zhongza Microblock on the east side. The plagiogranite in the ophiolitic melange was determined to have one of two U–Pb ages, respectively, 340 ± 3 Ma and 294 ± 3 Ma (Wang 1999), and the ocean ridge basalt has similar U–Pb ages, respectively, 362 ± 9 Ma and 296 ± 7 Ma (Zhan 1998), showing that Jinshajiang-Ailaoshan Ancient Ocean Basin was developed during Carboniferous and Early Permian. The calcic-alkalic moderately acidic volcanic rock system and Jiangda-Weixi Continental Margin Arc of Late Permian were developed on the eastern margin of Qamdo Microblock, showing that Jinshajiang Ancient Ocean Crust Plate began to subduct westward on a large scale at the end of Early Permian. The Rb–Sr isochron age of the syn-collision granite developed along the eastern margin of the continental margin arc is 227–255 Ma (Wang et al. 1999), showing that Jinshajiang Ocean Basin was closed and an arc-continent collision occurred during the early and middle Triassic Period. The main oceanic crust plate, even together with the arc volcanic-sedimentary materials, subducted obliquely and plunged beneath the Qamdo Microblock. The velocity disturbance profile of Sanjiang Orogenic Belt reveals that a subduction plate obliquely plunged beneath Qadom Microblock from east to west at an angle of about 45°. The subduction plate currently stays at a depth of 100–300 km underground, and its subduction front is beyond the west part of Qamdo Microblock and west of Tengchong (Zhong et al. 2001). Thus, it is inferred that the subduction plate might be at a shallower location on the eastern margin of Qamdo Microblock in the Paleocene. A relatively gentle high-speed (7.8–8.1 km/s) interlayer with the thickness of about 20 km is shown at the depth of 50–60 km in

Dali (Zhong et al. 2001). It is probably a local remnant of subducting mantle slices. The magma origin depth of Yulong porphyries is estimated to be greater than 38 km (Ma 1990); the formation depths of the other alkali-rich porphyries and lamprophyres are estimated to be about 60 km (Zhong et al. 2001). It is obvious that the depth of the magma source area of Yulong Porphyry Belt is roughly equivalent to that of the remnant parts of the subducting rock slices.

(2) Geochemical Evidence The Sr–Nd isotopic geochemical characteristics of the granite porphyry in Yulong Porphyry Belt show that the porphyry magma is originated from a mixed crust/mantle source in which mantle materials prevailed, and its crust/mantle ratio was estimated to be about 2/3–1/4 (Ma 1990; Wang et al. 1995). The following evidence shows that this mixed crust/mantle source might be a product of the metasomatic contamination of the overlying wedge-shaped mantle by the fluids from the subducting oceanic crust plate slices that were rich in LILEs, LREEs and Th. Boron is a light and easily soluble trace element that mainly occurs in seawater, ocean sediments and seabed altered rocks. However, in Yulong Porphyry Belt, boron occurs in large quantities in cryptoexplosive breccia and hydrothermal vein groups inside and outside the porphyry. The cryptoexplosive breccia is a product of the powerful explosion of the late-period magma and gas–liquid fluid in a lower-pressure environment (Rui et al. 1984; Wang et al. 2000). Its breccia is mainly composed of magma fragments, and the cementing matters mainly include tourmaline (up to 20–50%) and quartz. The hydrothermal veins are also mainly composed of tourmaline and quartz and are typically connected to the tourmaline-quartz breccia, showing that gas–liquid fluids extremely rich in boron were segregated from the calcic-alkalic porphyry magma at the late stage. Although the possibility of boron extraction from wall rocks by the uprising and invading porphyry magma can not be ruled out, the contribution of boron from oceanic sediments in the subduction zone to the porphyry magma may be a more reasonable explanation. Figure 4.24 Further Determine Composition of Magma Source Area of Yulong Porphyry Belt by Using Ratios of Main Elements Experimental studies in recent years show that the ratios w(CaO)/w(Na$_2$O) and w(Al$_2$O$_3$)/w(TiO$_2$) of a molten mass produced from a source rock mass consisting of different components can reflect the relative proportions of argillaceous rocks (clays) in the source rock (Douce and Johnston 1991; Holtz and Johannes 1995; Skjerlie and Johnston 1996). The monzonitic granite porphyry in Yulong Porphyry Belt is generally between mantle-derived basalt and argillaceous rock melt, near

their joint (Fig. 4.24). This shows that argillaceous rock materials might contribute to the formation of the porphyry magma. It seems that the 50–60 km-deep argillaceous rock materials participating in magmatic melting might be sediments from the boron-rich oceanic crust that plunged into the mantle through subduction. Compared with ocean ridge granites derived from the mantle, the rocks in Yulong Porphyry Belt are very rich in LILEs (Rb, Sr, Ba and K), LREEs and Th, are very poor in HREEs and Y and have largely equivalent high-field strength element contents (Nb, Ta, Zr, Hf, and Ti), which are represented by relatively flat curves in the NAP diagram. Strong Rb and K enrichments in the porphyry and their positive correlation imply that K-bearing minerals might exist in its magmatic source rocks. The porphyry is relatively poor in Ti, Y and HREEs, showing that its magmatic source rocks were relatively rich in water so that Ti-bearing minerals and Y- and HREE-rich minerals could be stable to become residual facies during the melting process of the magma. Potassium-rich lamprophyre and phlogopite harzburgite xenoliths have been found in the alkali-rich porphyry belt that was formed together with Yulong porphyry belt on the eastern margin of Tibet Plateau (Huang et al. 1997; Wang et al. 2000). This also shows that the magma source rocks of Yulon Porphyry Belt might contain water-bearing and K mineral-rich phlogopite. Its generation might be related to fluid metasomatism (Huang et al. 1996, 1997). In addition, due to the instability of LILEs (Rb, Sr, Ba, K), LREEs and Th during the water–rock reaction process, they often enriched this fluid during the dehydration process of the water-bearing subduction zone (Gill 1981; Tatsumi 1983), and this fluid participated in the metasomatic reactions with the materials in the mantle in the wedge-shaped zone, thus developing a water-bearing mantle portion rich in LILEs, LREEs and Th, i.e., island arc-type source rock (Tatsumi 1983, 1986). Since HFSEs are stable in water/rock reactions and hot water alterations, there was no HFSE enrichment in the island arc-type source rocks. Accordingly, we believe that the magma of Yulong porphyry belt is originated from a kind of island arc-type source rock related to subduction. Therefore, although it is occurred in an intracontinental environment, it has the same geochemical affinity as arc granite.

(3) Island Arc-type Source Rock + Lithospheric Strike-slip Fault—A New Diagenetic Mode of Ore-bearing Porphyry

Uyeda and Kanamori (1977) classified arcs into two extreme types: Mariana type and Chile type by subduction angle, stress state and characteristics of arc volcanic-magmatic products at the interface of

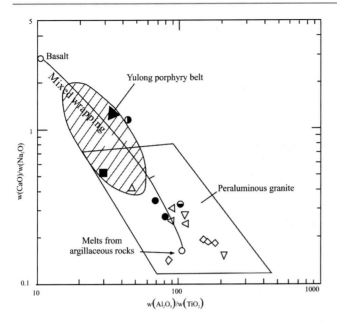

Fig. 4.24 Correlation between $w(CaO)/w(Na_2O)$ and $w(Al_2O_3)/w(TiO)$ of ore-bearing Porphyry in Eastern Tibet

convergence plates. The former features steep oceanic plate subduction angle, deep oceanic trench, volcanic island arc and back-arc spreading basin, and Kuroko-type deposits were often developed in the marine calcic-alkalic arc volcanic rock (Urabe and Sato 1978; Ohmoto 1983); the latter features gentle oceanic plate subduction angle, magmatic arc developed in the arc area due to strong compression and no back-arc spreading basins magmatic arcs, and porphyry copper deposits were often developed in the magmatic arc porphyrite (Sillitoe 1972). In eastern Tibet, the continental margin arc is a volcanic arc consisting mainly of calcic-alkalic island arc volcanic rock rather than a magmatic arc consisting mainly of granitic porphyry. Maybe, this is because the westward subduction angle of Jinsha River Oceanic Crust Plate is between the above two extreme types. In the Jiangda-Weixi Permian continental margin arc that extends hundreds of kilometers, neither typical Kuroko deposits nor typical Porphyry deposits have been found (Liu et al. 1993a, b; Hou 1993; Yang et al. 1993; Hou and Li 1999). However, the westward-subducting Jinsha River Oceanic Crust Plate provides a necessary island arc source area for the formation of the magma of Yulong porphyry belt through material and energy exchanges with the overlying wedge-shaped mantle, esp. the mantle metasomatism enrichment by the LILE-rich fluid from the subduction zone.

The convergence and collision of the Indian and Asian continents that started 65 Ma ago (Zhong 2001) developed the Himalayan Orogenic Belt and Gangdise syn-collision granite belt (55–36 Ma) in central Tibet, both of which are parallel to the collision zone. In the east of Tibet with an oblique or vertical collision zone, the northward movement of the Indian Continent and the lateral containment of Yangtze Continent led to great nearly SN shear strains, thus developing nearly SN right lateral strike-slip faults such as Sanjiang Fault System and Sagaing Fault Zone and a series of NW left lateral strike-slip faults such as Xianshuihe Fault and Honghe Fault. In western Yunnan, the left lateral strike-slip activities of the huge Honghe Fault controlled the development of the Red River alkali-rich porphyry belt and the development of the Ailaoshan metamorphic belt. In eastern Tibet, Chesuo Fault and Wenquan Fault, which constitute the Sanjiang Fault Zone, generated a series of NW folds and faults through early right lateral strike-slips and controlled the emplacement space and distribution characteristics of Yulong porphyry belt. The late-period stress relaxation and extension formed a tectonic framework of cutting and barrier alternating in the eastern margin of Qamdo Microblock. These strike-slip faults are not only basin margin faults of extensional and pull-apart basins but also crust-penetrating faults that induced partial melting in the mantle source area. Strike-slip faulting that cut lithosphere deeply led to pressure release and melting in the mantle source area and allowed the magma melt to move upward for emplacement. The partial melting in the mixed crust/mantle area, i.e., the island arc-type source area, developed Yulong Porphyry Belt, while the partial melting in the enrichment-type mantle source area developed the alkali-rich porphyry and high-potassium volcanic rock series (Huang and Wang 1996; Zhong et al. 2001). Due to the limited distribution of the island arc source area under subduction, Yulong porphyry belt is located only in eastern Tibet and parallel to the Late Permian arc volcanic belt. To sum up, we regard the diagenetic mode of Yulong porphyry belt as the coupled effects of the island arc source rock and lithospheric delamination strike-slip faulting (Fig. 4.25).

4.4.3.2 Large Basin System

In the huge Sanjiang Orogenic Belt, large basins have been located in tectonically active areas for a long time. Complex and changeable basin properties, rapid multi-source sediment accumulations, tectonic hydrothermal fluids in orogenic belts on both sides and their lateral migrations, complexly coupled basin and mountain and massive mixing of multi-source fluids finally led to large-scale mineralizations, various complex mineralization types and multi-source diplogene of metal assemblages in the basin system. Among these basins and large-scale mineralizations, Lanping Basin is the most typical.

A comprehensive analysis of the tectonic-sedimentary-fluid mineralization system in the large-scale Lanping Basin

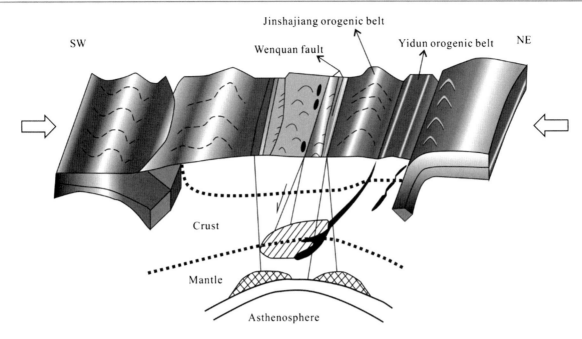

Fig. 4.25 Diagenetic mode of Yulong ore-bearing Porphyry

system and its basin evolution process reconstruction study show that the basin mainly experienced intracontinental rift-depression basin evolution (T2-3-J2-3), foreland basin evolution (K) and strike-slip pull-apart basin evolution (Cenozoic). Although mineralizations were developed to different degrees at every important evolution stage of Lanping Basin, they mainly occurred in the intracontinental rift-depression basin formed under large-scale lithosphere delamination and the strike-slip basin formed by the Himalayan intracontinental orogeny and ended with episodic drainage depositions and explosive mineralizations of Himalayan fluids.

Evolution and Mineralizations of Intracontinental Rift-Depression Basin

The intracontinental rift evolution stage is one of the important mineralization periods of Lanping Basin. The rift-depression basin and its mineralizations were controlled by the large-scale lithospheric delamination in the Late Triassic. At this stage, the lower crust was greatly delaminated, the hot mantle material upwelled in large quantities and the large basin extended, forming a central axis fault and a series of small contemporaneous faults. Driven by the upper uplift source, the hydrothermal fluid migrated vertically along the contemporaneous fault and brought about strong hydrothermal activities along the contemporaneous fault zone, thus developing hot water-deposited siliceous rocks and more than one hundred hot water-deposited alkali metal ore bodies that are widely scattered. Among them, there are two large-sized silver deposits and three medium-sized copper-silver polymetallic deposits, which are

represented by Heishan-Huishan silver–lead–zinc deposit, Yanzidong silver–copper–lead–zinc deposit, Xiawuqu-Dongzhiyan silver–copper deposit and Dongzhiyan-Hexi Strontium deposit. The developments of the deposits were controlled by specific strata. The ore-forming fluids mainly migrate vertically along the contemporaneous faults and are rich in Cl-, F-, Pb, Zn, Sb, Cu, Ag and other ore-forming materials. Although these medium–low-temperature hydrothermal sedimentary polymetallic deposits have been strongly reformed in the Himalayan period, there are still a lot of hydrothermal sedimentary characteristics (Fig. 4.26).

As the rift basin disappeared and the Lanping Basin entered the depression basin evolution stage in the Middle Jurassic, it received a large amount of high-porosity red bed deposits, and favorable thermal fluid migration channels were developed therein during late tectonic activities, thus locally forming structural traps conducive to mineralization enrichment.

Evolution and Mineralization of Foreland Basin

Lanping Basin, which entered the foreland basin development stage in Late Cretaceous, received large amounts of terrigenous coarse clastic materials from the orogenic belts under the great uplift and napping actions of the orogenic belts on the east and west sides. Although the lenticular and nodular Cu–Fe polymetallic stockwork ore bodies developed in the interstices of these coarse terrigenous materials were not of commercial significance, these high-porosity accumulation layers provided a main activity place for ultrahigh-pressure fluids when tectonic changes occurred later and their initial mineralization provided an important

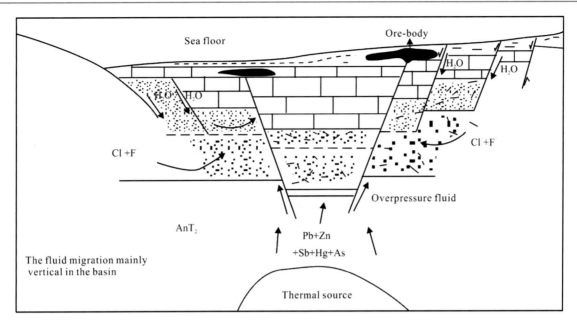

Fig. 4.26 Schematic diagram of rift basin and fluid mineralization (Wang et al. 2000, Scientific Research Report)

material source for large-scale mineralizations in the Himalayan Period.

Figure 4.27 shows the fluid activity course within the orogenic belt and foreland basin system. Oliver (1986) first proposed the viewpoint that hydrothermal fluids migrate laterally in a foreland basin's orogenic belt structure. This explains the spatial distribution pattern of various minerals in foreland basins. Luo and Du (1999) also believe that the deposits in Lanping Basin such as Jinding lead–zinc deposit exactly follow this mineralization mode. Although Jinding deposit was formed in the Himalayan Period and hosted in a strike-slip basin, massive amounts of data prove that this type of lateral migration did exist at the foreland basin stage. A lead isotope study of some lead–zinc deposits in Lanping Basin by Mo et al. (2000) also confirmed that the ore-forming materials are originated not only from Jinshajiang Orogenic Belt on the east side but also from Lancangjiang Orogenic Belt on the west and that fluids in the basin might migrate from the orogenic belts on both sides of the basin to its center. Due to the counterthrust nappings of the orogenic belts on both sides, the fluid pressure in the basin was equal to or slightly larger than the vertical pressure (Dominigne Granls 1997), thus resulting in lateral fluid migration under high-pressure stress without the opportunity of vertical upwelling. Because of the coarse sediment particles in the foreland basin, the porosity increased and fluids were preserved in the strata in the form of formation water, which provided large amounts of material reserves for Himalayan mineralization. Therefore, sedimentary sandstone-type copper deposits were formed only in some areas at this stage, such as the primary sedimentary deposit in Baiyangchang.

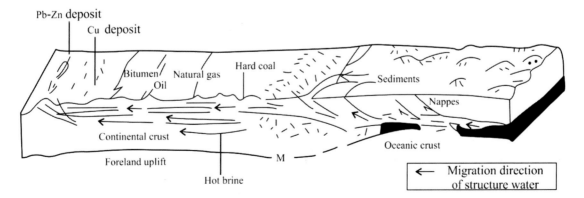

Fig. 4.27 Intrabasinal fluids might move from orogenic belts on both sides to center

Evolution and Mineralization of Strike-Slip Basin

The tensional and compressional strike-slip basin evolution period, which started from Paleocene, is the most important large-scale metallogenic period of Lanping Basin. The fluid evolution in the basin was more intense and active. Given the influence of the orogenic belts on both sides of the basin and the intensification of deep crustal tectonic activities, deep active ore-forming materials entered basin fluids through deep overpressure hydraulic or pneumatic fragmentation, which led to explosive mineralization.

At the Paleocene tensional strike-slip stage, the further delamination of lower crustal materials has led to large-scale and universal tensional strike-slips in Lanping Basin, and a hot uplift area was developed in the lower part of the basin, which acted as a "heat engine" to drive fluid activities in the basin (see Fig. 4.8). When the uplift area was cut due to a contemporaneous fault activity, the fluid rich in ore-forming materials in the area became part of the basin fluid, and a large-scale mixing of deep brine and shallow formation water occurred.

Summary of Basin Fluidization and Mineralization

In Lanping Basin that experienced complex property changes, there are superimposed basins of various properties. Since the Middle Jurassic, the very thick red bed deposits have served as a superior reservoir for the basin fluids. However, due to frequent tectonic activities and continuous activities of the contemporaneous inherited faults, good large-scale reservoir space was formed in the basin in the Eocene Epoch, such as Lanping Jinding Dome and Sanshan-Baiyangping fault traps. Thus, although the basin had good space for oil and gas generation, it had no good oil and gas storage space. Large quantities of oil and gas served as the transportation medium of metallatic minerals, and their residues were dispersed in various fissures and between rock-mineral grains, such as a large quantity of asphaltene found along fissures and beddings in Jinman copper deposit, through whose shrinkage a large number of chalcopyrite emulsion droplets were bled out, crude oil inclusions in Jinding lead–zinc deposit, etc.

The Himalayan ore-forming fluids are large-scale mixtures of deep fluids (rich in Co, Ni, Sb, As, Hg, Ba, Sr, Au, Ag, Cu, Pb, Zn, etc. and CO_2) and shallow fluids (rich in CH_4, SO_4^{2-}, Cl^-, Ca^{2+}, H_2S), etc. A distribution and metallogenic mechanism analysis of the extensive alkali metal deposits and Cu–Ag–Co polymetallic deposits in the basin show that the metallic matter and a large amount of CO_2 in the ore-forming fluids were mainly from deep fluids, while CH_4, SO_4^{2-}, Cl^-, Ca^{2+}, H_2S, etc. were mainly from shallow fluids. In the setting process of ore-forming fluids, basinal fluids ubiquitously involved degassing. The fact that H_2S gas is settled in place first is very important to the mineralization of alkali metal

deposits (such as Jinding Pb–Zn deposit). H_2S gas first settled in structural traps and occupied mineralization space, thus causing a series of strong activities to change the trap structure, such as triggering fracturing and mechanical slumping of brittle rocks in the napping dome (many such events are found in Jinding Mining Area), cryptoburst and fracture in fault traps (Sanshan-Baiyangping area). In addition, the setting of CO_2 gas is more important for Cu–Ag polymetallic deposits. It not only leads to cryptoburst and fracture in mineralization space, but also changes the physical and chemical properties of fluids, such as changing the pH value of the fluid to neutral-weakly alkaline. Under this condition, many silver-bearing tetrahedrites or silver tetrahedrites have been developed as the main copper-silver hosted minerals.

Given various structural types and multiple evolution stages of Lanping Basin, the ore-forming fluids of the basin have four characteristics: multi-level, multi-direction, multi-source and multi-migration mode. Specifically, they are represented by four fluid migration models: ① vertical migration model of moderately deep and deep fluids during tension and subsidence (T_3-J); ② extensive and large-scale lateral fluid migration and convection model under the Cretaceous single thrust condition; ③ fluid passive setting, accumulation and zonal migration model in the order of solubility and sedimentation under Paleocene-Early Eocene tensional strike-slip conditions; ④ lateral and vertical fluid migration-convection model under the counterthrust compression strike-slip conditions in Late Eocene and Oligocene, in which intrabasinal overpressure fluids were set in the uplift storage or fault trap (Fig. 4.28).

Large amounts of deep elements participating in mineralization show the existence of deep brine. With regard to sources of ore-forming fluids, there are the following views: ① the intrabasinal formation water became chemically very active through fluid/solid interactions; ② the fluids from the orogenic belts were partially migrated through huge pores of old red beds and were mixed into the deposits water in the basin under counterthrusts or a single thrust, forming a unique mixture rich in CH_4 and saturated $NaCl/CaCl_2$ hot brine; ③ high-temperature and high-pressure fluids were upwelled due to the flow removal of the mantle or crust and were mixed with the moderately deep and shallow fluids in the red beds in the basin into ore-forming fluids, which were rich in CO_2 gas.

4.5 Preliminary Discussion of Archipelagic Arc-Basin Metallogenic Theory

There are four huge metallogenic systems and 11 metallogenic systems in the Sanjiang Tethys mineralization domain, which were mainly controlled by two major tectonic processes, i.e., MABT evolution and collision orogeny.

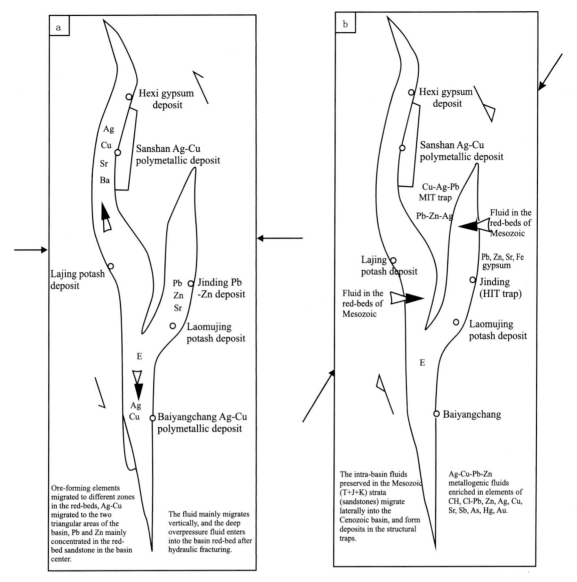

Fig. 4.28 Uplift storage or fault trap mineralization model of intrabasinal overpressure fluids

The basic framework and structural evolution of the MABT developed in Paleozoic and Early Mesozoic have led to three important metallogenic events in the Sanjiang area, i.e., early Paleozoic, late Paleozoic and late Triassic metallogenic events. The early Paleozoic mineralization episode is mainly occurred in a tensional environment at the edge of the block and within it at the early development stage of the MABT and is characterized by SEDEX Pb–Zn mineralization; the late Paleozoic mineralization episode is mainly occurred at the main development stage of the MABT and is characterized by seabed VMS mineralization; the late Triassic mineralization episode is mainly occurred in environments such as volcanic-magmatic arc, inter-arc rift basin, back arc-basin and superposed rift extension basin developed at the late development stage of the MABT, and is

characterized by porphyry Cu deposits, VMS, and SEDEX polymetallic deposits.

The basic framework and tectonic evolution of the MABT mainly resulted in two major metallogenic systems, i.e., continental margin split and convergence ones. There are two main mineralization modes: seabed SEDEX Pb–Zn mineralization and porphyry-type Cu–Au mineralization. As the island arcs evolved from immature ones (oceanic arcs) into mature ones, VMS mineralization evolved along the following route: Cu → Cu–Zn → Cu–Pb–Zn → Zn–Pb–Cu → Pb–Zn. Because of the change of the subduction plate angle, different sections of the island arcs have different stress states. VMS-type deposits occur in a tensile arc, while porphyry-type copper deposits occur in a compressive arc. In the Sanjiang area, multiple tectonic mineralizations could often occur in the same mineralization belt or even in the same deposit. Late porphyry Cu(–Mo–Au) deposits were developed under many early VMS-type deposits. For example, in Lancang Old Plant Lead Deposit in Changning-Menglian Belt, a porphyry Mo(–Cu) deposit has been found under an early Carboniferous (the volcanic tuff zircon SHRIMP age is 323 Ma, Huang 2009) VMS-type Cu (–Pb–Zn) deposit; in Yangla copper deposit in Jinshajiang Belt, Indosinian porphyry/skarn Cu(–Mo) mineralization is superimposed on the SEDEX Cu(–Pb–Zn) deposit formed from Late Devonian to Early Carboniferous, thus forming a composite deposit.

Given the volcanic-magnatic island arcs generated by the plate subduction in the late development period of the MABT, VMS-type Cu, Pb and Zn deposits and porphyry-type Cu(–Mo–Au) deposits were widely developed therein. The arc-continent and arc-arc collisions at the late stage (T_3) led to no obvious mineralization, but the plate break-off, remelting, and crustal extension after the collisions resulted in strong A-type granite magma emplacement and rift basin development in the Sanjiang area, thus inducing magmatic Sn deposits and hydrothermal Pb–Zn–Ag mineralization.

Based on previous research findings, we systematically dissected four ophiolitic melange belts, five island arc or continental-margin-arc volcanic-magmatic rock belts and the blocks between them through several rounds of scientific and technological research and prospecting action plans. It is found that the framework of the Sanjiang Tethys orogenic belt is not a simple trench-arc-basin assemblage (one trench, one arc and one basin) formed by a Tethys Ocean subduction. Rather, it was completed mainly by subductions and closures of a series of small ocean basins formed by the previous continental margin break-off and realized by archipelagic arc-basin orogenic processes induced by ocean basin subductions (including those of back arc-basins), such as ocean crust subduction or obduction, arc-arc collision, arc-microblock collision and continent–continent collision. Different genetic types of deposits were formed in different island arc or continental-margin-arc belts, blocks (microblocks) or basins, joint belts or ophiolitic melange belts. Thus, the archipelagic arc-basin metallogenic theory was put forward.

4.5.1   Temporal and Spatial Structure and Mineralization Pattern of Multi-Arc-Basin-Terrane (MABT)

In Chap. 3, we discussed in detail the basic characteristics and evolution of the Sanjiang MABT as well as its temporal and spatial structure. Here, the mineralizations of the arc and basin systems are further summarized.

The original plate tectonics theory and the global Tethyan tectonic framework believe that the Tethyan Orogenic Belt was formed mainly by the subduction of the Tethyan Ocean, it has the typical structural framework consisting of one trench (ocean trench), one arc (island arc) and one basin (back arc-basin) that were developed in sequence from ocean to continent, Cyprus-type VMS deposits were developed in trenches, and volcanogenic massive sulfide (VMS) deposits, porphyry copper deposits and epithermal Au deposits were developed in the island arc or continental margin arc (Fig. 4.29a).

The systematic dissection of several ophiolitic melange belts and various types of island arcs and basin systems in the Sanjiang Tethys orogenic belt shows that Sanjiang Tethys has gone through a long, continuous and complex process involving occurrence, development, shrinkage and extinction at least from Paleozoic to Mesozoic. Paleo-Tethys is not the inheritance and development of the original Tethys and Mesozoic Tethys is not Paleo-Tethys reopened after extinction. Some Tethys Ocean crusts can be merged with the subsequent Indian Ocean. According to a study of the ophiolitic melange belts and arc-basin system in the Sanjiang area and their evolutions, the basic tectonic framework of Sanjiang Orogenic Belt consists of four main arc-land collision zones (Ganzi-Litang, Jinshajiang-Ailaoshan, Lancangjiang and Nujiang-Changning-Menglian), five island or continental margin arc volcanic-magmatic rock belts (Dege-Xiangcheng, Jiangda-Weixi-Lvchun, Zaduo-Jinggu-Jinghong, Leiwuqi-Lincang-Menghai and Bomi-Tengchong) and five blocks in between (Zhongza-Shangri-La, Qamdo-Lanping, Leiwuqi-Zuogong, Baoshan-Zhenkang and Chayu-Gaoligong) and is similar to the spatial configuration of a series of island countries in Southeast Asia today. The Sanjiang Tethys orogenic belt was not formed by a simple Tethys Ocean subduction. It was completed mainly by subductions and closures of a series of small ocean basins and was realized by archipelagic arc-basin orogenic processes induced by ocean basin subductions (including those of back arc-basins), such as ocean crust subduction or obduction, arc-arc collision, arc-microblock (block) collision, and continent-continent collision; different genetic types of deposits were formed in different island arc or continental margin arc belts, blocks (microblocks) or basins, joint belts, or ophiolitic melange belts (Fig. 4.29b). The tectonic framework of the complex MABT developed from late Paleozoic to early Mesozoic in the Sanjiang area controls the temporal and spatial distribution and mineralization zoning of deposits therein.

4.5.1.1   Yidun Arc-Basin System and Mineralizations

The Yidun Arc-basin System is mainly composed of Ganzi-Litang Ophiolitic Melange Belt and Yidun Volcanic-magmatic Island Arc Belt on the west side. The existing

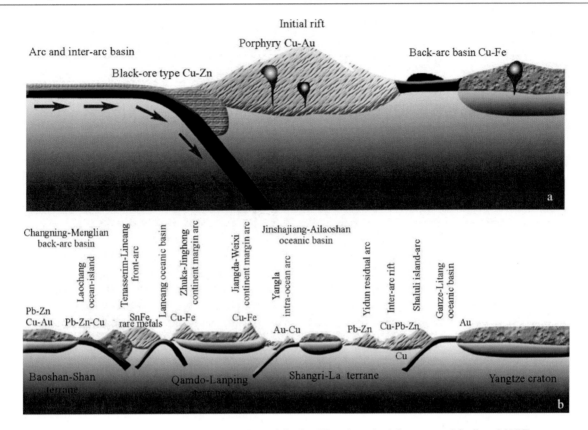

Fig. 4.29 Mineralization model comparison of traditional plate subduction "Trench-arc-basin" system and Sanjiang MABT

data show that on the continental crust that had extended and thinned for a long time, Ganzi-Litang Back Arc Ocean Basin opened in early Permian or earlier and began to subduct westward under Zhongza-Shangri-La Block in late Triassic. On the west side there occurred a typical spatial configuration framework consisting of Yidun island arcs (inner and outer arcs), inter-arc rift basin and back arc-basin system, in which bimodal volcanic rocks, subduction island arc volcanic rocks and intrusive rocks were developed. The ocean basin closure and arc-continent collision at the end of Late Triassic and the post-collision extension from Jurassic to Early Cretaceous led to the occurrence of a series of intrusive rocks in the collision and post-collision extension environment. At different stages and locations of the arc-basin system during its generation and evolution, VMS Pb–Zn–Ag deposits, porphyry/skarn-type Cu(–Mo–Au) deposits, epithermal polymetallic deposits and magmatic hydrothermal W–Sn–Bi–Mo polymetallic deposits were developed.

4.5.1.2   Jinshajiang Arc-Basin System and Mineralizations

The Jinshajiang Arc-basin System is mainly composed of the Jinshajiang-Ailaoshan ophiolitic melange belt and the

Jiangda-Weixi-Lvchun continental margin arc volcanic rock belt on the west side. According to the available research findings, Jinshajiang-Ailaoshan Back Arc Ocean Basin opened in the Carboniferous Period and began to subduct westward in the late Middle Permian after the Early Devonian above the Early Paleozoic metamorphic soft basement began to rift and subside; the subduction between ocean plates developed the Zhubalong-Yangla-Dongzhulin intra-oceanic volcanic arc at the central axis of the back arc ocean basin from the late Middle Permian to Late Permian as well as the Xiquhe-Xueyayangkou-Jiyidu-Gongnong back arc-basin (ocean crust basement) on its west side; the ocean crust on the west side of the back arc ocean basin subducted westward under Qamdo-Lanping Block and a typical spatial configuration framework consisting of the Jiangda-Deqin-Weixi-Lvchun continental margin arc and its back arc-basin (continental crust basement) was developed from the late Middle Permian to the Late Permian, thus forming subduction-type intra-oceanic arc volcanic rocks, island arc volcanic rocks and intrusive rocks. The ocean basin closure, arc-continent collision and post-collision extension in Late Triassic developed arc volcanic rocks, bimodal volcanic rocks, and intrusive rocks in a collision and post-collision extension environment. At different stages and different locations of the arc-basin system during its generation and evolution, VMS Cu–Pb–Zn deposits, SEDEX massive sulfide Fe–Ag deposits and skarn-type Fe–Cu deposits were developed.

4.5.1.3   Taniantaweng Arc-Basin System and Mineralizations

With the Jitang-Chongshan-Lancang remnant arc as the frontal arc, Taniantaweng Arc-basin System is composed of Changning-Menglian Ophiolitic Melange Belt, Lincang-Menghai Island Arc Volcanic Rock Belt on its east side, Lancangjiang Ophiolitic Melange Belt, and Zaduo-Zhuka-Yunxian-Jinghong Continental Margin Arc Volcanic Rock Belt on its east side. According to the available research findings, Lancangjiang Back Arc Ocean Basin opened in the Carboniferous Period after the Early Devonian above the Early Paleozoic metamorphic soft basement began to rift and subside, and Changning-Menglian Ocean was the inheritance and development of the original Tethys ocean; from the late Middle Permian on, the main body subducted eastward under Lancang Remnant Arc Block and Qamdo-Lanping Block, thus developing a spatial configuration framework from late Middle Permian to Late Permian which consisted of Lincang-Menghai Island Arc, Zaduo-Zhuka-Yunxian-Jinghong Continental Margin Arc and its back arc-basin (arc basement) and in which subduction-type continental margin arc volcanic rocks and intrusive rocks were developed. The ocean basin closure and arc-arc collision in Early and Middle Triassic and the post-collision extension in Late Triassic have developed arc volcanic rocks, bimodal volcanic rocks, and intrusive rocks in a collision and post-collision extension environment. At different stages and different locations of the arc-basin system during its generation and evolution, VMS Cu–Pb–Zn deposits, skarn-type Fe–Cu deposits, etc. were developed.

4.5.1.4   Boshulaling-Gaoligong Arc-Basin System and Mineralizations

The arc-basin system is mainly composed of Luxi-Santaishan Ophiolitic Melange Belt and Boshulaling-Gaoligong Magmatic Arc and Bomi-Polongzangbu Ophiolitic Melange Belt on its west side. The available research findings reveal that Tethys Ocean above the Proterozoic metamorphic basement and the Early Paleozoic passive marginal basin, which is represented by the Luxi-Santaishan ophiolite melange, subducted southwestward under Tengchong-Lianghe Block from Carboniferous to Permian, thus developing Boshulaling-Gaoligong Magmatic Arc and its back arc-basin (continental crust basement). The ocean's further subduction in the Mesozoic Era developed a spatial arc-basin system configuration framework consisting of Boshulaling-Gaoligong Magmatic Arc, Polongzangbo (Jiali) Inter-arc Ocean Basin and Chayu (South Gangdise)

Magmatic Arc, respectively, from east to west. The ocean basin closure and arc-arc or arc-continent collision at the end of Late Triassic (possibly to Early Jurassic) developed a series of island arc intrusive rocks and volcanic rocks in multi-collision environment, esp. the superimposed transformations of the Boshulaling-Gaoligong arc-basin system by the subduction of the Yarlung Zangbo Back Arc Ocean Basin from Middle Jurassic to Late Cretaceous and the arc-continent collision in the Paleogene Period resulted in large-scale magmatic hydrothermal vein-skarn Sn–W, Fe, Pb–Zn and rare metal mineralizations.

4.5.1.5   Edges of Stable Blocks and Mineralizations

The main bodies of the split blocks embedded in the MABT such as Zhongza-Shangri-La, Qamdo-Lanping-Pu'er, Leiwuqi-Zuogong, Baoshan-Zhenkang and Chayu-Gaoligongshan ones, which were above the Proterozoic metamorphic basement and the Early Paleozoic fold basement, experienced an evolution or transformation process consisting of relatively stable Devonian-Early Permian neritic shelf plateau, Middle and Late Permian back arc-basin and Middle and Late Triassic back arc foreland basin. In addition to VMS and SEDEX deposits in passive continental margin rift basins, the counterthrust structure framework developed along the nearly north-south back arc foreland basin's central axis in the Middle and Late Triassic after the Early Triassic ocean basin extinction and arc-continent or arc-arc collision, along with the drainage and convergence of hydrothermal fluids toward the center of the basin, provided a strong materials foundation for the later large-scale mineralizations of the intracontinental basin in the Cenozoic Era.

4.5.2   Evolution and Metallogenic Mechanism of Multi-Arc-Basin-Terrane (MABT)

Through dissecting and studying several ophiolitic melange belts and various arc-basin systems in the Sanjiang Tethys orogenic belt, it is found that the southwest margin of the unified Pan-Cathaysian Continent Group was formed due to the restriction of the eastward subduction of Tethys Ocean at the end of Early Paleozoic, Tanggula-Taniantaweng Remnant Arc split off like the Japanese Archipelago from Kangdian Continental Margin Arc constituted the Late Paleozoic–Mesozoic frontal arc at the southwest margin of Pan-Cathaysian Continent; according to records, an extensive area behind the frontal arc covering Yidun Arc-basin System (P-T$_3$), Zhongza-Shangri-La Block, Jinshajiang Arc-basin System (D-T$_3$), Qamdo-Lanping-Pu'er Block, Taniantaweng Arc-basin System (D-T$_3$), Leiwuqi-Zuogong Block and Early Paleozoic Lancang-Chongshan Remnant Arc Block saw a geological evolution history involving

MABT development, back arc spreading and arc-arc or arc-continent collision from Late Paleozoic to Mesozoic; the island arc orogeny in Triassic finalized the main body of the Pan-Cathaysian Continent and caused it to become an integral part of Eurasia, and in some areas of the Pan-Cathaysian Continent, rift basins were developed against the post-collision crustal extension background in Late Triassic.

Running through the whole process of the Tethys lithosphere evolution from Late Paleozoic to Mesozoic, the archipelagic arc-basin mineralizations in the Sanjiang area occurred in the Southeast Asia-like continental margin system characterized by Late Paleozoic–Mesozoic archipelagic arc-basin tectonic systems and were closely associated with a series of alternating processes including island or continental margin arc evolutions, back arc rift valley and ocean basin spreadings and extinctions, and activations and transformations of split blocks sandwiched between them, i.e., the complex structural-paleogeographic framework consisting of MABT in Late Paleozoic–Mesozoic in the Sanjiang area controls the temporal and spatial distribution of the deposits therein, and they feature obvious zoning, segmentation, diversity, and superposition. According to the evolution-mineralization relationship of the Sanjiang Tethys MABT, the mineralization or evolution process can be further divided into five stages: continental margin split period, ocean basin spreading period, ocean crust subduction period, arc-arc/continent collision period and post-collision extension period (Figs. 4.30 and 4.31).

4.5.2.1 Metallogenic Mechanism During Continental Margin Split Period

At the end of Early Paleozoic, the South China Ocean Basin was destroyed, and the folds turned into mountains, and multiple back arc ocean basins such as the Qilian Ocean Basin, Northern Qaidam Margin Ocean Basin and east and west Kunlun ocean basins disappeared one after another, thus developing a subduction collision orogenic system. So far, Yangtze was connected with Cathaysia, Sino-Korea Block, Qaidam, Tarim, Indosinian Continent, etc. to form the unified Pan-Cathaysian Continent Group. Between the southwest margin of Pan-Cathaysian Continent Group and the northern margin of Gondwana Continent was the vast Tethys Ocean. The Sanjiang Tethys tectonic belt occurred and evolved exactly against this global tectonic background. It is located at the junction of Pan-Cathaysian Continent Group and Gondwana Continent Group and originated from the continental margin arc active zone at the southwest margin of Pan-Cathaysian Continent Group.

The southwest margin of the unified Pan-Cathaysian Continent Group was formed at the end of the Paleozoic Era due to the restriction of the eastward subduction of Tethys Ocean from Early Paleozoic to Middle Devonian. Tanggula-Taniantaweng Remnant Arc split off like the Japanese Archipelago from Kunlun Frontal Arc and Kangdian Continental Margin Arc constituted the Late Paleozoic-Mesozoic frontal arc including Sanjianga at the southwest margin of Pan-Cathaysian Continent. The basin

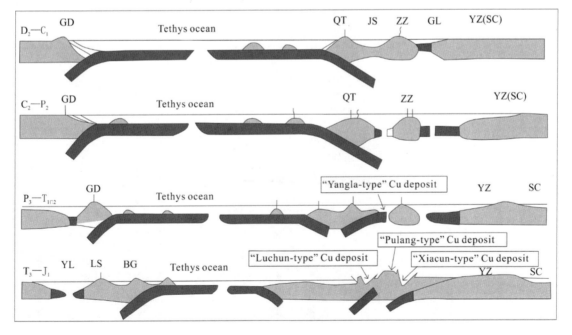

Fig. 4.30 Evolution stages and mineralizations of Sanjiang Tethys MABT. GD—Gangdise; LS—Lhasa; YL—Yarlung Zangbo; BG—Bange-Jiali Belt; QT—Qiangtang; JS—Jinshajiang; ZZ—Zhongza; GL—Ganzi-Litang; YZ (SC)—Yangtze (Sichuan)

Fig. 4.31 Spatiotemporal
evolution and metallogenesis of
the Paleo-Tethys in Sanjiang, SW
China

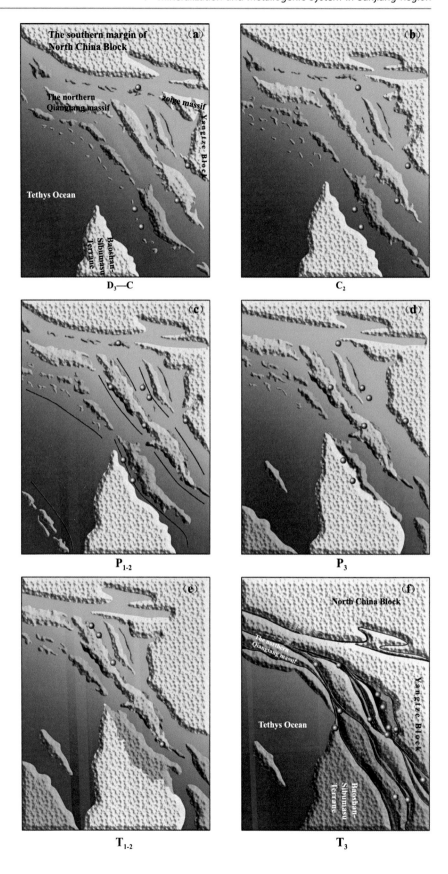

extension behind the frontal arc developed a series of back arc rift basins, including Ganzi-Litang Rift Basin at the east margin of Zhongza-Shangri-La Block, Jinshajiang-Ailaoshan Rift Basin between the west margin of Zhongza-Shangri-La Block and the east margin of Qamdo-Lanping-Pu'er Block and Langcangjiang Rift Basin between the west margin of Qamdo-Lanping-Pu'er Block and the frontal arc. The existing data show that the strong, rapid and short-duration back arc spreading was unfavorable for the formation of industrial grade deposits in the back arc rift basins but was favorable for the initial enrichment and formation of source beds or source rocks, thus providing an important material source for later mineralizations (including the Cenozoic intracontinental convergence and transformation process), such as Tuoding Cu deposit in Devonian might be related to this. The important mineralizations in this period, which are characterized by the occurrence of large-scale SEDEX Pb-Zn deposits, mainly occurred in the passive continental margin rift basins far away from the back arc spreading area, thus developing products enriched with crust-derived Pb and Zn materials by the hydrothermal fluid circulation system in the continental margin extension environment, such as the huge Huize Pb–Zn deposit in Devonian at the western and northwestern margins of Yangzi Continent, the large-sized Bafangshan-Erlihe Pb–Zn deposit in Fengxian County and Fengshu-Zhaojiazhuang Pb–Zn Ore Belt in Xunbei.

4.5.2.2 Metallogenic Mechanism During Ocean Basin Spreading Period

From Late Devonian to Early Permian, a series of back arc rift basins behind Jitang-Chongshan-Lancang Front Arc expanded into back arc ocean basins due to the restriction of the further eastward subduction of Tethys Ocean. The continuous spreading and eastward subduction of Tethys Ocean represented by Nujiang-Changning-Menglian Ocean caused Leiwuqi-Zuogong Block, Qamdo-Lanping-Pu'er Block and Zhongza-Shangri-La Block behind the frontal arc to split off from Yangtze Continent, thus developing Lancangjiang back arc spreading Ocean Basin, Jinshajiang-Ailaoshan back arc spreading Ocean Basin and Ganzi-Litang back arc spreading Ocean Basin successively from west to east, which constituted the basic framework of the Sanjiang Tethys MABT.

It is found through a systematic dissection of the mineralization environments and deposit types of the important metallogenic belts in the Sanjiang area that although there are three important back arc spreading ocean basins of different scales in the region, such as Lancangjiang Ocean Basin, Jinshajiang-Ailaoshan Ocean Basin and Ganzi-Litang Ocean Basin, the mafic-ultramafic rocks constituting the ocean crust are mostly quasi-oceanic ridge ones because the limited ocean basins formed by back arc spreading feature short-duration development and low ocean crust maturity

(Mo et al. 1993). This is the fundamental reason why it has no large-scale Cyprus-type VMS Cu and Cu–Ni deposits. Relatively speaking, the new ocean crust and large-scale quasi-oceanic ridge volcanic activities caused by back arc seabed spreading only occasionally developed small-scale magmatic liquation-type Cr–Pt sulfide deposits, such as Xumai Chromite–platinum group deposit in Jinshajiang ophiolite; deep ore-forming materials such as Au–Cr–Cu materials in the back arc ocean basins were more favorable for initial enrichment and formation of source beds or source rocks, which were kept and involved in ophiolitic melange belts, thus providing important material sources for later mineralizations (including the Cenozoic intracontinental convergence and transformation process), such as large or super-large deposits in the Jinshajiang-Ailaoshan ophiolitic melange belt including Laowangzhai Au deposit, Donggualin Au deposit, Jinchang Au deposit and Daping Au deposit and large deposits in the Ganzi-Litang ophiolitic melange belt including Cuo'a Au deposit and Shala Au deposit. The important mineralizations during this period mainly occurred in the oceanic island volcanic basins and their passive continental margin rift basins that were related to spreading ocean basins and are characterized by the occurrence of VMS deposits. The former were developed in the oceanic island basic-moderately basic volcanic rocks of expanding ocean basins by the hydrothermal fluid circulation system in a spreading ocean basin ridge environment, such as Cu–Pb–Zn deposits in the Changning-Menglian oceanic island ophiolitic melange; the latter were developed in passive continental margin rift basins related to spreading ocean basins by the hydrothermal fluid system in a continental margin extension environment, such as the large-sized Dapingzhang Cu polymetallic deposit in Carboniferous in the Jinggu-Jinghong continental margin arc zone on the western margin of Pu'er Block.

Only one Cyprus-type VMS Cu deposit, i.e., Tongchangjie Cu deposit, is known to exist in the Sanjiang Tethys orogenic belt. It was developed in the picrite with a large amount of olivine accumulation crystals that erupted out of the seabed due to magma chamber rupture under the Changning-Menglian spreading ridge. This zone, which is the Paleo-Tethys Ocean represented by Nujiang-Changning-Menglian Ocean, has very mature accumulations of ocean crust and mantle materials, so it provided the conditions for developing Cyprus-type VMS deposits. This is exactly the reason why no Cyprus-type VMS deposits exist in small back arc spreading-induced ocean basins.

4.5.2.3 Metallogenic Mechanism During Ocean Crust Subduction Period

From Middle Permian to Late Permian, the tectonic geological setting of the Sanjiang Tethys orogenic belt changed greatly on the basis of its early tectonic framework consisting of a series of small back arc ocean basins and microblocks arranged alternately, except that the main body of Ganzi-Litang Back Arc Ocean Basin was still in a spreading environment. Driven by the subductions of a series of back arc ocean basins behind the Jitang-Chongshan-Lancang frontal arc, the formation and evolution process of the MABT in the Sanjiang Tethys orogenic belt started when the alternating subducting island arcs, continental margin arcs and back arc-basins were formed. Among them, Ganzi-Litang Back Arc Ocean Basin began to subduct in early and middle Late Triassic.

Because of being generally restricted by the continuous eastward subduction of Tethys Ocean, Wulanwula-Lancangjiang Back Arc Ocean Basin subducted westward under Jitang-Lancang Remnant Arc Block so as to develop the Leiwuqi-Lincang-Mengdao island arc and subducted eastward under Qamdo-Lanping-Pu'er Block to form the Zaduo-Zhuka-Yunxian-Jinghong continental margin arc and the back arc-basin (continental crust basement) on its east side, thus developing the Taniantaweng continental margin arc-back arc ocean basin-back arc-basin system. The subduction between ocean crusts at the central axis of Jinshajiang-Ailaoshan Back Arc Ocean Basin developed the Zhubalong-Yangla-Dongzhulin intra-oceanic arc and the Xiquhe-Xueyayangkou-Jiyidu-Gongnong back arc-basin (ocean crust basement) on its west side; its westward subduction under Qamdo-Lanping-Pu'er Block developed the Jiangda-Deqin-Weixi-Lvchun continental margin arc and the back arc-basin (continental crust basement) on its west side, thus developing the Jinshajiang intra-oceanic arc-back arc ocean basin-continental margin arc-back arc-basin system. Due to the long-distance effect of the eastward subduction of Tethys Ocean and the restriction of the Jinshajiang back arc spreading ocean basin, the formation and spreading periods (mainly from Permian to Early and Middle Triassic) of the Ganzi-Litang back arc ocean basin, which is adjacent to Yangtze Continent on the easternmost side of Sanjiang Tethys, is obviously later than those of Jinshajiang and Lancangjiang back arc ocean basins on the west side (mainly from Late Devonian to Permian). The Ganzi-Litang back arc ocean basin subducted westward under Zhongza-Shangri-La Block in the early and middle Late Triassic, thus developing Yidun island arcs (inner and outer arcs), inter-arc rift basin and back arc-basin successively from east to west, which constituted the Yidun island arc (inner arc)-inter-arc rift basin-island arc (inner arc)-back arc-basin system.

As the main parts of the MABT of the Sanjiang Tethys orogenic belt, the island arc or continental margin arc volcanic-magmatic rock belts were products of the subductions of a series of small back arc ocean basins; the formation and development of the belts is also a process of great adjustments, great exchanges and recombinations of crust-mantle material components in the orogenic belt and also a process of enrichment of Cu–Au–Ag–Pb–Zn polymetallic ore-forming materials into deposits. In the traditional plate structure pattern and the framework of its trench-arc-basin metallogenic theory, it is established that its main mineralization types and tectonic environments are porphyry copper deposits in island arc zones and VMS deposits in back-arc-basins. However, during the evolution process of the MABT of the Sanjiang Tethys orogenic belt, there are multiple island or continental margin arc belts, together with more complex volcanic-magmatic arc types including intra-oceanic arc (equivalent to Mariana Arc), island arc (equivalent to Ryukyu Islands) and continental margin volcanic arc (Andean-type). All of the subduction-type volcanic-magmatic arc belts are important metallogenic belts in the Sanjiang area.

By tectonic environment and mineralization type, the volcanic-magmatic arc belts of the Sanjiang Tethys orogenic belt can be divided into five types: ① VMS deposit in an intra-oceanic arc volcanic basin, such as the large-sized Yangla Cu deposit in the Jinshajiang ophiolitic melange belt, which was developed in basic-moderately basic volcanic rocks by the hydrothermal fluid circulation system formed by mutual subductions of ocean crusts; ② VMS deposit in an intra-arc rift basin, such as the super-large Gacun Ag–Pb–Zn deposit and the large-sized Gayiqiong Ag–Pb–Zn deposit in the northern section of the Yidun island arc belt, which were developed in moderately basic and moderately acid bimodal volcanic rocks by the hydrothermal fluid circulation system in a tectonic environment of rift basin extension between volcanic arcs (inner and outer arcs); ③ Epithermal Au–Ag–Hg polymetallic deposit in a back arc-basin, such as the medium-sized Nongduke Au–Ag polymetallic deposit and the large-sized Kongmasi Hg deposit in the Yidun island arc belt, which were developed in moderately acid volcanic rock series by the volcanic hydrothermal cryogenic fluid circulation system formed under extensional rifting on the side of the volcanic arc adjacent to the continent; ④ Porphyry copper deposits, such as the large-sized Pulangte Cu deposit and the large-sized Xuejiping Cu deposit in the southern section of the Yidun island arc belt, which were developed in the island arc-type epithermal-ultraepithermal intrusive rock complex (porphyry) by the magmatic hydrothermal fluid circulation system in the compressive island arc tectonic environment; ⑤ Skarn-type Fe–Cu or Cu–Pb–Zn deposit, such as Hongshan Fe–Cu deposit in the southern section of the Yidun island arc belt and medium- and large-sized Cu–Pb–Zn deposits in the Bomi-Tengchong continental margin arc belt, such as Diantan deposit, Dadongchang deposit and Dakuangshan deposit, which were developed in an island arc-type mesogenic-hypogenic intrusive rock complex by the magmatic hydrothermal fluid circulation system in a compressive island arc tectonic environment.

The island arc or continental margin arc zones in the Sanjiang area have different types and scales of mineralization, but compared with the huge porphyry copper metallogenic belts in the Pacific Rim island arc belt and the large-scale Au metallogenic belts in its back arc-basins, none of the Leiwuqi-Lincang-Menghai island arc, Zaduo-Yunxian-Jinghong continental margin arc and Jiangda-Weixi-Lvchun continental margin arc has had any large-scale porphyry copper deposit or epithermal Au mineralization. Fundamentally, this is because the rapid subductions of a series of small back arc ocean basins led to the fast-paced development and formation of the multiple alternating immature island arcs or continental margin arcs and their back arc-basins; even though there are VMS deposits and porphyry copper deposits in the Yidun island arc belt, they are obviously segmented in space. The former occurred in the extensional volcanic arc in the northern part of the island arc belt, while the latter occurred in the compressive magmatic arc in the southern part of the island arc. A comprehensive analysis shows that the vertical tearing and differential subduction actions of the ocean crust plates of the small back arc ocean basins are the main reason for the segmented occurrence of VMS sulfide deposits and porphyry copper deposits in the same island arc belt.

4.5.2.4 Metallogenic Mechanism During Arc-Arc/continent Collision Period

During the Early and Middle Triassic, Sanjiang Tethys entered the arc-arc/continent collision stage after a series of small back arc ocean basins behind the frontal arc subducted and closed and the ocean crust destroyed except that the main body of the Ganzi-Litang back arc ocean basin was still in a spreading environment. The beginning of the stage is marked by the development of multiple alternating collision-type island arcs and the formation of the foreland basin (peripheral or back-arc). Among them, the Ganzi-Litang back arc ocean basin entered the arc-continent collision stage after its subduction and closure in the late Triassic.

Because of being generally restricted by the continuous eastward subduction of Tethys Ocean, Wulanwula-Lancangjiang Back Arc Ocean Basin extincted and the Leiwuqi-Lincang-Mengdao island arc collided with the Zaduo-Zhuka-Yunxian-Jinghong continental margin arc in Early and Middle Triassic, with the result that an ophiolic melange belt and the Qamdo-Lanping-Pu'er back arc foreland basin on its east side were developed. The Jinshajiang-Ailaoshan back arc ocean basin extincted and the Jiangda-Deqin-Weixi-Lvchun continental margin arc collided with Zhongza-Shangri-La Block, thus developing an ophiolitic melange belt and the Qamdo-Lanping-Pu'er back arc foreland basin on its west side. In the late Late

Triassic, the Ganzi-Litang back arc ocean basin extincted and the Yidun island arc collided with Yangtze Plate, thus developing an ophiolitic melange belt and the marginal foreland basin on its east margin. To sum up, the Sanjiang Tethys orogenic belt was not the result of the eastward subduction of Tethys Ocean, but was formed through an island arc orogeny consisting of subductions and extinctions of a series of multiple alternating back arc ocean basins and arc-arc or arc-continent collisions at the southwest margin of Pan-Cathaysian Continent.

The series of collision orogenies involving arc-arc, arc-continent and continent-continent collisions after the Sanjiang Tethys back arc ocean basin subducted and destroyed, as fallouts of the subduction of the back-arc ocean crust, are not only a formation process of the ophiolitic melange belts and collision-type volcanic-magmatic arc belts, but also a process of great adjustments and recombinations of material components. These collision-type arc volcanic-magmatic belts are also important metallogenic belts in the Sanjiang area. They developed many and various minerals and deposits of complex types. By tectonic environment and mineralization type, the deposits of the Sanjiang Tethys orogenic belt can be classified into three types: ① Skarn-type Fe–Cu and Cu–Pb–Zn deposits, such as Hongshan Fe–Cu deposit and Dongzhongda Cu polymetallic deposit in the Yidun island arc belt, Jiaduoling Fe–Cu deposit and Renda Cu–Fe deposit in the Jiangda-Weixi continental margin arc belt, the large-sized Narigongma Cu–Au–Ag polymetallic deposit in the Kaixinling-Zaduo continental margin arc belt, Changdonghe Cu–Pb–Zn deposit in the Lincang-Mengdao island arc belt, etc., each of which was developed in an island arc-type mesogenic-hypogenic intrusive rock complex by the magmatic hydrothermal fluid circulation system under a compressive island arc tectonic environment; ② magmatic hydrothermal vein-type (greisen-Shi Ying vein type) Sn–W polymetallic deposits, such as Bulangshan Sn deposit and Mengsong Sn deposit in the Zaduo-Jinggu-Jinghong continental margin arc belt, each of which was developed in an island arc-type mesogenic-hypogenic intrusive rock complex by the post-magmatic stage hydrothermal fluid circulation system in a compressive island arc tectonic environment; ③ sedimentary Sr or Cu deposit in a foreland basin, such as the large-sized Hexi Sr deposit and a sandstone-type Cu deposit in the Upper Triassic, which were developed through convergence of large quantities of ore-forming materials into Qamdo-Pu'er Basin as a typical back arc foreland basin because of being restricted by the counterthrust framework of the collision-type continental margin arcs on both sides; in addition, the convergence process provided important source beds or rocks for later mineralizations (including the Cenozoic intracontinental convergence and transformation process), such as the Hexi-Lanping-Sanshan-Baiyangping Pb–Zn–Ag–Cu polymetallic enrichment area and the Yunlong-Weishan Sb–Hg–Au–As polymetallic enrichment area.

4.5.2.5 Metallogenic Mechanism During Post-collision Extension Period

After a series of arc-arc or arc-continent collision orogenies in the Sanjiang Tethys orogenic belt, post-collision crust extension environments were developed in the island arc belts or continental margin arc belts and their edge zones in the tectonic background of continuous collisions and convergences generally at the continental margin probably because of inversions or disconnections of subducting back-arc oceanic crust plates. In these environments there occurred mineralizations and corresponding deposit types that were induced by dominant factors such as the influence of the superimposed rift basin, activities of the potassic magma from the crust or from both the crust and mantle and the related hydrothermal fluid system, thus constituting the important Cu–Pb–Zn–Au and W–Sn–Bi–Mo polymetallic deposit belts in the Sanjiang area. This is also a special mineralization process different from the traditional plate tectonic pattern and the trench-arc-basin metallogenic theory.

In the Jiangda-Deqin-Weixi continental margin arc belt and the Yunxian-Jinghong continental margin arc belt respectively on the east and west sides of the Qamdo-Lanping-Pu'er back-arc foreland basin, there occurred a series of nearly NS superimposed graben-type rift basins under post-collision extensions in the Late Triassic, in which bimodal volcanic rock series and potassic intermediate-acid volcanic rock series composed of high-potassium basic rocks and intermediate-acid rocks were developed. By mineralization environment and deposit type, deposits in the two belts can be divided into two types: ① VMS-type Cu–Pb–Zn polymetallic deposits related to the bimodal volcanic rock series, such as Luchun Cu–Pb–Zn polymetallic deposit, Laojunshan Cu–Pb–Zn polymetallic deposit, Hongpo Niuchang Cu–Au deposit, Zuna Ag–Pb–Zn deposit, etc. in the Jiangda-Deqin-Weixi continental margin arc belt and Sandashan Cu deposit, Minle Cu deposit, etc. in the Yunxian-Jinghong continental margin arc belt, which were developed due to activities of the sea-floor potassic volcanoes derived from both crust and mantle materials and their hydrothermal fluid systems. ② exhalative-sedimentary Ag-rich siderite deposits related to the intermediate-acid volcanic rock series, such as Zhaokalong Ag-rich Siderite deposit, Chugezha Ag-rich Siderite deposit, etc. in the Jiangda-Deqin-Weixi continental margin arc belt, whose minerals occur in high-potassium intermediate-acidity tuff rich in siderite and which were developed as a result of the activities of the sea-floor potassic volcanoes derived from crust materials and their hydrothermal fluid systems.

In the Yidun island arc belt on the eastern margin of Zhongza-Shangri-La Block and the Lincang-Menghai island arc belt on the west side of the South Lancangjiang ophiolitic melange belt, there occurred a series of potassic intermediate-acidity intrusive complexes in the form of stock or batholith under post-collision extensions from Late Triassic to Early Jurassic, in which there occurred a series of mineralizations and numerous magmatic hydrothermal vein-type (greisen-quartz vein) W–Sn–Bi–Mo polymetallic deposits and epithermal vein-type Ag–Au deposits due to dominant factors including activities of potassic intrusive magmas derived from crust materials or both crust and mantle materials and their hydrothermal fluid systems formed after the magmatic stage. Such deposits include Lianlong Greisen-quartz Vein-type W–Sn–Ag polymetallic deposit, Xiuwacu Greisen-quartz Vein-type W–Mo deposit, Xiasai Hydrothermal Vein-type Super-large Ag–Pb–Zn polymetallic deposit, etc. in the Yidun island arc belt, Tiechang Greisen-quartz Vein-type Sn–W polymetallic deposit, Haobadi Greisen-quartz Vein-type Sn–W polymetallic deposit, etc. in the Lincang-Menghai island arc belt and Lailishan Greisen-quartz Vein-type Large Sn–W deposit, large Baihuanao greisen-quartz vein-type Sn–W and rare metal (Nb–Ta–Rb) deposits, etc. in the Bomi-Tengchong continental margin arc belt.

4.5.3 Preliminary Discussion on Archipelagic Arc-Basin Metallogenic Theory

4.5.3.1 Foundation of Archipelagic Arc-Basin Metallogenic Theory

In the Sanjiang Tethys tectonic-mineralization domain in the east of Qinghai-Tibet Plateau, which is the epitome of the complex evolution of geological processes and mineralizations in Chinese mainland and has experienced superpositions and transformations imposed by the Late Paleozoic-Mesozoic Tethys accretionary orogeny and the Cenozoic continental collision orogeny, there occurred large-scale multi-phase mineralizations and huge metallic mineral accumulations leading to the occurrence of economic deposits so that it is one of China's most important polymetallic mineralization provinces or enriched areas that is composed mainly of nonferrous and precious metal deposits. According to the traditional trench-arc-basin metallogenic theory, Cyprus-type VMS deposits were developed in trenches marked by ophiolitic melange and VMS, porphyry copper deposits and epithermal gold deposits were developed in island arcs or continental margin arcs. In the Sanjiang area, however, there are up to four ophiolitic melange belts with shear zone gold deposits rather than Cyprus-type copper deposits, and neither porphyry copper deposits nor epithermal gold deposits have been formed in the

Jinshajiang and Lancangjiang arcs so far. Though there are VMS and porphyry copper deposits in the Yidun Island arc, they have obvious zoning. The former occurred in the northern tensional arc, and the latter in the southern compressive arc. Therefore, the traditional trench-arc-basin metallogenic theory cannot explain the mineralization pattern of Sanjiang Paleo-Tethys. A systematic dissection of the important metallogenic belts in the Sanjiang area shows that the Paleozoic-Mesozoic mineralization characteristics in this area result from the archipelagic arc-basin tectonic background and the important constraints of the mineralization environment created by it. The short-duration development of the limited ocean basin formed by the back arc expansion is the fundamental reason for the absence of Cyprus-type copper deposits; the subtraction orogeny of the back-arc-basins and the fast-paced formation and immature development of the multiple island arcs are the inherent reasons why some arcs lack porphyry copper deposits or epithermal gold deposits; the vertical tearing and differential subduction of the oceanic crust plate during its subduction are the main reasons for the zoned occurrence of VMS and porphyry copper deposits in the same island arc belt. Especially, the VMS or SEDEX deposits developed in large quantities in the post-collision extension environments in the volcanic arc belts cannot be explained by using the single trench-arc-basin metallogenic theory according to the traditional plate tectonic model. In this archipelagic arc-basin tectonic setting, VHMS is the most important mineralization type, with a Dapingzhang-type deposit in a marginal basin (D-C), a Yangla-type deposit in an oceanic arc (P_{1-2}), a Gacun-type deposit in an inter-arc rift basin (T_3) and a Luchun-type VMS deposit in a post-collision extension basin (T_3); porphyry copper deposits are the second important mineralization type, which was mainly developed in fast-paced and immature volcanic-magmatic arcs while VHMS was mainly developed in mature magmatic arcs with long-term stable development; SEDEX polymetallic mineralizations mainly occurred in the basins (such as Lanping-Pu'er Mesozoic Basin) developed on stable blocks under the post-collision extension (T_3) in the MABT and mainly served as important protores for Himalayan mineralizations and the materials foundation for the superimposed mineralizations.

Many regional geological surveys and mineral exploration evaluations show that the complex MABT tectonic-paleogeographic framework of the Sanjiang area controls the temporal and spatial distributions of the deposits therein. The mineralization processes of its large and super-large deposits feature obvious zoning, segmentation, diversity and superposition. The main deposit types here include VMS type, porphyry type, SEDEX type, tectonic hydrothermal type, etc. The metal assemblages involved include Pb–Zn, Pb–Zn–Cu–Ag, Cu and Sr–Ba. As the

MABT evolved, the framework became complex. These new findings have greatly deepened the understanding of the archipelagic arc-basin tectonic model, providing an important foundation for establishing the archipelagic arc-basin metallogenic theory.

4.5.3.2 Definition of Archipelagic Arc-Basin Metallogenic Theory

Sanjiang is one of the most complex orogenic belts in the world. It has experienced not only the tectonic evolution of Tethys but also the strong transformations of the Indian-Eurasian plate collisions and plateau uplifts, featuring complex geological structures, strong magmatic activities, active ore-forming fluids and various complex mineralizations. With the tectonic evolution, different metallogenic systems developed here. A systematic dissection of multiple ophiolitic melange belts and multiple arc-basin systems and a comparison with the arc-basin tectonic systems in Southeast Asia showed that most of the basin prototypes represented by these ophiolitic melange belts only had the characteristics of small ocean basins, back arc-basins and island arc marginal seas, with short-duration developments. However, these arc-basin systems have different origins and evolution processes, so the archipelagic arc-basin tectonic model was put forward. The Archipelagic arc-basin tectonic system refers to a complex tectonic system consisting of the frontal arc and the behind-frontal arc island arcs, volcanic arcs, sea ridges, island chains, seamounts, microblocks, back-arc oceanic basins, inter-arc-basins and marginal sea basins. The model emphasizes that the orogenies involved were mainly realized by the oceanic crust subduction or uplift, arc-arc collision, arc-microblock collision and continent-continent collision caused by the back-arc-basin subduction.

Archipelagic Arc-basin Metallogenic Theory refers to a deposit development pattern formed in a specific structural, tectonic and time evolution sequence during the continent convergence accompanied by the tectonic evolution of the MABT at the edge of the continent. Its main points are as follows: Sanjiang Paleo-Tethys was not a simple trench-arc-basin system, but a complex tectonic system composed of a series of multiple alternating arcs, small blocks and small ocean basins; during the evolution, new arcs were constantly born and small ocean basins constantly became extinct; the joining together orogeny of blocks was inherently not caused by the subduction of Tethys Ocean, but by the subductions and closures of a series of small ocean basins; different metal deposits were developed in different ocean basins, arcs and microblocks. The archipelagic arc-basin metallogenic theory reveals the genetic mechanisms of the deposits in the various metallogenic belts during this mineralization period, involving metallogenic mechanisms, patterns, systems and types.

4.5.3.3 In-Depth Study of Archipelagic Arc-Basin Metallogenic Theory

A tectonic analysis of the MABT can further deepen the understanding of the ore-forming geological background and mineralization pattern of the orogenic system. Based on the MABT's tectonic pattern, a summary of the archipelagic arc-basin metallogenic theory was given; a comprehensive evaluation model (Fig. 4.32) for different tectonic environments and metallogenic systems was put forward by combining the MABT's pattern with the metallogenic system theory, process theory and transformation theory (Pan et al. 2003), e.g., the Dapingzhang-type VMS deposit in the continental marginal rift basin, the Yangla-type VMS deposit in the oceanic arc, the Xiacun-type deposit in the inter-arc rift basin, the Pulang-type porphyry Cu deposit in the compressive arc and the Jiaduoling-type skarn-type Fe–Cu deposit in the continental margin arc, These theoretical models have achieved good results in the study of mineralization regularity, mineral exploration and prediction and evaluation of the Sanjiang Tethys orogenic belt.

Then, an in-depth study on the coupled evolution and mineralization processes of the marginal MABT in the early Paleozoic MABT of Qinling, Qilian Mountains and Kunlun Mountains, the Mesozoic MABT of Gangdise in Tibet, and even those of the continental margins on both sides of the ancient Asian Ocean might enrich and improve the archipelagic arc-basin metallogenic theory and be of great significance to the study of the ore-forming geological background of the orogenic system. The following aspects need to be further studied: ① determine the ocean basin attributes represented by each ophiolitic melange belt (back-arc ocean basin, inter-arc-basin or marginal sea basin, etc.) and restore the spatial-temporal structure and evolution of the formations, subductions and closures of these ocean basins; ② the basement properties (oceanic crust, continental crust, accretion wedge, etc.), mechanical properties

(tensile arc, neutral arc, compressive arc) and arc-making evolution (initial arc to mature arc) of volcanic-magmatic arcs; ③ deepen geostructure background studies on metallogenic processes of accretionary orogeny belts (deep crust-mantle processes during the arc-arc, arc-continent and arc oceanic island (seamount) collage collision orogenies, esp. collage collision orogenies that made ore-forming materials of magmatic and volcanic arcs reactivate and migrate again are favorable for large-scale mineralizations).

4.6 Preliminary Discussion on Intracontinental Tectonic Transition Metallogenic Theory

A deposit is essentially a product of metallic ore-forming material supernormal enrichment via a magmatic system, magmatic hydrothermal system and hydrothermal system. A mineralization system generally requires a specific tectonic setting and a specific geological environment. In essence, its being driven by abnormal heat energy and/or tectonic stresses from lithosphere scale to crustal scale is related to processes caused by continental lithosphere convergence, splice, delamination, spreading, etc., such as compression, extension, strike-slip and shear. Thus, a complete metallogenic theory needs to clarify three key scientific problems: ① dynamic background and driving mechanism of mineralizations; ② basic types and development mechanisms of metallogenic systems; ③ huge metal accumulation processes and metallogenic mechanisms of metallogenic systems.

Fine dating of the important mineralization events and a study of the temporal and spatial distribution of the important metallogenic belts in the Sanjiang area show that distinctive metallogenic systems and large deposits were developed in the tectonic transition environments therein. Through a detailed dissection and a comprehensive study of

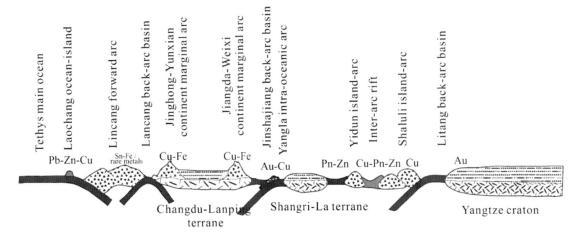

Fig. 4.32 Temporal and spatial structure evolution and mineralization evaluation model of Sanjiang Tethys MABT

its main metallogenic systems, an intracontinental tectonic transition metallogenic theory was put forward. According to the theory, the metallogenic systems in the Sanjiang area mainly including the crust-mantle magmatic hydrothermal metallogenic system, metamorphic fluid metallogenic system and basin fluid metallogenic system, against an intracontinental tectonic transition background, were driven by the abnormal thermal energy from the upwelling deep aesthenosphere materials and the shallow strike-slip, shear and compression tectonic stresses. The core of the intracontinental tectonic transition metallogenic theory is elaborated below.

4.6.1 Tectonic Transition and Mineralization Driving

4.6.1.1 Tectonic Transition Environment for Mineralizations

According to the intracontinental tectonic transition metallogenic theory, the mineralizations were developed in an intracontinental environment and the metallogenic systems were developed in a tectonic transition environment. In the Sanjiang area, the intracontinental tectonic transition environment is manifested as different scales, different main types and different forms. On the lithosphere scale, the Paleozoic-Mesozoic MABT realized the time-space tectonic transition from oceanic lithosphere to continental lithosphere in the Sanjiang area through back-arc-basin shrinkage, arc-arc collision and arc-continent collision in Paleozoic and Mesozoic and continental collision and intracontinental convergence in Cenozoic; on a large regional scale, along with the large-scale collisions between Indian and Asian continents since 65 Ma ago, large-scale tectonic transition systems were developed in the Sanjiang area, thus leading to wide-angle rotations and short slips of spliced blocks (Wang et al. 2006, 2008) and crustal shortening (Wang et al. 2001; Wang and Buchifel 1997; Liu 2006; He et al. 2009) and realizing the adjustment and transformation of the stresses and strains caused by large-scale collisions (Dewey et al. 1988; Wang et al. 2001); on the mineralization belt scale, the tectonic transition marked a large-scale thrust-napping system, a strike-slip fault system and a ductile-brittle shear system induced the extremely strong regional magma-fluid mineralization process (Hou et al. 2006). For example, the activities of the potassic mantle-derived magma controlled by a large-scale strike-slip fault system resulted in the formation of the 1000 km-long alkali-rich porphyry belt (Chung et al. 1998; Guo et al. 2005); during the compressive-to-tensile torsion transition period (40–30 Ma), pulsating epithermal emplacements of the ore-bearing porphyry at three stages resulted in three porphyry magmatic hydrothermal metallogenic systems (40 Ma, 36 Ma and

32 Ma) (Hou et al. 2006a). In Lanping Basin, the convergence collision and crustal shortening that began 65 Ma ago led to large-scale thrust-napping structures and a large number of structural traps and migration and convergence of regional basin fluids. The compressive-to-tensile torsion transformation 40–30 Ma ago generated a large number of strike-slip fault systems and strike-slip pull-apart basins as well as large quantities of regional fluid drainage and metal deposits, thus developing basin fluid metallogenic systems (Hou et al. 2008a). In the Ailaoshan Belt, large-scale strike-slip-shear actions (23–38 Ma) developed the Ailaoshan giant shear zone, and the ductile shear zone and the brittle-ductile transition portion controlled the development of the gold fields and deposits.

4.6.1.2 Drive of Abnormal Thermal Energy for Mineralizations

A comprehensive analysis of the intracontinental transition mineralizations in the Sanjiang area shows that they have four important characteristics: ① they were mainly developed in the discontinuous potassic igneous province with a peak age of 35 ± 5 Ma and were closely related to mantle-derived or crust/mantle-derived magmatic activities; ② the ultimate sources of the ore-forming materials (metals, fluids and gases) were closely related to deep materials, esp. mantle-derived magma; ③ the formations of the alkali-rich porphyry related to Cu–Mo–Au mineralization, the alkaline rock-carbonatite related to REE and the lamprophyre related to Au deposits in the shear zone were closely related to the deep asthenosphere activities; ④ the mineralizations mainly occurred 40–21 Ma ago, and among them, the porphyry Cu–Mo–Au mineralization, REE mineralization, hot brine Pb–Zn–Ag–Cu mineralization and partial shear zone Au mineralization mostly occurred 35 ± 5 Ma ago. These characteristics imply that the magma-hydrothermal-mineralization process in the intracontinental transition environment in the Sanjiang area was controlled by the unified deep process, and the abnormal thermal energy caused by asthenosphere upwelling might be the dynamic mechanism driving the formation and development of the metallogenic system.

Seismic tomography shows that in the lower part of the intracontinental transition orogenic belt (97°–99° E) at the eastern margin of the plateau there occurred an obviously low-velocity asthenosphere upwelling body from the depth of 450 km, which was necked at 200–250 km and continued to flow upward, causing the overlying lithosphere thin to 70–80 km (Liu et al. 2000). In the lower part of the Tengchong modern volcanic area, the underplating asthenosphere materials even thermally eroded the lithospheric mantle locally (Zhong et al. 2001). Spatially, the asthenosphere upwelling body is not in the shape of a large mushroom, but in the shape of a zonal vertical tile plate, which spreads along the NNW intracontinental transition orogenic belt (Zhong et al. 2001).

The formation of the asthenosphere upwelling body might be related to opposite block subductions. Of course, the NE intracontinental subduction of Lhasa Block (Wang et al. 2001) might also function as a subduction plate of the Indian continent when the Indian continent thrusted northeastward and subducted and converged obliquely. Its convergence with Yangtze Block and its intracontinental subduction also might induce the development of the asthenosphere upwelling body.

The asthenosphere upwelling provides a reasonable explanation not only for the formation of the alkali-rich porphyry (Zhong et al. 2001), but also for the development of the whole discontinuous potassic igneous province (Guo et al. 2005; Hou et al. 2006b). Under the asthenosphere tectonic thermal erosion and the injection of small melts, partial melting occurred in the crust-mantle transition zone that was once subjected to the strong metasomatism of the pristine oceanic crust plate fluid. The melting of the phlogopite peridotite in the lower part of the transition zone created Au-bearing syenite magma while the melting of the amphibolite eclogite in the lower crust created Cu-bearing adakitic-like magma (Hou et al. 2005). The emplacements of these porphyry magmas along the strike-slip faults and their intersections with the basement faults and the magmatic fluids segregated in the local tension and stress release environment developed a porphyry magma-hydrothermal metallogenic system (Hou et al. 2003a, b). The asthenosphere upwelling also caused the enriched mantle containing deep crustal cycle materials to melt into the CO_2-rich silicate melt mass, which formed into syenite-carbonatite due to unmixing (Hou et al. 2006b) and derived an REE-rich ore-forming fluid, thus developing a carbonatite magma-hydrothermal REE metallogenic system (Hou et al. 2009).

4.6.1.3 Drive of Tectonic Stresses for Mineralizations

In the Sanjiang intracontinental tectonic transition environment, there were at least three different tectonic deformations delivering tectonic stresses driving the formation and development of the metallogenic systems.

(1) Thrust nappe structural system, which, as a thin-skinned structure caused by collision orogeny and crustal shortening, was developed mainly in Lanping and Yushu's Tuotuohe Area. Through a series of thrust faults, the Mesozoic strata were cut into tectonic slices stacked in turn and were pushed over the sedimentary strata of the foreland basin. In the Lanping area, the thrust nappes advanced from both sides of the basin to the center, forming a counterthrust nappe structural system covering the main body of Lanping Basin. In the Yushu area, the Carboniferous thrust rock plate thrusted northward to form an uplift belt, and a front fold-thrust belt was formed due to the strong compression by Cenozoic in the Jiezha-Xialaxiu Triassic back-arc foreland basin (Hou et al. 2008a). The large-scale thrust nappe structure is characterized by episodic thrusts. The early and late episodes occurred respectively 55–50 Ma and 40–37 Ma ago (Li et al. 2006) and correspond to the main collision period (65–41 Ma) and late collision period (40–26 Ma) of Qinghai-Tibet Plateau, respectively (Hou et al. 2006a, b).

Controlled by a series of large thrust nappe structural systems, an MVT-like Pb–Zn polymetallic mineralization belt with a length of more than 1000 km has begun to appear (Hou et al. 2008a). Its representative deposits include Chaqupacha Pb–Zn deposit in the Tuotuohe area, Dongmozhazhao and Mohailaheng Pb–Zn deposits in the Yushu area and Jinding Super-large Pb–Zn deposit and Baiyangping Large Ag–Pb–Zn deposit in the Lanping area. These thrust nappe tectonic systems, as the main driving mechanism of the basin fluid metallogenic system, played three important roles: ① regional compression and thrust processes drove long-distance lateral migration and convergence of basin fluids (Oliver 1992; Deming 1992; Garven 1993); ② the deep detachment and decollement belt of the thrust nappe structure provided the optimal channel for large-scale basin fluid migration (Hou et al. 2008a); ③ the main thrust faults provided the communication means for vertical fluid migration in the basin, while various shallow thrust structures provided main places for fluid convergence (He et al. 2009).

(2) Large-scale Strike-slip Fault System The large-scale strike-slip fault system in the Sanjiang area is composed of multiple faults, including Jiali-Gaoligong Fault, Gongjue-Mangkang Fault Batang-Lijiang Fault (north section), Honghe Fault (south section), Kunlun Fault, Xianshuihe Fault and Xiaojiang Fault (from west to east and from north to south). They controlled not only the distributions of the Cenozoic alkali-rich intrusive rock belt and carbonate-syenite complex belt (Chung et al. 1998; Wang et al. 2001; Hou et al. 2003a, b, 2006a, b), but also the development of a series of strike-slip pull-apart basins (Liu et al. 1993a, b).

An analysis of the regional stress field shows that the Sanjiang area was in a compressive torsion state 40 Ma ago and changed into a tensile stress state in the Miocene (17–24 Ma) (Wang et al. 2001). The strike-slip fault system occurred 42 Ma ago and was highly active 30–40 Ma ago (Liang et al. 2008). The large-scale strike-slip fault system developed in the compressive-to-tensile torsion transformation period (40–30 Ma), as the tectonic driving mechanism of the magmatic hydrothermal metallogenic system, played three roles: ① the deep crust shear strains induced by the strike-slip faults would cause the crust- and mantle-derived melt to accumulate in the shear zone and rise along the diapir in the vertical shear zone (Sawyer 1994), with the vertical tensile torsion in the upper crusta providing an important channel for upward magma intrusion; ② The stress relaxation during the compressive-to-tensile torsion transformation period led to the pulsating emplacement of the lower magma, thus forming a large-scale, chronically active and continuously replenished magma chamber, so as to provide enough metals and fluids for the metallogenic system; ③ the stress relaxation and strike-slip pull-apart in the shallow crust led to pulsating segregation, massive exsolution and episodic drainage of the ore-forming fluids in the magmatic chamber, thus inducing the magmatic hydrothermal metallogenic system (Hou et al. 2003a, b).

(3) Large-scale Shear System This system was developed in its entirety along the Ailaoshan-Honghe fault zone, thus forming the famous large-scale Honghe Shear Zone. This shear zone might be developed at the tectonic boundary between Yangtze Block and the Sanjiang orogenic belt as the deep system of the large-scale strike-slip fault zone and experienced an early left lateral strike-slip and a late right lateral strike-slip (Tapponnier et al. 1990). The left lateral strike-slip-shear process started 42 Ma ago and continued until 23 Ma ago (Liang et al. 2008). Along the large shear zone, not only lamprophyre aged 35 ± 5 Ma was developed in large quantities and large-scale gold mineralizations occurred simultaneously to form the Ailaoshan gold deposit belt (Hu 1995).

The large-scale shear system, as the tectonic stress-based driving mechanism of the Au metallogenic system, had three functions: ① as a translithospheric fault zone with a history of multiple activities and reactivations, the large-scale shear system is a wide-angle inversion fault system and a thrust nappe shear zone (Li et al. 1999). It not only has the basic characteristics of large scale, deep cutting and good structural network connectivity, but also caused the Paleozoic Au-bearing ophiolite melange to suffer from intense metamorphism and transformation. In addition, it featured a high heat flow state due to the mantle-derived magma emplacement and provided the conditions of strong fluid activities and sufficient mineral supplies required by the orogenic-type gold metallogenic system (Barley and Groves 1992; Goldfarb et al. 2005; Groves et al. 2005); ② metamorphic fluids with rich CO_2 and low salinity (<6%) were often generated through metamorphism by large-scale shears (Kerrich et al. 2000), thus providing enough ore-forming fluids for the orogenic-type gold metallogenic system; ③ the secondary structures in the large-scale shear system, esp. the transformation portion or intersection zone between the ductile shear zone and brittle inversed fault, provided important space for ore-forming fluid convergence and gold mineral accumulation (Hu 1995; Sun et al. 2009).

4.6.2   Metallogenic Systems and Typical Deposits

The activities of the crust- and mantle-derived magma in the intracontinental tectonic transition environment, large-scale strike-slip, shear, thrust and nappe structures and fluid activities induced by them developed at least three important metallogenic systems: ① the magmatic hydrothermal Cu–Mo–Au metallogenic system (Rui et al. 1984; Tang et al. 1995; Hou et al. 2003a, b, 2009); ② the orogenic-type Au metallogenic system related to the shear zone (Hu 1995; Xiong et al. 2006; Hou et al. 2007a); ③ the Pb–Zn–Ag–Cu brine fluid metallogenic system controlled by the thrust nappe structure (Hou et al. 2008a) (Table 4.1).

4.6.2.1   Magmatic Hydrothermal Cu–Mo–Au Metallogenic System

The magmatic hydrothermal Cu–Mo–Au metallogenic system developed in the intracontinental tectonic transition environment is strictly controlled by the large-scale strike-slip fault system. The acidic granitic porphyry developed Cu or Cu–Mo deposits, such as the Yulong porphyry Cu deposit belt (Hou et al. 2003a, b); the syenite porphyry or monzonitic porphyry developed Au or Au–Cu deposits, such as Beiya Au deposit (Xu et al. 2007). Cu-bearing porphyry is geochemically similar to adakite (Hou et al. 2003a, b; Jiang et al. 2006) and Au-bearing porphyry is characterized by high potassium and rich alkali. Both of the two magmatic systems are characterized by high f_{O_2} (Liang et al. 2006) and originated from the rich mantle or crust/mantle transition zone (Chung et al. 1998; Hou et al. 2003a, b, 2005; Jiang et al. 2006). The molybdenite Re–Os dating data show that the porphyry Cu–Mo and Cu–Au metallogenic systems had three mineralization peaks (40 Ma, 36 Ma and 32 Ma; Hou et al. 2006a), equivalent to the diagenetic ages (41–27 Ma). These porphyry copper deposits in the intracontinental environment are very similar to the typical porphyry copper

Table 4.1 Environments and deposit types of late collision transition mineralizations

Tectonic environment	Rock assemblage	Stress state of mineralization environment	Metallogenic system	Type of deposit	Metal assemblage	Example
Strike-slip fault zone	Monzogranite porphyry	Compressive torsion stress of intracontinental transition	Magma-hydrothermal	Porphyry-type	Cu–Mo	Yulong
	Potassium-rich syenite porphyry	Compressive torsion stress of intracontinental transition	Magma-hydrothermal	Porphyry-type	Au(Cu)	Beiya
	Carbonate-syenite complex	Compressive torsion stress of intracontinental transition	Magma-hydrothermal	Complex type	REE	Haoniuping
Shear zone	Shear tectonic zone	Compressive torsion stress of ductile/brittle transition	Metamorphic-hydrothermal	Metamorphic hydrothermal type	Au	Ailaoshan
Overthrust zone	Sandstone formation	Structural trap in front zone	Regional brine	Jingding type	Zn–Pb	Jinding
	Carbonatite formation	Pop up structure in front belt	Regional brine	MVT-like type	Pb–Zn	Sanshan
	Continental clastic rock	Secondary fault in front zone	Regional brine	Hydrothermal vein type	Ag–Pb–Zn	Fulong Factory

deposits in the magmatic arc environment in terms of alteration zoning, mineralization characteristics, mineral assemblage and fluid system (Rui et al. 1984; Tang et al. 1995; Hou et al. 2003a, b, 2007b), indicating that both of them have similar mineralization processes.

4.6.2.2 Orogenic-Type Au Metallogenic System

The orogenic-type Au metallogenic system developed in the intracontinental tectonic transition environment is strictly controlled by the large-scale shear tectonic belt. The Ailaoshan mineralization belt is a typical product of the metallogenic system. Dependable dating data and deposit distribution characteristics show that the Ailaoshan mineralization belt was controlled by the Ailaoshan shear zone, the gold fields and deposits were controlled by the intersection of the NW brittle shear zone and the nearly EW thrust fault zone, individual deposits or ore bodies are controlled by brittle-ductile shear zones of different lithologic characters (Hu 1995). The mineralization ages mainly range from 35 to 40 Ma (Sun et al. 2009). The main gold-bearing formation is the Paleozoic ophiolite melange, which suffered greenschist facies metamorphism. The main ore bodies include the Au-bearing quartz veins filling fracture zones and the Au-bearing tectonic altered rocks in metasomatic surrounding rocks (Hu 1995). The available hydrogen, oxygen and sulfur isotopic data show that the ore-forming fluids were mainly metamorphic fluids containing meteoric water, while S and Au came from ore-bearing metamorphic mafic formations (Hu 1995; Xiong et al. 2006). The He–Ne–Xe isotopic evidence shows that mantle gas contributed to the ore-forming fluids (Hu et al. 1999; Sun et al. 2009).

4.6.2.3 Regional Pb–Zn Polymetallic Brine Metallogenic System

In the Sanjiang area, the regional brine Pb–Zn polymetallic metallogenic system was mainly developed in a tectonic transition zone and controlled by the large-scale regional thrust nappe and strike-slip pull-apart basin. According to the ore-controlling patterns and mineralization characteristics of the thrust nappe structures, three deposit models can be identified: Jinding-type Pb–Zn deposit developed in the "structural dome + lithologic trap" assemblage in the front zone of the thrust nappe structure system (Wang et al. 2007; Xue et al. 2007), Hexi-Sanshan-type Pb–Zn–Ag deposit controlled by the decollement structure in the front zone and Fulong Factory-type Ag–Pb–Zn–Cu deposit developed in a secondary or translation fault of the main thrust fault zone (Hou et al. 2008a). The distribution of these deposits in the region was controlled by the thrust nappe structure, but their locations were controlled by the extension structure. Most of the deposits and ore bodies here are vein-like and a few are plate-like shape; some are located in the continental siliceous clastic rock formations and some in carbonate rock formations, showing strata-bound characteristics, but the mineralizations are characterized by open space filling (Hou et al. 2008a). The ore-forming fluids show basin brine characteristics, with relatively low temperatures (mostly between 80 and 190 °C). The metal assemblages of the deposits might be related to the properties of the rocks experienced during the migration-convergence process of the ore-forming fluids (He et al. 2009). Its overall characteristics are different from those of MVT, SEDEX, Ireland and Laisvall Pb–Zn deposits (Xue et al. 2007; Hou et al. 2008a).

It is obvious that the intracontinental tectonic transition mineralization featuring Different metallogenic systems (magma-hydrothermal metallogenic system, metamorphic-hydrothermal metallogenic system and regional brine metallogenic system), super-large-scale metal accumulations and large and super-large deposits of different metal assemblages cannot be covered by the existing mineralization models or explained by any existing metallogenic theory.

4.6.3 Metallogenic Systems and Development Mechanisms of Large Deposits

A collision and continuous subduction of a continent inevitably would inevitably lead to dramatic lithosphere shortening and strain, which would be inevitably adjusted through tectonic transitions, as through faulting. Such a tectonic transition adjustment zone was mainly developed on a flank of a forward collision zone, such as the eastern margin of Qinghai-Tibet Plateau (Dewey et al. 1989), and was characterized by large-scale strike-slip fault system (shear and escape), thrust nappe structure system (internal deformation) and block rotation. In the tectonic transition adjustment zone, the internal deformation in the block, as an important way to adjust the collision strain, often forms a thrust nappe structural system or a compressive fold zone. For example, Lanping Basin had characteristics of foreland basins in Paleogene and Neogene (Wang et al. 2001) and the regional compressional torsion since Eocene makes the Mesozoic stratigraphic

system before collision overlap the foreland basin, forming a thrust nappe structural system (He et al. 2009). Driven by the gravity or compressional force of the orogenic belt, the regional fluids (or the basin brine) traveled a long distance along the detachment and decollement zone under the thrust nappe structure toward the foreland basin and leached the metallic materials in the surrounding rocks during their migration process, into low-temperature mineralizing fluids with high salinity and rich metal elements (Hou et al. 2008a). During the thrust-strike-slip transformation period, these ore-forming fluids were rapidly drained under compressive and tensile torsion stresses and converged into the secondary structures through the main thrust fault or a strike-slip fault (such as a structural dome, tensile structure, recoil structure, recoil structure, interlayer decollement structure), thus developing the Pb–Zn polymetallic fluid metallogenic system as well as various Pb–Zn–Cu–Ag deposits such as MVT-like Pb–Zn deposits in carbonatite formations, Zn–Pb deposits in clastic rock formations and Cu–Ag deposits in red beds (Fig. 4.33; Hou et al. 2008a).

Continuous intracontinental subduction might induce the upwelling of the asthenospheric materials under SCLM (Zhong et al. 2001), thus providing necessary heat energy for partial melting of SCLM, while the strike-slip faults cutting the lithosphere deep accelerated the decompressional melting of SCLM (Fig. 4.33; Hou et al. 2005), thus forming igneous provinces (belts) dominated by mantle-derived magma and crust/mantle-derived magma in the tectonic transition zone. These potassic magma originated from

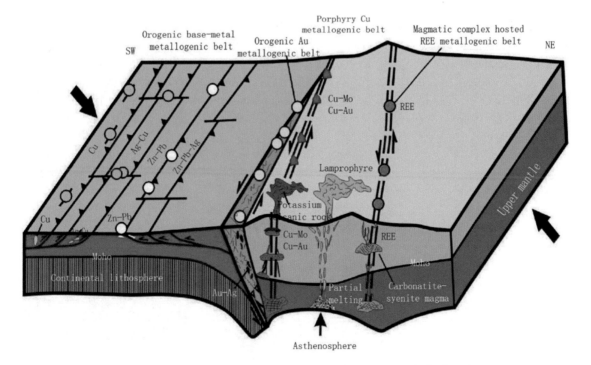

Fig. 4.33 Intracontinental Tectonic transition and development of different metallogenic systems in Sanjiang Area

SCLM enriched mantle or crust-mantle transition zone (Chung et al. 1998; Wang et al. 2001; Hou et al. 2003a, b; Guo et al. 2005; Jiang et al. 2006) may rise along the ductile shear zone deep in the strike-slip fault system in the deep crust, showing dike-like ascending emplacement; Large magma chambers were developed in the shallow crust, which were transported by strike-slip faults (Richard 2003). In the local tension and stress release environment, the ore-forming magma fluid was segregated from the felsic magma to develop the porphyry magmatic hydrothermal metallogenic system (Hou et al. 2003a, b). The Cu-rich fluid was segregated from the monzogranite porphyry magma from the crust-mantle transition zone to develop the porphyry copper deposits (Hou et al. 2007b); the Au-rich fluid was segregated from the syenite porphyry magma from the enriched mantle to develop the porphyry gold deposits (Xu et al. 2007). The SCLM's enriched portion that suffered the deep crust materials cycle melted to produce the CO_2-rich silicate melt (Hou et al. 2006b), from which the REE-rich ore-forming fluid was derived due to its unmixing property during its rising process (Niu and Lin 1994; Xie et al. 2009; Hou et al. 2009), thus developing the REE deposits related to the carbonate-alkaline rock complex (Fig. 4.33; Yuan et al. 1995; Yang et al. 2001; Hou et al. 2008b).

In the deep system of a large-scale strike-slip fault zone, there mainly occurred ductile and ductile-brittle shear zones, accompanied by mantle-derived magma emplacement and granulite-greenschist facies metamorphism. This strong shear and metamorphism process produced the CO_2-rich metamorphic ore-forming fluid, which rose along the ductile shear zone and developed vein-type Au–As deposits in deep ductile-brittle transformation structures and Au–Sb deposits in shallow brittle fissures (Fig. 4.33; Goldfarb et al. 2001; Sun et al. 2009).

4.6.4 Definition of Intracontinental Tectonic Transition Metallogenic Theory

In the Sanjiang area, the large-scale Cenozoic mineralizations developed many large and super-large deposits (commonly known as "late bloomers"), which occurred in the dynamic background of intracontinental transformation. All of its MABT, collision orogenic belt and metallogenic belts show the abnormally strong regional magma-fluid mineralization process induced by the tectonic transition. In Lanping Basin, the large-scale thrust-napping process that started 65 Ma ago developed a large number of structural traps, and migrations and convergences of large amounts of fluid during the collision and compression period as well as fluid drainages and metal depositions during the stress release period resulted in the development of Jinding- or Baiyangping-type Pb–Zn and silver deposits. At the eastern margin of the plateau, the magmatic activities controlled by the large-scale strike-slip

Fig. 4.34 Cenozoic strike-slip-shear action and metallogenic systems in Sanjiang Orogenic Belt

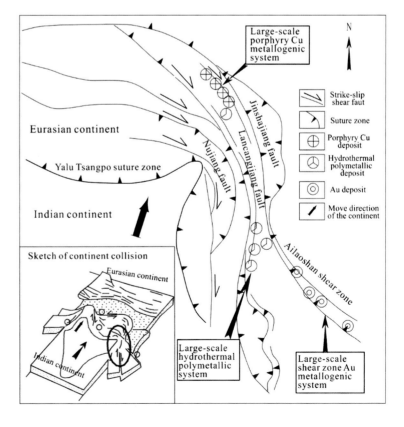

fault system developed a 1000 km-long large-scale alkali-rich porphyry belt; during the compressive-to-tensile torsion transformation period (40–30 Ma), pulsating epithermal emplacements of the ore-bearing porphyry at three stages resulted in three porphyry hydrothermal metallogenic systems (40 Ma, 36 Ma and 32 Ma), thus developing the porphyry mineralization belt represented by Yulong porphyry Cu deposits and Beiya porphyry Au deposits. In the Tengliang area, the syn-collision granite emplacement (50–65 Ma) controlled by the Gaoligong strike-slip fault system resulted in the Sn polymetallic superimposition mineralization that developed the large Sn deposit cluster superimposed on the Hercynian VMS, which is represented by Lailishan Sn deposit. In the Ailaoshan belt, the large-scale strike-slip-shear process (23–38 Ma) developed the huge Ailaoshan shear zone. The strongly sheared ophiolite melange zone controlled the development of the gold deposit belt, and the ductile-brittle-ductile shear belt transformation site controlled the development of the gold fields and deposits. Based on these systematic studies, we put forward the intracontinental tectonic transition metallogenic theory. The main points are as follows: 65 Ma ago, the Indian and Asian continents collided and uplifted to develop the Qinghai-Tibet Plateau, and with the intensification of the collision, the middle and south sections of the Sanjiang area on the eastern edge of the Qinghai-Tibet Plateau underwent great rotation, great strike-slip, great napping and great fluid migration (Fig. 4.34); large-scale strike-slip faults cut the lithosphere and induced copper-bearing crust/mantle magma activities, thus developing a large porphyry copper deposit system (such as Yulong copper deposit); a large-scale shear led to the formation of a ductile shear zone up to hundreds of kilometers in length, as well as the gold activation to form a large-scale shear zone gold deposit system (such as Zhenyuan copper deposit); the large-scale napping and strike-slip pull-apart developed a series of hydrothermal activity centers and the hydrothermal Ag polymetallic deposit systems (such as Lanping Jinding Pb-Zn deposit and Baiyangping Ag polymetallic deposit). This tectonic transition along with the stress-strain adjustment caused by a large-scale collision induced unusually strong regional tectonic-magmatic-fluid mineralizations. We call this process the intracontinental tectonic transition metallogenic theory.

References

Dewey JF, Shackleton RM, Chengfa C, Yiyin S (1988) The tectonic evolution of the Tibetan Plateau. Philos Trans R Soc Lond A Math Phys Eng Sci 327:379–413

Goldfarb R, Baker T, Dube B, Groves D, Hart C, Robert F (2005) Epigenetic lode gold deposits. In: Hedenquist JW, Thompson JFH, Goldfarb RJ, Richards JP (eds) Economic geology 100th anniversary volume. Society of Economic Geologists Inc., Littleton, Colorado, pp 407–450

Hou ZQ, Hou LW, Ye QT (1995) Tectono-magmatic evolution and volcanic origin of massive sulfide deposits in Yidun Island arc, Sanjiang area. Seismological Press, Beijing, pp 4–134

Hou ZQ, Khin Z, Qu XM, Ye QT, Yu JJ, Xu MJ, Fu DM, Yin XK (2001a) Origin of the Gacun volcanic-hosted massive sulfide deposit in Sichuan, China: fluid inclusion and oxygen isotope evidence. Econ Geol 96(7):1491–1512

Hou ZQ, Khin Z, Qu XM, Ye QT, Yu JJ, Xu MJ, Fu DM, Yin XK (2001b) Origin of the Gacun volcanic-hosted massive sulfide deposit in Sichuan, China: Fluid inclusion and oxygen isotope evidence. Econ Geol 96(7):1491–1512

Hou ZQ, Ma HW, Zaw K, Zhang YQ (2003a) The Yulong porphyry copper belt: product of large-scale strike-slip faulting in Eastern Tibet. Econ Geol 98:125–145

Hou ZQ, Yang YQ, Wang HP, Qu XM, Lü QT, Huang DH, Wu XZ, Yu JJ, Tang SH, Zhao JH (2003b) Collision-orogenic progress and mineralization system of Yidun Terrain. Geological Publishing House, Beijing, pp 1–345 (in Chinese with English abstract)

Huang ZL, Wang LK (1996) Discussion on geochemistry and genesis of lamprophyre in Machangqing Gold deposit, Yunnan. J Mineral Petrol 16(2):82–89

Jiang YH, Jiang SY, Ling HF, Dai BZ (2006) Low-degree melting of a metasomatized lithospheric mantle for the origin of Cenozoic Yulong monzogranite-porphyry, East Tibet: geochemical and Sr-Nd-Pb-Hf isotopic constraints. Earth Planet Sci Lett 241:617–633

Kerrich R, Goldfarb R, Groves D, Garwin S, Jia YF (2000) The characteristics, origins, and geodynamic settings of supergiant gold metallogenic provinces. Sci China (ser D) 43:1–68

Li XZ, Liu WJ, Wang YZ (1999) Tectonic evolution and mineralization of Tethys in sanjiang area, southwest China (general). Geological Publishing House, Beijing, pp 23–167

Liu ZQ, Li XZ, Ye QT (1993a) Division of tectonic magmatic belt and mineral distribution law in sanjiang area. Geological Publishing House, Beijing, pp 1–110

Liu ZQ, Li XZ, Ye QT et al (1993b) Dividing of tectono-magmatic belts and distribution of the ore deposits in Sanjiang region. Geological Publishing House, Beijing, pp 1–246 (in Chinese)

Mo XX, Lu FX, Shen SY (1993) Sanjiang Tethys volcanism and related mineralization. Geological Publishing House, Beijing, pp 1–267 (in Chinese with English abstract)

Ohmoto H (1983) Systmatics of sulfur and carbon isotopes in hydrothermal deposits. Econ Geol 67:551–579

Oliver J (1992) The spots and stains of plate tectonic. Earth Sci Rev 32:77–106

Pan GT, Hou ZQ, Xu Q, Wang LQ, Du DX, Li DM, Wang MJ, Mo XX, Li XZ, Jiang XS (2003) Archipelagic orogenesis, metallogenic systems and assessment of the mineral resources along the Nujiang-Lancangjiang-Jinshajiang area in southwestern China. Geological Publishing House, Beijing, pp 1–420 (in Chinese with English abstract)

Pearce JA (1984) Trace element discrimination diagrams for the tectonic interpretation of granitic rocks. J Petrol 25(4):956–983

Tapponnier P, Lacassin R, Leloup PH (1990) The Ailao Shan/Red River metamorphic belt: tertiary left-lateral shear between Indochina and South China. Nature 343:431–437

Xie YL, Hou ZQ, Yin SP, Tian SH, Xu WY, Simon D (2009) Continuous carbonatitic melt-fluid evolution for a REE mineralization system: evidence from inclusions in the Maoniuping REE deposit in the western Sichuan, China. Ore Geol Rev 36:89–105

Xu MJ, Fu DM (1993) Gacun Ag-Pb-Zn deposit, Sichuan province. Chengdu University of Science and Technology Press, Chengdu, pp 1–180

Xu XW, Cai XP, Zhong J, Song BC, Peters SG (2007) Formation of tectonic peperites from alkaline magmas intruded into wet sediments in the Beiya area, western Yunnan, China. J Struct Geol 29:1400–1413

Yang KH, Hou ZQ, Mo XX (1992) Basic characteristics and main types of volcanic massive sulfide deposits in Sanjiang area. Geol Miner Deposits, 35–44

5.1 Division of Metallogenic Belts

5.1.1 Principles of Metallogenic Division

The metallogenic belts (or regions) of the Sanjiang orogenic domain are divided under the ideas of systematic, hierarchical and correlated metallogenic geological-tectonic unit, based on the spatial-temporal structure of the MABT evolution → intracontinental convergence strike-slip → tectonic transformation orogenic process. Then, the evolutionary-spatial configuration of MABT in the continental margin of Sanjiang orogenic belt and the metallogenic process (generation-development-migration) during the intracontinental convergence strike-slip tectonic transformation orogeny are restored, and the coupling relationship between the evolutionary history of metallogenic-structure environment and ore deposits types in the Sanjiang orogenic domain is revealed. The guiding principles of these works are: (1) plate tectonic-geodynamic theories; (2) geological records of sedimentation, volcanic formation, intrusive activities and metamorphic deformation in different tectonic-metallogenic unties; (3) spatial-temporal structure analysis of metallogenic-tectonic environment of relatively stable blocks and basins of different scales and volcanic-magmatic and ophiolitic mélange belts of different periods; (4) metallogenic regularity research, mineral exploration and prediction evaluation requirements; (5) the basic division principles of sedimentary space-time type composition and existence state caused by major tectonic events in special areas; (6) the reasonable location and properties of ore-bearing geological bodies in Sanjiang region which are screened and stripped by using regional metallogenic and geophysical methods.

5.2 Division Scheme of Metallogenic Belts

Metallogenic belt refers to the region or belt where deposits (or ore spots) are concentrated. In other words, the spatial distribution of mineral resource is controlled by certain geological structures, sedimentary formations and magmatic activities, which is generally related to tectonic units and/or tectonic-magmatic belts. Therefore, mineralization has common characteristics. According to the aforementioned division scheme of tectonic unites and tectonic-magmatic belts, combined with the distribution of deposits (ore spots), the Sanjiang5.1 region is divided into 10 important metallogenic belts. However, the metallogenic belt is equivalent to the third or fourth tectonic unit or tectonic-magmatic belt (Table , Fig. 5.1).

I. Gantzê-Litang Au metallogenic belt
II. Dege-Xiangcheng Cu-Mo-Pb–Zn-Ag polymetallic metallogenic belt
III. Jinshajiang-Ailaoshan Au-Cu-PGEs metallogenic belt
IV. Jiangda-Weixi-Lvchun Fe-Cu-Pb–Zn polymetallic metallogenic belt
V. Changdu-Lanping-Pu'er Cu-Pb–Zn-Ag polymetallic metallogenic belt
VI. Zaduo-Jinggu-Jinghong Cu-Sn polymetallic metallogenic belt
VII. Leiwuqi-Lincang-Menghai Sn-Fe-Pb–Zn polymetallic metallogenic belt
VIII. Changning-Menglian Pb–Zn-Ag polymetallic metallogenic belt
IX. Baoshan-Zhenkang Pb–Zn-Hg and rare metals metallogenic belt
X. Tengchong-Lianghe Sn-W and rare metals metallogenic belt.

In a tectonic unit or tectonic-magmatic belt, due to the long-term, complex and multi-stage characteristics of geological evolution, the present tectonic units usually contain or preserve special deposits of different ages. Therefore, different deposit types and series formed in different evolution periods or stages can be formed as assemblage in the same metallogenic belt. On this basis, the complex multi-stage, multi-genesis, multi-source and multi-type

W. Li et al., *Metallogenic Theory and Exploration Technology of Multi-Arc-Basin-Terrane Collision Orogeny in "Sanjiang" Region, Southwest China*, The China Geological Survey Series, https://doi.org/10.1007/978-981-99-3652-6_5

Table 5.1 Summary for the major deposits in 10 metallogenic belts, Sanjiang region, SW China

Belt No.	Deposit type	Typical deposits	Ages	References
I	Orogenic Au deposits	Gala, Cuo'a, Xionglongxi, Ajialongwa	Zircon fission track: 80–120 Ma	Huan et al. (2011) Zhang et al. (2012) Deng and Wang (2016) Zhang et al. (2015)
		Suoluogou	Biotite K–Ar from lamprophyre: 26.4–26.7 Ma	
II (Northern section)	VMS Pb–Zn-Ag polymetallic deposits	Gacun, Gayiqiong	Zircon U–Pb and whole-rock Rb–Sr: 210–230 Ma	Hou et al. (2004) Xue et al. (2014)
	Skarn or hydrothermal vein-type Sn polymatallic deposits	Xiasai, Xialong, Cuomolong, Lianlong, Zhalong, Jiaogenma, Shaxi	Ar–Ar ages of K-feldspar; Rb–Sr isochronal of intrusions: 73–105 Ma	Qu and Hou (2002) Lin (2010) Wang et al. (2014)
II (Southern section)	Porphyry-skarn Cu deposits	Pulang, Xuejiping, Langdu, Lannitang, Chundu	Zircon U–Pb and Molybdenite Re-Os: 200–230 Ma	Li et al. (2006, 2010, 2011)
	Porphyry-skarn Mo-Cu polymetallic deposits	Xiuwacu, Relin, Hongshan, Tongchanggou, Donglufang	Zircon U–Pb and Molybdenite Re-Os: 73–90 Ma	Li et al. (2013) He et al. (2018) Zhang et al. (2020, 2021)
III	VMS + Porphyry-skarn (superimposed by later magmatic events)	Yangla, Tuoding, Tongjige	Zircon U–Pb: 290–303 Ma Molybdenite Re-Os: 214–239 Ma	Wang et al. (2010) Deng et al. (2014) Li et al. (2015)
	Porphyry-skarn Au deposits	Beiya, Machangqing	Molybdenite Re-Os: 30–35 Ma	Deng et al. (2014) Li et al. (2016a, b, c)
	Orogenic Au deposits	Donggualin, Jinchang, Laowangzhai, Daping	Ar–Ar of lamprophyre and alteration minerals: 32–26 Ma	Li et al. (2010), Deng and Wang (2016)
IV	VMS deposits	Laojunshan, Zhaokalong, Luchun, Hongponiuchang, Chugezha, Dingqinnong,	Molybdenite Re-Os and Zircon SHRIMP: 200–250 Ma	Li et al. (2010) Feng et al. (2011) Deng et al. (2016)
V	Porphyry Cu/Au deposits	Yulong, Mamupu, Bada, Gegongnong	Zircon U–Pb and Goethite (U-Th)/He: 35–42 Ma	Yang et al. (2008) Deng et al. (2017)
	MVT deposits	Jinding, Lanuoma, Zhaofayong, Sanshan, Changdong, Luoboshan	Apatite fission track, paleomagnetic data and zircon SHRIMP: 21–35 Ma	Yalikun et al. (2017) Leach and Song (2019)
VI	VMS deposits	Dapingzhang, Sandashan, Minle	Chalcopyrite/ Molybdenite Re-Os; Zircon U–Pb 410–440 Ma	Li et al. (2010) Lehmann et al. (2013)
VII	Greisen W-Sn deposits	Saibeinong, Xiaya, Duila	Late Yanshanian	Li et al. (2010)
	Greisen Sn-W, Cu (Pb–Zn) deposits	Tiechang, Haobadi, Changdonghe	Triassic	Li et al. (2010)
	BIF (Banded iron formation)	Huimin, Damenglong	Meso-proterozoic	Shen et al. (2008)
VIII	VMS + Porphyry Cu-Pb–Zn polymetallic deposits	Laochang, Tongchangjie, Jinla	Zircon U–Pb: 323 Ma; Molybdenite Re-Os: 43–45 Ma	Li et al. (2009, 2010) Chen et al. (2010)
IX	Skarn Pb–Zn polymetallic deposits	Luzhiyuan, Hetaoping, Jinchanghe, Heiniuwa, Fangyangshan	Calcite Sm–Nd; Sulfide/quartz/K-feldspar Rb–Sr 130–140 Ma	Wang et al. (2014) Xu et al. (2019a, b, 2020)
X	Skarn deposits	Diantan, Dadongchang, Dakuangshan, Huiyao	Late Jurassic-Early Cretaceous	Li et al. 2010 Deng et al. (2014) Cao et al. (2016) Tang et al. (2018)
	Greinsen Sn-W polymetallic deposits	Xiaolonghe, Lailishan, Tieyaoshan, Laopingshan, Baihuanao	Late Cretaceous-Paleogene	

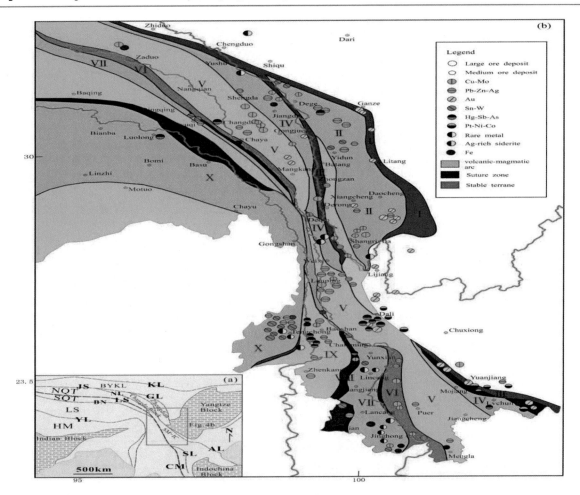

Fig. 5.1 a Tectonic framework of the Eastern Tethys; **b** distributions of major ore deposits and metallogenic belts in the Sanjiang region. Metallogenic belts: I: Ganzi-Litang Au metallogenic belt, II: Dege-Xiangcheng Cu-Pb–Zn-Ag polymetallic metallogenic belt, III: Jinshajiang-Ailaoshan Au-Cu-Pt metallogenic belt, IV: Jiangda-Weixi-Lvchun Fe-Cu-Pb–Zn polymetallic metallogenic belt, V: Changdu-Lanping-Pu'er Cu-Pb–Zn-Ag polymetallic metallogenic belt, VI: Zaduo-Jinggu-Jinghong Cu-Sn polymetallic metallogenic belt, VII: Leiwuqi-Lincang-Menghai Sn-Fe-Pb–Zn polymetallic metallogenic belt, VIII: Changning-Menglian Pb–Zn-Ag polymetallic metallogenic belt, IX: Baoshan-Zhenkang Pb–Zn-Hg and rare metals metallogenic belt, X: Tengchong-Lianghe Sn-W and rare metals metallogenic belt

metallogenic belts are formed in the Sanjiang region (Fig. 5.2).

5.3   Regional Metallogenic Models for the Major Sanjiang Metallogenic Belts

5.3.1   Ganzi-Litang Au Metallogenic Belt (I)

This belt is located in the eastern margin of the Sanjiang region and is developed on the Indosinian (Triassic) Paleo-Tethys combination zone and the Ganzi-Litang oceanic ridge volcanic-ophiolite mélange belt. This belt extends NW-SE for about 800 km from Zhiduo in the northwest, through Ganzi and Litang, and then to Muli in the southeast (Fig. 5.1). The Au mineralization is closely related to the

Paleo-Tethys closure and the Cenozoic strike-slip faulting. The Au deposits are mainly distributed in the middle and southern sections of this belt and include many large and medium-scale ones, e.g., the Gala, Cuo'a, Xionglongxi, Ajialongwa and Suoluogou deposits (Zhang et al. 2012, 2015; Table 5.1).

5.3.2   Geological Evolution and Mineralization

The Ganzi-Litang suture zone, which coincided spatially with the gold belt, was formed by the Zhongzan-Shangri-La microcontinent rifting from the western margin of the Yangtze block to form a small oceanic basin in the Permian-Middle Triassic, and then the oceanic crust subducted westward under the Zhongzan-Shangri-La microcontinent in the Late Triassic, resulting in the closure of the

Fig. 5.2 Comprehensive section of regional tectonics-metallogenic model in the Sanjiang MABT orogenic belt

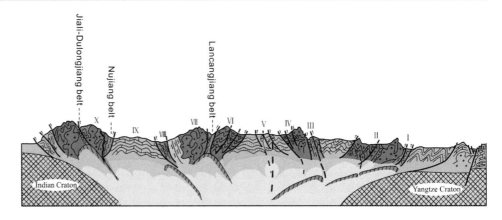

oceanic basin and Yidun arc-Yangtze plate collision. The Dege-Xiangcheng (Yidun) island arc-basins, intra-arc-basins and back-arc-basins were formed to the west of the Ganzi-Litang suture zone. During the development of Ganzi-Litang oceanic basin, the Permian-Late Triassic MORB-type gabbro, diabase dykes, serpentinite, tholeiitic/picritic basalts, rift-type alkaline basalt, mafic-ultramafic cumulates, radiolarian chert and flysch sandstone and slate were widely developed, which constitute an ophiolitic mélange belt (Yan et al. 2020). The Yanshanian (Jurassic-Cretaceous) collisional orogeny along this suture zone occurred after the closure of the Ganzi-Litang oceanic basin at the end of Late Triassic, followed by the superimposed Himalayan strike-slip shearing, accompanied by the formation of many orogenic Au deposits, such as the Gala and Cuo'a orogenic Au deposits (Table 5.1; Hou et al. 2004; Li et al. 2010; Deng and Wang 2016).

Orogenic Au deposits in this belt are mainly controlled by the main fault zone and its secondary fault zones (esp. in fault bends), as well as in the middle section of the suture zone. Spatially, the mineralization occurs in intensely altered fault zone as vein type or altered rock type. Four gold-bearing alteration zones have been discovered (silicic, pyrite, sericite and carbonate) and present some unique characteristics in different ore hosts, which can be used as important prospecting indicators in this region (Huan et al. 2011; Deng and Wang 2016).

5.3.3   Regional Metallogenetic Model

The tectonic evolution of the Ganzi-Litang ophiolitic mélange belt resulted in the gold mineralization in this ore belt. During the formation of the Ganzi-Litang ocean basin, the Au, Fe and other ore-forming elements from the lower crust or mantle were initially enriched in the basic volcanic and sedimentary rocks, and this process continued until the closure of oceanic basin at the end of the Late Triassic (Fig. 5.3a). The Yanshanian orogenic activities formed a

nappe-shear zone along the suture zone. With the evolution of metamorphic fluids, as well as the meteoric water-dominated hydrothermal activities, the volcanic-sedimentary rocks were fragmented, deformed and metamorphosed. The ore-forming elements were mobilized, migrated along and deposited in the nappe-shear zones (Li et al. 2010; Fig. 5.3b).

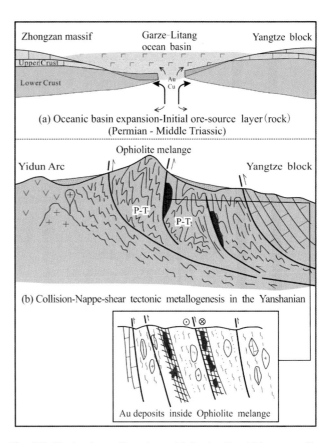

Fig. 5.3 Regional metallogenic model for the Ganzi-Litang metallogenic belt

5.4 Dege-Xiangcheng Cu-Mo-Pb–Zn-Ag Polymetallic Metallogenic Belt (II)

This belt is located near (west of) the Ganzi-Litang suture zone. It is an important ore belt developed in the Late Triassic Yidun island arc, which is consistent with the Changtai-Xiangcheng magmatic belt. It is an NNW-trending tectonic-volcanic belt with a length of 500 km from north to south and a width of 90–150 km from east to west (Fig. 5.1). The arc magmatic activities in the island arc orogenic period and the post-orogenic tectonic activities are intense in this belt. There are abundant mineral resources in the belt, with Pb-Zn-Ag deposits dominating the northern section (e.g., the Gacun, Gayiqiong, Xiasai, Xialong, Cuomolong etc.; Table 5.1) and Cu-(Mo) deposits (e.g., the Pulang, Xuejiping, Langdu, Chuncu, Lannitang porphyry Cu deposits and Hongshan, Xiuwacu, Relin, Donglufang, Tongchanggou porphyry-skarn Mo polymetallic deposits; Table 5.1; Li et al. 2011; Leng et al. 2018) dominating the southern section.

5.4.1 Geological Evolution and Mineralization

The Dege-Xiangcheng island arc was formed by the Late Triassic west-dipping subduction of the Ganzi-Litang Paleo-Tethys Ocean. Along the island arc belt, the regional uplift may have resulted in the tectonic paleogeographic pattern and various lithological lithofacies combinations. The belt experienced multi-stage of arc formation (compression setting), intra-arc rifting (extension setting) and arc reformation (compression setting). Recent studies have shown that the formation of these structures may be related to multiple subduction rollbacks and extensional settings (Li et al. 2017).

The Middle-Lower Triassic marine clastic rocks are inter-calated with carbonates and chert, and the Upper Triassic flysch is inter-calated with mafic-felsic volcanics and carbonates. Magmatic rock assemblages of three tectonic stages were developed in the Paleozoic: (1) Late Permian-early Late Triassic: volcanic passive margin (rifting) was developed, forming intraplate alkaline-transitional basalt and/or rhyolite assemblage. In particular, the Permian mafic volcanic rocks are geochemically comparable to the Emeishan flood basalt (Song et al. 2004); (2) Middle-Late Triassic: The Dege-Xiangcheng island arc began to develop, forming the andesite-dominated calc-alkaline volcanic assemblage and intermediate-felsic porphyry intrusions (Hou et al. 2004); and (3) Late Triassic: the Cuojiaoma-Daocheng batholith (dominated by calc-alkaline monzogranite and granodiorite) was formed. The Yanshanian-Himalayan (Cenozoic) tectonics superimposed the Cenozoic post-collisional granite, granite porphyry and related Cu-(Mo) polymetallic mineralization.

During the formation and development of Dege-Xiangcheng island arc, due to the steeper subduction angle and faster subduction velocity in the northern section, the extensional metallogenetic environment was formed in the northern section, while compressional metallogenetic environment was formed in the southern section of the Dege-Xiangcheng island arc (Li et al. 2010). Two tectonic environments mainly controlled three deposit types and ore metal assemblages: (1) in the northern section, extensive and multi-cycle bimodal volcanism (basalt and rhyolite) was developed, accompanied by the formation of Kuroko-type VMS deposits. The ore bodies mainly consist of a stratiform/lenticular upper part and a stockwork lower part. The ore metals in these deposits include Zn-Pb-Cu-Ag, and the mineralization ages are concentrated in the Late Triassic, such as the Gacun and Gayiqiong (Sichuan 210–230 Ma) (Table 5.1; Hou et al. 2001, 2003, 2004; Wang et al. 2013a; Dang et al. 2014; Xue et al. 2014); (2) in the southern part, a series of porphyry Cu polymetallic deposits was developed, such as the Pulang, Xuejiping, Chundu, Lannitang and Langdu (Li et al. 2006, 2011, 2017; Leng et al. 2010); and (3) in the Yanshanian post-collision extensional setting, the thickened lower crust may have delaminated and partially melting, forming extensive felsic magma emplacement into the Indosinian porphyries, consequently the Indosinian-Yanshanian intrusive complexes. Thus, there are many porphyry-skarn and granite-related quartz vein-type Mo polymetallic deposits distributed in this region, such as the Hongshan Cu-Mo (Li, et al. 2013; Peng et al. 2016; Gao et al. 2020), Xiuwacu W-Mo deposits (Liu et al. 2016; Zhang et al. 2017, 2020, 2021), Tongchanggou, Donglufang and Relin deposits (Liu et al. 2016; Li et al. 2017; He et al. 2018).

5.4.2 Regional Metallogenetic Model

Based on the coupling relationship between magmatic activities and mineralization, the Dege-Xiangcheng metallogenetic belt can be divided into northern section and southern section. Controlled by different tectonic backgrounds, VMS deposits were formed in the north extensional arc, while porphyry Cu polymetallic deposits were developed in the south compressive arc (Fig. 5.4a, b).

Subsequently, in Late Yanshanian and Himalayan periods, collisional orogeny and strike-slip activities were developed in this region, respectively, which resulted in the formation of Mo polymetallic and structured altered deposits (Fig. 5.4c).

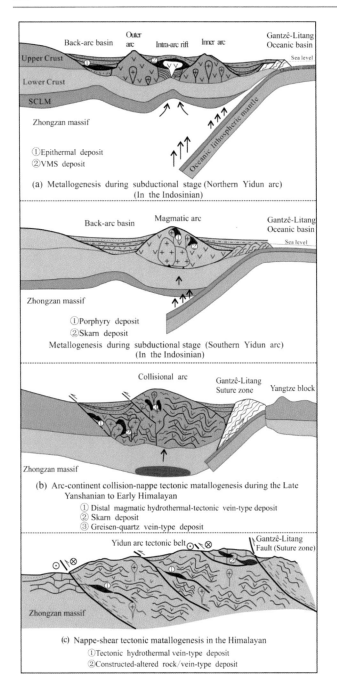

Fig. 5.4 Regional metallogenic model for the Dege-Xiangcheng metallogenic belt

5.5 Jinshajiang-Ailaoshan Au-Cu-PGEs Ore Belt (III)

This belt is developed in the Jinshajiang-Ailaoshan suture zone and the MORB-type ophiolites of the late Paleozoic (Lai et al. 2014a, b; Xu et al. 2019a, b). It extends NW-SE from Zhiduo in the northwest, through Yushu and Batang and to Ailaoshan in the southeast. The belt is 20–40 km

wide and thousands of kilometers long. The belt can be divided into Jinshajiang (mainly Cu) and Ailaoshan (mainly Au) sub-belts (Fig. 5.1).

5.5.1 Jinshajiang Sub-Belt

This sub-belt is mainly composed of fold-thrust belt in the western margin of Zhongzan-Shangri-La block and the low-grade metamorphic rocks in the Jinshajiang suture zone.

5.5.1.1 Geological Evolution and Mineralization

The Jinshajiang suture zone is sandwiched between the Jinshajiang and Yangla-Ludian fault zones. It is mainly composed of ophiolitic mélange, including: (1) MORB-transitional basic rocks formed during the oceanic basin opening process and Carboniferous-Lower Permian deep-water turbidites (Li et al. 2002); (2) intra-oceanic arc basaltic-andesitic rocks were formed during the west-dipping subduction of the oceanic basin, associated with the large-scale mineralization of the Yangla copper deposit (Table 5.1; Zhu et al. 2010; Yang et al. 2011; Li and Liu 2015). Furthermore, it is also revealed that ore-bearing felsic intrusive rocks have the characteristics of mantle-crust mixed magma source at the Yangla deposit (Zhu et al. 2010; Yang et al. 2011); (3) late Indosinian-Himalayan syn-collisional intermediate-felsic granitoids and the alkaline porphyries (strike-slip faulting stage) are closely related to Cu-Au ore deposits, such as Yangla, Tuoding and Tongjige porphyry-skarn mineralization in this period (Deng et al. 2014). The fold-thrust belt in the western margin of the Zhongzan-Shangri-La block is characterized by the development of imbricate nappes, detachment faults and klippe structures, accompanied by strong molybdenite mineralization, rheological fold, flow cleavage and dynamic metamorphism. The belt (over 600 km long) has many nonferrous metal deposits, such as the Tuoding Cu deposit (Mou et al. 1999).

The sub-belt experienced a complex multi-stage geological evolution. The Jinshajiang oceanic basin was formed during the middle Late Devonian to Early Permian and began to subduct westward in the late Early Permian, forming the Zhubalong-Yangla-Dongzhulin intra-oceanic arc and the Xiquhe-Gongnong back-arc-basin (oceanic crust basement) in the western margin of the arc. Strong sinistral shearing/napping from east to west occurred during the arc-continental collision and subsequent intracontinental orogeny after oceanic basin closed in the Late Triassic basin. Since the Late Triassic, under the background of collision between North China and Yangtze block together with large-scale strike-slip faulting between blocks, the development of various intermediate and felsic magmatic rocks and dikes as well as Cu-Pb-Zn mineralization had been superimposed (Ying et al. 2006; Deng et al. 2014). The distribution regularity of deposits in the Jinshajiang sub-belt could be summarized as: the VMS-type Cu deposits are mainly formed in the intra-oceanic arc volcanic rocks in the Jinshajiang ophiolite, while the Pb-Zn deposits are mainly hydrothermal vein type and occur in the thrust nappe and shear-detachment zones of the fold-thrust belt.

Regional Metallogenetic Model

The Jinshajiang oceanic basin may be formed in the Hercynian period, and the VMS mineralization occurred in the Middle-Upper Devonian submarine mafic volcanic-sedimentary sequences. Together with the deep-water turbidites, it formed the initial ore source bed and provided ore-forming materials for the sedimentary-hydrothermal superimposing deposits (such as the Yangla Cu deposit) under the magmatic hydrothermal events of the Early-Middle Triassic (Fig. 5.5a). The ophiolite provided initial ore-forming materials for the late tectonic altered-type Cu-Au-Pb-Zn mineralization.

The intracontinental collision and nappe events occurred during the Yanshanian to Himalayan period in response to

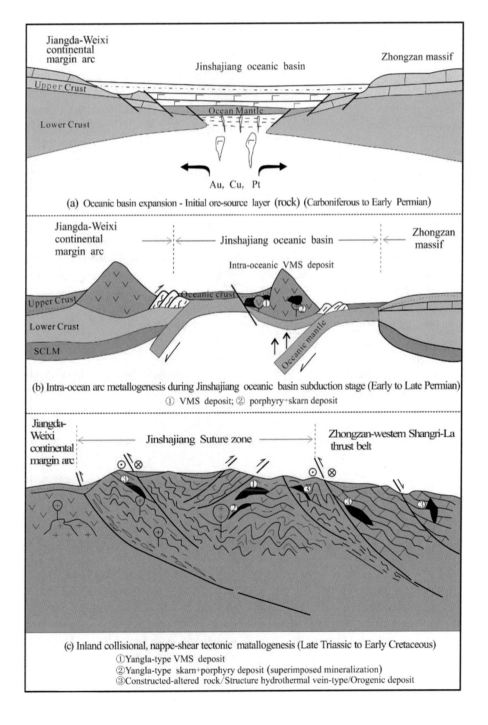

Fig. 5.5 Regional metallogenic model for the Jinshajiang Cu-Pb–Zn metallogenic sub-belt

the India-Asia collision (Deng et al. 2016). The tectonic event may have formed the alkaline magmatic rocks that emplaced into the mélange belt, which also formed the skarn-/porphyry-type Cu-Pb-Zn polymetallic mineralization (Fig. 5.5b). The thrust nappe, strike-slip shear-detachment tectonism of the Himalayan period complicates the super-imposition and transformation of ore deposits and ore bodies and forms hydrothermal vein-type Cu-Pb-Zn polymetallic deposits (Fig. 5.5c).

5.5.2   Ailaoshan Sub-Belt

This sub-belt is a complex nappe structure composed of the Ailaoshan basement nappe belt, Jinping slip unit and low-grade metamorphic rocks in the western margin of the Yangtze block (Fig. 5.1).

Geological evolution and mineralization

The sub-belt has experienced multi-stage geological processes. The Ailaoshan oceanic basin, which was formed by extension in the Late Devonian-Early Permian, has subducted westward and closed during the late Early Permian. Foreland depression basin was formed and molasse formation was accumulated during the subsequent collision process in the Late Triassic. There were strong nappe and sinistral strike-slip faulting westward during the late Yanshanian to Himalayan period (Li et al. 2010; Xu et al. 2019a, b, c, 2020). With the occurrence of large-scale strike-slip structure, alkali-rich porphyry intruded along the strike-slip zone and its secondary faults, forming porphyry-skarn Cu-Au deposits and Au polymetallic mineralization (e.g., the Beiya and Machangqing porphyry-skarn Au polymetallic deposits; Table 5.1; Yang et al. 2020a, b, c), and orogenic (also named structural-controlled alteration type) gold mineralization was also developed in the shear zone (e.g., the Donggualin, Jinchang, Laowangzhai and Daping deposits; Table 5.1).

There are three regional NNW-trending thrust faults are developed in this sub-belt, namely Red River (Red River), Ailaoshan and Jiujia-Mojiang. Among them, the Red River fault zone has undergone multiple tectonic phases and later (after 17 Ma) transformed into dextral strike-slip (Leloup et al. 2001; Li et al. 2010). There are four stages closely related to the formation of gold deposits: (1) the (quasi-) oceanic ridge-type volcanic-ophiolite belt during the formation of the oceanic basin is mainly distributed in the front nappe belt on the hanging wall of Jiujia-Mojiang fault zone, with intermittent distribution of more than 200 km, and forms the ophiolite mélange belt together with the deep-water turbidite sediments during Devonian-Early

Permian (Lai et al. 2014a, b); (2) the island arc basalt-andesite assemblage formed during the westward subduction and closure of the oceanic basin, including the Permian basalt on the western side of the Ailaoshan and Mojiang fault zones; (3) the intermediate-felsic intrusions formed in the syn-collision stage, which were concentrated on the volcanic belt of Permian, formed the felsic intrusive batholith with banded distribution; and (4) formed under the collision-nappe tectonic background during the Yanshanian-Himalayan orogeny, the Ailaoshan granites, alkaline porphyries, strongly granitic rocks and lamprophyre intrusions have close temporal-spatial genetic link with gold mineralization (Xu et al. 2019a, b, c, 2020).

The mineralization of Jinshajiang-Ailaoshan alkali-rich porphyry belt is characterized by metal zonation. The northern part is dominated by Cu-(Mo-Au) mineralization (e.g., the Yangla, Tongding, Tongjige Cu deposits), and the southern part is transformed into Au-(Cu/-Mo) mineralization, such as Machangqing and Beiya super-large porphyry-skarn-type Au deposits with proven Au reserves of over 387 t (Table 5.1; Li et al. 2016a, b, c; Hou et al. 2017; Li et al. 2021; Wang et al. 2018). Most of these deposits have magmatic hydrothermal mineralization systems (Li et al. 2016a, b, c). The ore bodies are mainly distributed in the inter-layer detachment planes between different lithologies and the detachment faults connected with the Jinshajiang-Ailaoshan main faults control the morphology, occurrence and size of the ore bodies, as well as the distribution of alteration zones. Typical orogenic Au deposits in Ailaoshan include the Donggualin, Laowangzhai, Jinchang and Daping (Zhang et al. 2011, 2019; Deng et al. 2015; Zhou et al. 2016).

Regional Metallogenetic Model

This sub-belt has experienced several significant geological events since Hercynian (Xia et al. 2016). With the fragmenting of Ailaoshan oceanic basin in the Hercynian period, the deep-water turbidites and ophiolites formed the initial source bed of gold mineralization, with the zircon U-Pb isotopic ages of 400–300 Ma (Hou et al. 2003; Li et al. 2010; Fig. 5.6a). Some ore-bearing strata may have resulted in new ore-forming rocks (e.g., pyroxene diorite) under the remelting and reconstruction of subduction zone during the Late Hercynian-Early Indosinian, with zircon U-Pb isotopic ages of 285–200 Ma (Hou et al. 2003; Li et al. 2010; Fig. 5.6b). In response to the Himalayan orogeny events, dominantly sinistral strike-slip movements and widespread lamprophyre intrusions were developed in the Ailaoshan sub-belt. Accompanying with strike-slip nappe movements, meteoric water infiltration and deep fluid mixing process extracted the ore-forming elements from the source rocks and then transported into secondary shear tectonic zones and

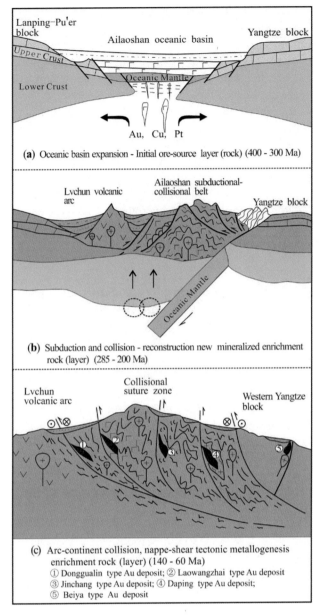

Fig. 5.6 Regional metallogenic model for the Ailaoshan Au metallogenic sub-belt

nappes, which resulted in the forming some ore deposits at the top of the tectonic convergence zones (Fig. 5.6c).

5.6 Jiangda-Weixi-Lvchun Fe-Cu-Pb–Zn Polymetallic Ore Belt (IV)

This belt is located in the volcanic arc of the Jiangda-Weixi-Lvchun continental margin between the Jinshajiang suture zone and Changdu-Lanping-Pu'er (Changdu- Lanping-Simao) block and is distributed along Jiangda-A'dengge-Nanzuo-Mojiang- Lvchun (Fig. 5.1).

5.6.1 Geological Evolution and Mineralization

The formation, development and mineralization of continental margin arc are closely related to the subduction of Jinshajiang-Ailaoshan oceanic basin and arc-continent collision in Paleo-Tethys. This belt is mainly composed of three secondary volcanic belts: (1) the Early-Late Permian Jiangda-Weixi-Lvchun arc related to subduction, which was formed by west-dipping subduction of the Jinshajiang-Ailaoshan oceanic basin beneath the Changdu-Lanping-Pu'er block (Fig. 5.7a); (2) the Early-Middle Triassic syn-collision magmatic belt related to the Jinshajiang oceanic basin closure, where intermediate to felsic magmatic rocks are formed (Fig. 5.7b); (3) the Late Triassic post-collisional rift-type magmatic belt related to crustal extension (Wang et al. 2001, 2002; Fig. 5.7c).

Most important deposits (e.g., Laojunshan, Zhaokalong, Luchun, Chugezha, Hongponiuchang) were formed in the Late Triassic, many of which were modified/superimposed by the following Himalayan mineralization (Fig. 5.7d). There are some Fe, Cu and Pb-Zn-Ag deposits in this belt, together with some Late Triassic large gypsum deposits (e.g., Lirenka). The major deposit types include: (1) VMS deposits occurred in the Late Triassic rift-related volcanic rocks, such as the large-scale Laojunshan Pb-Zn deposit and the Zhaokalong Ag-Fe polymetallic deposit (Table 5.1; Feng et al. 2011; Li et al. 2016a, b, c; G.S. Yang et al. 2020a, b, c); (2) SEDEX deposits were hosted in Upper Triassic sedimentary rocks, such as the large Zhuna Ag-Pb-Zn polymetallic deposit; (3) porphyry- and skarn-type deposits were related to (ultra)-hypabyssal intrusions, as represented by the Triassic Jiaduoling Fe-Cu deposit; and (4) hydrothermal vein-/altered rock-type polymetallic deposits, such as the large Lirenka Pb-Zn deposit and Longbohe IOCG deposit in the Jinping terrane, which could be compared with the Sin-Quyen IOCG deposit in Northwestern Vietnam (Cui et al. 2006; Halpin et al. 2016; Liu and Chen 2019).

5.6.2 Regional Metallogenetic Model

During the formation of continental marginal arc, the Jinshajiang oceanic basin began to subduct westward in the late Early Permian and subducted under the Changdu-Lanping-Pu'er block, forming the Jiangda-Deqin-Weixi-Lvchun subduction-collision type continental margin arc in the eastern margin of the block. The magmatism is mainly calc-alkaline intermediate-basic to felsic, forming volcanic rocks and syn-collisional granitoids such as granodiorite, monzogranite, plagio-granite, diorite porphyrite and quartz diorite porphyrite (Li et al. 2010), which resulted in the

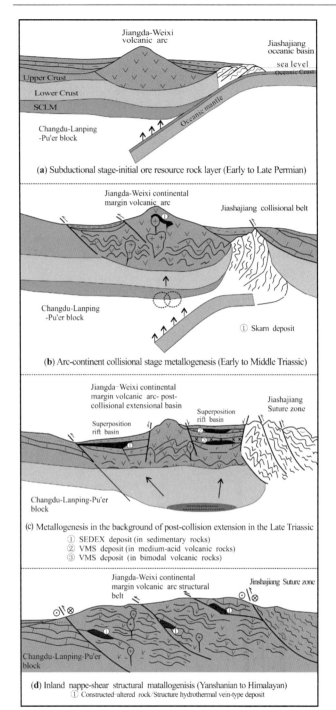

Fig. 5.7 Regional metallogenic model for the Jiangda-Weixi-Lvchun polymetallic metallogenic belt

formation of many porphyry- and skarn-type deposits (Fig. 5.7a, b).

During the post-collision extension in the Late Triassic, the continental margin changed from compression to extension, forming the superimposed rift basin and the associated VMS mineralization. Different degrees of basin development in certain sections and different stages of basin

evolution controlled certain types of deposits. SEDEX-type Pb-Zn-Ag deposits were developed in the rift basin of the northern part of the belt. In the rift basin of the middle part of the belt, VMS-type deposits of the bimodal volcanic mantle felsic section were formed in the early stage of the basin extension. Lirenka-type gypsum deposits occurred in the shallow marine molasse in the late stage of basin evolution (Fig. 5.7c).

During the intracontinental collision orogenic stage from the Middle-Late Yanshanian to Himalayan, thrust nappe and strike-slip structures were developed in this ore belt. This may further upgrade and enlarge the existing deposits and lead to the formation of ultrahydrothermal altered rock-type polymetallic deposits inside the fault zone and its secondary fault fissures (Fig. 5.7d).

5.7 Changdu-Lanping-Pu'er Cu-Pb–Zn-Ag Polymetallic Ore Belt (V)

This belt is developed in the Changdu-Lanping-Pu'er Mesozoic back-arc-basin located between the Jiangda-Weixi-Lvchun Fe-Cu-Pb-Zn and Zaduo-Jinggu-Jinghong Cu-Sn ore belts. Two important deposit types have been discovered: porphyry Cu-Au-(Mo) deposits in the Yulong-Mangkang region and Pb-Zn-Ag polymetallic deposits in the Changdu-Lanping basin (Fig. 5.1).

5.7.1 Yulong-Mangkang Porphyry Cu-Au(-Mo) Sub-Belt

This sub-belt is located in the eastern part of the Changdu basin, adjacent to the northern segment of the Jiangda-Weixi arc. The famous Yulong super-large Cu deposit is developed in this belt (Spurlin et al. 2005; Yang and Cooke 2019; Yang et al. 2020a, b, c). The Cu-Au-(Mo) mineralization is associated with the Himalayan felsic porphyry intrusions, e.g., the Mamupu and Bada deposits (Table 5.1).

Geological Evolution and Metallogenesis

The porphyry Cu-Au-(Mo) deposits in the Yulong-Mangkang region are mainly hosted in Early-Middle Triassic clastic and carbonate rocks, inter-layered with felsic volcanics and Late Triassic-Cretaceous continental red beds. Most of the folds and faults in Yulong-Mangkang region are arranged in echelon, showing dextral shear. The Himalayan felsic porphyry intrusions were emplaced into Early-Middle Triassic volcaniclastic rocks. Porphyry Cu-(Mo) deposits in the region are related to monzogranite porphyries, such as the super-large Yulong porphyry Cu-(Mo) deposit (Table 5.1; Chen et al. 2020; Huang et al. 2020). Meanwhile, the porphyry Au-(Ag) deposits are mainly related to monzonitic and syenite porphyries, such as the Mamupu, Gegongnong and Gegongnong deposits (Zhang et al. 2012). The Cu-(Mo) and Au-(Ag) deposits developed in the Yulong-Mangkang region show porphyry-epithermal-type characteristics.

Regional Metallogenetic Model

The Yulong-Mangkang region is an important Himalayan tectonic-magmatic belt formed by the transitional rifting along the Jinshajiang strike-slip fault. During the India-Asia collision, pull-apart basins were formed around the steps of the Jinshajiang strike-slip fault zone, and the extensional setting led to the emplacement of alkaline (incl. potassic) magma, forming the Yulong-Mangkang porphyry belt. The porphyry intrusions not only provide the ore-forming fluids and materials, but also the essential heat for driving the circulation of ore-forming fluids (Fig. 5.8a, b; Li et al. 2010).

5.7.2 Changdu-Lanping Pb–Zn-Ag Sub-Belt

This sub-belt is developed in the Changdu-Lanping transtensional pull-apart basin resulted from the movement of Jinshajiang strike-slip fault. This sub-belt is located in the south of the Yulong-Mangkang porphyry Cu-Au-(Mo) sub-belt and west of the Weixi magmatic arc. The

Changdu-Lanping sub-belt thins out in the center, and widens toward both ends (Fig. 5.1).

Geological Evolution and Metallogenesis

The Lanping-Pu'er basin in the mass has experienced three evolution stages in the block: (1) Basement formation stage. The Late Paleozoic Paleo-Tethys basin was opened, and the Proterozoic-Paleozoic metamorphic basement was exhumed outside the boundary fault of the Lanping-Pu'er basin. The Jinshajiang Ocean, which was a branch of Paleo-Tethys, separated the Western Yangtze block to the east from the Lanping-Pu'er basin to the west. The subduction of the Jinshajiang Ocean to the west and the Lancangjiang Ocean to the east led to the mutual subduction of both sides of the Lanping-Pu'er basin, resulting in the collision between the Baoshan-Sibumasu Terrane and Yangtze block in the Late Permian. During the Early-Middle Permian, strong orogenic activities may have resulted in the general absence of Middle-Lower Triassic strata in this region; (2) intercontinental basin evolution stage. This sub-belt experienced intracontinental extension, and the sedimentary cover was the alternating deposits of sea-river facies from Upper Triassic to Lower Jurassic since the Late Triassic. Until the Middle Jurassic, the Lanping region changed from intracontinental rift to intracontinental depression basin, which was filled with continental sediments from Middle Jurassic

Fig. 5.8 Regional metallogenic model for the Changdu-Lanping-Pu'er metallogenic belt (Fig. b modified after Sillitoe 2010; Fig. c modified after Leach and Song 2019)

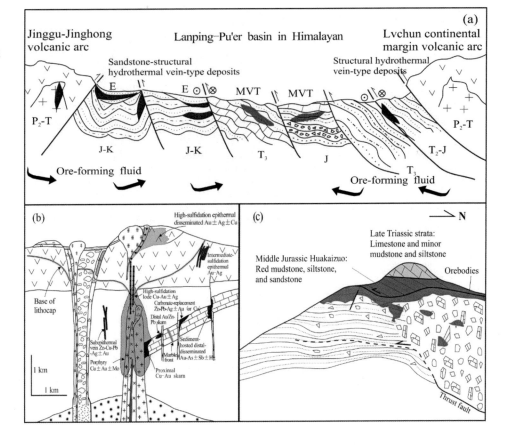

to Cretaceous; and (3) Basin strike-slip and uplifting stage. The Lanping basin experienced the evolution of basin development and strike-slip pull-apart basin opening in the Cenozoic. The sediments have the characteristics of closed lake basin environment, mainly developing gypsum-bearing red beds (Xue et al. 2007; Bi et al. 2019).

The Changdu basin basement is composed of Precambrian Jitang group and Youxi group. The Paleozoic sequence is scattered in this region, mainly Early Ordovician and Devonian-Permian clastics and carbonates. The Mesozoic sedimentary cover consists of Triassic pelagic carbonates with volcanic inter-beds, Jurassic pelagic carbonates and continental red beds and Cretaceous continental clastics. The orogenic belts on both sides of the Changdu basin were formed by megathrust event (the Jinshajiang and Lancangjiang fold belt) as the Paleozoic and Mesozoic strata were thrusted beneath the Proterozoic metamorphic basement in the eastern part of the Changdu basin in the Middle-Late Eocene, forming west-dipping nappe structures. In the western part of the basin, the Late Triassic and Jurassic continental red beds were thrusted over the Proterozoic-Early Paleozoic and Triassic clastic strata and Hercynian-Indosinian granitoids (Tang et al. 2006). From the Late Eocene to Oligocene, transtensional pull-apart basins were developed in this area, arranged in an echelon pattern (Bi et al. 2019).

The Pb-Zn-Ag polymetallic deposits are developed in the Changdu-Lanping basin (Fig. 5.8a) and fall into two genetic types: (1) Mississippi Valley-Type (MVT) deposits are closely related to the nappe structures, such as the super-large Jinding Pb-Zn deposit, Lanuoma, Zhaofayong, Sanshan, Changdong, Luoboshan (Song et al. 2017; Bi et al. 2019; Table 5.1; Fig. 5.8c); (2) sediment-hosted Cu-Ag deposits (e.g., Baiyangchang Cu-Ag deposit) are spatially associated with halokinetic structures and discordant breccia bodies interpreted as pre-evaporite grabens, which play an important role in controlling fluid flow in the basin and mineralization in reduced rocks, similar to the Proterozoic rock-hosted deposits in the Katangan Cu deposits (Leach and Song 2019). The above mineralization was mainly caused by the nappe and strike-slip faulting movements during the Himalayan orogeny.

5.8 Regional Metallogenetic Model

Under regional compression, substantial amount of fluid was discharged from the orogenic belts on both sides of the Changdu-Lanping basin (the Jinshajiang and Lancangjiang structure belt) and mixed with deep metamorphic and/or magmatic hydrothermal fluids. The ore-forming fluids may have entered the secondary faults in front of the nappes, forming vein-type and/or altered rock-type low-medium-

temperature hydrothermal deposits (100–300°C; Li et al. 2010). Alteration types include silicic, argillite, carbonate and barite.

In addition, the strong intracontinental convergence and extrusion during Late Yanshanian-Himalayan period triggered the transition of the eastern margin of Changdu Basin, leading to partial melting of the lower crust and upper mantle, which formed the Yulong-Mangkang felsic (some alkalic) porphyry Cu-Au-(Mo) ore sub-belt (Fig. 5.8b).

5.9 Zaduo-Jinggu-Jinghong Cu-Sn Polymetallic Ore Belt (VI)

The belt is located between the Lancangjiang suture/fault zone and the Changdu-Lanping-Pu'er basin (Fig. 5.1). It is related to the late Early Paleozoic-Early Mesozoic Zaduo-Jinggu-Jinghong magmatic belt. This belt contains massive Cu polymetallic sulfide deposits, including the Dapingzhang, Sandashan and Minle VMS deposits.

5.9.1 Geological Evolution and Mineralization

The oldest exposed sequences are the Middle-Upper Silurian andesite, rhyolite and sedimentary rocks formed in back-arc-basins (Lehmann et al. 2013). The Carboniferous strata are mainly composed of clastic rocks and mafic volcanic rocks in the continental marginal zones. The Permian consists of island arc volcanic-sedimentary formation (clastic and carbonate rocks, basalt to rhyolite). The Middle-Lower Triassic comprises an arc volcanic-sedimentary suite, including (basaltic)-andesite-rhyolite and pyroclastic rocks. A set of K-rich bimodal volcanic suite (basalt and rhyolite) was formed in Yunxian-Wenyu-Minle area (northern part of the belt) during the Late Triassic (Mo et al. 1993). The Middle-Late Silurian-Late Triassic volcanic-sedimentary activities are most closely related to the Cu polymetallic mineralization (Li et al. 2010). The earliest magmatic intrusion occurred in Caledonian period. Zircon U-Pb and whole-rock Re-Os dating of dacite-rhyolite in the Dapingzhang copper deposit show that the rock- and ore-forming ages are uniformly ca. 420 Ma. The pluton intruded mainly into the Permian-Triassic island arc volcanic-sedimentary rocks, including mainly diorite, granodiorite, monzogranite and granite porphyry. The granitoids show I-type characteristics (Mo et al. 1993).

Mineralization is closely related to magmatism and can be divided into two types: (1) VMS deposits are related to Late Silurian-Early Carboniferous volcanic passive margin magmatism and Late Triassic post-collisional rift-related magmatism, such as large Dapingzhang and the medium Sandashan and Minle Cu polymetallic deposits (Table 5.1;

Dai et al. 2004; Zhu et al. 2011; Lehmann et al. 2013); (2) Greisen-quartz vein-type Sn polymetallic deposits, which are closely related to the Indosinian island arc magmatism. These deposits may have been formed by the post-magmatic replacement-filling effect and are distributed in the southernmost part of the belt, such as the Bulangshan and Mengsong (Li et al. 2010).

5.9.2 Regional Metallogenetic Model

In the Late Caledonian, accompanying the expansion of the Southern Lancangjiang ocean basin, volcanic activities were developed in the volcanic passive margin on the western edge of the Lanping-Pu'er terrane (Jinggu-Jinghong area), forming the VMS-type deposits in the Late Silurian-Early Carboniferous submarine spilite-keratophyre series (Fig. 5.9a; Li et al. 2010).

At the end of Early Permian, the Lancangjiang ocean began to subduct eastward beneath the Lanping-Pu'er terrane, and the passive margin of the Lanping-Pu'er terrane changed to be an active margin. In the Triassic, arc-arc collision and accretion occurred, forming the Permian-Triassic island arc calc-alkaline intermediate-felsic volcanic-sedimentary sequences, together with extensive Indosinian collisional-related granitoid emplacement. The ore-forming fluids accumulated in the late-stage intrusive activity may have interacted with the surrounding rocks to form post-magmatic hydrothermal quartz vein-type Sn-W polymetallic deposits, together with the ore-related skarnization, greisenization, silicification and carbonation. The ore-forming intrusions are generally monzogranite, granite porphyry and other volatile-rich granites (Fig. 5.9b).

In the early Late Triassic, the volcanic arc changed from extrusion to post-collisional extension, forming a rift basin along this belt. In the basin, the volcanism is dominantly bimodal and may have brought in large quantities of ore-forming materials. Meanwhile, the ore-forming materials interact with the materials in the seawater to form the hydrothermal mineralization system. Under relatively reducing conditions, VMS deposits occur in the bimodal assemblage of felsic volcanic rock series (Fig. 5.9c; Li et al. 2010).

5.10 Leiwuqi-Lincang-Menghai Sn-Fe-Pb–Zn Polymetallic Ore Belt (VII)

Located in the west of Lancangjiang fault/suture belt (Fig. 5.1), the main body of the Leiwuqi-Lincang-Menghai is a Sn-Fe-Pb-Zn polymetallic ore belt developed on the Leiwuqi-Dongdashan and Lincang-Menghai magmatic arcs. The ore belt can be divided into two sub-belts: the Leiwuqi-Zuogong

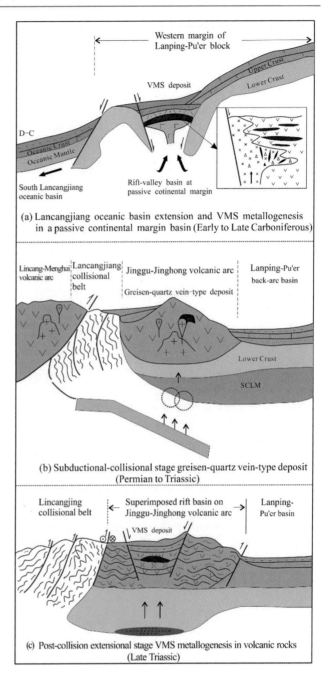

Fig. 5.9 Regional metallogenic model for the Zaduo-Jinggu-Jinghong polymetallic metallogenic belt (Fig. b modified after Leach and Song 2019)

Sn-W and Pb-Zn sub-belt in the north and the Lincang-Menghai Sn-Fe-Cu-rare metal sub-belt in the south.

5.10.1 Leiwuqi-Zuogong Sub-Belt

The main body of Leiwuqi-Zuogong is superimposed on the fold-thrust belt in the eastern margin of the Zuogong block and the Leiwuqi-Dongdashan magmatic arc. In addition to the small-/medium-sized Sn deposits related to granite, many Ag-Au-Cu mineral occurrences have been found in this sub-belt.

5.10.1.1 Geological Evolution and Mineralization

The Precambrian Jitang Group and the lower Paleozoic Youxi Group constitute the main body of the Taniantaweng terrane, which was in a passive continental margin setting during the Carboniferous-Permian. These strata are overlain by Middle-Upper Triassic fluvial-shallow marine carbonate-clastic rocks with minor volcanic inter-beds or locally overlain by Jurassic pelagic/terrestrial clastic-carbonate rocks. Magmatic rocks include metamorphosed mafic volcanic rocks of Precambrian Jitang Group and Lower Paleozoic Youxi Group, Carboniferous-Permian mafic volcanic rocks and Mesozoic-Cenozoic intermediate-felsic intrusive rocks. Among them, the Mesozoic-Cenozoic intermediate-felsic intrusive rocks are closely ore-related. Rock types of the granitoids include granodiorite, monzogranite, albite granite and granite porphyry. Strong structural deformation and tight folding resulted in nappe structures which are west-trending or south-trending in the southern section and north-trending in the northern section.

The ore metal assemblages are mainly Sn-W in Leiwuqi-Binda, Ag-Sn in Chaya-Jitang-Zuogong-Meiyu and Ag-Au in Tiantuo-Dongdacun. The main deposit types are Greisen-quartz vein-type W-Sn deposits formed by post-magmatic metasomatism (e.g., the Saibeinong, Xiayu and Duila Sn-(W) deposits; Table 5.1; Qiu et al. 2011) and hydrothermal vein-type Ag(-Au) deposits. The former is related to the late Yanshanian-Himalayan felsic granitoids, while the latter is mainly controlled by the Leiwuqi-Zuogong main fault and the NNW- and/or NNE-trend secondary faults. Furthermore, these two kinds of mineralization can coexist spatially.

5.10.1.2 Regional Metallogenetic Model

The Paleo-Tethys arc volcanic-sedimentary series provided material basis for the later tectono-magmatic mineralization in this sub-belt. On the geochemical background of host rocks in the Precambrian Jitang Group, Lower Paleozoic Youxi Group and Upper Triassic strata, the elements of W, Sn, Ag, B, As, Sb, Bi, Hg and Pb could be several to tens of times higher than the Clarke values. Therefore, these strata are related to Sn-W (and/or Au-Ag) mineralization, and these volcanic-sedimentary sequences are important source beds in this sub-belt (Li et al. 2010).

During the Late Yanshanian-Himalayan period, strong tectonic deformation occurred in the continent-continent collision stage, forming a series of NW-trending nappes, strike-slip shears and related NNW-trending faults. Magmatic activities are developed along the NW-trending main fault zone, forming many intermediate-felsic granitic stocks

including granodiorite, two-mica granite, albite granite and granite porphyry. The late Yanshanian-Himalayan granites are characterized by Sn-W and Ag-Au metallogenetic specialization. This is probably because the magma of the granites is mainly derived from the remelting of the continental crust with high metal background values and absorbed metal elements (e.g., Sn, W, B, Ag, Au and/or their combinations) which are eventually enriched in the high-middle temperature hydrothermal fluid. Therefore, in the late Yanshanian-Himalayan intracontinental convergence-compression setting, strong nappe, strike-slip shear and other tectonic processes, as well as the contemporaneous magmatic intrusion activities, are the main factors of tectonic-magmatic mineralization, forming the composite metallogenetic zone in this sub-belt (Fig. 5.10)

5.10.2 Lincang-Menghai Sub-Belt

It is mainly developed in the Lincang-Menghai magmatic arc, which is about 500 km long and 20–70 km wide (Fig. 5.1).

5.10.2.1 Geological Evolution and Mineralization

The Lincang-Menghai sub-belt mainly distributes migmatite, migmatitic gneiss, leptynite, schist, marble and other middle-deep metamorphic rocks of the Precambrian Lancang/Damenglong, Chongshan and Ximeng Groups. The protolith is a set of marine intermediate-basic volcanic rocks and sedimentary rocks. The Silurian-Carboniferous sequence is mainly composed of carbonate and clastic rocks, while the Permian-Triassic sequence is composed of intermediate-felsic volcanic and pyroclastic rocks formed in the active continental margin. The Precambrian intermediate-high-grade metamorphic rocks are important host rocks for Fe and Sn-(W) deposits in this sub-belt. Magmatic rocks include the aforementioned intermediate-felsic volcanic rocks formed in the Precambrian Groups, Permian-Triassic intermediate-mafic/felsic volcanic rocks and widespread granitoids. The Triassic granitoids mainly include the Lincang granitic batholith, extending up to 370 km long and 50 km wide (Wang et al. 2010). Furthermore, there are also Zhibenshan granite at Caojian, Yunlong County of Yunnan Province and many small Yanshanian stocks at Menghai. From Changning to Ximeng, a series of crust-remelted granites is exposed along the Lincang-Menghai fault, and its secondary faults occurred mainly in Indosinian, late Yanshanian and Himalayan, respectively.

The Indosinian granitoids are spatial-genetically closely related to Sn-(W) and Cu-(Pb-Zn) mineralization, while the late Yanshanian-Himalayan granites are spatial-genetically closely related to rare metal mineralization. There are mainly three types of ore deposits in this sub-belt:

Fig. 5.10 Regional metallogenic model for the Zaduo-Jinggu-Jinghong Cu-Sn polymetallic metallogenic belt

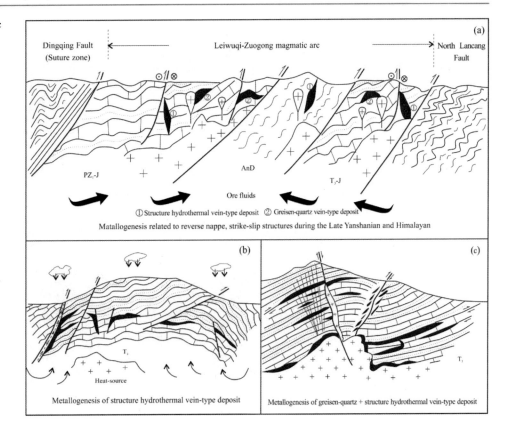

(1) post-magmatic hydrothermal Greisen-quartz vein-type Sn-(W) and Cu-(Pb-Zn) polymetallic deposits, which are mainly distributed in the Lincang granitic batholith and its surrounding rocks, such as Tiechang, Haobadi, Changdonghe Sn-W deposits (Liao et al. 2013); (2) modified VMS-type Fe deposits (e.g., the Huimin and Damenglong; Table 5.1; Shen et al. 2002; Xu and Yin 2010), which are mainly distributed in the middle-deep metamorphic rocks of the Precambrian Lancang and Damenglong Groups. During the ore-forming process, Fe deposits experienced multi-stage and strong superimposed transformation and finally formed the banded iron formation (BIF) deposits, which mainly inhibited the formation of Proterozoic VMS deposits; and (3) rare and rare earth element deposits were developed in the oxidized zone of the intrusions.

5.10.2.2   Regional Metallogenetic Model

The Damenglong and Lancang Groups were formed in the back-arc-basin in Rodinia convergence period. During the process, the Huimin Fe deposit and other Fe deposits/prospects were formed by direct exhalation deposition in volcanic rocks during the back-arc rift volcanism (Fig. 5.11a). The mineralization was controlled both by volcanic and sedimentary factors. Subsequently, the Early

Paleozoic Caledonian regional metamorphism (incl. migmatization) and the Late Paleozoic island arc magmatism resulted in the formation of modified VMS-type Fe deposits in the Precambrian metamorphic rocks.

During the Late Yanshanian-Himalayan, the collisional setting and strike-slip activities controlled the magmatic intrusions of crust-remelted-type muscovite/biotite/two-mica granites, which superimposed the early-formed (e.g., Paleozoic) intrusions. At the same time, the early ore deposits were superimposed and modified. Meanwhile, the intrusions were often accompanied by Sn-(W) mineralization. In the intrusions, the supergene alteration zone developed rare earth metals deposits, together with monazite, phosphyttrite deposits (Fig. 5.11b, c).

5.11   Changning-Menglian Pb–Zn-Ag Polymetallic Ore Belt (VIII)

This belt is developed on the (quasi-)oceanic ridge ophiolite belts and is located in the western part of the Lincang-Menghai magmatic arc. It extends N-S-trending from Changning to Menglian, with a length of about 270 km and width of 10–20 km (Fig. 5.1).

Fig. 5.11 Regional metallogenic model for the Leiwuqi-Zuogong Sn (W) and Ag (Au) polymetallic metallogenic sub-belt

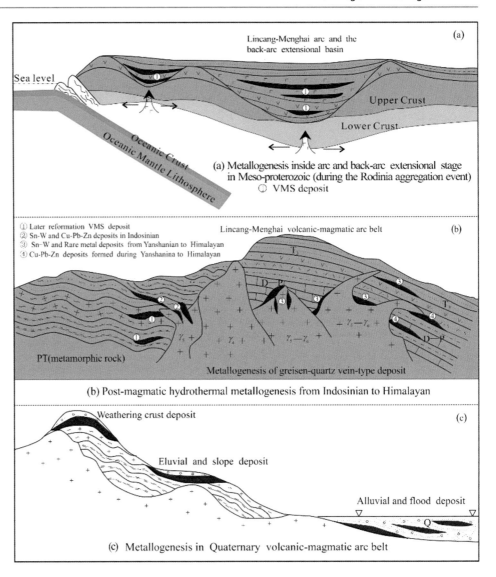

5.11.1 Geological Evolution and Mineralization

The Changning-Menglian suture zone, which contains the remnants of Proto- and Paleo-Tethys, experienced the expansion of Proto-Tethys Ocean in the Early Paleozoic, the subduction of Paleo-Tethys and the formation of island arc in the Early-Late Paleozoic. The main collision occurred in the Late Permian-Early Triassic, with late-collision orogeny and basin-mountain inversion in the Late Triassic (Li et al. 2010; Wang et al. 2018). According to the sedimentary records and magmatic activities, a large amount of magmatic age data have been reported: the zircon U-Pb age of Changning-Menglian ophiolite is 270–264 Ma (Jian et al. 2009a, b), the zircon U-Pb age of Ganlongtang plagioclase amphibolite is 331–349 Ma in Lincang, and the K-Ar isochron age from the gabbro in Tongchangjie ophiolite is 385 Ma (Zhang et al. 1985; Cong et al. 1993).

Previous studies have also documented the oceanic crust remnants in the Nantinghe area (454–439 Ma; Wang et al. 2013a, b), the oceanic type (O-type) high-Mg adakitic tonalite (about 468 Ma) in Niujinshan (Wang et al. 2016) and sodic cumulative plagioclase (471 Ma) in Mangnahe (Liu et al. 2017). Meanwhile, a number of retrograded eclogite outcrops (Li et al. 2017) were found in Kongjiao, Nakahe, Genhenhe and other places in the north of Mengku town and Shuangjiang (Yunnan Province) through 1:50000 regional geological survey. There are two types of protoliths of the retrograded eclogites: (1) E-MORB-like tholeiitic basalts and (2) OIB-like alkaline basalts (Sun et al. 2017). The formation ages of these protoliths are mainly 801 Ma, 429–463 Ma and 254 Ma (Li et al. 2017). These studies indicate that there are also Early Paleozoic ophiolites in the Changning-Menglian suture zone, in addition to the Late Paleozoic ophiolites. There are mainly two deposit types in

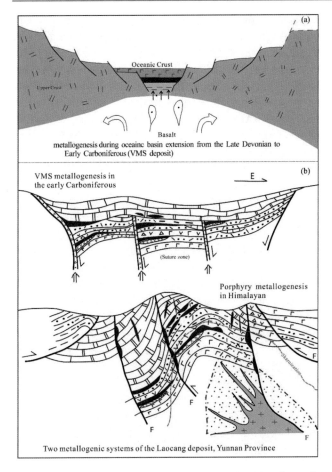

Fig. 5.12 Metallogenetic model for the Changning-Menglian poly-metallic ore belt (**a**) and the Laochang deposit (**b**) (modified after Li et al. 2009)

this belt: (1) VMS type, such as the Laochang Pb-Zn-Ag deposit, the Tongchangjie and Jinla deposits; (2) porphyry Mo-(Cu) deposits related to the Himalayan tectonic-magmatic activities, such as the Mo-(Cu) ore bodies super-imposed onto the existing Pb-Zn-Ag deposit in Laochang (Table 5.1).

5.11.2 Regional Metallogenetic Model

The regional mineralization is mainly related to the volcanism in the oceanic basin expansion stage in the Hercynian and the tectono-magmatism of the continental convergent in Late Yanshanian-Himalayan.

During the Hercynian Changning-Menglian oceanic basin expansion stage, the ore fluids formed in the course of submarine volcanism was directly exhaled and deposited in the volcanic-volcaniclastic rocks. The mineralization occurred related to the mafic volcanic eruption and produced the Cu, Pb, Zn and Ag polymetallic ore bodies with massive, stratiform and reticulate veins, which was corresponding to the main metallogenic stage of the deposits (Fig. 5.12a).

During the intracontinental convergence stage of the Late Yanshanian-Himalayan period, extensional setting, structure-excavation and corresponding plutonism of meta-morphic core complex in the Changning-Gengma fold-thrust belt were developed. The concealed porphyries may be related to the Mo-Cu mineralization, which are superim-posed on the Hercynian VMS-type ore bodies. This process has modified and upgraded the ore bodies by various ways

Fig. 5.13 Metallogenic model for the Baoshan-Zhenkang nonferrous and rare metals ore belt

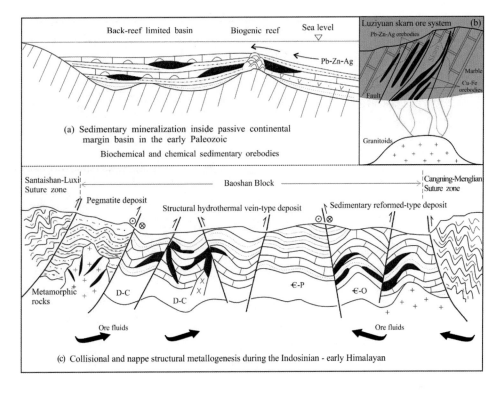

and controlled their shape and location, such as the Laochang Pb-Zn-Mo-Cu deposit (Fig. 5.12b).

5.12 Baoshan-Zhenkang Pb–Zn-Hg and Rare Metal Ore Belt (IX)

This belt is located in the west of Changning-Menglian ore belt, extending from Liuku and Baoshan in the north to Zhenkang in the south. This belt is about 230 km long and 50 km wide (Fig. 5.1).

5.12.1 Geological Evolution and Mineralization

In the Baoshan-Zhenkang block, a strong rifting process of the passive continental margin was developed from the late Precambrian to the early Paleozoic, and thick-bedded sub-flysch sequences were deposited in the Sinian-Cambrian. This region began to transform into a stable sedimentary setting, with the development of the littoral-neritic clastic rocks and impure carbonate rocks in the Ordovician. Subsequently, the sedimentation turned into calcareous sandy and argillaceous formation, which rapidly transitioned upward to impure carbonate inter-calated siliceous sedimentation during the Devonian-Carboniferous. In the early stage of the passive continental margin rift, a series of tectonic-magmatic events occurred in this region, as recorded by the Zhibenshan granite (Rb-Sr isochron age: 645 Ma) and the Pinghe granite (Rb-Sr isochron age: 529.9 Ma) (Zhang et al. 1990). The Paleozoic multi-layered carbonatites are important Pb-Zn-Ag polymetallic ore bodies. Then in the late Yanshanian-early Himalayan, the Pb-Zn-(Ag), Fe, Cu, Hg and Sb mineralization related to intermediate-felsic magmatism were superimposed on the Paleozoic ore bodies. In addition, the Late Cretaceous-Early Himalayan intermediate-felsic magmatism also formed rare earth metal (REM) deposits, such as the Huangliangou Bi-Ni-Nd deposit, and the Luziyuan, Hetaoping, Jinchanghe and Heiniuwa Pb-Zn polymetallic deposits (Table 5.1; Wang et al. 2014; Xu et al. 2019a).

5.12.2 Regional Metallogenetic Model

Regional mineralization is closely related to the evolution process of fold belt of Santaishan-Luxi on the west side and the Changning-Menglian suture zone on the east side. During the Paleozoic, the Baoshan-Zhenkang terrane was in a passive continental margin setting, with Proto-Tethys and Paleo-Tethys developed (Li et al. 1999). During the Early Paleozoic, in the sub-flysch formation which contained volcanic rock, siliceous and carbonate rocks in the passive

continental marginal (rift) basin of the central-east Baoshan-Zhenkang terrane, the ore-forming materials were deposited in the Cambrian-Ordovician bioclastic/reef limestone and in the carbonate-clastic transition interface by means of the (bio)-chemical process. Furthermore, the Early Paleozoic evolution may also have provided metals for the later Pb-Zn-(Ag), Fe and Cu mineralization that are of sedimentary-modified, skarn and tectonic-hydrothermal deposits (Fig. 5.13a).

Accompanying with the tectonic-magmatic hydrothermal activities, the infiltrated meteoric water or formation water was heated in the crust, which circulated in the sequence. The fluids leached and extracted the ore-forming materials and then formed ore fluids which migrated upward along the fault systems of the Baoshan-Zhenkang ore belt. When the ore fluids migrated to the inter-layer fracture zones in the Paleozoic strata, or the structural detachment spaces on the fold axis, or the contact interfaces between the carbonates and clastic rocks, the early-formed deposits were reworked and upgraded to form sedimentary-modified-type Pb-Zn-(Ag) polymetallic deposits. The hydrothermal vein-type Hg-Sb-As polymetallic deposits controlled by structures could also be formed in the new ore-host space (Fig. 5.13b).

During the Yanshanian-Himalayan period, the intermediate-felsic magmatism superimposed on the early-formed skarn Pb-Zn-(Ag) polymetallic deposits or resulted in stratiform Pb-Zn-Cu-Fe-(Ag) skarn mineralization, such as Luziyuan and Fangyangshan (Xu et al. 2021a, b). Some post-magmatic hydrothermal fluids rich in alkali, volatiles and rare metals produced the ore-bearing pegmatite veins by metasomatism-filling along the inner or outer-contact zones of intrusions and their surrounding rocks, and pegmatite-type rare metal deposits were developed (Fig. 5.13b).

5.12.3 Tengchong-Lianghe Sn-W and Rare Metals Ore Belt (X)

This belt is located in the southern part of the Bomi-Tengchong magmatic belt and extends northward to Chayu. The belt is connected to the Southeast Asian Sn belt in the south, forming a Sn belt of 2500 km long. Many Sn polymetallic deposits, including two large (Xiaolonghe and Lailishan) and several medium (e.g., Tieyaoshan, Laopingshan) ones, and one large rare metal deposit (Baihuanao) is distributed in this belt (Fig. 5.1; Li et al. 1999).

5.12.4 Geological Evolution and Mineralization

The metamorphic basement of the magmatic belt is composed of the Proterozoic Gaoligongshan Group. Late

Carboniferous-Permian glaciofluvial deposits and Middle Triassic carbonate rocks are the main components of the sedimentary cover. The Proterozoic Gaoligongshan Group is composed of greenschist-amphibolite facies metamorphic volcanic-sedimentary sequence, showing strong migmatization and ductile shear deformation features. The Carboniferous Menghong Group is sandy slates inter-calated with calc-silicates, which are the main surrounding rocks of granites and Sn polymetallic deposits in the belt.

On the basis of regional metamorphism (including migmatization) in the Precambrian, Caledonian and Variscan, intensive granitic magmatism was developed from the Late Mesozoic to the early Cenozoic, forming the Sn ore belt in western Yunnan Province. The Sn polymetallic mineralization has close genetic-temporal-spatial relation with the Mesozoic and Cenozoic granites. Three nearly parallel, N-S-trending magmatic-metallogenetic sub-belts (i.e., Eastern, Central and Western) were recognized from east to west, with their ages gradually decreasing from east to west (Wang et al. 2013a; Chen et al. 2014, 2015). (1) The eastern sub-belt is distributed in the eastern Tengchong, containing Sn, Fe, Pb-Zn polymetallic deposits related to Late Jurassic-Early Cretaceous granitoids. The ore bodies are mainly developed in fractured zones of strongly altered surrounding rocks (skarns), such as Diantan, Dadongchang, Dakuangshan and Huiyao skarn deposits (Table 5.1; Dong et al. 2005; Deng et al. 2014; Tang, et al. 2018). (2) Central sub-belt is distributed between the Lianghe and Tengchong terranes and is dominated by Sn-W polymetallic deposits of Late Cretaceous granitoids. Ore-related alteration mainly includes greisenization (main ore-forming stage) and silicification. The Sn-W mineralization was superimposed on the early skarn-altered rocks. The ore bodies are distributed along the fractures of altered rocks, forming veins and stockworks and lenticular Sn-W ore bodies, such as the large Xiaolonghe Sn deposit and medium Tieyaoshan W-Sn deposit (Wang et al. 2013a; Chen et al. 2014, 2015). (3) Western sub-belt is located in the west of the Lianghe terrane and contains Early Paleogene granitic Sn-W and rare metals deposits, such as the large Lailishan Sn-W deposit and large Baihuanao rare metal deposit (Shen 2002).

5.12.5 Regional Metallogenetic Mechanism

The northward thrust of the Indian plate beneath the Eurasia during the Late Mesozoic-Early Cenozoic may have resulted in a large arc-shaped thrust belt in the Gongshan-Ruili region, accompanied by the formation of a mylonite belt with a width of hundreds of meters and a transition zone from ductile-brittle to brittle deformation along the Tengchong-Lianghe suture zone (Ma et al. 2021). This tectonic thermal event may trigger the partial remelting of lower crust (e.g., metamorphic basement) with geochemical anomaly of rare, nonferrous metals and volatiles. There are good geochemical barriers and ore-forming spaces in the structural superimposed and intersected space, which are conducive to the evolution of ore-forming fluids. Therefore, the ore-forming elements are eventually concentrated in the favorable structural traps (Fig. 5.14).

The mineralization is spatially and temporally associated with the Mesozoic-Cenozoic granitoids in this ore belt, forming three nearly parallel, N-S-trending tectonic-magmatic ore sub-belts. Furthermore, the mineralization is mainly controlled by the metasomatic alteration and (along the fracture zone) post-magmatic hydrothermal fluids.

Fig. 5.14 Regional metallogenic model for the Tengchong-Lianghe Sn-W and rare metals metallogenic belt

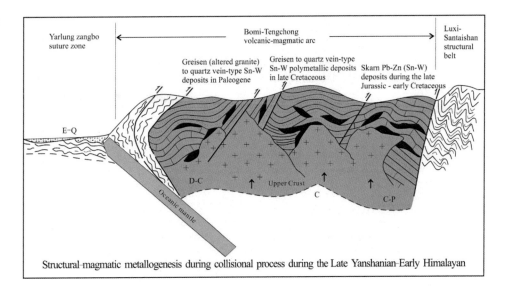

References

Bi XW, Tang YY, Tao Y, Wang CM, Xu LL, Qi HW, Lan Q, Mu L (2019) Composite metallogenesis of sediment-hosted Pb-Zn-Ag-Cu base metal deposits in the Sanjiang Collisional Orogen, SW China, and its deep driving mechanisms. Acta Petrologica Sinica 35 (5):1341–1371 (in Chinese with English abstract)

Cao HW, Zhang W, Pei QM, Zhang ST, Zheng L (2016) Trace element geochemistry of fluorite and calcite from the Xiaolonghe Tin deposits and Lailishan Tin deposits in Western Yunnan, China. Bull Mineral Petrol Geochem 35(5):925–935 (in Chinese with English abstract)

Chen J, Xu JF, Wang BD, Kang Z, Li J (2010) Origin of Cenozoic alkaline potassic volcanic rocks at KonglongXiang, Lhasa terrane, Tibetan Plateau: products of partial melting of a mafic lower-crustal source? Chem Geol 273:286–299

Chen XC, Hu RZ, Bi XW, Li HM, Lan JB, Zhao CH, Zhu JJ (2014) Cassiterite LA-MC-ICP-MS U/Pb and muscovite 40Ar/39Ar dating of tin deposits in the Tengchong- Lianghe tin district, NW Yunnan, China. Miner Deposita 49:843–860

Chen XC, Hu RZ, Bi XW, Zhong H, Lan JB, Zhao CH, Zhu JJ (2015) Petrogenesis of metaluminous A-type granitoids in the Tengchong-Lianghe tin belt of southwestern China: evidences from zircon U-Pb ages and Hf-O isotopes, and whole-rock Sr-Nd isotopes. Lithos 212:93–110

Chen XY, Liu JL, Burg JP, Tang Y, Wu WB, Yan JX (2020) Structural evolution and exhumation of the Yulong dome: constraints on middle crustal flow in southeastern Tibetan Plateau in response to the India-Eurasia collision. J Struct Geol 137:104070. https://doi.org/10.1016/j.jsg.2020.104070

Cong BL, Wu GY, Zhang Q, Zhang RY, Zhai MG, Zhao DS, Zhang WH (1993) The tectonic evolution of western Yunnan Paleo-Tethys structural rocks China. Sci China (d Series) 23 (11):1201–1207 (in Chinese with English abstract)

Cui YL, Qin DX, Chen SG (2006) Geochemistry of Meta-volcanic rocks from the Longbohe copper deposit, Yunnan Province China. Acta Mineralogica Sinica 26(4):395–408 (in Chinese with English abstract)

Dai BZ, Liao QL, Jiang SY (2004) Isotope Geochemical and Mineralization age of the Dapingzhang copper-polymetallic deposit in Lanping-Pu'er Basin Yunnan Province. J Nanjing Univ (nat Sci) 40(6):674–683 (in Chinese with English abstract)

Dang Y, Chen MH, Mao JW, Xue ZQ, Li YB, Xin TG, Ma CH, Li QQ (2014) Geochemistry of ore-forming fluid of Gacun-Youre ore district in Baiyu County Sichuan Province. Acta Petrologica Sinica 30(1):221–236 (in Chinese with English abstract)

Deng J, Wang QF (2016) Gold mineralization in China: Metallogenetic provinces, deposit types and tectonic framework. Gondwana Res 36:219–274

Deng J, Wang QF, Li GJ, Li CS, Wang CM (2014) Tethys tectonic evolution and its bearing on the distribution of important mineral deposits in the Sanjiang Region SW China. Gondwana Res 26 (2):419–437

Deng J, Wang QF, Li GJ, Hou ZQ, Jiang CZ, Danyushevsky L (2015) Geology and genesis of the giant Beiya porphyry-skarn gold deposit, northwestern Yangtze Block, China. Ore Geol Rev 70:457–485

Deng J, Wang QF, Li GJ (2016) Superimposed orogeny and composite metallogenetic system: case study from the Sanjiang Tethys belt SW China. Acta Petrologica Sinica 32(8):2225–2247 (in Chinese with English abstract)

Deng J, Wang CM, Zi JW, Xia R, Li Q (2017) Constraining subduction–collision processes of the Paleo-Tethys along the Changning-Menglian Suture: new zircon U-Pb ages and Sr-Nd-Pb-Hf-O isotopes of the Lincang Batholith. Gondwana Res 62:75–92

Dong FL, Hou ZQ, Gao YF, Zeng PS, Jiang CCX, Du AD (2005) Re-Os isotopic dating of molybdenite from Datongchang copper-lead -zinc deposit in Tengchong area, western Yunnan. Mineral Deposits 24(6):663–668 (in Chinese with English abstract)

Feng JR, Mao JW, Pei RF, Li C (2011) Ore-forming fluids and metallogenesis of Nanyangtian tungsten deposit in Laojunshan, southeastern Yunnan Province. Mineral Deposits 30(3):403–419 (in Chinese with English abstract)

Gao X, Yang LQ, Yan H, Meng JY (2020) Ore-forming processes and mechanisms of the Hongshan skarn Cu–Mo deposit, southwest China: insights from mineral chemistry, fluid inclusions, and stable isotopes. Ore Energy Resour Geol. https://doi.org/10.1016/j.oreoa.2020.100007

Halpin JA, Tran HT, Lai C-K, Meffre S, Crawford AJ, Zaw K (2016) U-Pb zircon geochronology and geochemistry from NE Vietnam: a 'tectonically disputed' territory between the Indochina and South China blocks. Gondwana Res 34:254–273

He WY, Xie SX, Liu XD, Gao X, Xing YL (2018) Geochronology and geochemistry of the Donglufang porphyry-skarn Mo-Cu deposit in the southern Yidun Terrane and their geological significances. Geosci Front 9:1433–1450

Hou ZQ, Qu XM, Xu MJ, Fu DM, Hua LC, Yu JJ (2001) The Gacun VHMS deposit in Sichuan Province: from field observation to genetic model. Mineral Deposits 20(1):44–56 (in Chinese with English abstract)

Hou ZQ, Gao YF, Qu XM, Rui ZY, Mo XX (2004) Origin of adakitic intrusives generated during mid-Miocene east-west extension in southern Tibet Earth Planet. Sci Lett 220:139–155

Hou ZQ, Zhou Y, Wang R, Zheng YC, He WY, Zhao M, Evans NJ, Weinberg RF (2017) Recycling of metal–fertilized lower continental crust: origin of non–arc Au–rich porphyry deposits at cratonic edges. Geology 45(6):563–566

Hou ZQ, Yang YQ, Wang HP, Qu XM, Lü QT, Huang DH, Wu XZ, Yu JJ, Tang SH, Zhao JH (2003) Collision-orogenic progress and mineralization system of Yidun Terrain. Beijing: geological publishing house 1–345 (in Chinese with English abstract)

Huan WJ, Yuan WM, Li N (2011) Study on the mineral electron microprobe evidence of the formation conditions and fission track of gold deposits in Ganzi-Litang Gold Belt, West Sichuan Province. Geoscience 25:261–270

Huang ML, Gao JF, Bi XW, Xu LL, Zhu JJ, Wang DP (2020) The role of early sulfide saturation in the formation of the Yulong porphyry Cu-Mo deposit: Evidence from mineralogy of sulfide melt inclusions and platinum-group element geochemistry. Ore Geol Rev 124:103644

Jian P, Liu DY, Kröner A, Zhang Q, Wang YZ, Sun XM, Zhang W (2009a) Devonian to Permian plate tectonic cycle of the Paleo-Tethys Orogen in southwest China (I): geochemistry of ophiolites, arc/back-arc assemblages and within-plate igneous rocks. Lithos 113(3–4):748–766

Jian P, Liu DY, Kröner A, Zhang Q, Wang YZ, Sun XM, Zhang W (2009b) Devonian to Permian plate tectonic cycle of the Paleo-Tethys Orogen in southwest China (II): insights from zircon ages of ophiolites, arc/back-arc assemblages and within-plate igneous rocks and generation of the Emeishan CFB province. Lithos 113(3–4):767–784

Lai CK, Meffre S, Crawford AJ, Zaw K, Halpin JA, Xue CD, Salam A (2014a) The Central Ailaoshan ophiolite and modern analogs. Gondwana Res 26:75–88

Lai C-K, Meffre S, Crawford AJ, Zaw K, Xue C-D, Halpin JA (2014b) The Western Ailaoshan Volcanic Belts and their SE Asia connection: a new tectonic model for the Eastern Indochina Block. Gondwana Res 26:52–74

Leach DL, Song YC (2019) Sediment-hosted zinc-lead and copper deposits in China. Soc Econ Geologists Special Publication 22:325–409

Lehmann B, Zhao XF, Zhou MF, Du AD, Mao JW, Zeng PS, Henjes-Kunst F, Heppe K (2013) Mid-Silurian back-arc spreading at the northeastern margin of Gondwana: the Dapingzhang dacite-hosted massive sulfide deposit, Lancangjiang zone, southwestern Yunnan, China. Gondwana Res 24:648–663

Leloup PH, Lacassin R, Tapponnier P, Harrison TM (2001) Comment on "Onset timing of left-lateral movement along the Ailao Shan-Red River shear zone: 40Ar/39Ar dating constraint from the Nam Dinh area, northeastern Vietnam." J Asian Earth Sci 20:95–99

Leng CB, Zhang XC, Wang SX, Qin CJ, Wang WQ (2010) The geochemical characteristics of Xuejiping Porphyry Cu deposit, Southwest Yunnan Province China. Mineral Deposits 29(S1):205–206

Leng CB, Cooke DR, Hou ZQ, Noreen JE, Zhang XC, Chen WT, Martin D, Brent IAM, Yang JH (2018) Quantifying exhumation at the giant pulang porphyry Cu-Au deposit using U-Pb-He dating. Econ Geol 113:1077–1092

Li WC, Liu XL (2015) The metallogenetic regularity related to the tectonic and petrographic features of Pulang porphyry copper orefield, Yunnan, and its ore-controlling characteristics. Earth Sci Front 22(4):53–66 (in Chinese with English abstract)

Li XZ, Du DX, Wang YZ (1999) The basin range transition and mineralization: examples from the Qamdo-Pu'er Basin and Jinshajiang-Ailaoshan orogenic belt in southwestern China. Tethyan Geol 22:1–16 (in Chinese with English abstract)

Li WC, Li LH, Yin GH (2006) Different data-processing methods for geochemical data from southern part of Sanjiang region in southwestern China. Mineral Deposits 25(4):501–510 (in Chinese with English abstract)

Li F, Lu WJ, Yang YZ, Chen H, Shi ZL, Liu SW, Xin R, Luo SL (2009) Mineralizing texture and metallogenetic model of the Laochang polymetallic deposits in Lancang Yunnan. Geol Explor 45(5):516–523 (in Chinese with English abstract)

Li WC, Zeng PZ, Hou ZQ, White NC (2011) The Pulang copper deposit and associated felsic intrusions in Yunnan province, Southeast China. Econ Geol 106:79–92

Li WC, Yu HJ, Yin GH (2013) Porphyry metallogenetic system of Geza arc in the Sanjiang region, southwestern China. Acta Petrologica Sinica 29(4):1129–1144 (in Chinese with English abstract)

Li GJ, Wang QF, Huang YH, Gao L, Yu L (2016a) Petrogenesis of middle Ordovician peraluminous granites in the Baoshan block: Implications for the early Paleozoic tectonic evolution along East Gondwana. Lithos 245:76–92

Li H, Xi HS, Sun HS, Kong H, Wu QH, Wu CM, Gabo RJAS (2016b) Geochemistry of the batang group in the Zhaokalong Area, Yushu, Qinghai: implications for the Late Triassic Tectonism in the Northern Sanjiang Region China. Acta Geol Sin 90(2):704–721

Li WC, Wang JH, He ZH, Dou S (2016c) Formation of Au-polymetallic ore deposits in alkaline porphyries at Beiya, Yunnan, Southwest China. Ore Geol Rev 73:241–252

Li WC, Yu HJ, Gao X, Liu XL, Wang JH (2017) Review of Mesozoic multiple magmatism and porphyry Cu–Mo (W) mineralization in the Yidun Terrain, eastern Tibet Plateau. Ore Geol Rev 90:795–812

Li WC, Pan GT, Zhang XF, Wang LQ, Zhou JX (2021) Tectonic evolution and multi–episodic metallogenesis of the Sanjiang Paleo-Tethys multi-arc-basin-terrane system, SW Tibetan Plateau. J Asian Earth Sci 221(2021):104932. https://doi.org/10.1016/j.jseaes.2021.104932

Li DM, Wang LQ, Xu TR, Diao ZZ, Chen KX, Lu YF, Wei JQ, Zhou ZX (2002) The Cu-Au metallogenesis and exploration in Jinshajiang tectonic belt. Geological Publishing House Beijing, 1–259 (in Chinese with English abstract)

Li WC, Pan GT, Hou ZQ, Mo XX, Wang LQ, Ding J, Xu Q, Li XZ, Li DM, Du DX, Jiang XS, Hu YZ, Lü QT, Yang WG, Lu YX, Fan YH, Yang XH, Shen SY, Xu QD, Zhu QW, Zhou YJ, Li XM, Guo YS, Zhang YF, Luo JL, Ren ZJ, Zeng PS, Yang YQ, Wang MJ, Yin GH (2010) Archipelagic-Basin, collision orogenic theory and prospecting techniques along the Nujiang-Lancangjiang-Jinshajiang area in Southwestern China. Geol Publish House Beijing 1–491 (in Chinese)

Liao SY, Wang DB, Tang Y, Yin FG, Sun ZM, Sun J (2013) LA-ICP-MS U-Pb age of two-mica granite in the Yunlong tin-tungsten metallogenetic belt in Three River region and its geological implications. Acta Petrologica Et Mineralogica 32(4):450–462 (in Chinese with English abstract)

Lin Q (2010) Geological features and prospecting potential for the Xialong Ag-Pb-Zn deposit in Batang Sichuan. Acta Geologica Sichuan 30(4):447–449 (in Chinese with English abstract)

Liu L, Chen T (2019) Geology, mineralization styles and age of ore-hosting rocks of the Proterozoic Longbohe-Sin Quyen Fe-Cu belt: implications for regional metallogeny. Ore Geol Rev 111:103013

Liu XL, Li WC, Zhang N, Lai AQ, Li Z, Yang FC (2016) Metallogenetic system of the Yanshanian porphyry Mo polymetallic deposit in the Xiangcheng-Lijiang suture zone, western margin of Yangtze block SW China. Acta Petrologica Sinica 32(8):2281–2302 (in Chinese with English abstract)

Liu GC, Sun ZB, Zeng WT, Feng QL, Huang L, Zhang H (2017) The age of Wanhe ophiolitic mélange from Mengku area, Shuangjiang County, western Yunnan Province, and its geological significance. Acta Petrologica Et Mineralogica 36(2):163–174 (in Chinese with English abstract)

Ma P-F, Xia X-P, Lai C-K, Cai K-D, Cui Z-X, Xu J, Zhang L, Yang Q (2021) Evolution of the Tethyan Bangong-Nujiang Ocean and its SE Asian connection: perspective from the early cretaceous high-mg granitoids in SW China. Lithos 106074

Mo XX, Lu FX, Shen SY, Zhu QW, Hou ZQ, Yang KH (1993) Sanjiang Tethyan volcanism and related mineralization. Geol Publish House Beijing 1–267 (in Chinese with English abstract)

Mou CL, Wang J, Yu Q, Zhang LS (1999) The evolution of the sedimentary basin in Lanping area during Mesozoic-Cenozoic. J Mineral Petrol 19(3):30–36 (in Chinese with English abstract)

Peng HJ, Mao JW, Hou L, Shu QH, Zhang CQ, Liu H, Zhou YM (2016) Stable isotope and fluid inclusion constraints on the source and evolution of ore fluids in the hongniu-hongshan Cu Skarn deposit, Yunnan Province, China. Econ Geol 111:1369–1396

Qiu JQ, Qiangba ZX, Ding XG (2011) Division and characteristics of metallogenetic belt and prospecting area in Leiwuqi, Eastern Tibet. Acta Mineralogica Sinica S1:385–386 (in Chinese)

Qu XM, Hou ZQ (2002) 40Ar/39Ar age of the Panyong Pillow Basalt: implication for the evolution relationship between the Jinshajiang and Garze-Litang Suture Zones. Geol Rev S1:115–121 (in Chinese)

Shen GF (2002) Weathering crust of baihuanao granite: a potential superlarge-scale Rb, Cs, Y, Sc, quartz and albite ore deposit. Bull Mineral Petrol Geochem 21(3):182–184 (in Chinese with English abstract)

Shen SY, Feng QL, Wei QR, Zhang ZB (2008) New evidence for the original Tethyan Island-arc volcanic rocks in the Southern Segment of South Lancangjiang Belt. J Mineral Petrol 28(4):59–63 (in Chinese with English abstract)

Song XY, Zhou MF, Cao ZM, Robinson PT (2004) Late Permian rifting of the South China craton caused by the Emeishan mantle plume? J Geol Soc 161(5):773–781

Song YC, Hou Z, Liu YC, Zhang HR (2017) Mississippi Valley-Type (MVT) Pb-Zn deposits in the Tethyan domain: a review: Geol China 44:664–689 (in Chinese with English abstract)

Spurlin MS, Yin A, Horton BK, Zhou JY, Wang JH (2005) Structural evolution of the Yushu-Nangqian region and its relationship to syncollisional igneous activity, east-central Tibet. Geol Soc Am 117:1293–1317

Sun ZB, Li J, Zhou K, Zeng WT, Duan XD, Zhao JT, Xu GX, Fan YH (2017) Geochemical characteristics and geological significance of retrograde eclogite in Mengku Area, Shuangjiang County, Western Yunnan Province China. Geoscience 31(4):746–756 (in Chinese with English abstract)

Tang JX, Zhong KH, Liu ZC, Li ZJ, Dong SY, Zhang L (2006) Intracontinent orogen and metallogenesis in Himalayan Epoch: Changdu Large composite basin, eastern Tibet. Acta Geol Sin 80 (9):1364–1376 (in Chinese with English abstract)

Tang WL, Xu JF, Chen JL, Tan RY, Huang WL (2018) Implications of the geochronology and geochemistry of Diantan A-type granites in the Tengchong area. Geochemical 47(1):1–13 (in Chinese with English abstract)

Wang LQ, Li DM, Guan SP, Xu TR (2001) The evolution of the luchun-hongponiuchang superimposed rifting basin, Deqin County Yunnan Province. J Mineral Petrol 21(3):81–89 (in Chinese with English abstract)

Wang LQ, Li DM, Guan SP, Xu TR (2002) The Rb-Sr age determinations of the "bimodal" volcanic rocks in the Luchun-Hongponiuchang superimposed rift basin, Deqin Yunnan. Sedimentary Geol Tethyan Geol 22(1):65–71 (in Chinese with English abstract)

Wang CM, Deng J, Zhang ST, Yang LQ (2010) Metallogenetic province and large-scale mineralization of volcanogenic massive sulfide deposits in china. Resour Geol 60(4):404–413

Wang BD, Wang LQ, Pan GT, Yin FG, Wang DB, Tang Y (2013a) U-Pb zircon dating of Early Paleozoic gabbro from the Nantinghe ophiolite in the Changning-Menglian suture zone and its geological implication. China Science Bulletin 58(8):920–930

Wang QF, Deng J, Li CS, Li GJ, Yu L, Qiao L (2014) The boundary between the Pu'er and Yangtze blocks and their locations in Gondwana and Rodinia: constraints from detrital and inherited zircons. Gondwana Res 26(2):438–448

Wang DB, Luo L, Tang Y, Yin FG, Wang BD, Wang LQ (2016) Zircon U-Pb dating and petrogenesis of Early Paleozoic adakites from the Niujingshan ophiolitic mélange in the changning-menglian suture zone and its geological implications. Acta Petrologica Sinica 32(8):2317–2329 (in Chinese with English abstract)

Wang X, Yang L, Deng J, Li HJ, Yu HZ, Dong CY (2018) Identification of multistage hydrothermal mineralization in the Beiya gold deposit: Evidence from geology, petrography, fluid inclusion, H-O-S Isotopes. Acta Petrologica Sinica 34(5):1299–1311 (in Chinese with English abstract)

Wang BD, Wang LQ, Yin FG, Wang DB, Tang Y (2013b) Longmu Co-Shuanghu- Changning-Menglian Suture zone: residual of uniform early Paleozoic Tethys Ocean. Acta Geologica Sinica (English Edition) 87(Zl):72–75

Xia XP, Nie XS, Lai CK, Wang YJ, Long XP, Meffre S (2016) Where was the Ailaoshan Ocean and when did it open: a perspective based on detrital zircon U-Pb age and Hf isotope evidence. Gondwana Res 36:488–502

Xu D, Yin GH (2010) The technological model of exploration in Yangla Cu deposit. Mineral Deposits 29(S1):647–648 (in Chinese)

Xu J, Xia X-P, Cai K, Lai C-K, Liu X-J, Yang Q, Zhou M-L, Ma P-F, Zhang L (2020) Remnants of a Middle Triassic island arc on western margin of South China Block: evidence for bipolar subduction of the Paleotethyan Ailaoshan Ocean. Lithos 360:105447

Xu R, Deng M-G, Li W-C, Lai C-K, Zaw K, Gao Z-W, Chen Y-H, Niu C-H, Liang G (2021b) Origin of the giant Luziyuan Zn-Pb-Fe (-Cu) distal skarn deposit, Baoshan block, SE Tibet: constraints from Pb–

Sr isotopes, calcite C-O isotopes, trace elements and Sm–Nd dating. J Asian Earth Sci 205:104587

Xu J, Xia X-P, Lai C-K, Zhou M, Ma P (2019a) First identification of late permian Nb-enriched basalts in Ailaoshan Region (SW Yunnan, China): contribution from emeishan plume to subduction of eastern paleotethys. Geophys Res Lett 46. https://doi.org/10.1029/2018GL081687

Xu J, Xia X-P, Lai C, Long X, Huang C (2019b) When did the paleotethys Ailaoshan Ocean Close: new insights from detrital zircon U-Pb age and Hf isotopes. Tectonics. https://doi.org/10.1029/2018TC005291

Xu J, Xia X, Huang C, Cai K, Yin C, Lai C-K (2019c) Changes of provenance of Permian and Triassic sedimentary rocks from the Ailaoshan suture zone (SW China) with implications for the closure of the eastern Paleotethys. J Asian Earth Sci 170. https://doi.org/10.1016/j.jseaes.2018.10.025

Xu R, Chen W, Deng M-G, Li W-C, Chen F-C, Lai C-K, Sha J-Z, Jia Z, Liu W (2021a) Geology and COS-Pb isotopes of the Fangyangshan Cu-Pb-Zn deposit in the Baoshan block (SW China): implications for metal source and ore genesis. Ore Geol Rev 103992

Xue CJ, Zeng R, Liu SW, Chi GX, Qing HR, Chen YC, Yang JM, Wang DH (2007) Geologic, fluid inclusion and isotopic characteristics of the Jinding Zn-Pb deposit, western Yunnan, South China: a review. Ore Geol Rev 31(1–4):337–359

Xue ZQ, Chen MH, Mao JW, Dang Y, Li YB, Xin TG, Wang K, Gao Y, Wu XL (2014) Comparison of fine structures of ore-bearing belts in the gacun- youre ore district and its significance for prospecting. Geol Exolor 50(4):599–616 (in Chinese with English abstract)

Yalikun Y, Xue CJ, Symons DTA (2017) Paleomagnetic age and tectonic constraints on the genesis of the giant Jinding Zn-Pb deposit, Yunnan, China. Mineralium Deposita https://doi.org/10.1007/s00126-017-0733-9

Yan ST, Tan CH, Duan YH, Li H (2020) The discovery of the ocean island rock association in the middle part of the Garze-Litang Ophiolite Mélange belt and its implication for tectonic evolution. Acta Geol Sin 94(2):439–449 (in Chinese with English abstract)

Yang ZM, Cook DR (2019) Porphyry copper deposits in China. SEG Spec Pub 22:133–187

Yang ZM, Hou ZQ, Li ZQ, Song YC, Xie YL (2008) Direct record of primary fluid exsolved from magma: evidence from unidirectional solidification texture (UST) in quartz found in Qulong porphyry copper deposit Tibet. Mineral Deposits 27(2):188–199 (in Chinese with English abstract)

Yang TN, Zhang HR, Liu YX, Wang ZL, Song YC, Yang ZS, Tian SH, Xie HQ, Hou KJ (2011) Permo-Triassic arc magmatism in central Tibet: Evidence from zircon U-Pb geochronology, Hf isotopes, rare earth elements, and bulk geochemistry. Chem Geol 284:270–282

Yang GS, Wen HJ, Ren T, Xu SH, Wang CY (2020a) Geochronology, geochemistry and Hf isotopic composition of Late Cretaceous Laojunshan granites in the western Cathaysia block of South China and their metallogenetic and tectonic implications. Ore Geol Rev 117:103297. https://doi.org/10.1016/j.oregeorev.2019.103297)

Yang M, Zhao F, Liu X, Qing H, Chi G, Li X, Duan W, Lai C (2020b) Contribution of magma mixing to the formation of porphyry-skarn mineralization in a post-collisional setting: The machangqing Cu-Mo-(Au) deposit, Sanjiang tectonic belt SW China. Ore Geol Rev 122:103518

Yang Z, Zhang XF (2021) Multiphase intrusion at the giant Pulang porphyry Cu-Au deposit in western Yunnan (Southwestern China): comparison between ore-causative and barren intrusions. Mineral Petrol. https://doi.org/10.1007/s00710-020-00734-8

Yang ZM, Hou ZQ, Zhou LM, Zhou ZW (2020c) Critical elements in porphyry copper deposits of China. China Sci Bull 65. https://doi.org/10.1360/TB-2020c-0246 (in Chinese)

Ying HL, Wang DH, Fu XF (2006) Timing and lead and sulfur isotope composition of Xiasai granite and silver polymetallic deposit in Batang Sichuan Province. Mineral Deposits 25(2):135–146 (in Chinese with English abstract)

Zhang Q, Li DZ, Zhang KW (1985) Preliminary study on tongchangjie ophiolitic mélange from Yun County Yunnan Province. Acta Petrologica Sinica 1(3):1–14 (in Chinese with English abstract)

Zhang YQ, Xie YW, Wang JW (1990) Rb and Sr isotopic studies of granitoids in Tri-River Region. Geochimica 4:318–326 (in Chinese with English abstract)

Zhang Y, Sun XM, Shi GY, Xiong DX, Zhai W, Pan WJ, Hu BM (2011) SHRIMP U-Pb dating of zircons from diorite batholith hosting Daping gold deposit in Ailaoshan gold belt, Yunnan Province China. Acta Petrologica Sinica 27(9):2600–2608 (in Chinese with English abstract)

Zhang SM, Xiao YF, Gong TT, He JL, Wang Q, Zhang L, Sun JD (2012) Optimal selection assessment on geochemical anomalies at gegongnong, hengxingcuo, mamupu in the Yulong Metallogenetic Zone Tibet. Bull Mineral Petrol Geochem 31(4):354–360 (in Chinese with English abstract)

Zhang WL, Cao HW, Yang ZM, Xi XY, Liu WW, Peng SM, Zhang L (2015) Geochemical characteristics and genesis of lamprophyres of the Cenozoic from the Suoluogou gold deposit, Sichuan Province China. Bull Mineral Petrol Geochem 34(1):110–119 (in Chinese with English abstract)

Zhang XF, Li WC, Yin GH, Yang Z, Tang Z (2017) Geological and mineralized characteristics of the composite complex in Xiuwacu W-Mo mining district, NW Yunnan, China: constraints by geochronology, oxygen fugacity and geochemistry. Acta Petrologica Sinica 33(7):2018–2036 (in Chinese with English abstract)

Zhang HC, Chai P, Zhang HR, Hou ZQ, Chen SM, Sun YB, Peng Q (2019) Two-stage sulfide mineral assemblages in the mineralized ultramafic rocks of the laowangzhai gold deposit (Yunnan, SW China): implications for metallogenetic evolution. Resour Geol 69 (3):270–286

Zhang XF, Li WC, Yang Z, Gao X, Zhu J, Liu WD (2020) Mineralization significance of Aplite in the Xiuwacu W-Mo deposit, NW Yunnan, China: constrains by geochronology, oxygen fugacity and geochemistry. J Asian Earth Sci https://doi.org/10.1016/j.jseaes.2020.104555

Zhang XF, Li WC, Wang R, Zhang L, Qiu KF, Wang YQ (2021) Preservation of Xiuwacu W-Mo deposit and its constraint on the uplifting history of Eastern Tibetan Plateau. Ore Geol Rev https://doi.org/10.1016/j.oregeorev.2021.103995

Zhou XK, Wang JG, Liu YC, Lei HY, Li GJ (2016) Primary superimposed halo features and deep ore-prospecting in the Donggualin gold deposit Yunnan Province. Geol China 43 (5):1710–1720 (in Chinese with English abstract)

Zhu J, Li WC, Zeng PS, Yin GH, Wang YB, Wang Y, Yu HJ, Dong T, Deng YQ, Luo C (2010) Geochemical characteristics and tectonic significance of basic rocks in the Yangla ore district, northwest Yunnan province. Geol Explor 46:899–909 (in Chinese with English abstract)

Zhu WG, Zhong H, Wang LQ, He DF, Ren T, Fan HP, Bai ZJ (2011) Petrogenesis of the basalts and rhyolite porphyries of the Minle copper deposit, Yunnan: geochronological and geochemical constrains. Acta Petrologica Sinica 27(9):2694–2708 (in Chinese with English abstract)

Geological Prospecting Method of Sanjiang and Integration of Exploration Technologies

6

As a specific assemblage of rocks and minerals, ore deposit is the product under the joint action of structure, magma, fluid and mineralizing activities in a long geological time, which is in a limited distribution range and fixed in spatial output position, shape and occurrence. The material composition of ore bodies (ores) is different from that of surrounding rocks (nonore), and some are even diametrically opposite. Thus, the ore deposits (types) formed under different geological backgrounds and along with different geologic and tectonic evolutions take on different performance characteristics, hence different prospecting methods as well as technical integration of effective prospecting methods. As one of the most complex orogenic belts in the world, the Sanjiang area has gone through not only the tectonic evolution of Tethys, but the fierce transformation of India-Eurasian Plate collision and plateau uplift. Therefore, the Sanjiang area is characterized by the most complicated structure, the strongest magmatic activities, the most active mineralizing fluid and complicated and diverse mineralization, thereby becoming a world-known metallogenic belt of nonferrous and precious metals. But simultaneously, the Sanjiang area is also the most complicated area in topography and geomorphology. Given the complex geomorphological landscape in the deep cut mountains and canyons in the Sanjiang area as well as the coverage of vegetation and ice deposits, conventional prospecting methods are no longer useful, particularly the geophysical prospecting technologies. The vast number of geological workers in the Sanjiang area have been pursuing how to make use of mature methods of geophysical prospecting, geochemical prospecting and remote sensing, while actively introducing and innovating new methods and technologies. Many methods have been successfully applied in the Sanjiang area and played a crucial part in guiding the discovery and evaluation of some ore deposits. However, with the increasing difficulty of ore prospecting, conventional and single methods may fail to help ore prospecting and exploration. Therefore, while introducing new methods and technologies, consistent efforts shall be made to explore the adaptability of various methods to topography and geomorphology and the effectiveness of different types of ore deposits in use and innovate the integration of methods and technologies. Yunnan's basic geological work has been done well. All types of geological techniques and methods have been widely applied, new technologies and methods have been used early, and exploration has gone deep. Apart from the area-based basic survey work, different prospecting methods and exploration techniques have been applied to the metallogenic provinces and belts with different metallogenic backgrounds. In the gold mineralization belt (such as Ailaoshan Gold Deposit Belt), relevant surveys have been conducted, including a geochemical survey of 1:10,000—1:25,000 soil in 1: 200,000 stream sediment anomaly areas (belts) and a geochemical survey of 1:50,000 stream sediment or 1:50,000 soil in many polymetallic metallogenic zones of gold, copper, lead, zinc and silver; geochemical survey of 1:50,000 stream sediment and 1: 50,000 soil in some areas simultaneously for comparison; 1:100,000 aeromagnetic survey and 1:100,000 ground gravity survey in the prospecting target area where concealed deposits are predicted (for example, the gravity survey of potassium salt in Pu'er Basin exceeds 10,000 km^2, and the evaluation of Hetaoping copper-iron mine is 400 km^2); and 1:10,000: 50,000 high-precision magnetic survey in ore-concentrated areas of iron, nonferrous metals and precious metals with magnetic conditions. Besides, we have conducted remote sensing interpretation of different data sources, applied hyperspectral and PIMA methods and used the Transient Electromagnetics (TEM), EH-4 electrical conductivity imaging system and controlled source audio magnetotelluric sounding earlier. Thanks to decades of practice and application, we have probed into five sets of effective integrated technologies for ore prospecting (exploration) in the prospecting practice in the Sanjiang area and achieved fast and efficient prospecting and exploration evaluation.

The following integrated technologies have been brought forward in a targeted manner, respectively, the integrated

W. Li et al., *Metallogenic Theory and Exploration Technology of Multi-Arc-Basin-Terrane Collision Orogeny in "Sanjiang" Region, Southwest China*, The China Geological Survey Series, https://doi.org/10.1007/978-981-99-3652-6_6

technology of "porphyry ore deposit model + hyperspectral + PIMA + high-precision magnetic survey + IP" for porphyry copper deposits, the "metallogenic model + horizon + Transient Electromagnetics + Induced Polarization" for VMS ore deposits, the integrated technology of "shear zone + geochemical exploration" for structurally altered rock type (orogenic belt) gold deposits, the integrated technology of "structural trap + hydrothermal circulation center + multiple electrical methods" for hydrothermal vein type Pb–Zn-Ag polymetallic deposits, and the integrated technology of "metallogenic system + gravity + magnetism + multiple electrical methods" for prospecting skarn/porphyry concealed deposits. These integrated technologies of ore prospecting have been extensively used.

6.1 Integrated Technologies for Exploration of Porphyry Copper (Gold, Molybdenum) Ore Deposits

In terms of porphyry copper (gold, molybdenum) ore deposits, we have achieved a breakthrough in prospecting Pulang porphyry copper deposits through the integration of "porphyry ore deposit model + hyperspectral + PIMA + high-precision magnetic survey + IP".

6.1.1 Background of Ore Deposit Prospecting

Pulang copper deposit is a significant enrichment area of porphyry copper polymetallic resources. In the early 1970s, two medium-sized copper mines, Hongshan (skarn) and Xuejiping (porphyry), were evaluated. The search for porphyry copper deposits in this area stopped since the exposed rock bodies are mostly intermediate-basic porphyry and the old belief that porphyry is beneficial to iron formation rather than copper formation. Given the discovery of porphyry copper–gold deposits and gold ore deposits associated with basic porphyry in some foreign regions, in 1999, Yunnan Provincial Bureau of Geology and Mineral Exploration and Development set up Gaoshan Company in collaboration with Britain's Billiton Corporation, which performed geological exploration of risks mainly targeted at porphyry copper–gold deposits, conducted 1:100,000 aeromagnetic survey in Gezhongza area, Shangri-La, and delineated several copper mineralization bodies in Pulang copper–gold prospecting area. Therein, drilling verification has been performed on the Qiansui magnetic anomaly of Pulang porphyrite (porphyry) rock mass. Three boreholes have been drilled on Pulang porphyrite (porphyry). The ore discovery of the first hole is good, while that of the second and third holes are of poor. In the Qiansui borehole, there are chiefly thicker magnetic bodies composed of pyrite, pyrrhotite and magnetite. In 2001, the BHP Billiton Group decided to withdraw from the venture exploration in this area.

The Ganzi-Litang oceanic crust subducted westward to shape Yidun Island Arc. The difference of plate structure and the nonuniformity of subduction speed may have resulted in the different subduction angles caused by the tearing of plate during the westward subduction. The plate in the northern section of the ocean crust is at a steep subduction angle and a fast speed, and an extensional arc characterized by inter-arc rift and back-arc-basin, namely Changtai arc, takes shape, which results in the formation of VMS Pb–Zn-Cu deposit (Xiacun) and epithermal Au–Ag-Hg ore deposit. The plate in the southern section of the ocean crust is at a gentle subduction angle and a fast speed, and a compressive arc characterized by andesite volcanic rocks and intermediate-acid porphyry, namely Shangri-La Arc, takes shape (Fig. 6.1). Therefore, on the basis of the key scientific and technological projects of the Ministry of Land and Resources, the research project entitled "Comprehensive Research on Mineralization Law and Prospecting Direction of "Three Rivers" in Southwest China" arranged by China Geological Survey has always insisted that conditions are available for looking for porphyry and skarn copper (molybdenum and gold) in Pulang area.

The members of the project team conducted a great deal of research and learned that this area is not only covered with both the intermediate-basic porphyry as known before and the widely distributed and large-scale intermediate-basic porphyry. From their perspective, magma in this area is in a complete evolution sequence, and all sorts of magmatic rocks range from basic to neutral to acidic. They state that favorable conditions are available in this area for looking for large porphyry copper deposits: ① fierce magma differentiation has occurred to the large-scale mother magma of porphyry in this area, producing copper-containing acidic porphyry magma, which can be described as "fat mother and strong baby"; ② a large area of propylitization and potassium silication has developed from the porphyrite/porphyry complex rock mass with lithofacies differentiation, confirming that a magma-fluid system associated with porphyry copper deposits has developed in the late phase of porphyry evolution; ③ Among the three verification holes, good porphyry copper mineralization is discovered in one borehole, which demonstrates an enormous potential of copper resources. Therefore, despite the "two ups and two downs" of this area, the project team members still believe in the prospects of prospecting. In 2002, the members of the project team made a special report to China Geological Survey with new research knowledge and new progress made in field investigation, which was highly valued by the China Geological Survey, allocating one million (RMB) to continue exploration and outcrop exposure. Expected results were achieved through surface exposure and further

Fig. 6.1 Background for formation of Shangri-La Porphyry copper zone in Yidun Island arc zone

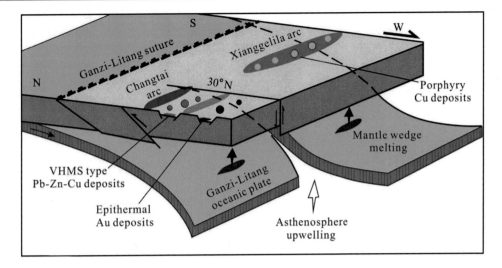

research. In 2003, the China Geological Survey included the prospecting of this area into a new round of major land and resources survey project and carried out large-scale explorations, launching the copper mine evaluation in Shangri-La area again.

6.1.2 Integrated Technologies for Exploration

The breakthrough of prospecting ideas has laid the foundation for making the breakthrough of prospecting, and the selection of prospecting methods (integration) is a key link to achieve the effectiveness and fast evaluation of prospecting. In Pulang area, we made the breakthrough of prospecting through the integrated technologies of "porphyry ore deposit model + hyperspectral + PIMA + high-precision magnetic survey + IP".

6.1.2.1 Porphyry Ore Deposit Model

The ore deposit model of porphyry copper deposit has become relatively mature. The wall rock alteration in the porphyry body i inside out, generally silicification (core)—potassium silicified zone (biotite-potash feldspathization zone)—phyllic zone (quartz-sericitization zone)—(argillization zone)—propylitization zone- (skarn) hornfelsization zone and the mineralization types also correspondingly vary from being free from ore core—sparse (dotted) disseminated—dense (veinlet) disseminated—contact metasomatism (skarn-type)—filling metasomatic (big vein type) in wall rocks, in line with a clear distribution law. Therefore, the prospecting of porphyry copper (molybdenum and gold) deposits must be conducted under the guidance of the metallogenic model of porphyry copper (molybdenum and gold) mines. The mineralization of Pulang copper deposit can be summarized as below.

(1) Ganzi-Litang Ocean Basin began to subduct westward in the Early Triassic, leaving the Shudu ophiolite melange only and foming the southwest porphyry belt (242.92–237.5 Ma), in which the Xuejiping porphyry copper mine (the isochron age of whole ore rock + biotite Rb–Sr is 224.6 Ma, Tan Xuechun 1985) was produced; large-scale subduction occurred in the Late Triassic. After a large number of dacitic volcanic eruptions broke out, large-scale subvolcanic rocks (e.g., quartz diorite porphyry) and intrusive rocks (e.g., quartz monzonite porphyry) (218–203 Ma) developed in succession, forming complex rock masses. The mineralizing $^{40}Ar/^{39}Ar$ age of Pulang porphyry copper deposit is 213–216 Ma; the Re-Os isochron age of molybdenite is 213 Ma ± 3.8 Ma, which is a typical ore deposit shaped in the Indosinian period.

(2) Quartz dioritic porphyrite is distributed in a large area, generally the mineralization of pyrite, chalcopyrite, etc. Industrial ore bodies are chiefly produced in quartz monzonite porphyry, and some vein ore bodies are produced in granodiorite-porphyry and other wall rocks.

(3) Copper, the main ore-forming metal, chiefly occurs in chalcopyrite, and a minute quantity occurs in bornite and covellite, distributed in porphyry (porphyrite) in the fine-vein disseminated and fine-grained—fine-grained sparse and dense disseminated form together with pyrite. A large vein-shaped ore body is shaped in the outer zone, which takes a "three-story" mode of porphyry copper granular disseminated, fine-meshed vein disseminated and large vein-shaped ore body.

(4) The alteration zoning of ore-bearing rock body in Pulang copper deposit is obvious, with strong silicification zone (local), potash feldspar biotitization zone, quartz-sericitization zone, propylitization zone (local illite-carbonatization zone) and skarn hornelialization

Fig. 6.2 Alteration zoning pattern of Pulang copper deposit in Shangri-La

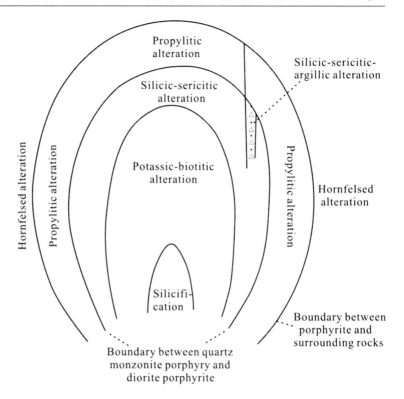

zone developed from the center. Among them, propylitization zone and hornelialization zone are developed, with large earth surface distribution area. Compared with the complete alteration zone of porphyry copper deposit, the argillic zone of this copper deposit is not developed, with the wide range of propylitization zone. The ideal model is shown as Fig. 6.2. The distribution of the main trace elements, like Cu, Mo, Au, Ag, Pb, Zn, W and Bi, in the rock have the characteristics of annular zoning with the rock mass as the center. While Cu and Mo (W, Bi) are in the inner zone, Cu and Au are distributed across the rock mass and surrounding rock, and Ag, Pb and Zn are in the outer contact zone.

(5) Ore body shape. In the plane, the ore body is in the shape of a "gourd", and it is wide in the south and narrow in the north and lies to the north. In the section, the ore body inclines eastward, and it is speculated that there is uplift on the east. Industrial ore bodies mainly occur in quartz monzonite porphyry. The middle and upper part of quartz monzonite porphyry show veinlet disseminated state, while the lower part of it present dense and sparse disseminated state, and the upper porphyrite has a vein ore body, forming a "three-story" mineralization style (Fig. 6.3).

(6) Metallogenic Temperature. The homogenization temperature of copper ore mostly concentrates in four ranges (160–180 °C, 180–210 °C, 280–320 °C, 300–430 °C) (multi-phase inclusions), and the average Th values are roughly 170 °C, 230 °C, 305 °C and 360 °C.

(7) Metallogenic Epoch. According to the Re-Os date of molybdenite and K–Ar date of mineralized porphyry, the activity time of potash selection (biotitization and potash feldspathization) biotite quartz monzonite porphyry mineralization in Pulang porphyry copper mine is roughly 235.4 Ma ± 2.4 Ma to 221.5 Ma ± 2.0 Ma, and the Re-Os age of molybdenite in quartz-molybdenite stage is roughly 213 Ma ± 3.8 Ma, both ages of which are very similar. The mineralization of Pulang porphyry copper deposit was completed in the Indosinian period.

(8) Mineralization Stage. According to the mineral assemblage and production characteristics of the ore deposits in the mining area, the metallogenic period of mineralization can be divided into ① Late magmatic mineralization. It refers to the mineralization of potassium-rich magmatic gas and liquid, which is carried out upward and outward from the rock mass without Tianshui, and forms biotite-potash feldspar-metal mineralization association with potassium. ② The post-magmatic hydrothermal metallogenic period: It is the most important metallogenic epoch, and all types of major mineralization are formed by this mineralization. From early to late, from inside to outside, from high temperature to low temperature, the mineralization can be divided into three stages. In the high temperature stage, quartz-biotite-potassium feldspar-metal sulfide association is formed in porphyry; in the medium

Fig. 6.3 Schematic diagram of output forms of different types of ores (bodies)

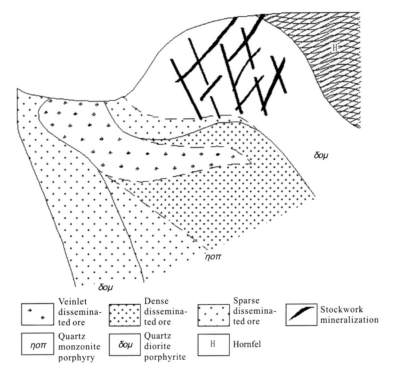

| | Veinlet disseminated ore | | Dense disseminated ore | | Sparse disseminated ore | | Stockwork mineralization |
| | Quartz monzonite porphyry | | Quartz diorite porphyrite | | Hornfel | | |

temperature stage, quartz-sericite-metal sulfide association was formed in porphyry; in the low-temperature stage, with weak mineralization, only a few veinlets of propylite -chalcopyrite-pyrite assemblage are produced, and no ore body is formed. ③ The super-gene period: The super-genesis in the shallow part near the west and southeast of KT1 ore body is strong, which causes the metal minerals chalcopyrite, pyrite and pyrrhotite to oxidize into limonite and malachite. The oxidation zone is 10–40 m in depth. The copper in the oxidation zone is partially leached, so the copper grade of ore body decreases, and the copper sulfate solution formed in the oxidation process forms carbonate copper mineral.

The model of Pulang copper deposit is shown as Fig. 6.4.

6.1.2.2   Extraction of Hyperspectral Data (Hyperspectral Images) and Alteration Information

Remote sensing hyperspectra are used to identify magmatic rock belts and main alteration types, infer concealed rock masses and delineate exposed rock masses.

In November 2003, Yunnan Geological Survey purchased hyperspectral data (hyperspectral images) of 4 scenes in Pulang area of Shangri-La County. Yunnan Geological Survey has successively cooperated with Professor Wang Runsheng from China Aero Geophysical Survey and Remote Sensing Center for Land and Resources (AGRS) and Professor Hu Guangdao from China University of Geosciences (Wuhan) to interpret and analyze, which

provide a basis for delineating and tracing the distribution of magmatic rock belts, the spreading characteristics and main alteration types of alteration belts in this area. The following is an introduction to the interpretation of Professor Hu Guangdao and other research groups. In the extraction of altered mineral information, the spectral angle mapping was adopted to identify altered minerals in "Gaochiping-Lannitang area" and "A're-Pulang area". The interpretation and analysis of A're-Pulang area are illustrated as follows: magmatic rocks in Pulang area are distributed in zone, three magmatic rocks zones can be delineated from north to south, including NW-trending Bidu-Zhuoma intermediate-acid magmatic rocks zone, NW-trending Disuga-Songnuo-Pulang intermediate-acid magmatic rocks zone, and the close NS-trending Chundu-A're intermediate-acid magmatic rocks zone (Fig. 6.5). By processing the hyperspectral data and extracting the alteration information of A're-Pulang-Langdujing, most altered areas are consistent with the known porphyry bodies (Figs. 6.6, 6.7 and 6.8).

6.1.2.3   PIMA Application

The ore-bearing porphyry bodies and alteration zoning can be rapidly delineated within complex rock bodies by using PIMA technology.

Mineral mapping is a spectral mapping technology developed on the basis of hyperspectral remote sensing. The spectral interval used for mineral mapping is in the infrared region. At present, short wavelength infrared has been applied to observe minerals containing water or OH and some sulfate minerals and carbonate minerals.

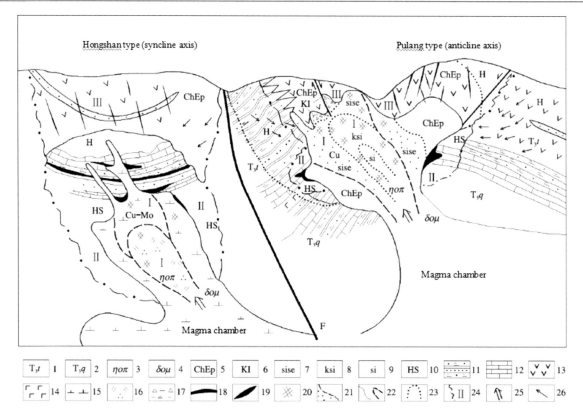

Fig. 6.4 Metallogenic model of copper deposits related to porphyrite in Shangri-La area. 1-Tumugou formation; 2-Qugasi formation; 3-quartz monzonite porphyry; 4-quartz dioritic porphyrite; 5-propylitization zone; 6-argillic zone; 7-Sericitization zone; 8-potassium silicification zone; 9-silicified nucleus; 10. hornstone skarnization zone; 11-sandstone slate; 12-limestone; 13-intermediate volcanic rocks; 14-basic volcanic rocks; 15-porphyrite; 16-quartz monzonite porphyry; 17-argillization; 18-skarn ore body; 19-vein ore body; 20-porphyry reticulated ore body; 21-disseminated ore body; 22-lithologic and lithofacies erathem of intrusive rocks; 23-alteration zoning erathem; 24-mineralization type erathem and serial number; 25-magmatic intrusions and rising direction of gas and liquid in the late period; 26-migration direction of mixed hydrothermal liquid

As a new concept, the purpose of mineral mapping is to determine the distribution and relative content changes of one mineral or some minerals on earth surface, and it can also be used to determine rock and mineral samples from deep. Given the high resolution of hyperspectral remote sensor, it is feasible to map out the distribution of target minerals from spectral images and determine their relative abundance at the same time. Here, the concepts of stratum, rock stratum or rock mass in the traditional geological significance are not taken into consideration, and the basic unit of mapping has been as small as a single mineral. The distribution of individual mineral may be controlled by primitive lithology and may also be caused by late geological processes (hydrothermal alteration, tectonic deformation, weathering and denudation, etc.) or modern human activities. As the mineral map can provide information related to various geological processes in the late period and even human activities after the formation of stratum and rock stratum, the mineral map can be used in the fields of prospecting, engineering geology, natural disaster

monitoring and environmental pollution investigation. In terms of mineral prospecting and exploration, the mineral map directly provides information on mineral distribution and abundance related to mineralization. With mineral information and the genetic model of ore deposits, the detailed investigation and exploration targets can be effectively determined. In addition, the rock mass erathem and mineralization range can be roughly delineated through a combination of high-precision magnetic survey and induced polarization (IP) method, realizing a breakthrough in rapid prospecting.

PIMA portable near-infrared spectrometer produced by Integrated Spectronics Pty Ltd (Australia) was used in rock (ore) samples test in Pulang copper deposit. The PIMA was applied by Lian Changyun and other researchers from the Development Research Center of China Geological Survey and Yunnan Geological Survey.

A total of 18 altered minerals were identified by PIMA, which are listed as follows in order of quantity: anhydrite, illite, magnesium-chlorite, hydrated kaolinite (halloysite),

Fig. 6.5 Geological Diagram of
A're-Pulang-Langdu area

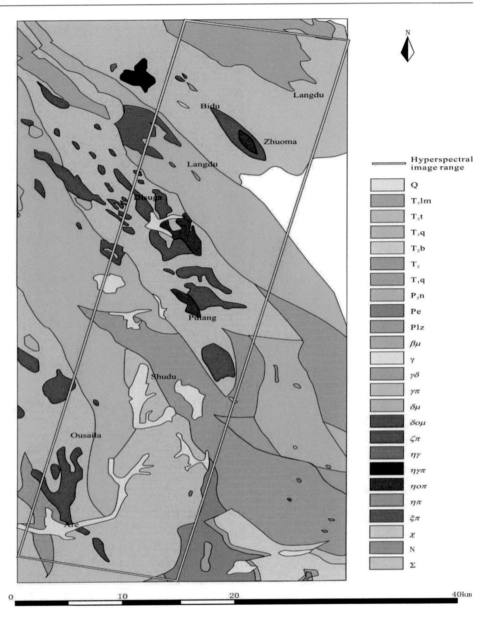

ferromagnesian-chlorite, hornblende, ferrochlorite, muscovite, biotite, montmorillonite, phengite, actinolite, tremolite, calcite, phlogopite, dolomite, tourmaline and kaolinite.

One principle for sampling is that the sampling is carried out at a certain dot spacing (generally 5 m dot spacing, which can be made denser to 1–2 m in areas with strong alteration) along the existing exploration line profile. The earth surface and profile measurement are based on exploratory trench and borehole core, respectively. Another sampling principle is to make the samples representative.

Core samples from 9 boreholes and rock samples from 2 earth surface long profiles were collected in Pulang porphyry copper deposit area. A total of 927 core samples were collected from these 9 boreholes. The collection of earth surface samples is mainly concentrated on the 0 exploration line and 0' exploration line. Part of two profiles were exposed, with samples taken from the exploratory trench; where the engineering disclosure was not carried out, sampling was conducted on the surface outcrop. The actual control length was 2249 m, and 356 samples were collected. In order to test the effectiveness of PIMA instrument in mineral identification and delineate alteration zoning, representative samples were selected from the samples tested by PIMA instrument for optical thin slice identification at the same time. Some projects were added according to the "cross" profile. A total of 460 optical thin slice samples of rock-mineral were collected for comparison and comprehensive identification.

Fig. 6.6 Hyperspectral image map of A're-Pulang-Langdu area

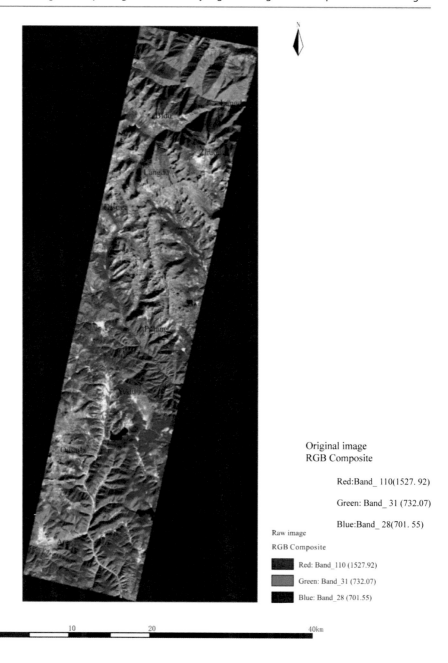

Original image
RGB Composite

Red:Band_ 110(1527. 92)

Green: Band_ 31 (732.07)

Blue:Band_ 28(701. 55)

Raw image

RGB Composite

Red: Band_110 (1527.92)

Green: Band_31 (732.07)

Blue: Band_28 (701.55)

According to the spectral measurement results, at least, the following three main alteration zones can be identified in Pulang porphyry copper deposit area:

(1) Potassic alteration zone: It is mainly characterized by abundantly developing biotite (sometimes phlogopite) and actinolite, and it is super-imposed with chlorite, illite, anhydrite and other minerals (Fig. 6.9). There are a lot of actinolite in this zone. This indicates that the denudation of porphyry metallogenic system is relatively light, for actinolite is formed by oxidation due to acid alteration near the earth surface. Generally speaking, there are many biotites in the Pulang porphyry

copper deposit area, which indicates that potassium alteration is an important thermal event in this area. According to the law of porphyry metallogenic system, the dispersed biotite alteration often occurs in the early stage of mineralization, with a relatively large range. After that, the potassic alteration zone was super-imposed and transformed by sericitization in the late period, so more chlorite and illite can be seen in ZK0608 borehole.

(2) Sericitization zone: It is characterized by developed sericite (mainly illite here), chlorite and anhydrite, which formed a halo around the potassic alteration zone. Illite is the main altered mineral in this zone, and

Fig. 6.7 Extraction of Altered Minerals from A're-Pulang-Langdu (SAM, Threshold = 0.2)

some places are distributed with muscovite and polysilicate muscovite. This zone tends to, but not always strictly controlled by porphyry bodies. The main ore body in Pulang copper deposit is located in this zone.

(3) Argillic zone: This zone is mainly composed of montmorillonite, illite and chlorite are also distributed. This zone is mainly located outside the sericitization zone. The appearance of large amount of montmorillonite in the periphery means that montmorillonite was formed in the late stage of mineralization due to faults or the infiltration of cold groundwater of breccia zone. Montmorillonite represents the last stage of mineralization of Pulang porphyry copper deposit.

Figure 6.10 shows the spatial variation characteristics of altered minerals depicted by spectral measurement results in Pulang porphyry copper deposit area. Figure 6.11 demonstrates the zoning of altered minerals inferred according to the spatial distribution and variation characteristics of altered minerals. From Fig. 6.11, the altered mineral zoning obtained by PIMA has obvious asymmetric characteristics. The potassic zone of altered center is located in the eastern part of the mining area, in which low-grade copper ore bodies are produced. Sericitization zone is located above and on the east side of the main ore body, and it is the main output part of the copper ore body. The narrow argillic zone, which is located in the west of the mining area, is marked by the increase of altered minerals, including montmorillonite.

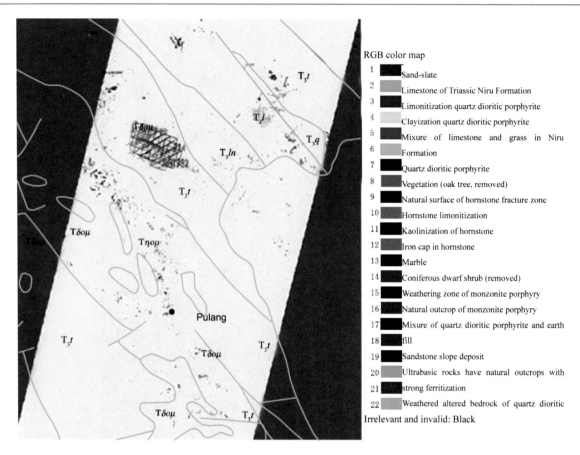

Fig. 6.8 Distribution characteristics of alteration zone in Pulang copper deposit and its vicinity

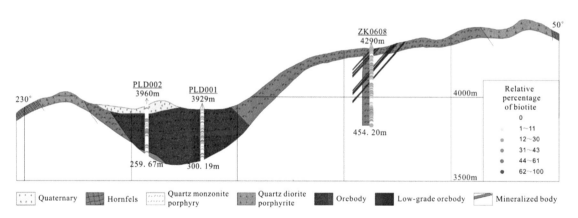

Fig. 6.9 Biotite distribution identified in profile of 0 exploration line of Pulang porphyry copper deposit area

Application of PIMA in ore prospecting. Based on the model of altered mineral zoning and the mineralization law of porphyry copper deposit, the prospecting prediction area of Pulang porphyry copper deposit area was delineated (Fig. 6.12), which provides guidance for the exploration of similar deposits in unknown areas.

According to the altered mineral zoning model established by us, further prospecting in Pulang porphyry copper deposit area should be confined to the annular range centered on ZK0608 borehole, that is, to increase exploration efforts in the periphery of potassic alteration zone and sericitization zone. Given characteristics of ore body formation, results of geological, geophysical and geochemical exploration revealed, and the occurrence of rock mass and ore body, it is suggested that further exploration of porphyry copper deposits in Area A and northward of Area A in Fig. 6.12 should be conducted in the near future.

Fig. 6.10 Spatial distribution characteristics of altered minerals measured by PIMA instrument in Pulang porphyry copper deposit area

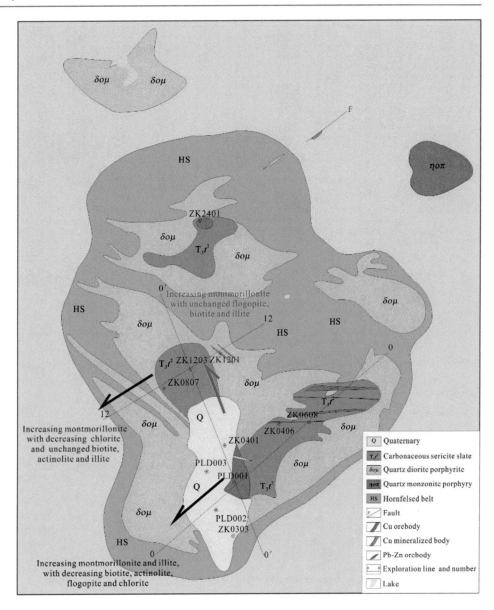

In terms of mining area evaluation, the prediction area map (Area A in Fig. 6.12) depicted by PIMA was drilled and verified without finding out the ore body shape and strike. The results show that the ore body extends to the northeast direction, and the lower part of the prediction area is the best ore-seeing position at present.

6.1.2.4 Magnetic and Electrical Measurement

On the basis of 1: 100,000 geochemical prospecting (water system) and 1: 50,000 geochemical prospecting (soil) survey, various geophysical and geochemical prospecting were carried out in the mining area, including 1: 10,000 IP intermediate gradient surface scanning, 1: 100,000 aeromagnetic survey, 1: 50,000 and 1: 10,000 high-precision magnetic survey, transient electromagnetic measurement

(TEM), amplitude-frequency IP survey, etc. The results of high-precision magnetic survey and IP survey were better (Fig. 6.13).

6.2 Integrated Technologies for Exploration of Volcanic-Associated Massive Sulfide Deposit (VMS) and Sedimentary Exhalative Deposit (Sedex)

The massive sulfide deposit discussed in this section generally refers to volcanic-associated massive sulfide deposit (VMS) and sedimentary exhalative deposit (Sedex). Although these two types of ore deposits are different in terms of formation conditions and ore deposit characteristics,

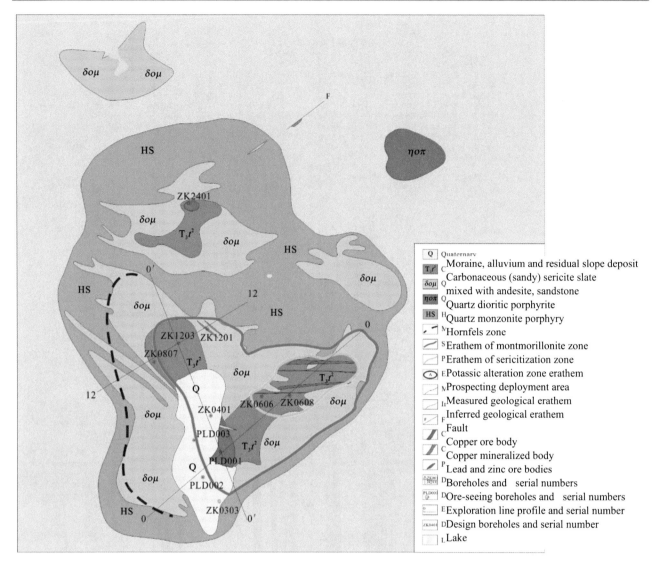

Fig. 6.11 Altered minerals zoning measured by PIMA in Pulang porphyry copper deposit area

their common features are that there are ore-bearing beds and stratiform ore bodies, which generally contain high content of sulfides and magnetic minerals. The model of this kind of ore deposit has been studied deeply, so model prospecting is still an important prospecting method. During the prospecting, ore deposits with different occurrence types have different emphasis, but "metallogenic model + horizon + Induced Polarization" is an effective prospecting method. For exposed VMS deposit and Sedex deposit, in addition to "metallogenic model + horizon + Induced Polarization", high-precision magnetic measurement of 1: 10,000 can be added to form an integrated technical method of "metallogenic model + horizon + Induced Polarization + magnetic method". For deep buried deposits, the integrated technology of "metallogenic model + horizon + Transient Electromagnetics + Induced Polarization" was adopted, which delivered satisfactory results.

6.2.1 Integrated Technologies of "Metallogenic Model + Horizon + Transient Electromagnetics + Induced Polarization Method" Have Achieved a Breakthrough in Prospecting of Dapingzhang Copper Polymetallic Deposit

For rift metallogenic system in Yunxian-Jinghong rift zone (Wang Baolu et al. 2001), the initial time of its development has always been a hot issue. It is generally believed that the system originated from Carboniferous. Yang Yueqing (2000) determined that the diagenetic age of quartz porphyry in Dapingzhang mining area was 236 Ma (Rb–Sr isochronous age). However, the recent Re-Os isochronous ages (428.8 Ma ± 6.1 Ma to 432.4 Ma ± 5.6 Ma) obtained in Dapingzhang ore body indicate that the ore deposit was formed before Middle Silurian. This is of great significance

Fig. 6.12 Schematic diagram of prospecting deployment area in Pulang porphyry copper deposit area. Interpretation and inference map of comprehensive profile of 26 exploration line in Pulang mining area

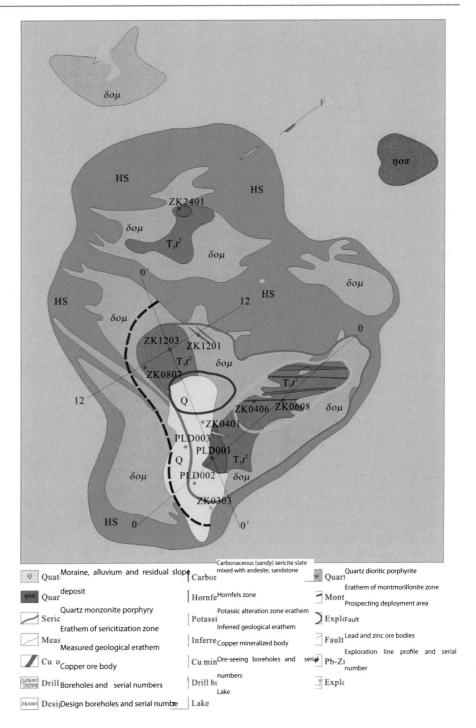

for guiding further discussion of Tethys tectonic evolution and regional prospecting. Dapingzhang Cu-Pb–Zn polymetallic deposit is a submarine volcanic-sedimentary exhalative deposit (VMS deposit), which is distributed along volcanic apparatus and volcanic depressions in Jiufang fault zone. The metal elements of the ore deposit show vertical zoning, that is, the massive ore body of the upper basin facies is Cu-Pb–Zn-Au–Ag symbiotic assemblage; the metal element of lower pipeline facies veinlet disseminated ore body are dominated by Cu.

According to the composition, intensity of molten lava and types of volcanic eruptions from early to late, the primitive volcanic eruption process can be basically divided into one eruption cycle and three eruption sub-cycles.

The first eruption sub-cycle is dominated by a large amount of submarine sodium-rich molten lava. At first,

Fig. 6.13 Interpretation and inference map of geophysical prospecting profile in Pulang deposit area

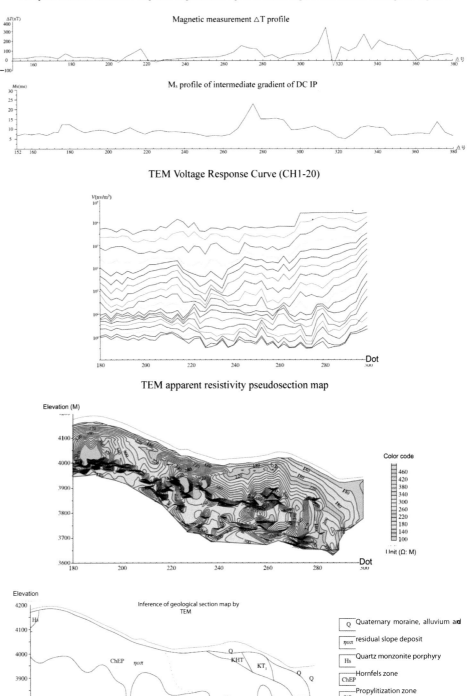

Interpretation and inference map of comprehensive profile of 26 exploration line in Pulang mining area

quartz keratophyric molten lava spouted, then the composition of the molten lava turned into keratophyric material. At the same time, basic spilitic molten lava spouted in a short time and finally ended in submarine volcanic deposition. VMS deposits were formed during the intermittent eruption period. The second eruption sub-cycle not only has strong volcanic eruption activity but also has volcanic eruptions of a certain scale. At first, the eruption of rhyolitic molten lava and intermediate-basic molten lava dominated, but the eruption weakened in the medium term. In the intermittent period, volcanic breccia and tuff were produced by associated eruption. In the late period, eruption and outburst tended to weaken. Some tuff and sedimentary pyroclastic rocks far away from volcanic channels were formed. In the third eruption sub-cycle, volcanic activity is coming to an end. At the beginning of this eruption cycle, strong intermediate molten lava erupted, and then, it turned to the eruption of acid molten lava after a short break. Consequently, volcanic outburst dominated. Volcanic breccia was formed in this process. It finally ended with tuffaceous-siliceous rocks and normal sedimentary rocks formed by tuff.

Metallogenic mechanism: In the intermittent period of submarine volcanic eruption, Cu, Zn, Pb, Ag, and Au carried by volcanic jet-hydrothermal solution contain halogen and sulfur; driven by magmatic hydrothermal solution, groundwater derived from seawater produces convection, and leach minerals from rocks. When the ore-bearing hydrothermal solution rises to the vicinity of the submarine eruption outlet, the solution boils and gasifies due to pressure release. Then, the solution is injected into sedimentary depressions to form massive ore bodies. Volcanic depressions are the best exhalative metallogenic environment. There is disseminated (volcanic) hydrothermal Cu mineralization in the volcanic channel of volcanic rocks or the contact zone inside and outside the edge of the late volcanic dome. Then Dapingzhang disseminated copper ore bodies were formed.

The ore body occurs in the spilite keratophyre series, and the upper stratiform and stratiform-like massive ore body and the lower veinlet ore body are combined to form a "double-bed structure". See Fig. 5.11 for the ore deposit model. The massive ore body mainly develops banded, striped and massive ore structures, and its ore metal minerals include sphalerite, chalcopyrite, pyrite, galena and silver tetrahedrite, with a total content of 83.83% (Yang Guilai). The massive ore body is characterized by rich sphalerite and chalcopyrite, and gangue minerals include quartz, calcite, sericite, chlorite and barite. The content of metal sulfide in disseminated ore is generally less than 35%, mainly including pyrite, chalcopyrite, trace sphalerite, galena, chalcocite, limonite, etc. The gangue minerals include quartz, calcite, sericite and chlorite. The alteration closely related to mineralization mainly develops silicification, chloritization and pyritization, etc. Three massive ore bodies are delineated. The average grade of ore bodies is Cu 2.14%-3.80%, Pb 0.52%-2.89%, Zn 2.60%-9.46%, Au 0.43×10^{-6}-2.15×10^{-6}, Ag 82.12×10^{-6}-158.73×10^{-6}. The average grade of disseminated ore body is Cu 0.92%, Pb 0.04%, Zn 0.21%, Au 0.52×10^{-6}, Ag 10.91×10^{-6}.

Breccia ore is developed in the ore body, and the composition of breccia is basically the same as that of massive ore. It is inferred that after the formation of massive ore body, the ore body at the top of volcanic neck breaks and enters into volcanic neck to form breccia ore.

According to the characteristics of this type of deposit, in the prospecting of Dapingzhang copper polymetallic deposit, we have adopted various geophysical and geochemical prospecting methods to explore its integrated prospecting technology.

6.2.1.1 Comprehensive Geophysical Prospecting Experiment of Profile

The electrical and magnetic properties of rocks (ore) exposed in the area (mainly sandstone, mudstone, limestone, quartz keratophyre, rhyolite porphyry and sulfide ore, etc.) were measured. The results show that while the charge rate of ore is the highest (>30%), the apparent resistivity of ore is the lowest (<72.5 $\Omega \cdot$ m). The ore has the characteristics of low resistivity and high polarization. The charging rate of tuff, dacite, mudstone and limestone containing little or no sulfide is extremely low, ranging from 3 to 4%. There are obvious differences in electrical properties between ore bodies and surrounding rocks, and there is a good geophysical premise for IP work in this area. The magnetic measurement of rock (ore) shows micro-magnetism or nonmagnetism, indicating extremely weak magnetism. There is not much difference between rock and ore in magnetic measurement, so it is obvious that magnetic measurement results show it does not have geophysical conditions.

We need to study the geophysical characteristics of ore deposits, understand the distribution range and burial depth of ore bodies as soon as possible, and guide the arrangement and construction of prospecting projects. So, firstly, we did comprehensive geophysical prospecting (IP sounding, spontaneous electric field, high-precision magnetic survey and charging survey) tests on geological exploration line 7 and exploration line 16. Then, Transient Electromagnetics tests were conducted on exploration line 1, 10, 16 and 57. Based on the tests, area IP intermediate gradient and self-electric measurement were carried out. It showed that the results of IP and Transient Electromagnetics were abnormal. IP intermediate gradient and self-electric measurement were in good agreement with ore bodies (Fig. 6.14).

6.2.1.2 Area Comprehensive Geophysical Prospecting

Through the comprehensive test of five geophysical prospecting methods on the profile, the results of IP, self-electricity and Transient Electromagnetics are better, which affirms the effectiveness of these three geophysical prospecting methods. First, IP method is selected as the main method and technology to carry out surface scanning work.

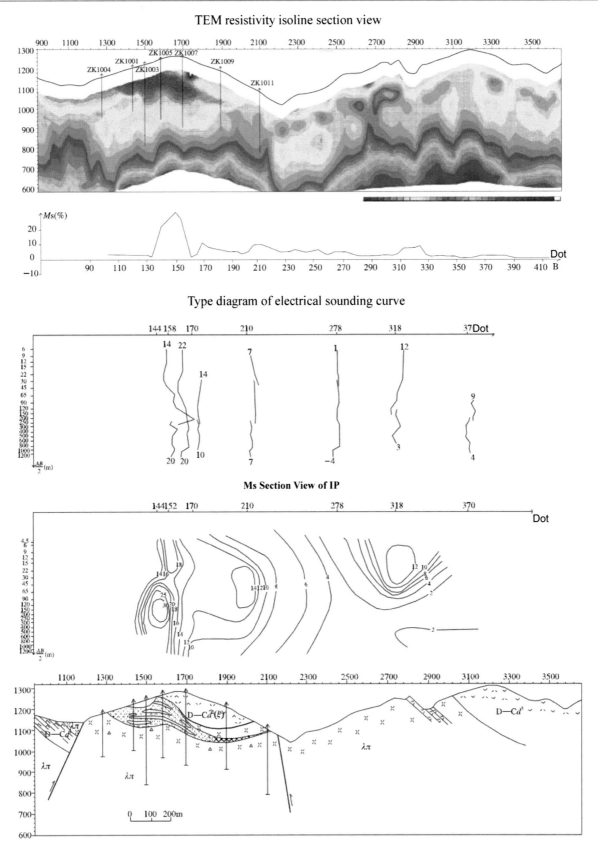

Fig. 6.14 Geological and geophysical prospecting model of exploration line 10 in Dapingzhang ore deposit

The mining area was scanned for 9 km² by 1: 10,000 IP middle gradient, and four IP intermediate gradient anomalies with M ≥ 10% were circled (Fig. 6.15). With engineering verification, industrial ore bodies and mineralized bodies are found in all projects constructed in anomalies. Physical property data and ore-seeing projects in the area show that the remaining abnormal values on ore bodies are all greater than 10%. Therefore, when there is a certain scale of anomaly, the residual value is generally greater than 10% and is ore-induced anomaly. This indicates that IP intermediate gradient has a certain effect on finding disseminated ore in Dapingzhang deposit, delineating the distribution range of mineralized body and guiding engineering layout.

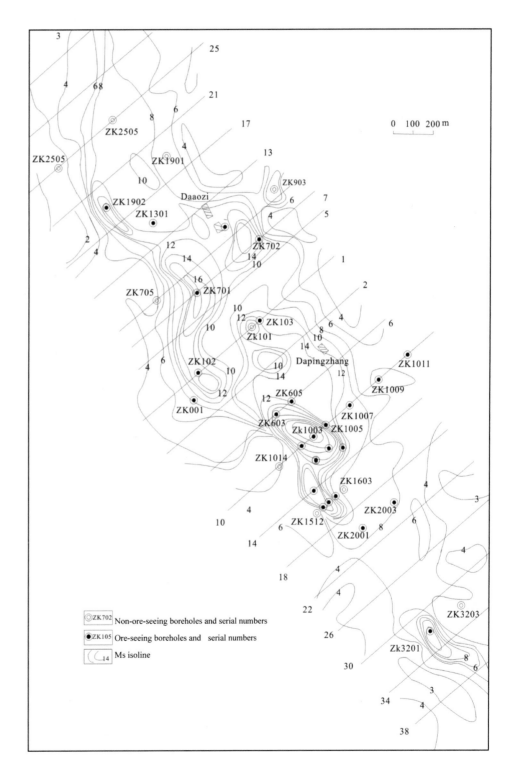

Fig. 6.15 IP intermediate gradient (Ms) anomaly diagram of Dapingzhang ore deposit

To sum up, the geophysical prospecting anomalies in this area are obvious. The ore-seeing rate of engineering verification is high, with remarkable prospecting effect.

Geophysical prospecting model: The plane distribution range and morphological characteristics of IP anomaly are in good agreement with copper polymetallic ore bodies exposed in outcrops or boreholes. The data of IP sounding and Transient Electromagnetics obtained from typical exploration profiles fully reflect the characteristics of low resistance and high polarization of copper polymetallic ore bodies. The geoelectric profiles of Ms, ρs and TEM are consistent with the exploration profiles, reflecting the spatial mode of occurrence of copper polymetallic ore bodies, as shown in Fig. 6.14. The geophysical prospecting model is characterized by the coincidence of IP intermediate gradient and TEM low resistivity anomaly. The residual abnormal value of IP intermediate gradient is greater than 10%. The Transient Electromagnetics has low resistivity, and the apparent resistivity is less than 200 Ω m. The combination of geophysical prospecting methods is effective under the specific metallogenic geological conditions and geophysical premise of Dapingzhang deposit area by measuring the area IP intermediate gradient and carrying out the area Transient Electromagnetics selectively.

After exploration, we put forward the integrated technology of "metallogenic model + horizon + Transient Electromagnetics + Induced Polarization" to find volcanic-associated massive sulfide deposit (VMS) similar to Dapingzhang copper polymetallic deposit. This is an effective integrated technology for prospecting.

6.2.2 Location Prediction of Ore Body in Luchun Zn-Cu-Pb (Ag) Polymetallic Ore Target Area

According to the geologic characteristics, ore deposit genesis, mineral symbiotic assemblage, mineral form of Zn-Cu-Pb component and physical properties of rocks and ores of Luchun Zn-Cu-Pb (Ag) polymetallic ore deposit, the geophysical prospecting methods such as high-precision

magnetic method, Transient Electromagnetics and amplitude-frequency Induced Polarization are used to predict the location of ore body and the ore deposit, and good results have been achieved.

6.2.2.1 High-Precision Magnetic Method

Determination of Geological Physical Properties

60 specimens of rock magnetic were collected and measured for parameters determination in Luchun deposit (Table 6.1). With strong magnetism, the magnetic susceptibility of the massive ore is (36,520–239,270) $4 \pi \times 10^{-6}$ SI, and the remanence is 1560×10^{-3}–$25,420 \times 10^{-3}$ A/m. With weak magnetism, the magnetic susceptibility of the surrounding rock is (98–1627) $4 \pi \times 10^{-6}$ SI in general, and its remanence is 50×10^{-3}–1260×10^{-3} A/m. The statistical table of magnetic parameters shows that there are obvious magnetic differences between the mineralized body and surrounding rock. Therefore, it is considered that the anomaly is ore-induced anomaly, which has the physical property for implementing high-precision magnetic survey.

Characteristics of High-Precision Magnetic Anomalies

The high-precision magnetic method was arranged in the whole mining area, with a total of 15 exploration lines, showing good magnetic survey results for the known outcrop areas of ore bodies and the areas covered by plants, slope deposits and ice deposits (Figs. 6.16 and 6.17).

The results of high-precision magnetic surface scanning in Luchun deposit showed that the intensity of magnetic anomaly in the north of exploration line P12 was large, with obvious magnetic gradient changes. The magnetic anomaly was consistent with the horizon and strike of mineralized body in the mining area. The middle ore section was basically connected with the mineralized bodies in the north ore section and the south ore section in the strike. The ore sections were not staggered or discontinued by the large displacement of the close east–west fault. The outcrop of the shallow mineralized bodies had a certain downward displacement due to the landslide. The south area of exploration line P12 is the extension section of this work, without any

Table 6.1 Statistical table of magnetic parameters of rocks and ores in Luchun deposit

Rocks	Number of specimens	κ/(4π × 10⁻⁶)SI		Jr /(10⁻³A·m ⁻¹)	
		Range of change	Average	Range of change	Average
Massive ore	10	36,520–239,270	133,920	1560–25,420	8460
Sericite slate	20	203–484	313	134–292	228
Chlorite slate	10	147–360	260	90–232	106
Rhyolite	10	98–205	150	50–120	77
Mineralized limestone	5	150–3020	679	70–1100	319
Basalt	5	224–1627	750	100–1260	540

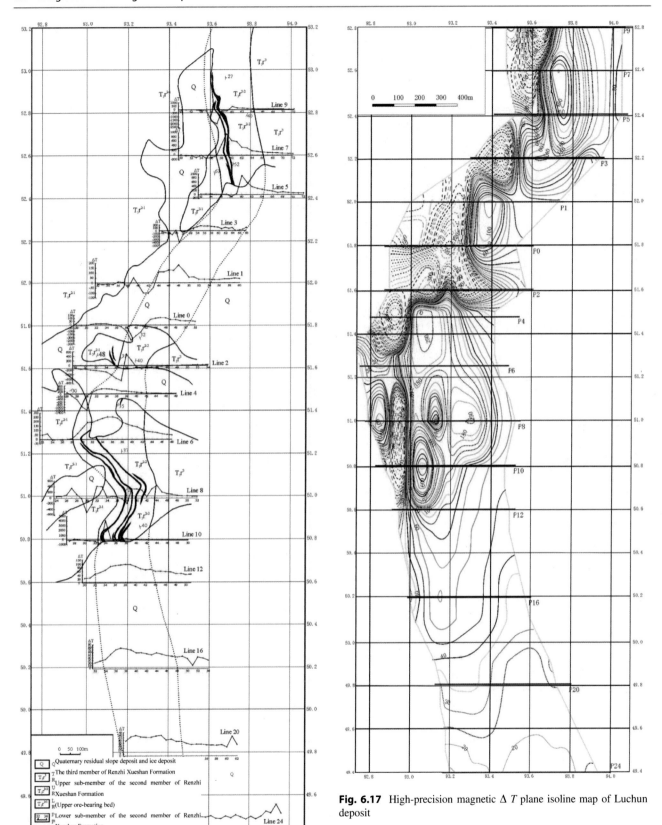

Fig. 6.16   High-precision magnetic ΔT profile plan of Luchun deposit

Fig. 6.17   High-precision magnetic Δ *T* plane isoline map of Luchun deposit

outcrop. Covering by vegetation and slope deposits, the extension section showed a low and slow magnetic anomaly. Magnetic survey anomaly showed that there were mineralized bodies beneath the overburden, which were connected with mineralized bodies in the south ore section in strike. The magnetic anomaly (mineralized) body in the upper ore-bearing bed extends steadily from north to south, with a length of 3600 m from north to south and a width of 60–200 m from east to west. Moreover, in the range of 2000 m between exploration line P11-P10, not only high positive magnetic anomalous zone appears on the mineralized bodies, but also obvious negative magnetic anomalous zone appear on the west side of the positive magnetic anomalous zone. The spatial "pairing" arrangement of positive and negative magnetic anomalous zones indicates that the mineralized bodies (stratiform ore bodies) tend to the east with good continuity in strike. In addition to the obvious magnetic anomaly characteristics in the upper ore-bearing beds, there were also magnetic anomalies in the lower ore-bearing beds on the west side of exploration line P10 and P8.

Analytical Continuation of Magnetic Anomaly

Through analytic continuation of anomalies, which is produced by magnets at a certain spatial position below the earth's surface, the anomalies are highlighted. When using the existing analytical continuation method of potential field to extend downward to the region near the top surface depth of the field source body, the potential field will have strong oscillation, and the vertical super-position body cannot be clearly distinguished. By introducing the relevant correction functions such as potential field frequency and buried depth with regularization factor, the field value is not singular when downward continuation passes field source body by computer processing. In this condition, it can reach any required depth below the field source body by downward continuation.

Upward Continuation of Magnetic Anomaly

Upward continuation is to calculate the abnormal value of a certain height above the ground according to the measured abnormal value on the ground. The purpose is to suppress the interference of shallow magnets and highlight the meaningful anomalies produced by deep-seated magnets. In fact, upward continuation is equivalent to improving the observation plane, and the anomaly curve obtained after continuation mainly reflects the anomaly characteristics of deep-seated magnets. By using upward continuation, deep-seated magnets can be found, and local anomalies and regional anomalies can be divided.

The upward continuation of magnetic anomaly in Luchun deposit has been carried out in several heights (0 m → 25 m 50 m → 100 m → 200 m → 300 m → 500 m) (Figs. 6.17,

6.18 and 6.19). According to the results, at the height from 0 to 50 m, there are many magnetic anomaly centers between exploration line P7 and P5, exploration line P1 and P0, exploration line P4 and P6, exploration line P6 and P8, exploration line P8 and P12 respectively in the north of exploration line P12. There are still low and slow anomalies in the south of exploration line P12. When upward continuation is carried out at the height from 100 to 300 m, the magnetic anomaly center in the north of exploration line P12 only appears between exploration line P7 and P5, exploration line P8 and P12. The magnetic anomaly in the south of exploration line P12 still shows a low and gentle anomaly. When upward continuation is carried out at the height of 500 m, the magnetic anomaly center between exploration line P7 and P5 moved southward to exploration line P5 and P3, and the magnetic anomaly center between exploration line P8 and P12 moved northward to exploration line P8 and P10. The low and slow anomaly south of exploration line P12 was not obvious.

The upward continuation results of magnetic anomaly in Luchun deposit show that magnetic anomalies and anomalous bodies are stable and continuous in a certain depth range along dip and strike. The maximum extension depth of magnetic anomalous bodies along dip direction lies between exploration line P5 and P3 in the north ore section and between exploration line P8 and P10 in the south ore section. This is also the center of deep magnetic anomalous bodies. The characteristics of dense isoline in the west and sparse isoline in the east show that the magnetic anomalous body inclines eastward. This is also consistent with the actual observation of geological profile of the ore deposit.

Downward Continuation of Magnetic Anomalies

Downward continuation is to calculate the anomaly value of a certain depth below the ground according to the measured anomaly on the ground. Downward continuation is equivalent to reducing the height of the observation plane, aiming at distinguishing super-imposed anomalies and highlighting anomalies caused by deep-seated magnets. By using downward continuation, we can explore the spatial occurrence and depth of deep magnets extending downward, thus realizing two-dimensional spatial recourse of anomalous bodies.

The magnetic anomalies of 15 magnetic side profiles in Luchun deposit are extended 250 m below the earth surface, and the magnetic anomalies of seven exploration line P9, P7, P1, P0, P2, P6 and P10 are fitted by computer in shape spatial occurrence (Figs. 6.1, 6.2, 6.20 and 6.21). It can be seen from the figure that the magnetic anomaly characteristics in the upper ore-bearing bed are obvious and tend to become larger when downward continuation is conducted to the depth range of 250 m in the north of exploration line P12. The anomalous bodies fitted by computer all show stable extension, and it is a plate-like body and tends to the

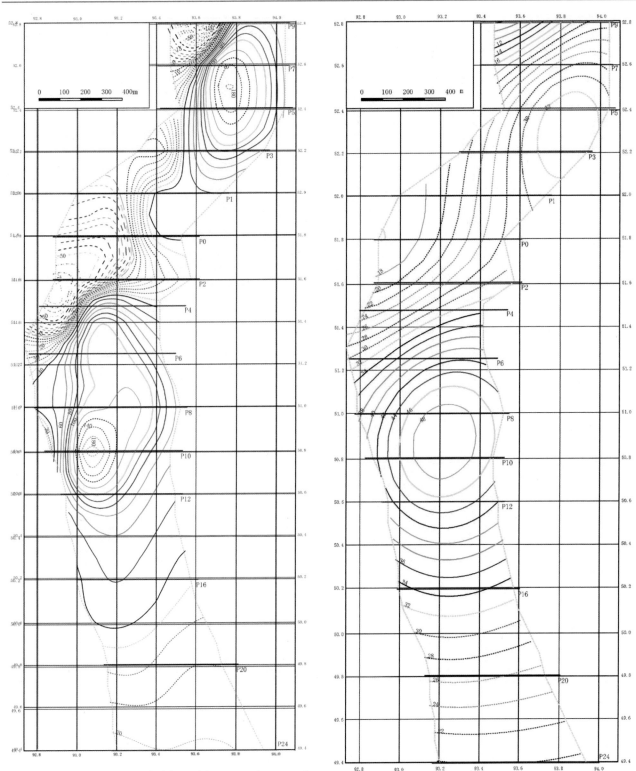

Fig. 6.18 Plane isoline map of high-precision magnetic survey in Luchun deposit (100 m upward continuation)

Fig. 6.19 Plane isoline map of high-precision magnetic survey in Luchun deposit (500 m upward continuation)

east. The ore body exposed between exploration line P0 and P2 has a certain downward displacement due to landslide. There are also anomalous bodies in the lower ore-bearing beds on the west side of exploration line P10. The low and gentle anomalies in the south of exploration line P12 extend downward to the depth range of 250 m, and the characteristics of magnetic anomalies are obvious in-depth ranges of 100–250 m below the overburden. The maximum

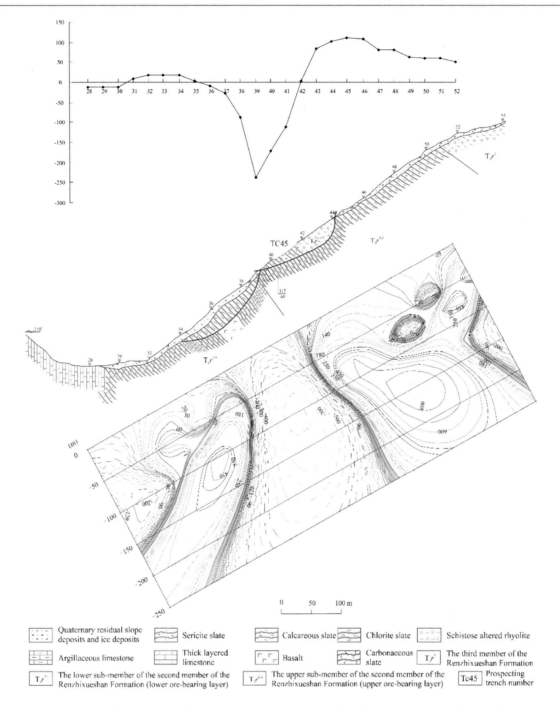

Fig. 6.20 Comprehensive map of geological profile and downward continuation profile of magnetic anomaly of exploration line P0 of Luchun deposit

downward continuation of magnetic anomalies and anomalous bodies in Luchun deposit is located in the area of exploration line P7, P5 and P3 in the north ore section and area of exploration line P6 and P8 in the south ore section. The downward continuation depth of plate-like magnetic bodies can be 300–400 m.

The downward continuation results of magnetic anomaly in Luchun deposit show that the magnetic anomaly in north of exploration line P12 and the anomalous body extending along the dip range of 250 m below the earth's surface are relatively stable and continuous. The anomalous body is a plate-like body with dip to the east. In Quaternary overburden area in

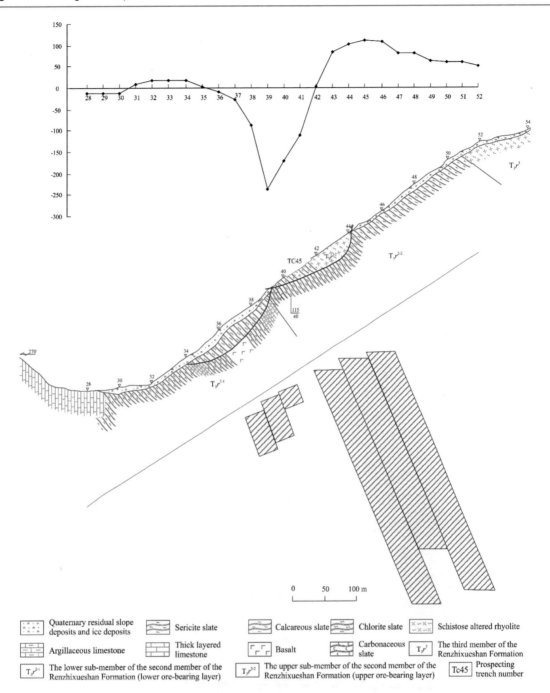

Fig. 6.21 Comprehensive map of geological profile and computer fitting profile of anomalous bodies of P0 exploration line of Luchun deposit

the south of exploration line P12, there are magnetic anomalies in depths from 100 to 250 m below the overburden. The maximum downward continuation of magnetic anomalies and anomalous bodies in Luchun deposit is located in the area of exploration line P7, P5 and P3 in the north ore section and in the area of exploration line P6 and P8 in the south ore section. The downward continuation depth of plate-like magnetic bodies can be 300–400 m, which is consistent with the location of magnetic anomaly center displayed by upward continuation. The spatial distribution of magnetic anomalies and anomalous bodies in Luchun deposit is consistent with the configuration of the measured geological profile of the ore deposit. It also corresponds to the spatial position of mineralized bed (bodies) in the profile one by one. In addition to the obvious magnetic anomalies and anomalous bodies in the upper ore-bearing beds, there are also magnetic anomalies and anomalous bodies in the lower ore-bearing beds on the west side of exploration line P10 and P8.

6.2.2.2 Transient Electromagnetics (TEM)

Arrangement of Transient Electromagnetics (TEM) Exploration Line

The ore body of Luchun deposit is rich in magnetite and Zn-Cu-Pb sulfide minerals, so it has the physical properties for Transient Electromagnetics (TEM) detection. Through TEM, the deep-seated spatial occurrence of Luchun mineralized body can be detected. Nine east–west direction exploration line profiles were arranged in Luchun deposit, which are exploration line P9, P5, P3, P0, P6, P8, P10, P12 and F from north to south. Among them, overlapping loops were arranged in exploration line P9, P0, P6, P10, and F

with a dot spacing of 50 m. Large fixed-loop sources were arranged on exploration line P5, P3, P8 and P12 with a dot spacing of 25 m and the control depth of 350–500 m.

Results of Transient Electromagnetics

The detection of nine geophysical prospecting profiles in Luchun deposit was conducted by overlapping loops method and large fixed-loop source method, with the controlled depth of 350–500 m. Effective results are shown in the known outcrop areas of ore bodies and the areas covered by plants and slope deposits (Figs. 6.16, 6.17, 6.22 and 6.23).

The TEM results for deep exploration of Luchun deposit show that the mineralized body in the north of exploration

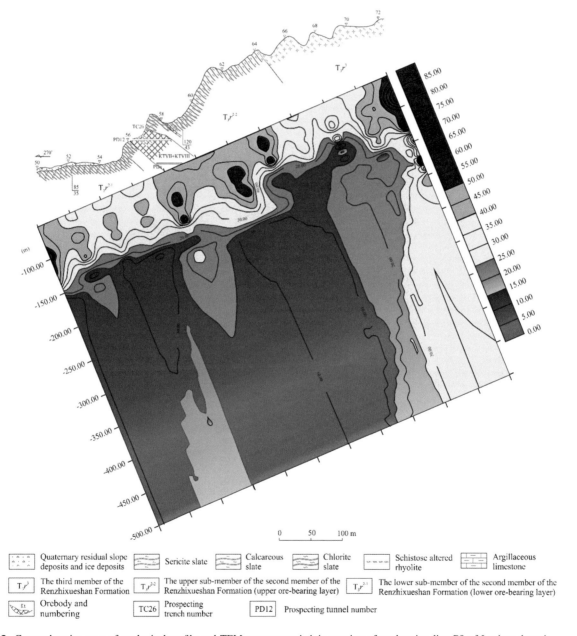

Fig. 6.22 Comprehensive map of geological profile and TEM apparent resistivity section of exploration line P9 of Luchun deposit

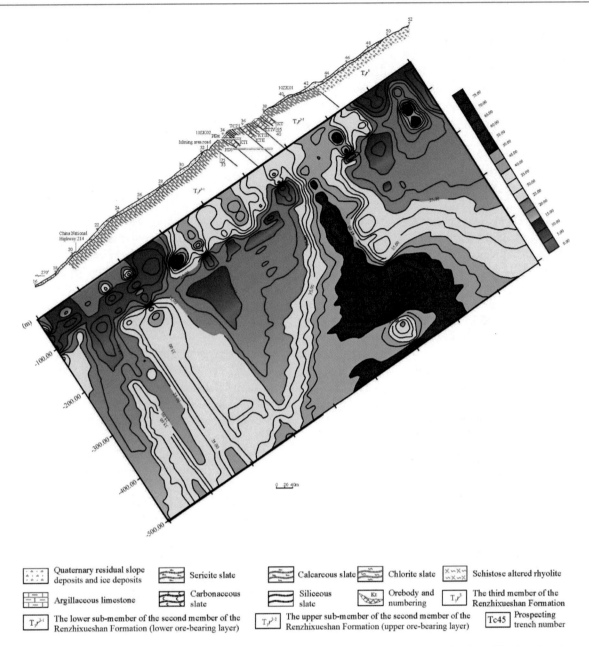

Fig. 6.23 Comprehensive map of geological profile and TEM apparent resistivity section of P10 exploration line of Luchun deposit

line P12 extends steadily and continuously to the deep, with obvious low resistance body and low apparent resistivity (ρs value: 0–25 Ω m). The stratiform mineralized bodies tend to extend up to 200–250 m along the dip. In Luchun deposit, the spatial distribution of low resistance body is consistent with that of magnetic anomalies and anomalous bodies and also roughly corresponds to the configuration of the measured geological profile of the ore deposit. There are two large-scale tubular low resistance bodies beneath the stratiform mineralized bodies in exploration line P9 in the north ore section and exploration line P10 in the south ore section in Luchun deposit, respectively, with apparent resistivity

value of 0–25 Ω·m. These two tubular low resistance bodies are presumed to be tubular mineralized bodies beneath the stratiform ore bodies. There are low resistance bodies beneath the vegetation and slope deposit overburden in the south of exploration line P12. The location of the low resistance bodies is basically consistent with the range of high-precision magnetic survey anomalies, and it is presumed that the low resistance bodies are mineralized bodies. The depth of the overburden varies from 50 m (east) to 120 m (west), and the low resistance bodies can connect with the mineralized bodies (beds) of exploration line P10 in space to the north.

6.2.2.3 Amplitude-Frequency Induced Polarization

Amplitude-frequency Induced Polarization is a technique used to detect the difference of electric field by taking advantage of the slowness of polarization process and adopting different frequency currents to excite polarized bodies to different degrees. As amplitude-frequency induced polarization can detect IP effect and measure apparent resistivity and amplitude-frequency effect simultaneously, it can be used to detect various objects suitable for resistivity method. In addition, amplitude-frequency induced polarization has a strong ability to find sulfides. Generally, sulfides can be found effectively only if their content is only 1% or even lower. Most nonferrous and precious metal deposits are closely related to sulfides, and they can be detected effectively by IP method.

There are a large number of strongly polarized sulfide minerals such as magnetite, pyrite, chalcopyrite, galena, sphalerite, chalcocite and hematite in Luchun zinc-copper-lead (silver) polymetallic ore deposit, which have the electrical conditions for carrying out Induced Polarization (dipole amplitude-frequency induced polarization detection). In 1998, the research group of Chengdu Institute of Geology and Mineral Resources conducted the "Research on Tectonic Evolution and Metallogenic Regularity of Copper and Gold Deposits in Jinsha River Junction Zone". In the same year, the project group (a national key scientific and technological research project in the Ninth Five-Year Plan) carried out "Comprehensive Demonstration Research on Rapid Positioning and Prediction of Important Copper Deposits (Bodies) Types". These two groups jointly carried out geophysical prospecting tracing of deep-seated ore body by amplitude-frequency Induced Polarization in Luchun deposit area. Two induced polarization profiles were arranged, with a total length of 1020 m.

The amplitude-frequency induced polarization profile is arranged near exploration line P8 in the south ore section and near P2 exploration line (Quaternary coverage which is the middle ore section area). The profile near the exploration line P8 is 500 m long, with an orientation of NEE-SWW, and passes through KTI, KT II, KT III and KT IV ore bodies in EW direction. The dot spacing between profiles is 20 m, with the sounding depth of 120 m. The data are averaged, that is, the amplitude-frequency effects of exploration line F6 and 0.3 are averaged arithmetically, and the apparent resistivity of exploration line $\rho6$ is averaged geometrically. The results show certain regularity.

It can be seen from Fig. 6.24 that the low resistivity anomalies D_1, D_2 and D_3 are in good agreement with the delineated ore bodies (KTI, KT II, KT III and KT IV) on the earth surface, and their messy state is probably related to goaf. Low resistivity anomalies D_1 and D_2 correspond to amplitude-frequency effect anomaly J_{1-1}, and low resistivity anomalies D_3 and D_4 correspond to amplitude-frequency effect anomaly J_{1-2}. All of them reflect large extension of ore body.

It is worth noting that there are strong amplitude-frequency effect anomalies J_{2-1} and J_{2-2} in the lower part of the west side of the exploration line profile, and there are D_5 low resistivity anomalies corresponding to them. Strong amplitude-frequency effect anomaly J_3 and D_6 low resistivity anomaly are found in the deep part. They are all in a closed state, indicating that they may be caused by deep "sac ore body". The anomaly is also reflected on the earth surface, showing that there are two beds outcrop of zinc-copper-lead (silver) polymetallic mineralized bodies with a width of 1.5–2.0 m in the measured geological profile. Based on metallogenic geological conditions, ore deposit genesis and mineralization clues, combined with deep-seated geophysical prospecting results, it is preliminarily inferred that it is a fractured tubular mineralized body near the exhalative channel.

The profile using amplitude-frequency Induced Polarization arranged near P2 exploration line is 520 m long, and its orientation is close to east–west direction. No ore body is exposed on the earth surface, with the dot spacing between profiles of 40 m. The data obtained after processing have a certain regularity.

In Fig. 6.25, it can be seen from the profile that there are obvious high amplitude-frequency effect values and low resistivity anomalies, which are just in the southward extension of KT V and KT VI ore bodies between exploration line P0 and P2 in the middle ore section, but the earth surface has been covered by Quaternary. The spatial position can correspond to the landslide body on geological profile and the finite extension anomalous body under magnetic anomaly downward continuation fitted by computer. It shows that the ore body exposed between exploration line P0 and P2 is a landslide body, and the normal ore-bearing bed, which is now covered by landslide body and Quaternary, is higher in elevation.

6.3 Location Prediction of Ore Body and Ore Deposit in Pb, Zn, Cu, Ag Polymetallic Ore Target Area with Hot Ditch in Gacun's Peripheral Area

The Pb–Zn–Cu–Ag polymetallic ore target area with hot ditch in Gacun's periphery refers to the range from the south of exploration line 32 to the north of exploration line 95 in Gacun deposit (Fig. 6.26). According to the analysis of regional metallogenic geological environment and local tectonic of the target area, the target area with hot ditch and

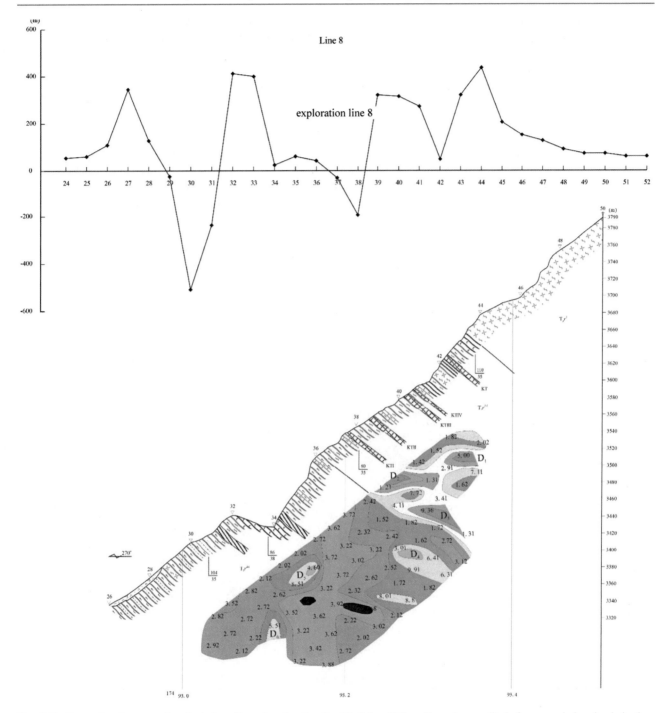

Fig. 6.24 Comprehensive map of geological profile and exploration line F6, 0.3 and P6 profiles using amplitude-frequency induced polarization in south ore section in Luchun deposit

Gacun ore deposit occurs in the same volcanic-sedimentary basin controlled by intra-arc rift. Their volcanic activity characteristics, sedimentary and tectonic characteristics are very similar, but they are located in local basins influenced by different volcanic activity centers. In terms of regional geophysical, geochemical (Pb, Au, Zn, Hg, Cu) and remote sensing anomaly characteristics, they are located in different anomaly centers in the same regional anomalous zone. Therefore, it is of practical significance to carry out comprehensive metallogenic prediction research in this target area.

Based on the comprehensive prospecting model established by predecessors (Hou Zengqian et al. 1992; Lv Qingtian et al. 1999) in Gacun Deposit area, the rapid

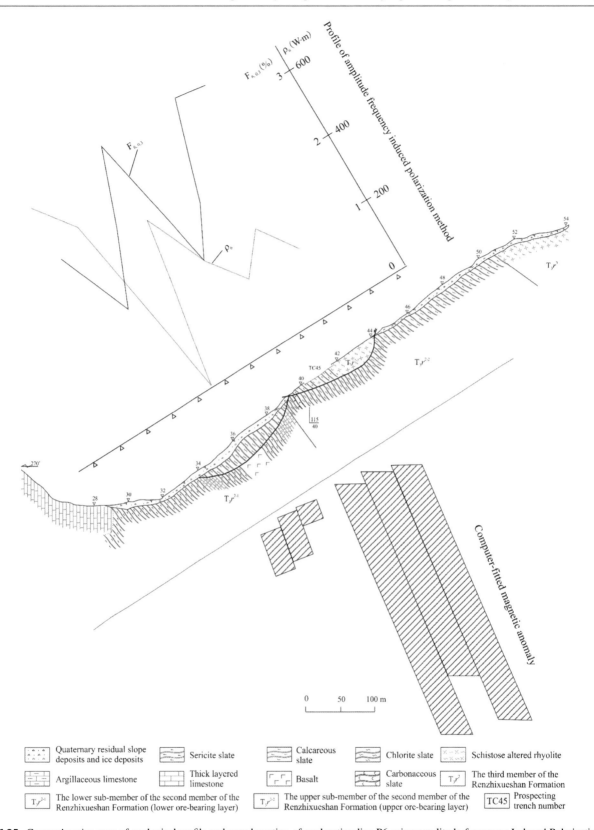

Fig. 6.25 Comprehensive map of geological profile and pseudosection of exploration line P6 using amplitude-frequency Induced Polarization in middle ore section in Luchun deposit

Fig. 6.26 Magnetic polarization reduction isoline map of Gacun and its peripheral target area with hot ditch

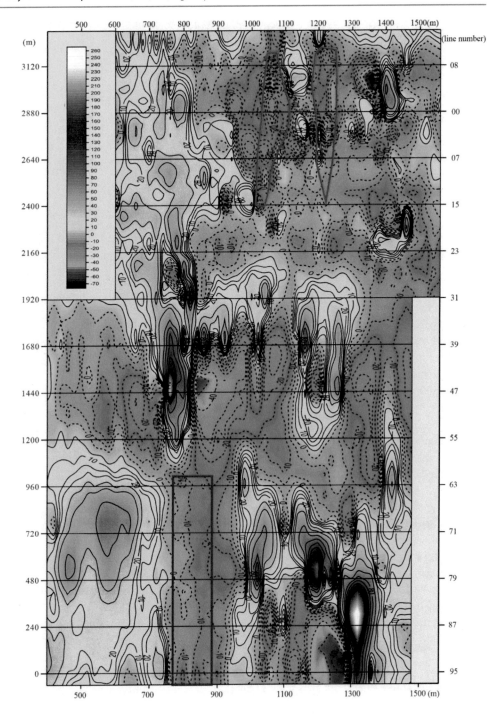

surface scanning work focusing on magnetic method, X-ray fluorescence and amplitude-frequency IP was carried out in the target area. Then, the ore body location prediction was carried out in the key abnormal areas using Controlled Source Audio-frequency Magnetotellurics (CSAMT) and Transient Electromagnetics with large detection depth. The measurement and inversion result of each method are analyzed below. In combination with the geological and physical property measurement results, the metallogenic prospect of this target area is comprehensively predicted.

6.3.1   Magnetic Measurement Results

The measurement results of magnetic method in known mining areas are as follows: ΔZ isoline map generally reflects the lithologic distribution characteristics of the mining areas; low and slow positive anomalies reflect andesite with relatively strong magnetism; large negative anomaly area reflect nonmagnetic or diamagnetic rhyolite, mineralized rhyolite, ore body, and barite rocks. Because the magnetism of the reticulated vein ore, which occupies the main body of the ore deposit, is equivalent to that of rhyolite, dacite and barite, so it is difficult to delineate the accurate position and form of the ore body by magnetic method. However, negative anomaly can still be used as a necessary condition and an important symbol to determine the existence of ore body, so as to narrow the space scope of prospecting. Compared with the known mining area, the measured results in unknown mining area have the following characteristics: except for a few anomalies, the amplitude and anomaly strike of positive anomalies are equivalent to those in known mining area. This means that the lithology composition of main body in unknown mining area is consistent with that in known mining area. The local high magnetic anomalies located in exploration line 47, 79 and 87 may be the volcanic activity center, and there are some basic volcanic rocks. The amplitude of negative anomalies in unknow mining area is similar to that of known mining areas. But the overall strike of anomalies tends to be north–south (possibly due to the large spacing between exploration lines), and many negative anomalous zones close the north–south direction are formed. The largest negative anomalous zone is the one from exploration line 39 to 95 with abscissa between 800 and 900. According to the analogy of known mining area and unknown mining area, prospecting work should focus on these negative anomalous zones. Controlled Source Audio-frequency Magnetotellurics (CSAMT) and Transient Electromagnetics (TEM) are used in negative anomaly region. The results will be described later.

6.3.2 Rapid Analysis of X-ray Fluorescence

The results of rapid analysis of X-ray fluorescence for soil in the target area with hot ditch show that there are many Zn anomalous zones (Fig. 6.27), most of which are consistent with the negative magnetic anomalous zone. This further illustrates the prospecting significance of this anomalous zone.

6.3.3 Transient Electromagnetics (TEM)

Six profiles, namely exploration line 87, 79, 71, 63, 55 and 47, were made using TEM in the target area with hot ditch. Figure 6.28 shows the resistivity inversion results of 400 m buried horizontal section. It can be clearly seen from the figure that there is a low resistivity anomalous zone from exploration line 63 to exploration line 87. The low resistivity anomaly center is located at 800 m of abscissa, and the anomaly axis tends to close north–south direction. If the range of anomalous body is delineated with resistivity of 60 Ω·m, the anomalous body is nearly 200 m wide, and its length from north to south is more than 700 m, with the

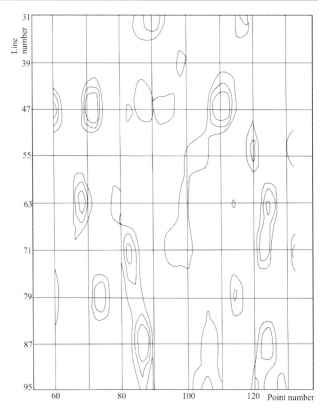

Fig. 6.27 X-ray Fluorescence Zn Anomaly Diagram of Target Region with Hot Ditch

anomalous body open to the south. The location of TEM low resistivity anomaly is basically consistent with that of low magnetic anomalous zone and high Zn anomalous zone.

6.3.4 Controlled Source Audio-Frequency Magnetotellurics (CSAMT)

CSAMT results in known mining areas are good, especially for massive ores with good electrical conductivity. With CSAMT, the form and occurrence of ore bodies can be better determined. Six profiles corresponding to Transient Electromagnetism were also made in the unknown area, namely exploration line 87, 79, 71, 63, 55 and 47. Good low resistance anomalies were also found from exploration line 63 to exploration line 87 (Fig. 6.29). The anomalous body is about 100 m wide and 700 m long (controlled by existing exploration lines). The anomalous body is not closed to the south and may be longer. The width and length of this anomalous body are basically equivalent to that of the anomalous body delineated by Transient Electromagnetics. The vertical extension of anomaly varies from one exploration line to another. According to the results of profile inversion, the extension of low resistivity anomaly on exploration line 63, 71 and 79 exceeds 400 m. The planar

Fig. 6.28 TEM Resistivity Plane Isoline Map of Target Region with Hot Ditch

low magnetism anomalies and low resistivity anomalies, only carbonaceous slate can cause it. However, according to earth surface geological observation and trench exploration disclosure, carbonaceous slate is ubiquitous in this area and generally distributed in planar, and not in belt. Carbonaceous slate extended vertically. In addition, CSAMT and TEM low resistivity anomalies do not appear in places where carbonaceous slate outcropped in known mining areas, so the anomaly will not be caused by carbonaceous slate. The possibility of fault exists. Because there is just a fault passing through the place where the anomalous zone occurs. The fault extends to the mining area and is the main ore-controlling fault in the mining area. Two main ore bodies in the mining area appear on both sides of the fault. There are wide cleavage zones on both sides of the fault. This can cause low resistance and low magnetic zone anomaly to a large extent. Therefore, there is no sufficient reason to rule out the possibility of fault at present. Even if it is a fault, both sides of the fault are favorable places for mineralization. It is very possible that this comprehensive anomaly is related to mineralization. Because the characteristics of this comprehensive anomaly are very similar to that of known deposits, and there are Zn element anomalies. Therefore, it is suggested that:

(1) The existence of conductive anomalous zone is further confirmed by using denser magnetic method, CSAMT and TEM (60 m × 20 m density of exploration grid) in the anomalous zone with hot ditch.

(2) Adding high-density electrical method and geochemical exploration to determine the anomaly nature and spatial form.

(3) Geological trench exploration, pit depth survey or shallow drilling in anomalous zone should be carried out to directly determine the property of anomalous body.

(4) The exploration scope of the periphery of Gacun should be expanded. A breakthrough in prospecting should be made in the periphery of Gacun.

position of the low resistivity anomaly found by CSAMT is basically consistent with that of the largest low magnetic anomalous zone (at 800–900 m of abscissa). Neither of the two anomalies is closed to the south. In addition, there is continuous high resistivity below 200 m at deep part in the eastern part of each profile (at 1000–1100 m of abscissa). This is probably the product of volcanic activity (either crater or concealed rock mass formed by volcanic activity).

To sum up, according to results of several geophysical prospecting methods and X-ray fluorescence Zn elemental analysis, we have confirmed the existence of anomalous zones at the same time. Is the anomalous zone an ore body or a fault? or carbonaceous slate or with other lithology? According to the geological tectonic map of the mining area and its periphery, the stratum corresponding to the anomaly position is volcanic complex of Gacun Formation of Upper Triassic, mainly including rhyolitic-dacite tuff, breccia, agglomerate, lava, phyllite, sand-slate, siliceous rock, dolomite, etc. Obviously, most of the above rocks cannot cause

6.4   Location Prediction of Ore Deposit and Ore Body in Nongduke Ag Polymetallic Ore Target Area

Nongduke Ag polymetallic ore target area is a newly discovered metallogenic prospect. The mineralized zone has been discovered and basically confirmed by the general survey and prospecting work of Team 403 of Sichuan Bureau of Geology and Mineral Resources in recent years. However, the mining area is completely covered, and the strike, scale, host and control conditions and genetic types of

Fig. 6.29 Inversion results map of CSAMT apparent resistivity in Gacun's peripheral target area with hot ditch

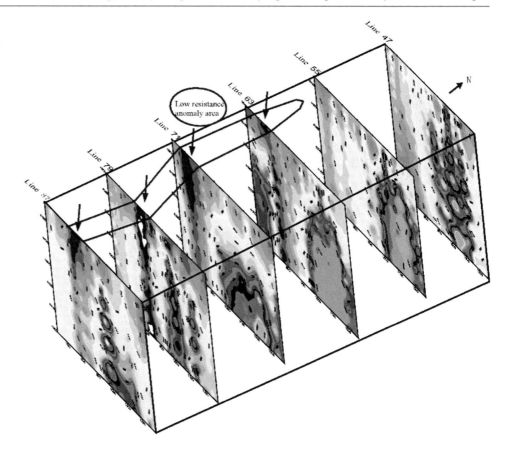

the ore deposit are unclear. In order to find out these problems quickly, Team 403 of Sichuan Bureau of Geology and Mineral Resources, Chengdu University of Technology and other units have carried out comprehensive evaluation and research focusing on geology and light geophysical prospecting. Geophysical prospecting methods include magnetic method, amplitude-frequency induced polarization, γ-ray energy spectrum. Rapid X-ray fluorescence analysis is also carried out.

Nongduke Ag Polymetallic Ore Target Area in Changtai, western part of Sichuan Province, is located in Changtai volcanic-sedimentary basin in the middle section of Yidun remnant arc, about 20 km southwest of Changtai. The regional stratum is developed with Triassic rhyolitic volcanic rock series of Miange Formation, sand-slate series of Lamaya Formation, sand-slate and basalt series of Qugasi Formation, which are formed in the rift zone of back-arc expansion environment (Fig. 6.30).

6.4.1 Physical Properties of the Target Area

A total of five specimens were collected in the target area, and detailed physical properties survey and lithoscopic identification were carried out. The results are shown in Table 6.2 and Fig. 6.31. It can be seen that the main

ore-bearing rocks (sericitization cataclasites) are characterized by low magnetism (diamagnetism), low resistivity and high polarization. These characteristics are mainly due to the large number of nonmagnetic metal minerals in the ores. The rhyolite in the same horizon with the ore body (specimens No.2 and No. 4 in the table) is characterized by low magnetism, high resistivity and high polarization. The reason for high polarization is that rhyolite contains a certain amount of metal minerals. Specimen No.1 collected from Qugasi Formation (sand-slate with conglomerate and basaltic tuff) on the east side of the mining area shows strong magnetism and polarizability. Although no obvious magnetic minerals are found under rock microscope, the physical property survey results show that there should be a certain amount of magnetic minerals in the ores. By analyzing the physical properties of ore and surrounding rock in this target area, it can be concluded that the best methods to search and trace this kind of ore body are electrical method (resistivity and polarizability) and magnetic method.

6.4.2 Results and Analysis of Magnetic Method

As this mining area is completely covered, it is a fast and effective method to carry out high-precision magnetic surface scanning along the mineralized zone. From the survey

Fig. 6.30 Outline geological map of Nongduke Ag polymetallic ore target area

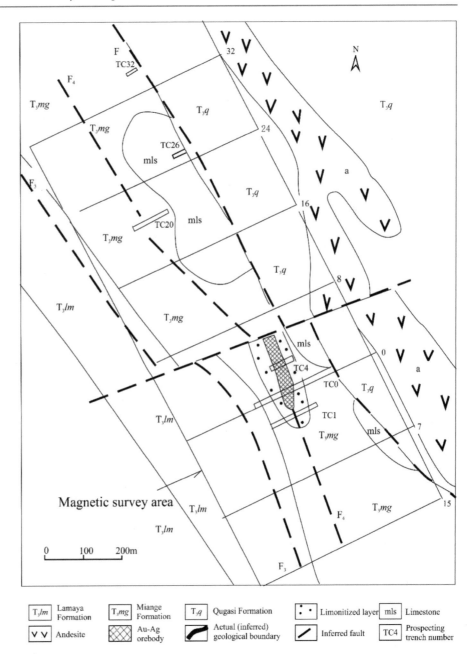

Table 6.2 Specimens Physical Properties of Nongduke Ag Polymetallic Ore Target Area in Baiyu County, Sichuan Province

Serial No.	Lithology	Magnetic susceptibility $4\pi \times 10^{-6}$SI	Residual magnetization 10^{-3}A·m^{-1}	Resistivity Ω·m	Polarizability %	Density g·cm^{-3}
1	Feldspar quartz sandstone	128.4	0.35	2014.95	4.46	2.609
2	Rhyolite	28.2	0.24	5743.67	2.59	2.568
3	Sericitization cataclasites (mainly mineralized rock)	−17.6	0.36	549.1	3.04	2.637
4	Rhyolite	9.6	0.84	3598.76	1.40	2.597
5	Carbonate-quartz vein	94	0.14	7794.6	3.23	2.742

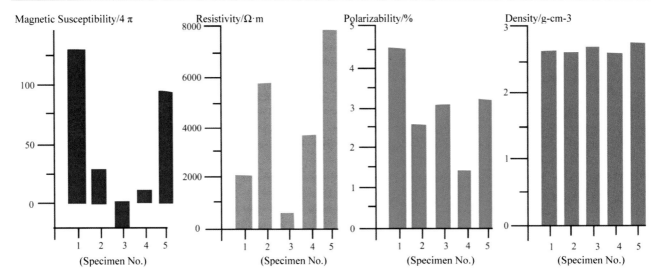

Fig. 6.31 Histogram of physical properties of Nongduke Ag polymetallic ore target area

results (Fig. 6.32), the magnetic survey better reflects the lithologic distribution and ore-controlling tectonic. According to the characteristics of magnetic field, the measured area can be divided into two anomaly areas with different characteristics by taking exploration line 12 as the erathem: southern anomaly area and northern anomaly area. The northern region can be divided into high, medium and low magnetic field regions, and the anomalies are generally NE strike except for local small anomalies. The magnetism in the southern region is relatively weak. The southern anomaly region can be divided into three types: high, medium and low magnetic fields, all of which show obvious ribbon anomaly characteristics. For example, the high magnetic anomaly in the east side of the survey area shows close north–south strike. This reflects the distribution of intermediate-acid volcanic rocks. However, the western part of the survey area is characterized by low magnetism and gentle field, which corresponds to volcanic-sedimentary rocks of Lamaya Formation and Miange Formation. But the anomaly strike is still close north–south, reflecting the basic strike of stratum and tectonic in the survey area. On the north and south sides of the southern anomaly area, the anomaly strike changes from close north–south to close east–west. This is probably caused by the east–west fault tectonic. According to the measurement results of specimens' physical property and the corresponding relationship between magnetic anomaly and mineralized body in exploration line 0, it is found that the magnetic anomaly transition zone (medium intensity anomaly) in the middle section of the southern anomaly area (between 45 and 55 dot) corresponds to the mineralized body. This magnetic anomaly transition zone represents the contact zone or fault zone between eastern volcanic rocks and western sedimentary rocks, and mineralization occurs along this zone. According to this

corresponding relationship, it is speculated that displacement and fault may occur on mineralized bodies by east–west faults in the north and south direction. To the north direction, the mineralized zone may correspond to the northern low magnetic anomaly area, so the NW or SW direction should be considered for the trace of extension of this mineralized zone.

6.4.3 Results of Amplitude-Frequency IP, γ-ray Energy Spectrum and X-ray Fluorescence Analysis

The results of soil analysis of amplitude-frequency IP, γ-ray energy spectrum and X-ray fluorescence completed by Chengdu University of Technology (Ge Liangquan et al.) show that the amplitude-frequency IP has low-frequency dispersion anomaly in this zone, and the comprehensive anomaly of γ-ray energy spectrum forms two main anomalous zones G-1 and G-2 in the survey area (Fig. 6.33). The transition zone of this magnetic anomaly also has this phenomenon in different degrees. The G-1 anomalous zone is about 900 m long and 40 ~ 60 m wide, with good continuity. It is distributed in the east-central part of survey area, spreading in NE direction. In the south of exploration line 0, the anomaly basically coincides with the transition zone of magnetic anomaly and corresponds to the contact zone between intermediate-acid volcanic rocks and sand-slate as well as the mineralized zone in exploration line 0, and 4. G-2 anomalous zone is located in the west-central part of exploration line 6–20, with a length of about 500 m and a width of 30 ~ 50 m, distributing in NW strike. G-2 anomalous zone basically corresponds to the low magnetic anomaly in the northern magnetic field area.

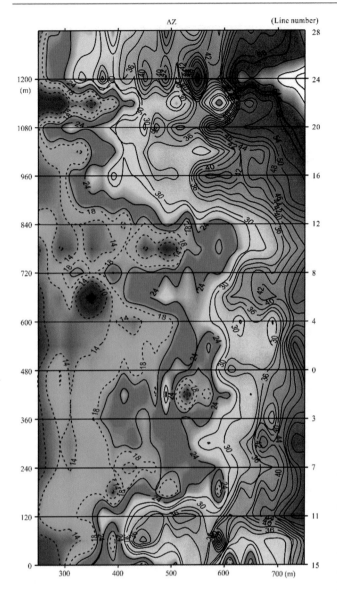

Fig. 6.32 High-precision Magnetic Δ Z Isoline Map of Nongduke Target Area

The background values of elements Zn, Pb and As in the south of exploration line 8 were significantly different from those in the north of exploration line 12. The background values in the north part were about one time higher than those in the south part. This reflects the parent characteristics of different soils. It is worth noting that the high background values of chalcophile elements (Zn, Pb, As, Hg) in the south part are located in the middle section of exploration line 2 and exploration line 6 and distributed in the NE direction. This may be related to mineralization and alteration.

Based on the results of existing geological survey, magnetic survey, amplitude-frequency induced polarization, γ-ray energy spectrum and X-ray fluorescence soil analysis, it can be preliminarily determined the following: The G-1 anomalous zone is located in the middle section of the

southern anomaly area and adjacent to the high magnetic anomalous zone and γ-ray energy spectrum on the east side. The G-1 anomalous zone is a reflection of mineralized bodies, with a length of about 300 m and a width of 30～50 m in the south of exploration line 12. The G-1 anomalous zone is staggered by two close east–west faults on the north and south sides, respectively. According to geophysical prospecting data, it is inferred that the Ag-Au polymetallic deposit should have medium scale or above. It is therefore recommended that:

(1) To further determine the spatial form of ore bodies by using high-density electrical method or other geophysical prospecting methods with large detection depth such as CSAMT and TEM for possible mineralized anomalies.
(2) Additional density is added to shallow exploratory trench to control. The possible mineralized zones are tracked continuously in the NW and SW directions of the survey area.

6.5 Ore Deposit and Ore Body Location Prediction of Qingmai Pb–Zn–Cu–Ag Ore Target Area

Xiangcheng Basin is an important volcanic basin in the southern section of Sanjiang Yidun Arc. For many years, Xiangcheng Basin has been concerned by ore deposit geologists. Because large and super-large volcanic-sedimentary exhalative volcanic-associated massive sulfide deposit and porphyry ore deposits have been discovered one after another in Zengke (located in Yidun Arc same as Xiangcheng Basin), Changtai Basin and Shangri-La magmatic arc in the south. These volcanic-sedimentary basins are similar to Xiangcheng Basin in sedimentary characteristics, tectonic and volcanic rock scale, so Xiangcheng Basin should also have geological conditions for producing such deposits. Despite a great deal of geological exploration being carried out over the years, there has been no big breakthrough in prospecting. In addition to the insufficient degree of prospecting, improper methods and ways are also important factors. Therefore, we have carried out comprehensive metallogenic prediction based on geophysical prospecting in Qingmai target area, which has the best metallogenic conditions in Xiangcheng volcanic rock basin.

Qingmai Pb–Zn–Cu–Ag ore target area is located in the southeast margin of Xiangcheng volcanic-sedimentary basin in the southern section of Yidun arc. Xiangcheng volcanic-sedimentary basin is tensional basins formed during the period of back-arc rift. The outcrop stratum in the area is Genlong Formation, Miange Formation and Lamaya

Fig. 6.33 Comprehensive Anomaly Diagram of γ-Ray Energy Spectrum of Nongduke Ag Polymetallic Ore Target Area

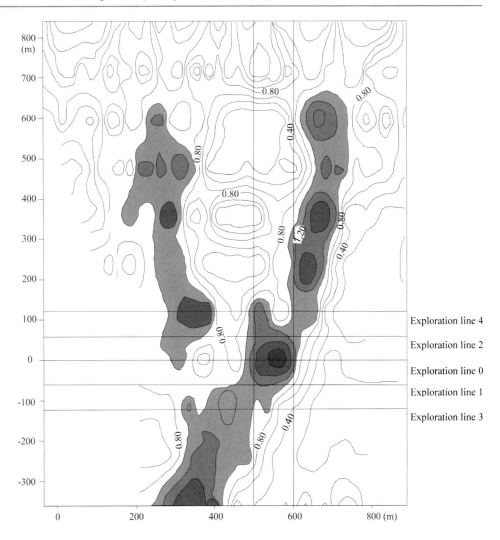

Formation of Upper Triassic. The Dege-Xiangcheng fault in the region passes through the middle of the basin, resulting in complex tectonic and strong deformation of the basin.

6.5.1 Characteristics of Regional Geophysical Prospecting, Geochemical Prospecting and Remote Sensing

On the regional aeromagnetic map (Fig. 6.34), Xiangcheng Basin corresponds to a moderate-intensity annular local positive anomaly. This indicates that the basin is a local intermediate-acid volcanic activity center. On the regional gravity map, Xiangcheng Basin is located in the gravity gradient zone (Fig. 6.35). Yangla block (high local gravity) represents the ancient Yangtze landmass in the west. It is low local gravity formed by a large area of Himalayan granite in the east. According to the research (Wu Xuanzhi et al. 1999), the periphery of rigid blocks is often favorable areas for mineralization. The results of regional remote sensing interpretation show (Wang Haiping 1999) that there

are many ore-showing annular tectonics and mineralized alteration image anomalies in Qingmai target area and its periphery of Xiangcheng Basin. This shows a good prospecting potential.

1: 200,000 geochemical prospecting results show that there are four multi-element composite anomalies and many single-element anomalies in the target area and its periphery. Among them, the anomalies related to the target area are the composite anomalies composed of Ba, Cu, Au and other elements. The composite anomalies are distributed in belt in SN direction along the Dougai-Yajin-Heida line, covering an area of about 50km². The Ba anomaly has the largest range and forms a concentration center near Heida. This shows a good prospect for searching for "black ore".

6.5.2 Geological Information of the Target Area

The Qingmai Pb–Zn–Cu-Ag ore target area is located in the southeast margin of Xiangcheng Basin (12 km south of Xiangcheng County), which is a sub-basin with an area of

Fig. 6.34 Aeromagnetic Pole
Reduction Anomaly Diagram of
Qingmai Target Area and its
Peripheral Area

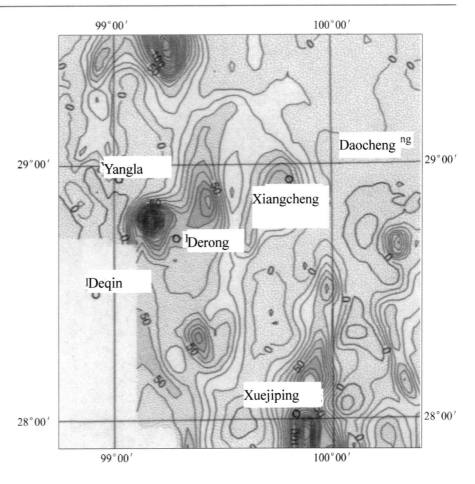

about 100 km². The stratum outcrop is mainly calc-alkaline
volcanic-sedimentary rock series of Miange Formation. The
stratum distribution in this area is controlled by Xiangcheng
fault with close north–south strike. The stratum generally
forms a westward inclined broken anticline near the target
area, which is located in the east wing of the broken anticline. At present, four ore occurrences and mineralized
points have been found. They mainly occur in the middle
and upper sub-cycles of the intermediate-acid volcanic-
sedimentary cycle. The ore-bearing surrounding rocks
include rhyolite, rhyolitic breccia lava, clinker tuff and
agglomerate. Although no ore bodies with industrial value
have been found in this target area, according to a lot of field
geological work achievements (Hou Zengqian et al. 1996), it
is found that there may be a "trinity" ore-bearing rock series
tectonic of volcanic rocks, ore bodies and exhalative sedimentary rocks in this target area. Typical signs of volcanic-
sedimentary exhalative deposits are found, such as "bimodal" rock assemblage, barite and volcanic apparatus. It is
predicted that there may be "Gacun" style ore body under
this target area (Fig. 6.36). Therefore, comprehensive geophysical prospecting prediction research has been carried out
in this target area.

6.5.3 Physical Properties of the Target Area

Physical property is the premise of geophysical prospecting
interpretation. In order to precisely study the physical
property characteristics of various rocks and ores in the
target area, the main rocks and ores were sampled when a
geophysical survey was conducted. The indoor physical
property determination and microscopic identification of
rocks and ores were carried out. A total of 20 specimens of
various types were collected in the target area, and the
physical property measurement results are shown in
Table 6.3 and Fig. 6.37.

The results of physical property measurement show that
obvious magnetic differences existed between rocks and ores
in this target area are basalt, massive ore and strongly
mineralized altered rhyolitic pyroclastic rock (specimens
No.2, No.3 and No.11). Other rocks such as rhyolitic tuff,
dacite tuff and sand-slate are nonmagnetic or weakly alkaline
except andesite (with magnetism). The resistivity of rocks
and ores in the target area depends more on the content of
sulfide and carbon and the degree of alteration. Most altered
rocks and carbonaceous slate (specimen No.12, No.13,
No.14, No.16 and No.17) have lower resistivity than

Fig. 6.35 Anomaly Diagram of Gravity Vertical Derivative in Qingmai Target Area and its Peripheral Area

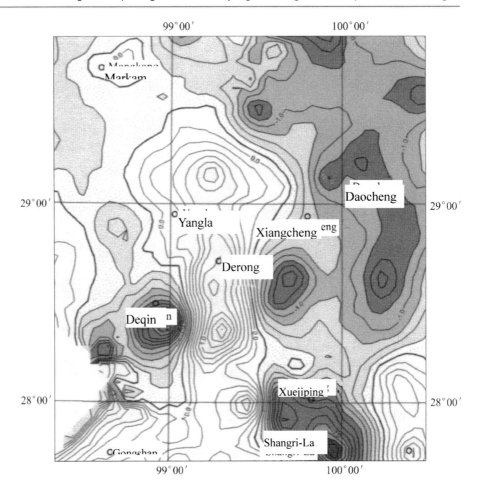

unaltered rocks. Of course, massive ores have extremely low resistivity. The polarizability is completely determined by the content of sulfide, and the polarizability of No. 1, No.2 and No.11 mineralized rocks is obvious and several times higher than that of ordinary rocks. There are obvious density differences between massive ore, strongly mineralized rock and ordinary rocks. To sum up, the geophysical prospecting methods for finding the same type ore deposits as the target area are magnetic method and electrical method mainly for measuring polarizability.

6.5.4 Comprehensive Geophysical Prospecting Survey Results and Analysis

According to the location of concealed ore bodies and known ore occurrence inferred by the geological data, and considering the topographic conditions, two areas are selected for geophysical prospecting surface scanning in the target area. The first area is centered on Mulanggong, covering an area of nearly 4km^2. The second area is from Yajin ore occurrence in the north to Yajinqiao in the south, covering an area of 0.2 km^2. Magnetic method, amplitude-frequency induced polarization and very low frequency electromagnetic method are used.

The measurement results and analysis are as follows.

(1) Survey Results and Analysis of Mulanggong Survey Area

There are 26 magnetic survey lines in the Mulanggong survey area. The spacing between survey lines ranges from 80 to 240 m, with the dot spacing of 5 m. The ΔZ anomaly diagram of the measured results after polarizing is shown in Fig. 6.38.

It can be seen from Fig. 6.38 that there is an obvious high magnetic anomaly in the middle and south of the survey area, respectively. The anomaly in the middle section of the survey area is large in scale and approximately equiaxed, with an average anomaly of 25nT. The anomaly contains several local high-value areas. The anomaly scale in the south section of the survey area is relatively small. It is similar to the anomaly caused by two spheres. From the measurement results of physical properties, the properties of the two high magnetic anomalies may be ore bodies

Fig. 6.36 Geological Map of Volcanic Sedimentation and Tectonic Inference in Qingmai Target Area. 1-Lanashan Formation; 2-Tumogou Formation; 3-marmarization limestone; 4-tuffaceous sandstone; 5-tuffaceous slate; 6-sedimentary tuff; 7-barite; 8-gypsum layer; 9-andesite; 10-rhyolitic tuff; 11-rhyolite; 12-silicon-rich rhyolite; 13-basalt; 14-agglomerate; 15-volcanic apparatus; 16-rhyolite pillow body; 17-known occurrences; 18-inferred ore body location; 19-scope of magnetic survey work area

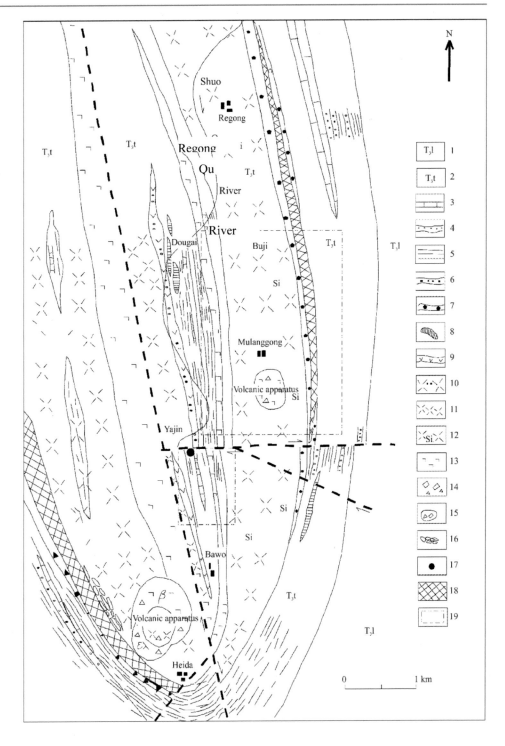

(mineralized bodies) or basalt bodies. In view of the fact that the property of anomalies cannot be distinguished only by magnetic method, we have made six frequency induced polarization profiles at the middle-high magnetic anomaly area (Fig. 6.39), and the expected high polarizability anomaly has not been found. The frequency dispersion rate (proportional to the polarizability) mostly changes around 2%. According to the experience of other places, when the frequency dispersion rate measured by this instrument is more than 10%, it can be determined as mineralization. There is only a local low resistivity anomalous zone, which is close NS (tend to be NW) strike, in the east of the profile. It is verified by the earth surface survey that this low resistivity anomalous zone is caused by irrigation canal. According to the measurement results of amplitude-frequency IP, it can be concluded that this anomaly is

Table 6.3 Determination of physical property of Yajin lead and zinc deposit in Xiangcheng

Serial No.	Lithology	Magnetic susceptibility $4\pi \times 10^{-6}$SI	Residual magnetization 10^{-3}A·m^{-1}	Resistivity Ω·m	Polarizability %	Density g·cm^{-3}
1	Compact massive ore deposit	129.4	0.13	33.6	35.9	3.903
2	VMS	981	13	55.8	42.6	3.542
3	Basalt	921	2.2	2232.6	2.8	2.840
4	Rhyolitic-dacite crystalline tuff	1.3	0.52	11,747.6	2.2	2.589
5	Quartz andesite	6.55	0.22	8049.9	3.2	2.626
6	Andesite-dacite	0.6	0.12	9067.8	3.4	2.577
7	Dacite crystalline tuff	3.4	0.15	2496.4	0.7	2.590
8	Andesite-dacite	134.5	3.1	4372.8	2.2	2.523
9	Andesite-dacite	269.5	21	2059.9	6.2	2.618
10	Strongly silicified andesite-dacite	14	0.20	58,020.6	0.0	2.585
11	Mineralized rhyolitic volcanic-sedimentary rocks	4040	240	1463.7	23.8	3.034
12	Dacitic altered crystalline tuff	123.6	0.43	1535.8	2.9	2.704
13	Silica sludge carbonaceous slate	281.2	0.092	626.2	3.1	2.679
14	Siltstone	120	0.2	1344.6	2.5	2.673
15	Microcrystalline limestone	35.6	0.29	7039.5	7.6	2.664
16	Andesite-dacitic crystalline tuff	238.2	0.99	1412.4	3.5	2.528
17	Andesite or basaltic-andesite	115.6	3.4	356.4	4.5	2.524
18	Andesite	424	2.7	2403.2	3.0	2.689
19	Rhyolitic tuff	345	0.72	1160.9	2.6	2.684
20	Dacite rhyolitic crystalline tuff	149.8	0.38	347.8	4.1	2.510

probably caused by basaltic rocks. However, considering that the survey area is covered with thick beds (about $30 \sim 50$ m) and all of them are dry boulder sediments, the grounding conditions are seriously affected. Therefore, the exploration depth is greatly reduced. The depth of the anomalous body was not reached. Therefore, it cannot be completely denied that there is a deposit in the survey area.

VLF electromagnetic measurements results show that there is no obvious anomaly at the corresponding position of high magnetic anomaly (Fig. 6.40). There are some linear anomalies with close NW strike in the map. After verification of earth surface topography and surface features, it shows that these anomalies are all caused by irrigation canals and topography. The absence of the expected high conductor anomaly may further indicate that the magnetic anomaly is caused by basalt or show that the VLF electromagnetic measurements cannot detect the target when the detection target is buried deeply. In this condition, the method is ineffective.

(2) Geophysical prospecting survey results and analysis in Yajin survey area.

A total of 14 profiles have been measured by magnetic method in Yajin survey area, with a line spacing of 50 m and a dot spacing of 5 m. The preprocessed ΔT isoline of the measurement results is shown in Fig. 6.41. It can be seen from the figure that there is an obvious high-value anomalous zone in the Yajin survey area. The anomalous zone has a width of 50–100 m, spreading in the northwest and extending more than 600 m. At the northern end of the anomalous zone, it is a known occurrence, and massive lead–zinc deposits with a thickness of $10 \sim 20$ cm are found in the outcrop in the valley of the anomalous zone. According to the above characteristics and geological data of Yajin deposits, combined with physical property measurement data, it can be judged that this anomalous zone is the reflection of ore zone. In order to further confirm the property of the anomalous zone, an amplitude-frequency IP

Fig. 6.37 Histogram of physical properties of rocks and ores in Qingmai target area

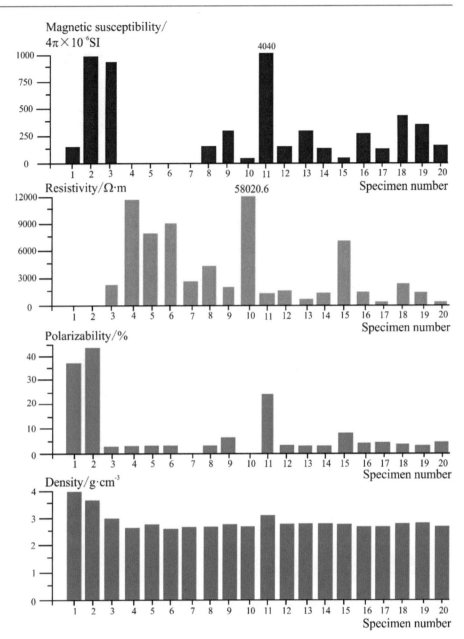

profile is made at the southern end of the anomalous zone. The results show that there is a strong IP anomaly, which is characterized by high polarization and low resistance. Frequency dispersion rate anomaly shows as bimodality of one high and one low. While resistivity anomaly is generally low resistivity, and the resistivity only slightly increases when it corresponds bimodality anomaly of frequency dispersion rate. Comparing IP anomaly with magnetic anomaly, it is found that IP bimodal anomaly does not completely correspond to high magnetic anomaly. But only corresponds to half of high magnetic anomaly, and the other half of bimodal anomaly corresponds to low magnetic anomaly. That is, the center of IP anomaly corresponds to the gradient zone of magnetic anomaly. The most reasonable explanation for this

feature is that the mineralized zone occurs in the contact zone of two lithologies, one of which has higher magnetism. According to the measurement results of physical properties, it can obviously be concluded that it is basaltic rock. If the width of mineralized zone is delineated according to the frequency dispersion rate of 5% and the length of mineralized zone is estimated by the extension length of magnetic anomaly, it can be estimated that the mineralized zone has a width of about 60 m and a length of at least 600 m (the two ends of anomalous zone are not closed) within the Yajin survey area, with considerable scale.

VLF electromagnetic measurements results also seem to verify the presence of mineralized zones (Fig. 6.42). According to the interpretation principle that the "trough" of

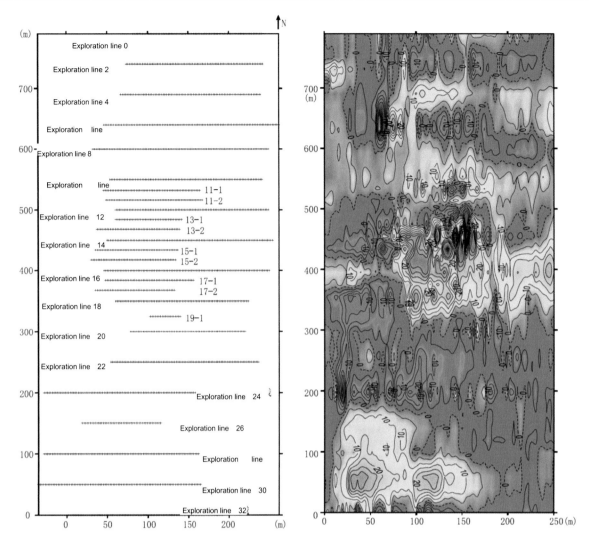

Fig. 6.38 High-precision magnetic ΔZ isoline map of Mulanggong survey area

vertical component corresponds to the plate-shaped good conductor, it can be seen that there is an intermittent distributed relatively low-value anomaly at the western edge of the survey area. The low-value anomaly is close to the western edge of the NW-strike high-value anomaly in the middle area and basically corresponds to the inferred mineralized zone.

6.5.5 Conclusions and Suggestions

(1) Based on the field observation, physical property survey and research of the known occurrence in this target area and comprehensive geophysical prospecting data interpretation of the two survey areas, it can be

concluded that the main mineralization of this target area occurs in the contact zone between basic basalt and volcanic-sedimentary rocks. While the mineralization of Gacun ore deposit in the north is mainly related to rhyolite. Therefore, there is no condition for prospecting a Gacun style deposit.

(2) Mulanggong high magnetic anomaly is caused by basalt. It is not ore-induced anomaly.

(3) The Yajin high anomaly is directly related to the ore belt, and the strike of the anomaly basically reflects that of the ore belt.

(4) The direction of further prospecting in this target area should be along the two sides and the north–south extension direction of the magnetic anomalous zone. Suggestions for future work:

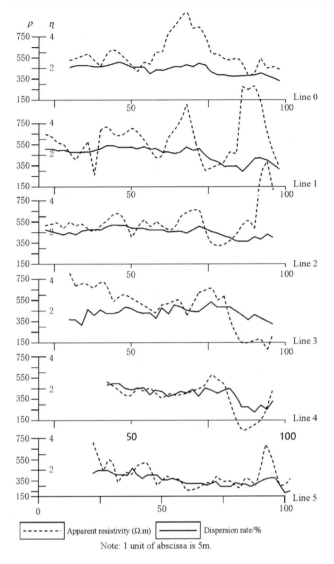

Fig. 6.39 Plan and profile of amplitude-frequency IP in Mulanggong magnetic anomaly area

- ① To identify the presence of ore bodies at depth in the contact zone of the periphery of Mulanggong anomaly by the electrical method with deep detection depth.
- ② To conduct pit depth and exploratory trench survey in Yajin ore zone to determine the thickness of rich ore body. At the same time, electrical methods with large detection depth and high resolution, e.g., high-density electrical method, shall be used to determine the deep-seated form of ore body.
- ③ In the north–south extension direction of the Yajin anomalous zone, magnetic method and light electrical method shall be used to track the strike of the ore zone.

6.6 Integrated Technologies for Exploration of Shear Zone Type Gold Deposits (Orogenic Gold Deposits)

For the prospecting of shear zone (ductile, brittle-ductile shear zone) type gold deposits, we adopted the integration of "ductile shear zone + geochemical prospecting anomaly" and realized the prospecting breakthrough of Zhenyuan gold deposits (Laowangzhai, Donggualin, Daqiaoqing, Langnitang gold deposits, etc.).

Zhenyuan gold deposit consists of six main ore sections: Laowangzhai, Donggualin, Langnitang, Daqiaoqing, Kudumu and Bifushan. It is a super-large ductile shear zone type (orogenic type) gold deposit located on the west side of Ailaoshan ductile shear zone (see Figs. 3.15 and 5.7). The Zhenyuan gold deposit was called Laowangzhai gold deposit in some literature in the early period. In recent years, many prospecting breakthroughs have been made in some ore sections (e.g., Langnitang ore section) and periphery (e.g., Shangzhai, Hepingyakou). The original exploration depth of the ore deposit is generally 300–500 m. According to the undulation of nappe tectonic in ductile shear zone and the metallogenic characteristics of strong–weak-strong–weak mineralization regularity along tectonic, it is predicted that there are still great prospecting potential in the deep parts of Donggualin and Langnitang ore sections.

During the chromite general survey in the 1970s, native gold was found in the heavy sand of Laowangzhai ultrabasic rocks, yet not much attention was given. In 1976, gold element analysis was supplemented to the geochemical prospecting duplicate samples during the general survey of chromite, and gold anomaly was circled. In 1983, ore bodies I and ore bodies II in Laowangzhai were found when mine inspection was carried out. Both ore bodies I and ore bodies II were related to altered ultrabasic rocks. Immediately, the search for ultrabasic rocks was taken as an important symbol of prospecting, and 1:50, 000 ground magnetic survey was deployed. 11 magnetic anomalies were circled, such as Baitushan, Laowangzhai, Suoshan and Shilihe Iron Works. No great progress was made after anomaly verification. The gold prospecting idea of "altered ultrabasic rocks" is in a dilemma.

From 1985 to 1987, soil chemical exploration of 1: 10,000 and 1: 25,000 was carried out in this mining area and its periphery, and 67 gold anomalies were delineated. It was found that the anomalies formed a beaded distribution in NW–SE direction and coincided with the main fault tectonic. It has been verified that gold mineralization has been found in gold anomalous concentration center and the tectonic coincidence area. For a time, gold prospecting made progress, but not big. Up to 1988, the amount of resources

Fig. 6.40 Isoline map of very low-frequency vertical component of Mulanggong. *Note* 1 unit is 5 m in ordinate

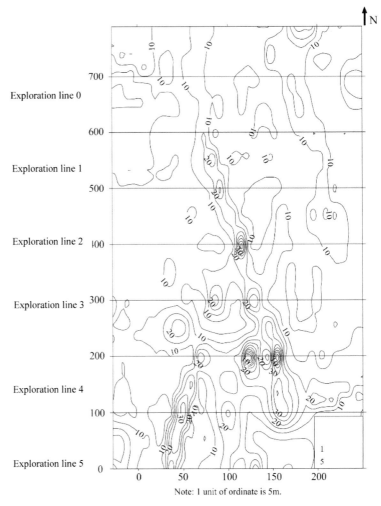

Note: 1 unit of ordinate is 5m.

obtained was less than 10t, and the exploration work stagnated again.

After 1991, when carrying out ore deposit evaluation, it was noted that strike-slip-shear and nappe tectonic are developed in this mining area. Although brittle-brittle-ductile tectonic is more common on the earth surface in this mining area, there are many shear structures such as boudin, folded bed, shell fold and mylonite. Mylonite zones are also common between brittle-brittle-ductile tectonic, and ductile shear zone occurring in bedding can also be seen between window lattice tectonics (Fig. 6.43). At the same time, it is found that most ore bodies also exist in brittle-ductile shear zones around rigid rock blocks (Figs. 6.44 and 6.45). The metallogenic model of gold deposits controlled by ductile shear zones was put forward for the first time. The ore deposit types are positioned as tectonic altered rock types or ductile shear zone type gold deposits related to ductile shear zones. At the same time, it shows that the ore deposit is generally controlled by Ailaoshan ductile shear zone. While the ore body is also controlled by brittle-ductile shear tectonic. The shear zone is in good agreement with geochemical anomalies of Au, As, Sb and other elements (Fig. 6.46).

Subsequently, according to the prospecting model of "ductile shear zone + geochemical prospecting anomaly", the exploration and evaluation work was redeployed. Regional remote sensing technology was used in combination with the earth surface identification and recourse control. So a large number of brittle-ductile shear tectonic have been delineated. Soon, the 1280 m long gold deposit in Donggualin was controlled. It is found that the gold deposit is located at the transitional position between shallow brittle shear and deep-seated ductile shear. The gold deposit probably extends to the deep ductile shear zone. Thus, the prospecting model is further deepened and perfected. As a result, major prospecting breakthroughs have been made in six main ore sections, namely Laowangzhai, Donggualin, Langnitang, Daqiaoqing, Kudumu and Bifushan. So the deposit scale has expanded from less than 10t to 104t. At present, the gold deposit reserves in this zone are still expanding. In the shear zone, there may be the isometric distribution of ore deposits and pod-like distribution of ore bodies. Therefore, attention should be paid to the prospecting work in Hepingyakou in the north section of the ore deposit, Bifushanyakou in the south section and its southern

Fig. 6.41 High-precision Magnetic Δ T Isoline Map of Yajin Target Area

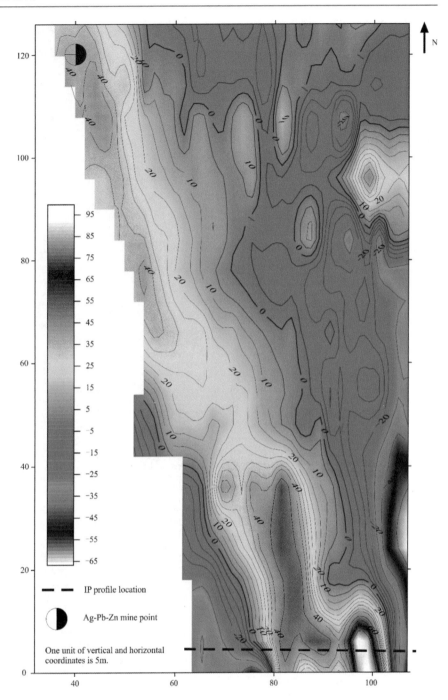

extension. Attention should be paid to the reappearance of ore sections (beds) extend to deep part in pod-shaped or pinchout form in Donggualin, Langnitang and Daqiaoqing.

The Zhenyuan gold deposit is a typical example of geochemical prospecting for gold. Its ore is controlled by ductile shear zone. This drives gold deposit prospecting in Yunnan. The discovery of many gold deposits, including Ailaoshan and Gaoligongshan metamorphic zones, draws lessons from the successful prospecting experience of Zhenyuan gold deposit.

6.7 Integrated Technologies for Exploration of Hydrothermal Vein Type Lead–Zinc Polymetallic Deposits

For the prospecting of hydrothermal vein type lead–zinc polymetallic deposits, we have adopted the integration of tectonic trap + hydrothermal circulation center + various electrical techniques.

Because the regional stress push and extrude from NE to SW, the relative movement directions of local stress fields

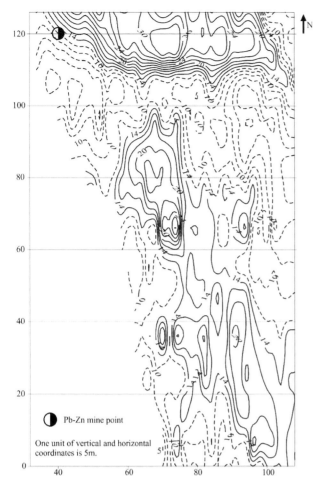

Fig. 6.42 Isoline Map of Very Low-frequency Electromagnetic Method (NZ) in Yajin Target Area

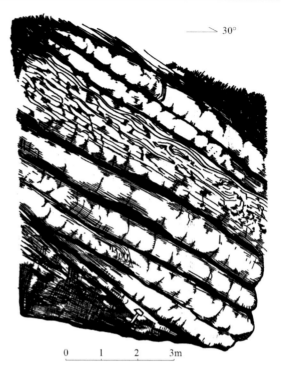

Fig. 6.43 Window Lattice Tectonic and Bedding Ductile Shear Zone in Thin Plate Argillaceous Limestone Fold Beds

on the NE and SW sides of anticline are inconsistent. The NE wing is in a relatively tensile state, which is easy to be mineralized. While the condition in SW wing is opposite.

6.7.1 "Thermal Cycle" Mineralization in Lanping Basin

Lanping-Pu'er basin has always been the focus of geological research and prospecting. After the evaluation of Lanping Jinding lead–zinc deposit, it has always been a dream of geologists to find the second "Jinding" deposit in the basin.

In the past 65 Ma, India and Eurasia began to collide and uplift, forming the Qinghai-Tibet Plateau. With the intensification of collision, large rotation, strike-slip, nappe and fluid migration occurred in the middle and south section of Sanjiang area, which is located in the eastern margin of Qinghai-Tibet Plateau (Fig. 6.47).

India pushed, extruded and collided northward. Firstly, a series of tectonic traps in Lanping Basin were formed. These

tectonic traps became the reservoir space of fluid (Fig. 6.48). The subsequent large-scale strike-slip pull-apart lead to the formation of a series of strike-slip pull-apart basins. At the same time, a series of dilatation centers, which provided channels for the rise of the subsequent magma, were also formed. Large-scale strike-slip faults cut the lithosphere and induce magmatic upwelling. These intrusions drive hydrothermal activity. Then, a series of nearly equidistant hydrothermal activity centers formed (Fig. 6.49). The hydrothermal fluid circulates repeatedly in the hydrothermal cycle center and extracts the metal components in the stratum. The hydrothermal fluid rises together with mantle-derived materials and forms different ore deposits in different environments. In some places, magma upwell to the earth's surface or shallow part, forming porphyry deposits (Yulong porphyry copper deposit in Tibet, etc.). In other areas, e.g., the eastern margin of Lanping Basin. Although magma does not reach the shallow part, and with large buried depth, it drives the thermal cycle or erupts into the tectonic depression (lake bottom) together with the shallow ore-bearing hot brine to precipitate and form an ore deposit (Lanping Jinding lead–zinc deposit). Or the magma precipitates in a series of shallow tectonic interfaces to form an ore deposit (Lanping Baiyangping lead–zinc-silver-copper polymetallic deposit). Or the magma spouts out of the earth's surface to form a hot spring precipitation, which constitutes the ore deposit (Yunlong Dalong lead–zinc-silver deposit). Jinding super-

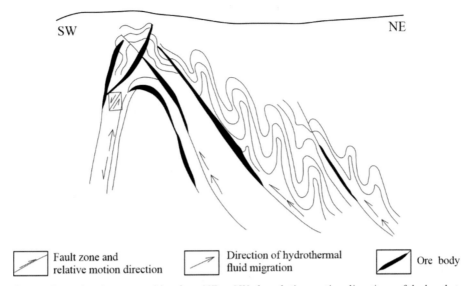

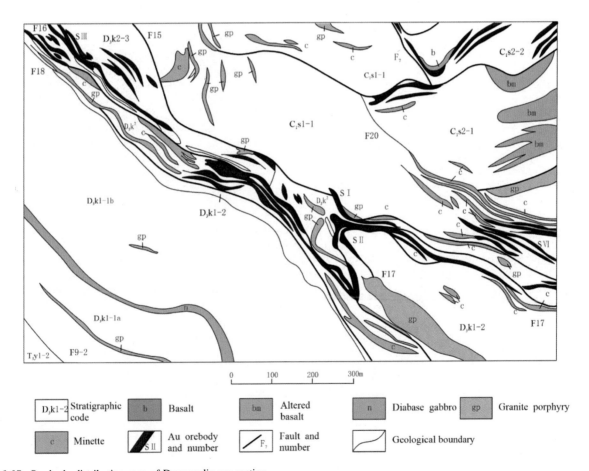

Due to the regional stress pushing from NE to SW, the relative motion directions of the local stress field on the NE and SW sides of the anticline are inconsistent. The NE wing is in a state of relative extension and easy to mineralization, while the SW wing is opposite.

Fig. 6.44 Tectonic-mineralization model of ore section

Fig. 6.45 Ore body distribution map of Donggualin ore section

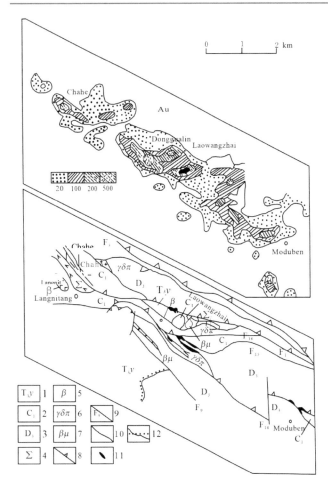

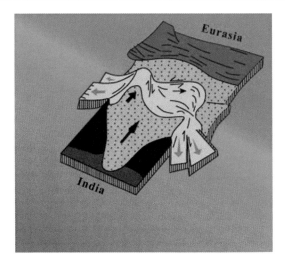

Fig. 6.47 Location of Sanjiang Tectonic and Formation Background of Strike-Slip Tectonic

Fig. 6.46 1: 10,000 soil survey in Laowangzhai gold deposit. Au anomaly corresponds to brittle-ductile shear tectonic 1-Upper Triassic-Wanshui Formation; 2-Lower Carboniferous; 3-Upper Devonian; 4-Ultrabasic Rocks; 5-Basalt; 6-Granite Diorite Porphyry; 7-Diabase; 8-Brittle-Ductile Shear Zone; 9-Fault and its Serial Number; 10-Geological Erathem; 11-Ore Bodies; 12-Unconformity Erathem. The prospecting breakthrough of Baiyangping silver-lead–zinc polymetallic deposit in Lanping has been made

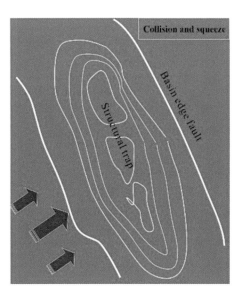

Fig. 6.48 Schematic Diagram of Tectonic Traps in Lanping Basin

large Pb–Zn ore deposit exists in the Lanping basin. During the strike-slip process of Bijiang fault, shallow hot brine mixed with deep-seated rising ore-bearing fluid. Then, the mixture exhaled and sedimented in the sunken lake, and the deposit formed. The ore-bearing hydrothermal fluid erupted into lake ditch for deposition. Next, the metasomatism and filling occurred along the tectonic interface, and the deposit formed. The ore deposit output on the side of the Bijiang strike-slip fault occurred at the bottom of the decollement zone in the footwall of the thrust nappe (Fig. 6.50).

The forming process of Lanping Baiyangping Pb–Zn-Ag–Cu polymetallic deposit is shown below. Driven by deep-source heat, the hydrothermal solution with high salinity circulated repeatedly and extracted the metal components in the stratum. Together with mantle-derived

materials, the hydrothermal solution precipitated in a series of shallow tectonic interfaces to form an ore deposit, which directly occurred in the overthrust zone and its hanging wall tectonic fracture zone (Fig. 6.51). At the Yunlong Dalong Pb–Zn mineralized spot, hydrothermal fluid directly erupted from the earth surface and formed hot springs. Pb–Zn-Ag mineralization is common in the precipitated sinter.

According to this metallogenic theory, we quickly narrowed the exploration focus to the hydrothermal circulation center. By adopting the integrated technology of "tectonic trap + hydrothermal circulation center + multiple electrical methods", the prospecting breakthrough of Baiyangping Cu-Pb–Zn-Ag polymetallic deposit has been realized.

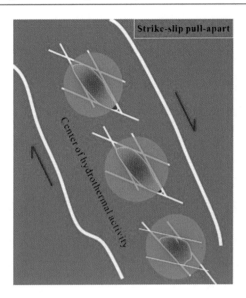

Fig. 6.49 Schematic Diagram of Equidistant Hydrothermal Activity Center

The submitted resources of Baiyangping Cu-Pb–Zn-Ag polymetallic deposit are Ag 4009.55 t, Pb–Zn 870,000 t and Cu 378,800 t.

6.7.2   Demonstration Research of the Exploration Technology Integration

In 1992, the exploration and evaluation were carried out for the Huishan lead–zinc deposits in and Heishan lead–zinc deposits, which were discovered by anomaly inspection. At that time, the genesis of these two deposits was unclear, and the prospecting focus and target were vein copper polymetallic ore bodies in red beds, and the prospecting effect was not obvious. The 1: 200,000 geochemical surveys in

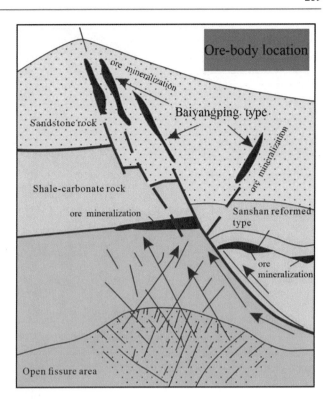

Fig. 6.51 Metallogenic model of Lanping Baiyangping Pb–Zn-Ag Deposit

Baiyangping area are processed by multiple methods, especially the SA fractal method. The survey results show that the lead–zinc geochemical anomalies are isometric (Fig. 6.52). After careful study, it is considered that the occurrence of this isometric anomaly is not accidental but related to the formation of regional strike-slip tectonic (Fig. 6.53). The isometric anomaly is the result of thermal cycle mineralization.

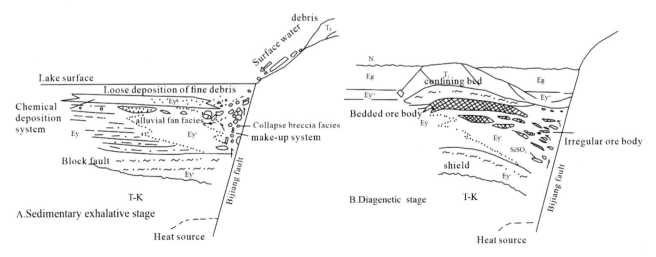

Fig. 6.50 Metallogenic Model of Lanping Jinding Pb–Zn Deposit

Fig. 6.52 Isometric Distribution of Lead–Zinc Anomaly in Lanping Baiyangping Area

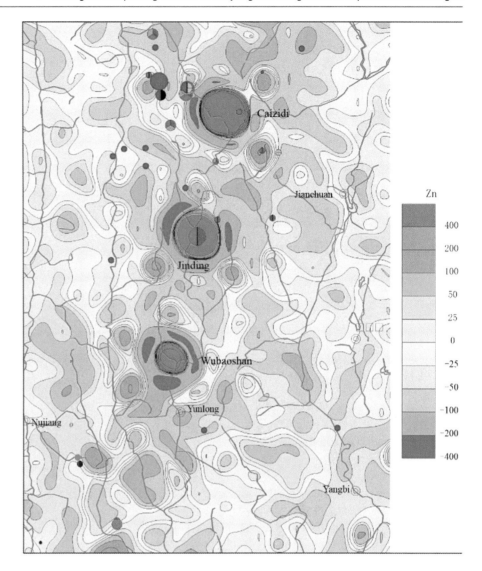

Later, we found new evidence of super-imposed mineralization of mantle-derived materials through further study. It is considered that large-scale mineralization in the basin mainly occurred in the Himalayan period, in which the fluids in deep part merged with brine in the basin to form isometric and multi-center cycle mineralization along a series of nappe fracture zones. Then, silver-lead–zinc polymetallic deposits formed in Baiyangping, Jinding and Baiyangchang.

In 1994, when evaluating Baiyangping ore deposit, it is found that the ore body is distributed along a group of faults close north–south direction. Immediately, the tectonic system, ore-controlling conditions and metallogenic regularity in Baiyangping area were studied. It was discovered that the regional ore-bearing stratum included Sanhedong Formation of Upper Triassic, Huakaizuo Formation of Middle Jurassic, Jingxing Formation of Lower Cretaceous, Baoxiangsi Formation of Paleogene, etc. The ore body output was mainly controlled by regional thrust nappe tectonic system (Fig. 6.54).

The study found that these thrust nappe tectonics have created many tectonic traps with different tightness. These tectonic traps become important ore-hosting spaces. Under the extrusion background, regional fluids migrated and converged to low stress areas. They leached and accumulated into mineralized materials along the way. Then, the fluid upwelled and excreted along thrust nappe fault zones, precipitated along a series of tectonic interfaces in tectonic traps, and eventually formed ore bodies. Within the ore deposit scale, the ore body groups occurring roughly in parallel also show the characteristics of approximately equidistant distribution along Huachangshan nappe-strike-slip fault zone, such as Yanzidong-Heishan Huishan ore section and Dongzhiyan-Xiaquwu ore section in the east ore zone (Fig. 6.55).

According to the above theoretical model, the integrated prospecting idea of "tectonic trap + hydrothermal circulation center + multiple electrical methods" is put forward. That said, a large-scale regional geological survey in the whole

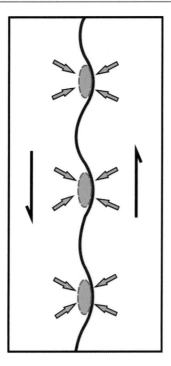

Fig. 6.53 Schematic Diagram of Formation Mechanism of Strike-Slip Tectonic and Hydrothermal Cycle Center

region should be carried out again. The survey focuses on the tectonic system, mainly including trap tectonic, nappe tectonic and strike-slip tectonic. The geochemical investigation of ore-forming fluids (fluid mapping) was arranged. The fluid inclusions were systematically studied, and the fluid inclusions were in Baiyangping Pb–Zn–Ag–Cu polymetallic deposit along the ore-bearing tectonic and the profile crossing the ore zone at certain intervals. So the properties, characteristics and state of ore-forming fluids were determined. The fluid inclusions in this area are mainly two-phase inclusions with rich fluid phase. The composition of the fluid inclusions is mainly $NaCl + H_2O$ system. For the fluid inclusion, its homogenization temperature is between 90 and 220 °C, and its upper limit of metallogenic temperature is about 280 °C. The salinity of the metallogenic fluid is generally between 5 and 15% (NaCl), with medium–low temperature and low salinity. The gas phase composition of fluid inclusions is mainly H_2O, followed by CO_2. The reduction parameter w $(H_2 + CH_4 + CO)/w$ (CO_2) is low, which indicates that mineralization is carried out in a relatively reduced environment. Electrical work mainly based on redox potential method is conducted. The results of electrical work show that the anomalies of negative potentials are distributed in parallel belt. It is inferred that they are caused by metal mineralization. Therefore, the earth surface engineering exposure of the anomalous zone was arranged. The earth's surface ore body was systematically controlled, and then, the control in deep part was carried out at a certain

engineering interval, thus realizing the breakthrough of prospecting. This method has been fully applied in the ore zone, and the evaluation has been started from Baiyangping and Fulongchang ore sections in the west ore zone to seven ore sections distributed along Huachangshan fault zone in the east ore zone, such as Dongzhiyan, Xiaquwu, Xinchangshan, Yanzidong, Huachangshan, Huishan and Heishan, with a large ore deposit scale. In the work, we pay attention to the favorable tectonic parts (a series of nappe structures) and also focus on the favorable metallogenic factors of lithofacies (bioclastic limestone and dolomitic limestone).

The new understanding of multi-center and equidistant hydrothermal circulation mineralization in Baiyangping area breaks through the traditional understanding that there are only intracontinental hot water jet deposits and sediment-hot brine transformation in Lanping Basin for many years and expands the prospecting ideas, which is of great significance to regional prospecting.

6.8 Integrated Technologies for Exploration of Skarn/porphyry Concealed Deposits

For the prospecting and exploration of skarn/porphyry concealed deposits, we have probed into the integrated technology of "metallogenic system + gravity + magnetism + multiple electrical methods" through the practical application of numerous deposits.

6.8.1 Discovery and Evaluation of Hetaoping Cu-Pb–Zn-Fe-Au Polymetallic Deposit in Baoshan. An Example of "Metallogenic System + Gravity + Magnetism + Multiple Electrical Methods" to Find Concealed Deposits

Baoshan Hetaoping is a polymetallic ore deposit integrating copper, lead, zinc, silver, gold and iron, which is a good example of applying multi-variate information for metallogenic prediction and prospecting concealed ore by combining gravity, magnetism and electricity method. In 1988, geochemical anomalies were discovered by 1: 200,000 geochemical exploration, and in 1989, the Hetaoping ore deposit was discovered during the anomaly inspection, but "only the trace ore was seen, but the ore deposit was not seen".

Through the secondary development and comprehensive analysis of geological, geophysical, geochemical and remote sensing data in this area, it is found that many "trace ore" in this area have internal genetic connections. They constitute a skarn (porphyry)/hydrothermal vein polymetallic metallogenic system which may be related to concealed rock mass

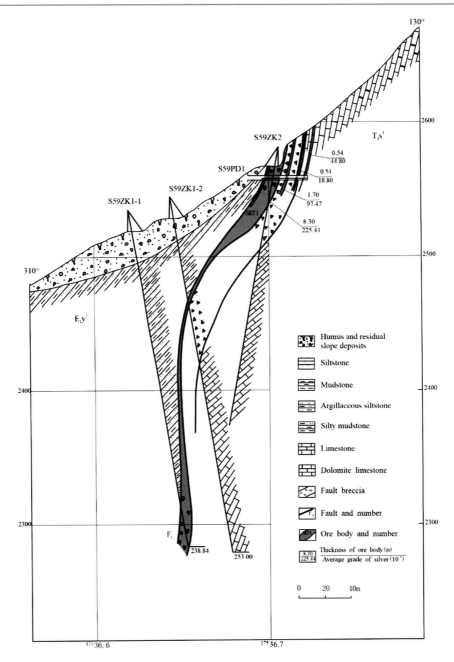

Fig. 6.54 Geological Profile of Exploration Line 59 in The Ore Section of Xiaquwu

and controlled by fault structure and fracture system, and the "ore deposit" is likely to be concealed at the top of concealed rock mass, especially the low gravity and high magnetic anomaly shown by 1: 500,000 gravity and magnetic field, which further confirms the possible existence of concealed granite rock mass in deep.

Regional geological research shows that the Lincang granite body, which was mainly uplifted during Indosinian, inclined to the north, and the magma uplifted gradually become newer to the north.

Lincang granite batholith shows obvious low gravity (negative anomaly), and gravity also shows obvious negative anomaly in Zhenkang and Baoshan Hetaoping to the west and north. Therefore, from the geological, gravity and magnetic anomalies, it is reflected that there may be concealed rock masses in the deep parts of Zhenkang and Hetaoping (Fig. 6.56).

According to the 1: 200,000 stream sediment survey in Yunnan Province, an obvious 1: 200,000 geochemical anomaly was circled in Baoshan Hetaoping area, and then, a 1: 50,000 stream sediment geochemical survey was arranged in the anomaly area, and the circled lead, zinc and silver geochemical anomalies were horseshoe-shaped, with local gold anomalies (Fig. 6.57). In subsequent geological

Fig. 6.55 Output distribution of ore bodies along faults in Baiyangping east ore zone

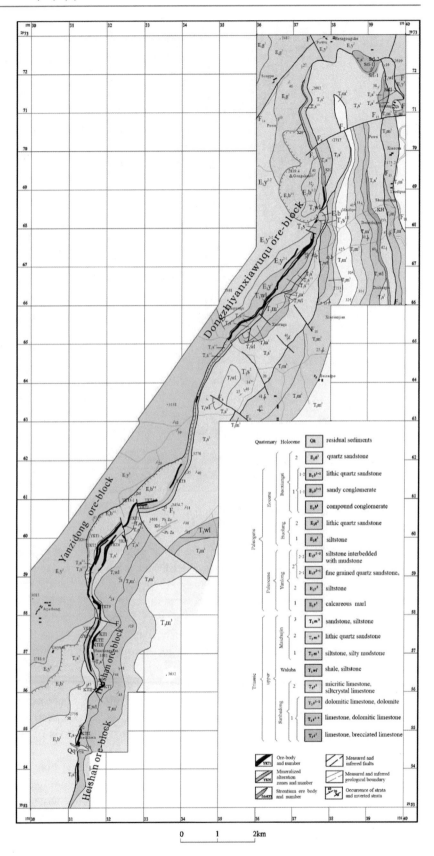

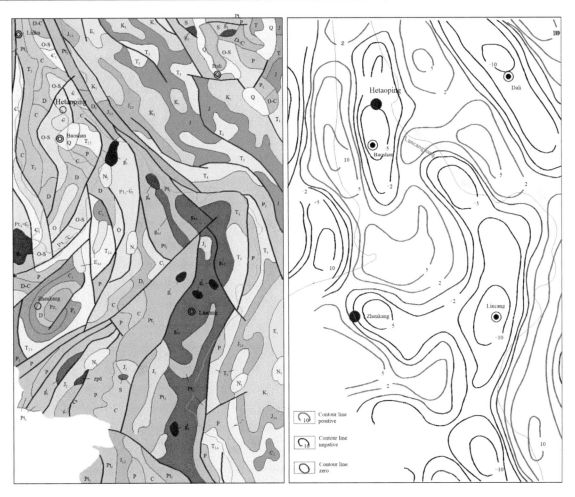

Fig. 6.56 Geological map and gravity isoline map of Hetaoping area

prospecting, we found ore bodies in Hetaoping. After sparse engineering control, the estimated lead and zinc resources are small scale. For a long time after that, we failed to get project approval for further exploration in this area, and even failed to get project approval for many years due to the small scale of the deposit, insufficient data for potential analysis and inadequate basis for project establishment.

After further study, it is considered that all the comprehensive geological elements such as geology, geophysical prospecting, geochemical prospecting and remote sensing in this area show promising metallogenic conditions and great resource potential. Nevertheless, what method can we use to make a breakthrough in the prospecting of concealed deposit? We have adopted the technical integration of "gravity + magnetism   (measurement) + electricity (method)" and achieved a breakthrough in prospecting.

(1) With 1: 100,000 gravity survey and 1: 50,000 quickly surface scanning of ground magnetic survey, we found that gravity, magnetic and geochemical anomalies are

horseshoe-shaped along Hetaoping anticline. According to the high local gravity and strong magnetic anomalies on the background of low gravity (Figs. 6.58 and 6.59), it is speculated that this is caused by magnetic anomalies. The geochemical anomalies correspond to most of the magnetic anomalies, and the anomalies of the main elements Pb, Zn, Cu, Ag and Cd overlap with each other. While the anomalies of the elements Au, Sb, As and Hg mostly appear outside the main element anomalies, showing obvious zoning, and the mineralization is related to the intrusion of granite located at the axis of Hetaoping anticline.

(2) Among the 13 magnetic anomalies delineated, 1: 10,000 high-precision magnetic survey and high-power IP intermediate gradient and electrical sounding are carried out in the anomalies with high local gravity and large magnetic anomaly area and high intensity. Among them, Jinchanghe, Shangchang and Hetaoping have good magnetic and electrical reflections. The area of magnetic anomaly in Jinchanghe is large, with a length

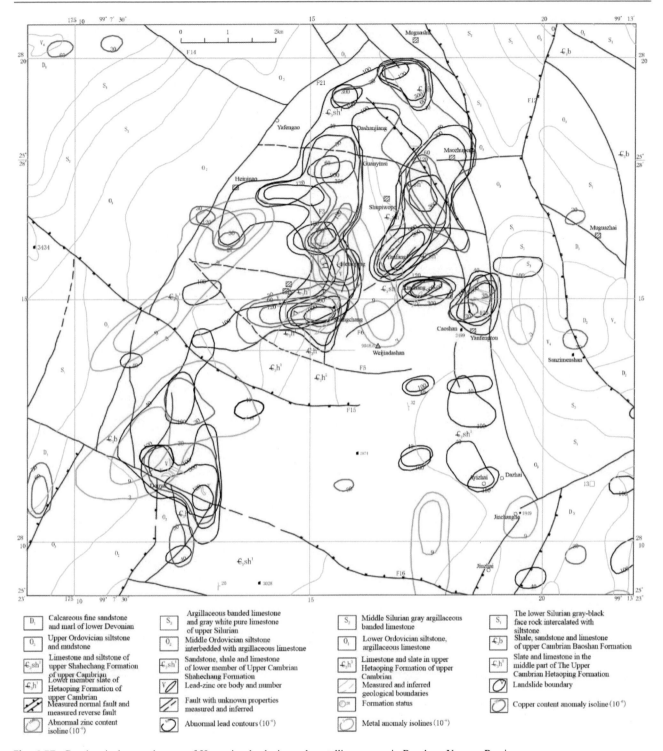

Fig. 6.57 Geochemical anomaly map of Hetaoping lead–zinc polymetallic ore area in Baoshan, Yunnan Province

of about 3 km and a width of about 2 km. The anomaly delineated by isolines of positive magnetic anomaly extreme value of 1031 nT and 400 nT shows NE direction. The negative anomaly is distributed in the NE direction of the positive anomaly, with its extreme value of 131 nT.

(3) In order to accurately determine the spatial morphology of the mineralized body, the downward continuation and horizontal derivative treatment of the electromagnetic anomalous body were carried out in Jinchanghe. The downward continuation was carried out with 100 m, 150 m, 250 m and 500 m, respectively, for

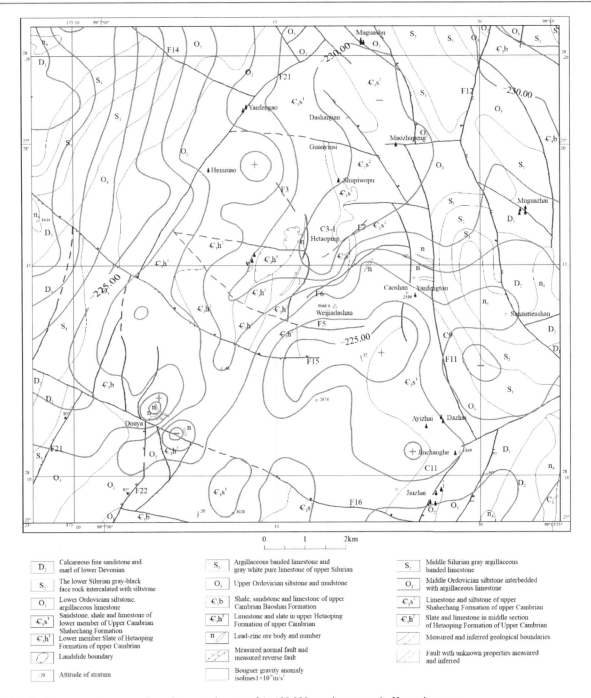

0   1   2km

D_1	Calcareous fine sandstone and marl of lower Devonian	
S_1	The lower Silurian gray-black face rock intercalated with siltstone	
O_1	Lower Ordovician siltstone, argillaceous limestone	
C_3s^1	Sandstone, shale and limestone of lower member of Upper Cambrian Shahechang Formation	
C_3h^1	Lower member Slate of Hetaoping Formation of upper Cambrian	
	Landslide boundary	
$/20$	Attitude of stratum	

S_1	Argillaceous banded limestone and gray white pure limestone of upper Silurian
O_1	Upper Ordovician siltstone and mudstone
C_3b	Shale, sandstone and limestone of upper Cambrian Baoshan Formation
C_3h^3	Limestone and slate in upper Hetaoping Formation of upper Cambrian
n	Lead-zinc ore body and number
	Measured normal fault and measured reverse fault
	Bouguer gravity anomaly isolines1×10^{-5}m/s^2

S_2	Middle Silurian gray argillaceous banded limestone
O_2	Middle Ordovician siltstone interbedded with argillaceous limestone
C_3s^2	Limestone and siltstone of upper Shahechang Formation of upper Cambrian
C_3h^2	Slate and limestone in middle section of Hetaoping Formation of Upper Cambrian
	Measured and inferred geological boundaries
	Fault with unknown properties measured and inferred

Fig. 6.58 Geology anomaly map and gravity anomaly map of 1: 100,000 gravity survey in Hetaoping area

continuation transformation, and the anomaly oscillated at -500 m (Fig. 6.60), indicating that the magnetic body was within 500 m. Using characteristic point method and tangent method to explain and infer the results with various methods, it is considered that the plane shape of the magnetic body is as shown in Fig. 6.61, with the top interface buried at 250–270 m, striking at a thick plate-like body with a length of 650 m and a thickness of 130 m and inclining to the northwest (Fig. 6.62).

In 2003, boreholes were arranged in the magnetic anomaly for verification. In the first borehole, we found concealed skarn-type copper-Fe ore body at 276 m, with Cu-Au at the top, Fe-Au at the middle and Fe at the lower part. The main Cu-Au ore body is 45.1 m thick, with grades of Cu 1.42% and Au 0.5%. The lower magnetite body is 308 m thick, and TFe grade is 31.84% (Fig. 6.62). On this basis, the evaluation of concealed ore deposit is carried out, and more than 100 boreholes have been constructed

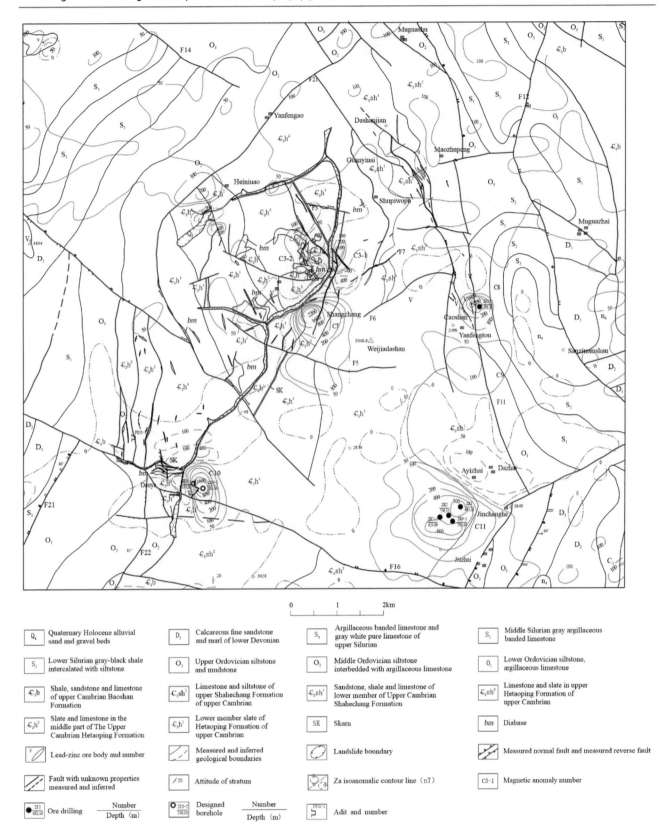

Fig. 6.59 Geology anomaly map of Hetaoping area and magnetic anomaly map measured by 1: 50,000 high-precision magnetic method

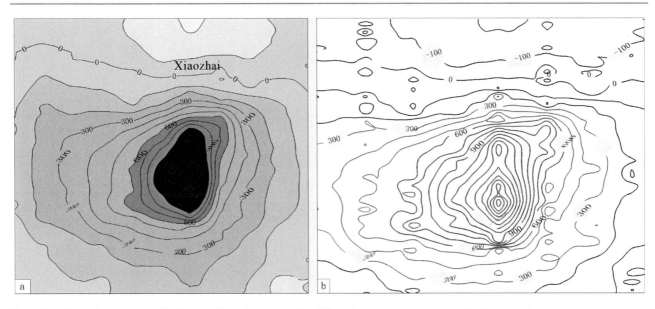

Fig. 6.60 Magnetic anomaly of Jinchang in Hetaoping area and its 500 m downward extension polarization anomaly

successively. The boreholes controlled Cu–Pb–Zn are large, while that controlled Fe is medium, and the prospect is large.

With theoretical understanding of Hetaoping deposit and the successful experience of Jinchanghe ore body positioning, a series of breakthroughs were made in mineral exploration in the whole region. New ore bodies were discovered in Hetaoping, Dachang'ao, Jinchanghe, Douya, Huangcaodi, Maozhupeng, Xinchang, Caoshan and Shangchang, showing the promising prospect of large-scale polymetallic resource bases.

6.8.2 The Prospecting Breakthrough of Deep Concealed Porphyry Molybdenum-Copper Deposit in Laochang Lead Deposit, Lancang "Metallogenic System + Gravity + Magnetism + Multiple Electrical Methods"

Laochang Pb–Zn-Ag deposit in Lancang, Yunnan Province is located in the southern section of Changning-Menglian Rift zone (Fig. 6.63). Laochang Pb–Zn-Ag deposit had a mining record in 1404 AD. At the beginning of 1980s, the exploration was completed, and the proven reserves of Pb and Zn were 672,700 t, and silver was 741t. Although the explanations of its genesis are not completely unified, the main viewpoints are as follows: the ore deposit is an early Carboniferous submarine volcanic eruption sediment-associated ore deposit (Yang Kaihui 1992), a composite genetic ore deposit of volcanic eruption sedimentation + late magmatic hydrothermal super-position (Wang Zengrun

1992; Li Lei 1996; Li Feng et al. 2000; Chen Wanyou 2002) and a VMS deposit + submarine eruption sedimentary deposit (Long Hansheng 2007).

With further development and prospecting in recent years, two genetic mechanisms, namely stratiform ore bodies with volcanic-associated massive sulfide deposit (VMS-type) and vein ore bodies with hydrothermal filling meta-somatism in fault zones, have been identified (Fig. 6.64). Meanwhile, many scholars have inferred and predicted that acid rocks (porphyry) may be hidden under the Sanjiang metallogenic zone, which has the potential for us to further search for copper polymetallic. For example, according to the Balkan model of massive sulfide, polymetallic metaso-matic vein, skarn and porphyry ore deposit (body) established by Ren Zhiji in 1991, it is pointed out that VMS deposit and vein Pb and Zn ore deposits in carbonate rocks are found in carbonate areas. Attention must be paid to searching for skarn-type and porphyry-type copper poly-metallic deposits. According to geophysical exploration, geochemical exploration and geological disclosure, it is proposed that there may be hidden rock masses in the deep part (Fig. 6.65). Zhao Zhifang et al. (2002) also inferred that it was caused by deep rock mass according to remote sensing hydrothermal ring.

Many prospecting methods and technologies have been adopted in the prospecting work of Lancang lead–zinc-silver deposit, especially in the prospecting project of crisis mines, including 1: 100,000 gravity survey, 1: 50,000 high-precision gravity survey, 1: 10,000 high-precision magnetic survey, high-frequency magnetotelluric method (EH-4) survey, Transient Electromagnetics (TEM) survey and so on.

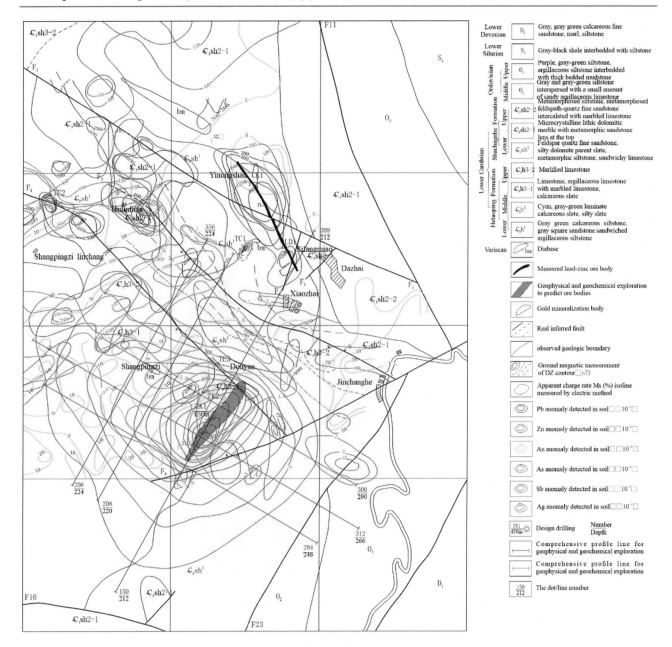

Fig. 6.61 Comprehensive interpretation of geophysics and geochemics of Jinchanghe in Hetaoping area

After practice and summary, the effective methods and technologies are integrated into "metallogenic system + gravity + magnetism + multiple electrical methods".

6.8.2.1 Metallogenic System

The metallogenic system of Lancang Pb–Zn-Ag deposit is a "quaternity" metallogenic system, including porphyry molybdenum-copper deposit, skarn-type Cu-Pb–Zn deposit, valcanic-associated massive sulfide deposits (pyrite chalcopyrite), hydrothermal vein Pb–Zn-Ag deposit and gold deposit, including volcanic eruption sedimentary metallogenic series and porphyry metallogenic series. Volcanic eruption sedimentary metallogenic series can be divided into eruption sedimentary massive lead–zinc-silver sulfide ore bodies (I and II ore body groups), eruption sedimentary massive brass-bearing pyrite ore bodies (V ore body groups) and eruption pipeline facies hydrothermal filling metasomatic disseminated and reticulated vein ore bodies. The porphyry metallogenic series can be further divided into hydrothermal filling vein-like Pb–Zn-Ag sulfide ore bodies

Fig. 6.62 Buried depth inference and borehole verification profiles of Jinchanghe electromagnetic anomalous body

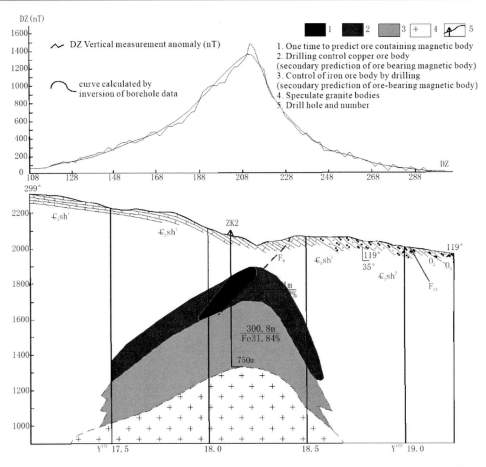

Fig. 6.63 Geotectonic zoning flowchart in western Yunnan (According to Li Feng 2009, revised)

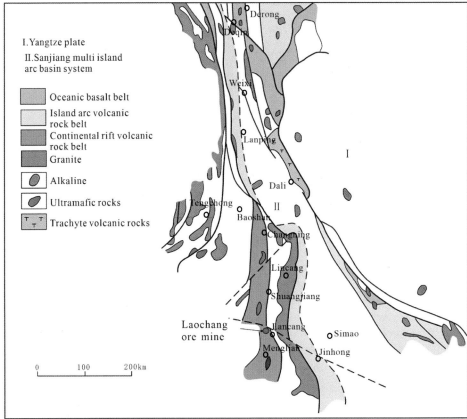

Fig. 6.64 Profile of Line 150A of Laochang Lead–Zinc-Silver Deposit (According to Li Feng 2009)

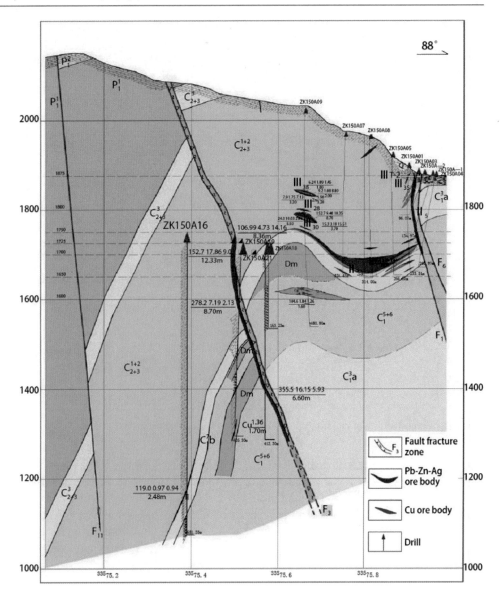

(III and IV ore body groups), skarn veinlet disseminated molybdenum (copper) ore bodies (VI ore body group) and porphyry molybdenum-copper ore bodies (Li Feng 2009).

In the epoch of two mineralization (series), many scholars have made in-depth research. The epoch of volcanic eruption sedimentary mineralization includes I, II and V ore body groups, which occur at the top of Lower Carboniferous volcanic-sedimentary rock series and in the transitional zone of eruption cycle. In the past, the lead isotope model ages made by Nonferrous 309 Team (1992) and Xue Bugao (2003) mainly concentrated in the range of 355–295 Ma, while the SHRIMP age made by Huang Zhilong using zircon from C1 tuff was 323 Ma. According to comprehensive analysis, the metallogenic age of this period should be between 323–295 Ma (Li Feng et al. 2009). Porphyry mineralization period, including III, IV, VI ore body groups, metallogenic epoch: 6 molybdenite model ages are 44.0–44.4 Ma, isochron ages are 43.78 Ma ± 0.78 Ma (Li Feng 2008), and Rb–Sr isochron ages of pyrite are 45 Ma (Huang Zhilong 2008, Communication). Therefore, the mineralization of granite porphyry should be between 43.78 ~ 45 Ma. Molybdenum-copper mineralization during porphyry mineralization is not limited to porphyry body itself but also forms diffusive mineralized halo in surrounding rocks above porphyry body, and the thickness of molybdenum-copper ore body in surrounding rocks in some sections is more than

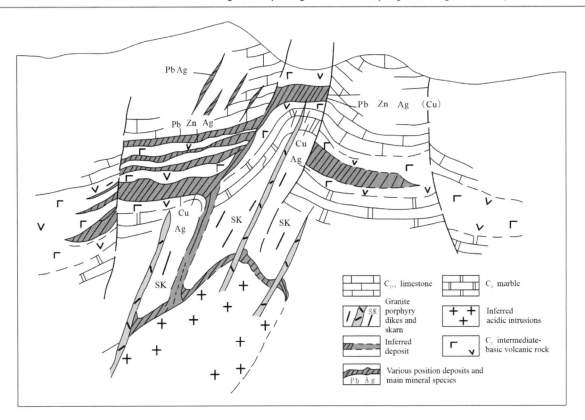

Fig. 6.65 Combined pattern diagram of Laochang silver-lead–zinc-copper deposit in Lancang (According to Ren Zhiji 1991)

200 m. Based on the metallogenic process and ore deposit characteristics, we can summarize the metallogenic model of "double metallogenic series" for the "quaternity" metallogenic system of Laochang Pb–Zn-Ag deposit in Lancang (Fig. 6.66).

6.8.2.2 Gravity Measurement

Because of the difference of rock density, gravity measurement can effectively infer the existence and distribution of concealed granitoid rock mass. Bouguer gravity anomaly changes obviously with strong directivity in Lancang area. Lancang lead–zinc-silver deposits are distributed in Bouguer gravity anomaly gradient zone.

The average Bouguer gravity anomaly of 1 km × 1 km shows that Bouguer gravity anomaly has two gravity high zones in the west and the east, and the north and south areas taking Mianxupu fault as the boundary show different characteristics: In the north area, it mainly shows a strip-shaped low gravity on the west side, while the east side is occupied by a large range of high gravity. The west side of the south area demonstrate a large-scale high gravity, and the east side of the south area is a complex

gravity anomaly area with two low gravity area sandwiched by one high gravity area (the residual gravity anomaly shows a unified negative gravity anomaly with two local low-value centers).

According to gravity measurement, it is inferred that there is a rock mass under the two concealed rock outburst (Fig. 6.67).

6.8.2.3 High-Precision Magnetic Survey

In porphyry copper deposits, especially in mining areas where rock mass has been exposed to the earth surface, magnetic anomalies are often distributed in ring-belt shape along the boundary of rock mass, from which the boundary of rock mass can be inferred.

In Lancang Pb–Zn-Ag mining area, a total 6km^2 of 1: 10,000 high-precision magnetic survey was completed, and three anomalies were delineated. However, there are massive sulfide stratiform ore bodies and vein ore bodies distributed along the structure at the top of the rock mass. Therefore, magnetic anomalies mostly reflect shallow anomalies, especially the vein ore body related to faults, and the anomalies are obvious.

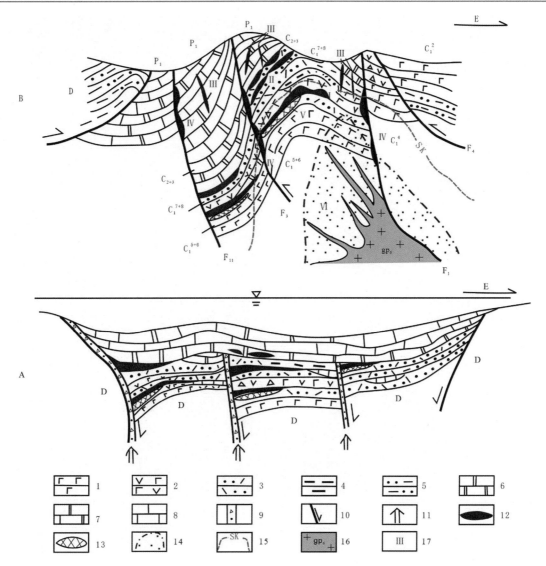

Fig. 6.66 Metallogenic model of "double metallogenic system" in Laochang, Lancang (According to Li Feng 2009, revised). A-Early Carboniferous massive volcanic eruption sedimentary metallogenic system; B-Himalayan porphyry metallogenic system. 1-Basaltic lava and tuff; 2-andesitic lava, pyroclastic rocks; 3-sedimentary tuff; 4-Carbonaceous siliceous shale and carbonaceous shale; 5-Middle-Upper Devonian gravel-bearing shale; 6-Middle and Upper Carboniferous dolomite; 7-Lower Permian dolomitic limestone; 8. Limestone interbed in volcanic rock series; 9-syngenetic fault; 10-epigenetic faults and their numbers; 11-direction of hydrothermal jet movement; 12-ore body; 13-copper-bearing pyrite ore body; 14-inferred porphyry Mo (Cu) mineralized area; 15-distribution area of silicified mineral; 16-concealed granite body; 17-Ore body number

6.8.2.4 Electrical Measurement

High-Frequency Magnetotelluric (EH-4) Measurement
In 2005, the Institute of Geological, Geophysical and Geochemical Exploration of Nonferrous Metals carried out EH-4 test profile along 150A and 152A exploration lines in the mining area, with a total length of 2280 m. In 2007, 16 EH-4 electromagnetic profiles with a total length of 23.08 km were arranged in the National Crisis Mine Replacement Resource Exploration Project. The main results are as follows: ① The downward extension of about 12 known faults is inferred; ② there are 12 plane anomalies and 272 profile anomalies, most of which correspond to I, II and III ore body groups, and the anomalies correspond well to mineralization; ③ it indicates the prospecting direction of the mining area and thinks that North Xiangshan, South Xiangshan-

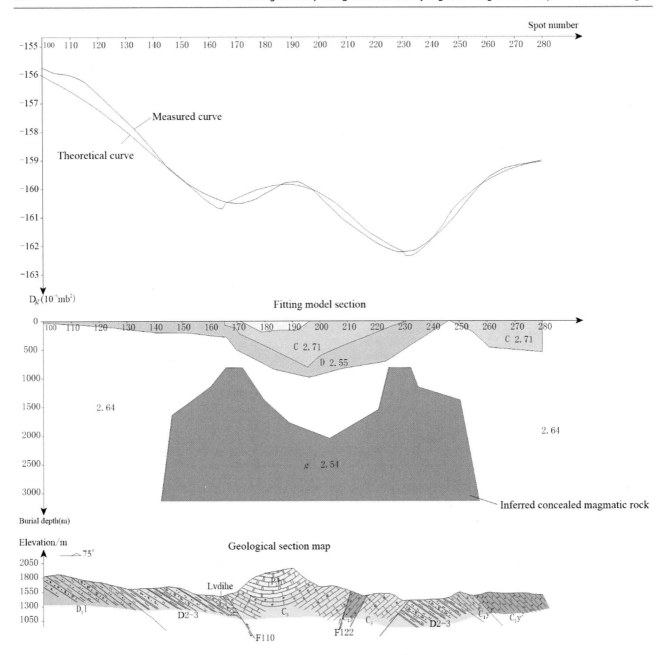

Fig. 6.67 Comprehensive interpretation profile of I-I 'in Laochang area, Lancang county, Yunnan province (According to Institute of Geophysical and Geochemical Exploration, Yunnan Geological Survey, 2009, revised)

Lianhuashan, Xiongshishan and Shangpingba-Shuishishan have further prospecting potential, while the west of Xiangshan has poor potential, so it is not suitable to do too much work; ④ the inferred electrical interface may be related to concealed rock mass (Fig. 6.68).

ZK153101, which is arranged according to EH-4 anomaly, is very good.

Transient Electromagnetics Measurement

In 2005, the Institute of Geological, Geophysical and Geochemical Exploration of Nonferrous Metals carried out a TEM survey, and the obtained TEM low resistivity anomalies mostly correspond to known ore bodies, which is of great significance for future prospecting.

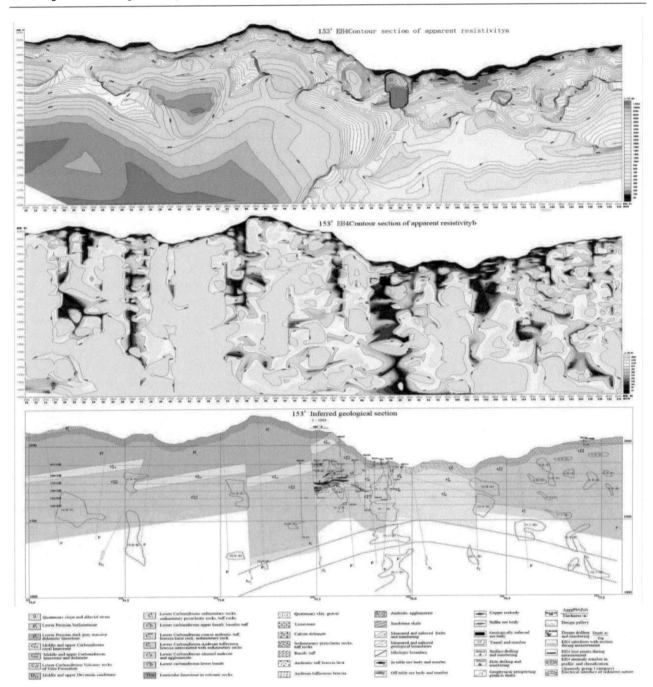

Fig. 6.68 Comprehensive Geophysical Prospecting Profile of Laochang (According to Geophysical and Geochemical Exploration Branch of Yunnan Nonferrous Geology, Geophysics and Chemistry Exploration Institute, 2009)

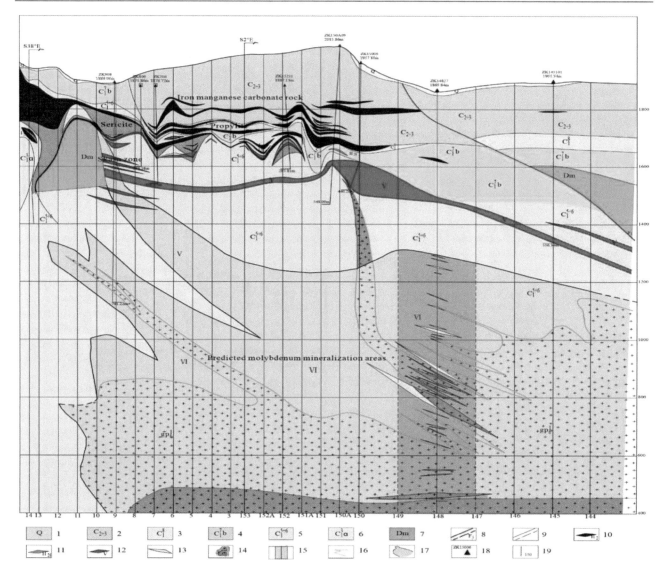

Fig. 6.69 Longitudinal profile of ore body in Laochang mining area (According to Li Feng 2009). 1-Quaternary slope alluvium, clay gravel, mud sand lead ore; 2-Middle and Upper Carboniferous limestone and dolomite; 3-Lower Carboniferous sedimentary rocks, sedimentary pyroclastic rocks and tuff; 4-Lower Carboniferous basalt, basaltic tuff; 5-lower Carboniferous trachyandesitic tuff, breccia lava and sedimentary rocks; 6-Lower Carboniferous almond-shaped andesite and agglomerate; 7-limestone lenses in volcanic rocks; 8-measured and speculated faults and number; 9-measured and speculated geological erathem; 10-on-balance-sheet ore bodies and their numbers in the lead–zinc deposit; 11-off-balance sheet ore bodies and their numbers; 12-copper ore body; 13-skarn; 14-Inferred granite porphyry body and molybdenum mineralization area; 15- molybdenum ore body controlled by borehole; 16-measured and speculated porphyry intrusion erathem; 17-measured and speculated ore bodies; 18-ore-seeing borehole and number; 19-Exploration lines and numbers

According to the technical integration of "metallogenic system + gravity + magnetism + electricity", a breakthrough has been made in prospecting in the mining area. In addition to further controlling the previously known vein-like and stratiform ore bodies and increasing the resources, there are 6 deep boreholes in deep porphyry bodies, such as ZK14827, which control the thick molybdenum-copper ore bodies, with a hole depth of 1417 m and delineated with w (Mo) $\geq 0.3\%$. The total ore-seeing length of molybdenum ore bodies reaches 696.25 m, with an average grade of 0.068%, of which 477.5 m has a grade of 0.082% (Fig. 6.69). Concealed porphyry molybdenum-copper deposits have great prospecting potential.

References

Chen BY (2002) Metallogeny of laochang Ag-Pb-Zn-Cu polymetallic deposit in Lancang, Yunnan. Changsha: Doctoral dissertation, University of Technology, Central South University, 1–20

Hou ZQ (1991) Ore fluid chemistry, thermal evolution history and ore-forming process of the Gacun Kuroko-type polymetallic deposit in western Sichuan. Mineral Deposits 4:313–324 (In Chinese with English Abstract)

Hou ZQ, Urabe T (1996) Hydrothermal alteration of ore-bearing volcanic rocks and mass chemical change in the Gacun Kuroko-type deposit, western Sichuan. Mineral Deposits 15:97–108 (In Chinese with English Abstract)

Li F, Zhuang FL (2000) Fluid inclusion characteristics and metallogenic significance of Dapingzhang Copper polymetallic deposit in Simao Yunnan Province. Geotecton Metallog 24(3):237–243

Li L, Duan JR, Li F (1996) Geological characteristics and multi-stage synchronic mineralization of laochang copper polymetallic deposit in Lancang. Yunnan Geol 15(3):246–256

Long HS, Jiang SP, Shi ZL (2007) Geological and geochemical characteristics of laochang Large Ag-Pb-Zn polymetallic deposit in Lancang Yunnan Province. Acta Mineralogica Sinica Z1:360–365

Lv QT, Hou ZQ, Zhao JH, Wu FX, Huang LJ (2001) Integrated prospecting model for Gacun volcanic-hosted massive sulfide deposit, Sichuan province. Mineral Deposits 20:313–322 (In Chinese with English Abstract)

Ren ZJ, Zhu ZH, Zhao ZS (1996) Terrane structure and mineralization in Yunnan province. Metallurgical Industry Press, Beijing, pp 1–50

Sillitoe RH (1979) Some thoughts on gold-rich porphyry copper deposits. Miner Deposita 14:161–174

Wang HP, Sun LR (1998) TM analysis of ground wave spectrum of copper deposit in the southern section of Yidun island arc and its application effect. Mineral Deposits 17:830–832 (In Chinese)

Wang ZR, Wu YZ (1992) Mineralization of Lancang River Rift in western Yunnan and genesis of laochang large cu-Pb-Ag deposit. Nonferrous Minerals Explor 4:207–215

Xue BG (2002) Geological characteristics and prospecting prospect of Xiaolongtan porphyry copper deposit Binchuan. Mineral Resour Geol 16(2):82–86

Yang KH (1992) Basic characteristics and genetic types of volcanic massive sulfide deposits in sanjiang area. Mineral Deposits 11(1):35–44

Yunnan Bureau of Geology and Mineral Resources (1990) Regional geology of Yunnan Province. Geological Publishing House, Beijing, pp 542–597

Zhao ZF, Lu YX (2002) Demonstration study on metallogenic prediction by remote sensing and GIS in Luziyuan area Zhenkang. Yunnan Geology 21(3):300–307

Printed in the United States
by Baker & Taylor Publisher Services